SOLAR MAGNETOHYDRODYNAMICS

GEOPHYSICS AND ASTROPHYSICS MONOGRAPHS

Editor

B. M. McCormac, *Lockheed Palo Alto Research Laboratory, Palo Alto, Calif., U.S.A.*

Editorial Board

R. Grant Athay, *High Altitude Observatory, Boulder, Colo., U.S.A.*
W. S. Broecker, *Lamont-Doherty Geological Observatory, Palisades, New York, U.S.A.*
P. J. Coleman, Jr., *University of California, Los Angeles, Calif., U.S.A.*
G. T. Csanady, *Woods Hole Oceanographic Institution, Woods Hole, Mass., U.S.A.*
D. M. Hunten, *University of Arizona, Tucson, Ariz., U.S.A.*
C. de Jager, *The Astronomical Institute, Utrecht, The Netherlands*
J. Kleczek, *Czechoslovak Academy of Sciences, Ondřejov, Czechoslovakia*
R. Lüst, *President Max-Planck Gesellschaft für Förderung der Wissenschaften, München, F.R.G.*
R. E. Munn, *University of Toronto, Toronto, Ont., Canada*
Z. Švestka, *The Astronomical Institute, Utrecht, The Netherlands*
G. Weill, *Service d'Aéronomie, Verrières-le-Buisson, France*

VOLUME 21

QB 521 PRI

MAIN LIBRARY
QUEEN MARY, UNIVERSITY OF LONDON
Mile End Road, London E1 4NS
DATE DUE FOR RETURN.

WITHDRAWN
FROM STOCK
QMUL LIBRARY

SOLAR MAGNETO-HYDRODYNAMICS

ERIC R. PRIEST
St. Andrews University, Scotland

D. REIDEL PUBLISHING COMPANY

A MEMBER OF THE KLUWER ACADEMIC PUBLISHERS GROUP

DORDRECHT / BOSTON / LANCASTER

Library of Congress Cataloging in Publication Data

Priest, Eric Ronald, 1943–
    Solar magneto-hydrodynamics.

    Bibliography: p.
    Includes index.
    1. Solar magnetic field. 2. Magneto-
hydrodynamics. 1. Title.
QB539.M23P74   523.7'2   82-5225
ISBN 90-277-1374-X   AACR2
ISBN 90-277-1833-4 (pbk.)

Published by D. Reidel Publishing Company,
P.O. Box 17, 3300 AA Dordrecht, Holland.

Sold and distributed in the U.S.A. and Canada
by Kluwer Academic Publishers,
190 Old Derby Street, Hingham, MA 02043, U.S.A.

In all other countries, sold and distributed
by Kluwer Academic Publishers Group,
P.O. Box 322, 3300 AH Dordrecht, Holland.

Reprinted with corrections.

All Rights Reserved
© 1982, 1984 by D. Reidel Publishing Company, Dordrecht, Holland
No part of the material protected by this copyright notice may be reproduced or
utilized in any form or by any means, electronic or mechanical
including photocopying, recording or by any information storage and
retrieval system, without written permission from the copyright owner

Printed in The Netherlands

To Clare

CONTENTS

PREFACE	xv
ACKNOWLEDGEMENTS	xix
CHAPTER 1. A DESCRIPTION OF THE SUN	1
1.1. Brief History	1
1.2. Overall Properties	3
1.2.1. Interior	4
1.2.2. Outer Atmosphere	6
1.3. The Quiet Sun	13
1.3.1. The Interior	13
A. The Core	13
B. A Model	13
C. Convection Zone	15
1.3.2. The Photosphere	19
A. Motions	19
B. Magnetic Field	22
C. A Model	24
1.3.3. The Chromosphere	26
1.3.4. The Corona	29
A. At Eclipses	29
B. In X-rays	30
C. Solar Wind	32
1.4. Transient Features	37
1.4.1. Active Regions	38
A. Development	38
B. Structure	42
C. Loops	42
D. Internal Motions	44
1.4.2. Sunspots	46
A. Development	46
B. Umbra	48
C. Penumbra	49
D. Motion	50
E. Solar Cycle	52
1.4.3. Prominences	56
A. Introduction	56
B. Properties	58

C. Development	58
D. Structure	59
E. Eruption	61
F. Coronal Transients	61
1.4.4. Solar Flares	64
A. Basic Description	65
B. Ground-Based Observations	67
C. Space Observations	70

CHAPTER 2. THE BASIC EQUATIONS OF MAGNETOHYDRODYNAMICS — 73

2.1. Electromagnetic Equations	73
2.1.1. Maxwell's Equations	73
2.1.2. Ohm's Law	75
2.1.3. Generalised Ohm's Law	76
2.1.4. Induction Equation	77
2.1.5. Electrical Conductivity	78
2.2. Plasma Equations	80
2.2.1. Mass Continuity	80
2.2.2. Equation of Motion	81
2.2.3. Perfect Gas Law	82
2.3. Energy Equations	84
2.3.1. Different Forms of Heat Equation	84
2.3.2. Thermal Conduction	85
2.3.3. Radiation	87
2.3.4. Heating	89
2.3.5. Energetics	90
2.4. Summary of Equations	91
2.4.1. Assumptions	92
2.4.2. Reduced Forms of the Equations	92
2.5. Dimensionless Parameters	93
2.6. Consequences of the Induction Equation	96
2.6.1. Diffusive Limit	96
2.6.2. Perfectly Conducting Limit	99
2.7. The Lorentz Force	101
2.8. Some Theorems	106
2.8.1. Cowling's Antidynamo Theorem	106
2.8.2. Taylor–Proudman Theorem	106
2.8.3. Ferraro's Law of Isorotation	107
2.8.4. The Virial Theorem	107
2.9. Summary of Magnetic Flux Tube Behaviour	108
2.9.1. Definitions	109
2.9.2. General Properties	109
2.9.3. Flux Tubes in the Solar Atmosphere	112
2.10. Summary of Current Sheet Behaviour	113
2.10.1. Processes of Formation	115
2.10.2. Properties	115

CHAPTER 3. MAGNETOHYDROSTATICS — 117

- 3.1. Introduction — 117
- 3.2. Plasma Structure in a Prescribed Magnetic Field — 119
- 3.3. The Structure of Magnetic Flux Tubes (Cylindrically Symmetric) — 121
  - 3.3.1. Purely Axial Field — 123
  - 3.3.2. Purely Azimuthal Field — 123
  - 3.3.3. Force-Free Fields — 125
    - A. Linear Field — 125
    - B. Nonlinear Fields — 125
    - C. Effect of Twisting a Tube — 126
    - D. Effect of Expanding a Tube — 127
    - E. A Tube of Non-Uniform Radius — 128
  - 3.3.4. Magnetostatic Fields — 129
- 3.4. Current-Free Fields — 130
- 3.5. Force-Free Fields — 133
  - 3.5.1. General Theorems — 134
  - 3.5.2. Simple Constant-α Solutions — 137
  - 3.5.3. General Constant-α Solutions — 140
  - 3.5.4. Non-Constant-α Solutions — 143
  - 3.5.5. Diffusion — 144
  - 3.5.6. Coronal Evolution — 145
- 3.6. Magnetohydrostatic Fields — 149

CHAPTER 4. WAVES — 153

- 4.1. Introduction — 153
  - 4.1.1. Fundamental Modes — 153
  - 4.1.2. Basic Equations — 154
- 4.2. Sound Waves — 157
- 4.3. Magnetic Waves — 157
  - 4.3.1. Shear Alfvén Waves — 159
  - 4.3.2. Compressional Alfvén Waves — 162
- 4.4. Internal Gravity Waves — 163
- 4.5. Inertial Waves — 166
- 4.6. Magnetoacoustic Waves — 168
- 4.7. Acoustic-Gravity Waves — 170
- 4.8. Summary of Magnetoacoustic-Gravity Waves — 173
- 4.9. Five-Minute Oscillations — 175
  - 4.9.1. Observations — 175
  - 4.9.2. Models — 177
    - A. Photospheric Ringing — 177
    - B. Wave Trapping — 178
  - 4.9.3. Wave Generation — 180
  - 4.9.4. Strong Magnetic Field Regions — 181
  - 4.9.5. The Future — 182
- 4.10. Waves in a Strongly Inhomogeneous Medium — 182
  - 4.10.1. Surface Waves on a Magnetic Interface — 183

4.10.2. A Twisted Magnetic Flux Tube	186
4.10.3. A Stratified Atmosphere	187

CHAPTER 5. SHOCK WAVES — 189

5.1. Introduction	189
5.1.1. Formation of a Hydrodynamic Shock	189
5.1.2. Effects of a Magnetic Field	193
5.2. Hydrodynamic Shocks	195
5.3. Perpendicular Shocks	197
5.4. Oblique Shocks	199
5.4.1. Jump Relations	199
5.4.2. Slow and Fast Shocks	202
5.4.3. Switch-Off and Switch-On Shocks	203
5.4.4. The Intermediate Wave	205

CHAPTER 6. HEATING OF THE UPPER ATMOSPHERE — 206

6.1. Introduction	206
6.2. Models for Atmospheric Structure	207
6.2.1. Basic Model	207
6.2.2. Magnetic Field Effects	211
6.2.3. Additional Effects	212
6.3. Acoustic Wave Heating	213
6.3.1. Steepening	213
6.3.2. Propagation and Dissipation	214
6.4. Magnetic Heating	217
6.4.1. Propagation and Dissipation of Magnetic Waves	218
6.4.2. Nonlinear Coupling of Alfvén Waves	220
6.4.3. Resonant Absorption of Alfvén Waves	224
6.4.4. Magnetic Field Dissipation	225
A. Order of Magnitude	226
B. Current Sheets	228
C. Current Filaments	233
6.5. Coronal Loops	234
6.5.1. Static Energy-Balance Models	235
A. Uniform Pressure Loops	236
B. Cool Cores	239
C. Hydrostatic Equilibrium	240
6.5.2. Flows in Coronal Loops	242

CHAPTER 7. INSTABILITY — 246

7.1. Introduction	246
7.2. Linearised Equations	248
7.3. Normal Mode Method	251
7.3.1. Example: Rayleigh–Taylor Instability	251

	A. Plasma Supported by a Magnetic Field	253
	B. Uniform Magnetic Field ($B_0^{(+)} = B_0^{(-)}$)	256
7.4.	Variational (or Energy) Method	257
	7.4.1. Example: Kink Instability	259
	7.4.2. Use of the Energy Method	264
7.5.	Summary of Instabilities	265
	7.5.1. Interchange Instability	265
	7.5.2. Rayleigh–Taylor Instability	267
	7.5.3. Pinched Discharge	268
	7.5.4. Flow Instability	271
	7.5.5. Resistive Instability	272
	7.5.6. Convective Instability	276
	7.5.7. Radiatively-Driven Thermal Instability	277
	7.5.8. Other Instabilities	279

CHAPTER 8. SUNSPOTS — 280

8.1.	Magnetoconvection	280
	8.1.1. Physical Effects	280
	8.1.2. Linear Stability Analysis	283
	8.1.3. Magnetic Flux Expulsion and Concentration	285
8.2.	Magnetic Buoyancy	291
	8.2.1. Qualitative Effect	291
	8.2.2. Magnetic Buoyancy Instability	293
	8.2.3. The Rise of Flux Tubes in the Sun	297
8.3.	Cooling of Sunspots	298
8.4.	Equilibrium Structure of Sunspots	299
	8.4.1. Magnetohydrostatic Equilibrium	299
	8.4.2. Sunspot Stability	306
8.5.	The Sunspot Penumbra	308
8.6.	Evolution of a Sunspot	309
	8.6.1. Formation	309
	8.6.2. Decay	311
8.7.	Intense Flux Tubes	314
	8.7.1. Equilibrium of a Slender Flux Tube	314
	8.7.2. Intense Magnetic Field Instability	316
	8.7.3. Spicule Generation	319
	8.7.4. Tube Waves	319

CHAPTER 9. DYNAMO THEORY — 325

9.1.	Introduction	325
9.2.	Cowling's Theorem	327
9.3.	Qualitative Dynamo Action	328
	9.3.1. Generation of Toroidal and Poloidal Fields	328
	9.3.2. Phenomenological Model	330
9.4.	Kinematic Dynamos	331

9.4.1. Nearly-Symmetric Dynamo	331
9.4.2. Turbulent Dynamo: Mean-Field Electrodynamics	332
9.4.3. Simple Solution: Dynamo Waves	334
9.4.4. Solar Cycle Models: The α–ω Dynamo	335
9.5. Magnetohydrodynamic Dynamos	338
9.5.1. Modified Kinematic Dynamos	338
9.5.2. Strange Attractors	339
9.5.3. Convective Dynamos	341
9.6. Difficulties with Dynamo Theory	342

CHAPTER 10. SOLAR FLARES — 344

10.1. Magnetic Reconnection	345
10.1.1. Unidirectional Field	346
10.1.2. Diffusion Region	348
10.1.3. The Petschek Mechanism	351
10.1.4. External Region	353
10.2. Simple-Loop Flare	357
10.2.1. Emerging (or Evolving) Flux Model	357
10.2.2. Thermal Nonequilibrium	360
10.2.3. Kink Instability	362
10.2.4. Resistive Kink Instability	365
10.3. Two-Ribbon Flare	366
10.3.1. Existence and Multiplicity of Force-Free Equilibria	368
10.3.2. Eruptive Instability	375
10.3.3. The Main Phase: 'Post'-Flare Loops	377

CHAPTER 11. PROMINENCES — 382

11.1. Formation	382
11.1.1. Formation in a Loop (Active-Region Prominences)	383
11.1.2. Formation in a Coronal Arcade	386
11.1.3. Formation in a Current Sheet	390
A. Thermal Nonequilibrium	391
B. Line-Tying	393
11.2. Magnetohydrostatics of Support in a Simple Arcade	395
11.2.1. Kippenhahn–Schlüter Model	395
11.2.2. Generalised Kippenhahn–Schlüter Model	398
11.2.3. The External Field	403
11.2.4. Magnetohydrodynamic Stability	404
11.2.5. Helical Structure	405
11.3. Support in Configurations with Helical Fields	406
11.3.1. Support in a Current Sheet	406
11.3.2. Support in a Horizontal Field	408
11.4. Coronal Transients	410
11.4.1. Twisted Loop Models	410
11.4.2. Untwisted Loop Models	413

11.4.3. Numerical Models	414
11.4.4. Conclusion	415

CHAPTER 12. THE SOLAR WIND 417

12.1. Introduction	417
12.2. Parker's Solution	418
12.3. Models for a Spherical Expansion	420
12.3.1. Energy Equation	420
12.3.2. Two-Fluid Model	421
12.3.3. Magnetic Field	423
12.4. Streamers and Coronal Holes	427
12.4.1. Pneuman-Kopp Model	427
A. Basic Model	427
B. Angular Momentum Loss	431
C. Current Sheet	431
12.4.2. Coronal Hole Models	432
12.5. Extra Effects	434

APPENDIX I. Units 436

APPENDIX II. Useful Values and Expressions 440

APPENDIX III. Notation 444

REFERENCES 448

INDEX 460

PREFACE

I have felt the need for a book on the theory of solar magnetic fields for some time now. Most books about the Sun are written by observers or by theorists from other branches of solar physics, whereas those on magnetohydrodynamics do not deal extensively with solar applications. I had thought of waiting a few decades before attempting to put pen to paper, but one summer Josip Kleczek encouraged an immediate start 'while your ideas are still fresh'. The book grew out of a postgraduate lecture course at St Andrews, and the resulting period of gestation or 'being with monograph' has lasted several years.

The Sun is an amazing object, which has continued to reveal completely unexpected features when observed in greater detail or at new wavelengths. What riches would be in store for us if we could view other stars with as much precision! Stellar physics itself is benefiting greatly from solar discoveries, but, in turn, our understanding of many solar phenomena (such as sunspots, sunspot cycles, the corona and the solar wind) will undoubtedly increase in the future due to their observation under different conditions in other stars.

In the 'old days' the solar atmosphere was regarded as a static, plane-parallel structure, heated by the dissipation of sound waves and with its upper layer expanding in a spherically symmetric manner as the solar wind. Outside of sunspots the magnetic field was thought to be unimportant with a weak uniform value of a few gauss. Recently, however, there has been a revolution in our basic understanding. High-resolution ground-based instruments have revealed a photosphere full of structure and with small-scale magnetic fields that are probably concentrated into intense kilogauss flux tubes. The chromosphere is now known to be made up of cool jets, and space experiments have shown the corona to be a dynamic, highly complex structure consisting of myriads of hot loops. At small scales in the corona, hundreds of X-ray bright points are seen where new flux is emerging from below the solar surface and causing mini-flares. Also, coronal heating is now thought to be magnetic, either via various wave modes or by direct current dissipation, and the solar wind has been found to escape primarily from the localised regions known as coronal holes, where the magnetic field lines are open. Many of these new features are dominated by the magnetic field. Indeed, much of the detailed structure we now see owes its very existence to the field, and so solar MHD is at a most exciting stage as we attempt to explain and model the magnetic Sun.

Magnetohydrodynamics (or MHD for short) is the study of the interaction between a plasma (or electrically conducting fluid) and a magnetic field. As the temperature of a material is raised, so it passes through the first three states of matter (solid, liquid and gas), and eventually it reaches the fourth state (plasma) when many electrons are no longer bound to the nuclei. A plasma, therefore, is an ionised gas, which behaves quite differently from the other states. Most of the matter in the universe is in this plasma state, such as the gas in a glowing fluorescent light tube, the Earth's ionosphere or the atmosphere of the Sun. Indeed, we on the Earth represent a tiny enclave of solid, liquid and gas immersed in the outflow of solar wind plasma, like a pebble in a stream of water. A magnetic field affects a plasma in several ways. It exerts a force, which is able, for instance, to support material in a prominence against gravity or propel it away from the Sun at high speeds. It provides thermal insulation, and so allows cool plasma to exist alongside hotter material, as in prominences or cool loop cores. It also stores energy, which may be released violently as a solar flare.

Solar MHD is an important tool for understanding many solar phenomena. It also plays a crucial role in explaining the behaviour of more general cosmical magnetic fields and plasmas, since the Sun provides a natural laboratory in which such behaviour may be studied. While terrestrial experiments are invaluable in demonstrating general plasma properties, conclusions from them cannot be applied uncritically to solar plasmas and have in the past given rise to misconceptions about solar magnetic field behaviour. Important differences between a laboratory plasma on Earth and the Sun include the nature of boundary conditions, the energy balance, the effect of gravity and the size of the magnetic Reynolds number (generally of order unity on the Earth and very much larger on the Sun).

The importance of mathematical modelling in our subject must be stressed. The full nonlinear equations of MHD (including thermal and diffusive effects) are so complex that they often need to be approximated drastically by focusing on the dominant physical mechanisms in any particular phenomenon. One begins with a simple model for a coronal transient or a dynamo, for instance, which perhaps admits analytical solutions to the equations. Then more and more effects may be added in an attempt to make the model more realistic. Thus the simple analytical model and the complex numerical computation can play the complementary roles of examining a new effect and attempting a more realistic simulation.

The overall structure of the book is as follows. It begins with two introductory chapters on solar observations (Chapter 1) and the MHD equations (Chapter 2). Then the fundamentals of MHD are developed in chapters on magnetostatics (Chapter 3), waves (Chapter 4), shocks (Chapter 5), and instabilities (Chapter 7). Finally, the theory is applied to the solar phenomena of atmospheric heating (Chapter 6), sunspots (Chapter 8), dynamos (Chapter 9), flares (Chapter 10), prominences (Chapter 11) and the solar wind (Chapter 12). The chapter on heating was placed after chapters 4 and 5 because of its close relationship to them. Appendices discuss the question of units and notation and list some expressions and constants which may be found useful. The building blocks of the solar magnetic field are flux tubes and current sheets. Their properties are developed at numerous points throughout the book and are summarised in Sections 2.9 and 2.10. Some of the chapter headings possess rather broad meanings. For instance, the chapter on sunspots includes sections on magnetoconvection, magnetic buoyancy and intense flux tubes, while the prominence chapter includes a discussion of coronal transients.

The notation that has been adopted for cylindrical polar coordinates is (R, ϕ, z), whereas that for spherical polars is (r, θ, ϕ). All quantities are measured in rationalised mks units, with the magnetic field in tesla (T) as far as most formulae are concerned (1 T = 10^4 G). In the text, however, magnetic field strengths are commonly quoted in G. Lengths in formulae are usually measured in metres, although in the text they are often quoted in megametres (1 Mm = 10^6 m).

I am extremely grateful to many people for their advice and help, notably colleagues in St Andrews (B. Roberts, M. Wragg, P. Browning, P. Cargill, T. Forbes, M. Gibbons, A. Hood), friends in Boulder (G. Athay, J. Christensen-Dalsgaard, P. Gilman, T. Holzer, A. Hundhausen, G. Pneuman, E. Zweibel), as well as T. G. Cowling, J. Heyvaerts, J. Hollweg, H. Spruit, P. Ulmschneider, N. Weiss and those who have so willingly supplied the figures. There are many others who have taught me what is written down here, although any errors or misconceptions are in no way their fault. My only hope is that this book may help others to understand a little more about the intriguing behaviour of solar magnetic fields and the mathematical language for its expression.

St Andrews, May, 1981   E. R. PRIEST

ACKNOWLEDGEMENTS

The author gratefully acknowledges permission to reproduce the following copyright figures: Fig. 1.20 (American Geophysical Union), Fig. 1.27 (Associated Book Publishers Ltd., London), Figs. 11.1 and 11.13 (Astronomy and Astrophysics Journal), Fig. 1.12 (S. Habbal), Fig. 8.4 (D. Galloway), Fig. 10.19 (Gordon and Breach), Fig. 10.7 (J. Heyvaerts), Fig. 11.11 (A. Milne), Fig. 11.15 (T. Mouschovias), Fig. 1.17 (R. H. Munro), Figs. 6.6, 1.26, 8.11 (E. N. Parker), Fig. 2.2 (R. Rosner), Figs. 1.30, 1.37, 9.4, 10.6 (Royal Society of London), Figs. 1.3a–c, 1.8, 1.13, 1.37, 10.18 (Association of Universities for Research in Astronomy, Inc., Sacramento Peak Observatory). The author is also most grateful to those who have supplied the many figures that are Reidel copyright, as indicated in the captions.

CHAPTER 1

A DESCRIPTION OF THE SUN

The Sun is an object of great beauty and fascination that has been studied with interest for thousands of years. During this century it has gradually become clear that much of the observed structure owes its existence to the Sun's magnetic field. This book is principally concerned with developments during the last couple of decades in our present understanding of the effect of this field on the solar atmosphere. To put them into perspective, however, we begin with a list of earlier developments, many of which took place before the solar magnetic field was discovered. Section 1.2 continues by describing the overall characteristics of the solar interior and atmosphere, and is followed by a more detailed account of both the quiet Sun (Section 1.3) and the active Sun (Section 1.4).

1.1. Brief History

2000 B. C.   Eclipses are recorded and predicted by the Chinese, and later (from 600 B. C.) by the Greeks.

350 B. C.   Theophrastus of Athens, one of the pupils of Aristotle, observes *sunspots* with the naked eye.

280 B. C.   Aristarchus of Samos suggests that the Earth is travelling around the Sun and estimates the distance to the Sun as 5 million miles.

190–125 B. C.   Hipparchus explains the motion of the Moon and Sun in terms of epicycles, with the Earth as the centre of the universe.

23 B. C.   From now until the Middle Ages, sunspots are observed systematically by the Chinese.

A. D. 140   Ptolemy publishes tables of the positions of the planets, and maintains that the Sun moves around the Earth, a belief held for the next 1400 years.

1530   Copernicus suggests that the six known planets revolve about the Sun in concentric circles.

1609   Kepler uses Brahe's observations of Mars to formulate his laws of planetary motion. He gives the Sun–Earth distance as 14 million miles.

1610   In the West, sunspots have been forgotten, until they are now observed by Galileo and others through recently invented telescopes. He is later tried for heresy, because his talk of sunspots contradicts the orthodox Christian teaching, due to Aristotle, that the Sun is without blemish. Some of those who do believe in sunspots regard them as planets, while others think they are the slag of the burning Sun or opaque clouds of smoke!

1666	Newton formulates the law of gravitation and applies it to the motion of planets around the Sun.
1770	Euler gives the Sun–Earth distance correctly as 93 million miles.
1814	Fraunhofer discovers most of the first 547 lines in the solar spectrum, which show up when sunlight is passed through a prism. At this time it is commonly thought that the Sun is inhabited!
1836	A systematic study of solar eclipses is begun.
1842	At an eclipse, *prominences* are rediscovered, after having been mentioned previously in medieval Russian chronicles and observed by Vassenius in 1733. Also the outer layers of the solar atmosphere (the *chromosphere* and *corona*) are clearly seen.
1843	Schwabe proposes the existence of an *eleven-year cycle* for the frequency of sunspot occurrence.
1851	The corona is photographed for the first time as a faint halo, visible around the Sun during an eclipse. Prominences are shown to be associated with the Sun rather than the Moon.
1852	Sabine, Wolf and Gautier find that the sunspot cycle is related to geomagnetic storms.
1858	Carrington discovers the latitude drift of sunspots during a cycle.
1859	Carrington and Hodgson observe a *solar flare* for the first time.
1861	Sporer discovers his law of sunspot distribution.
1868	At an eclipse, Secchi detects the emission line of a new element, which is given the name helium (after Helios, the Greek Sun god).
1869	Another new emission line, in the corona, is ascribed to a hypothetical element called coronium.
1869	Lane models the Sun as a gaseous sphere.
1874	The controversy which has raged throughout the 1860s about the appearance of the Sun's visible surface (the *photosphere*) is resolved when Langley gives a detailed description of its fine structure, called *granulation*.
1875	Secchi realizes that the form of the corona changes during the solar cycle.
1877	Secchi describes spicules as a burning prairie.
1889	Hale invents the spectroheliograph.
1908	Hale discovers that sunspots possess a strong magnetic field.
1909	The outward flow in sunspot penumbrae is observed by Evershed.
1930	Lyot invents his coronograph to view the corona without the aid of an eclipse.
1934	A theory for sunspots and an anti-dynamo theorem are put forward by Cowling.
1935	Hartmann conducts experiments on the flow of mercury in a magnetic field.
1938	The carbon–nitrogen and proton–proton chains are proposed by Bethe as an explanation for the source of solar energy.
1940	Edlén follows the suggestion of Grotrian and establishes that coronal emission lines arise from normal elements that are ionised by the extremely hot corona ($>10^6$ K).

QUEEN MARY
University of London

Self Service Issue Receipt

Customer ID: *****8251

Items that you have checked out

Title: Georg Wilhelm Friedrich Hegel
ID: 23029122554308
Due: 23/09/2019 23:59

Total items: 1
Account balance: £0.00
10/11/2017 10:59
Checked out: 1
Overdue: 0
Hold requests: 0
Ready for collection: 0

Thank you

Mile End Library
Queen Mary University of London

Extended Summer Vacation Loans
Ordinary Loans borrowed in February
May and June will be due back on 19 Sept

One Week Loans borrowed from February
May and August will be due back on Wednesday 11 Sept

PLEASE NOTE:
Summer Vacation Loans will be recalled
if they are requested by another reader
Please check often to ensure your item
hasn't been recalled or you may be away from college

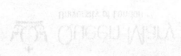

Queen Mary
University of London

QMUL Library Self Service

Customer ID: ******8521

Items that you have checked out

Title: Solar magneto-hydrodynamics
ID: 2312336364
Due: 19/09/2017 23:59

Total items: 1
Account balance: £1.00
06/07/2017 14:01
Checked out 1
Overdue: 0
Hold requests: 0
Ready for collection: 0

Thank you!

Mile End Library
Queen Mary, University of London

Extended Summer Vacation Loans
Ordinary Loans borrowed or renewed
from Saturday 13th May
will be due back on Friday 29th Sept

One Week Loans borrowed or renewed
from Saturday 3rd June
will be due back on Wednesday 27th Sept

PLEASE NOTE
Summer Vacation Loans will be recalled
if they are required by another reader
and must be returned
Please check or forward your e-mail
if you will be away from college

1941	Biermann proposes that sunspots are cool because of the inhibition of convection.
1942	Alfvén sets up a theory for *magnetic waves*. Also radio emission from the Sun is detected by radar.
1945	W. Roberts names and describes *spicules* in detail.
1948	Biermann and Schwarzschild propose that the outer solar atmosphere is heated by sound waves that propagate outwards from the convection zone; the effect of a magnetic field is not incorporated until much later by Osterbrock (1961).
1951	This year sees the beginnings of laboratory experiments on the magnetic containment of a plasma; the ultimate goal is to contain plasma for only about a second at 10^6 K, so that light atoms can fuse together and release energy.
1952	A great improvement in the measurement of solar magnetic fields comes with Babcock's invention of the magnetograph.
1956	The basic theory of magnetohydrodynamics is summarised by Cowling in his book on the subject.
1957	The first satellite observations of the interplanetary plasma are made. Also, Kippenhahn and Schlüter propose a model for prominence support.
1958	Parker predicts the existence of the *solar wind* and puts forward his model for it.
1960	Leighton discovers *5-minute oscillations* in the photosphere.
1962	Leighton, Noyes and Simon (1962) find that the *network* (first seen by Hale and Deslandres in the 1890s) outlines *supergranule cells*.

The last ten years have witnessed a further development in the theory of magnetohydrodynamics (as summarised in the remarkable book by Parker (1979a)) and in its application to solar phenomena. Also, high-resolution observations of the Sun from the ground have revealed new features of the photosphere and chromosphere in fine detail. These have been complemented by satellite observations of the transition region and corona directly. In particular, many exciting discoveries have been made with the series of Orbiting Solar Observatories and with the Skylab programme, which consisted of three manned missions of one, two and three months' duration between 25 May 1973 and 8 February 1974; they have revealed *coronal holes*, *X-ray bright points* and myriads of *coronal loop structures* in great detail. Finally, the Solar Maximum Mission of 1980 is greatly enhancing our understanding of the solar flare.

1.2. Overall Properties

The Sun is a fairly ordinary star of spectral type G2 V and absolute stellar magnitude 4.8, but of course its proximity to the Earth makes it unique, and its study is of central importance for understanding the behaviour of stars and of cosmical plasma in general. Its overall physical properties are as follows:

Age: 4.5×10^9 yr.
Mass: $M_\odot = 1.99 \times 10^{30}$ kg.
Radius: $R_\odot = 696\,000$ km ($= 6.96 \times 10^8$ m).

Mean density: 1.4×10^3 kg m^{-3} ($= 1.4$ g cm^{-3}).
Mean distance from Earth: 1 AU $= 1.50 \times 10^{11}$ m ($= 215 \, R_\odot$).
Surface gravity: $g_\odot = 274$ m s^{-2}.
Escape velocity at surface: 618 km s^{-1}.
Radiation emitted (luminosity): $L_\odot = 3.86 \times 10^{26}$ W ($= 3.86 \times 10^{33}$ erg s^{-1}).
Equatorial rotation period: 26 days.
Angular momentum: 1.7×10^{41} kg m^2 s^{-1}.
Mass loss rate: 10^9 kg s^{-1}.
Effective temperature: 5785 K.
1 arc sec ($\equiv 1''$) $= 726$ km.

Many of these numbers may be rather meaningless to the reader, so let us put them into perspective. The Earth has a mass of 6×10^{24} kg and a radius of 6000 km (6×10^6 m); so the Sun is 330 000 times more massive than the Earth and has a radius that is 109 times larger, while its surface gravity is 27 times greater. The mean density of the Earth is 5.5×10^3 kg m^{-3}, roughly equal to that of the Sun, while the atmospheric density at the surface of the Earth is only 1 kg m^{-3}; this makes the Sun's surface pressure only 0.2 times that of the Earth's sea-level atmosphere. The mean distance from the Earth to the Sun is 93×10^6 miles, which takes light 8 min to travel. The radiation emitted by the Sun amounts to about 1 kW m^{-2} at the surface of the Earth. Furthermore, the Sun's equator is inclined at about $7°$ to the plane of the Earth's orbit and the solar rotation gives an equatorial velocity of 2 km s^{-1}.

Traditionally, solar phenomena have been divided into two classes, quiet and active; for example, they have been well described by Kiepenheuer (1957), Gibson (1973) and Bruzek and Durrant (1977), which have been used as the main sources for the present chapter and in which detailed references may be found. The *quiet Sun* is viewed as a static, spherically symmetric ball of plasma, whose properties depend to a first approximation on radial distance from the centre and whose magnetic field is negligible. The *active Sun* consists of transient phenomena, such as sunspots, prominences and flares, which are superimposed on the quiet atmosphere and most of which owe their existence to the magnetic field. This division is retained here, by describing the quiet Sun in Section 1.3 and transient phenomena in Section 1.4, even though it is less than ideal. For instance, the quiet atmosphere is influenced markedly by the magnetic field; it is structured by the magnetic network above and around evolving supergranule cells (Section 1.3.2A) and the normal heating of the outer atmosphere may well be due to the magnetic field (Section 6.4).

1.2.1. INTERIOR

The Sun, like all stars, is such a massive ball of plasma that it is held together and compressed under its own gravitational attraction. It consists mainly of H (90%) and He (10%), mostly in an ionised state because of the high temperature; the remaining elements, such as C, N and O, comprise about 0.1% and are present in roughly the same proportions as on Earth, which suggests a common origin such as the interiors of older stars.

The *interior* of the Sun is shielded from our view; only its surface layers can be seen. However, models of its structure give a central temperature (1.6×10^7 K) and density (1.6×10^5 kg m^{-3}) that are high enough for thermonuclear reactions to take place. The central temperature is so high that the material there remains in a gaseous (plasma) state under a pressure 2.5×10^9 times that of the Earth's surface atmosphere; the central density is 30 times the mean density of the Earth! For every kilogram of H that is fused to form He, 0.007 kg is converted into energy, so that the great furnace in the solar core burns up 5×10^6 tonne s^{-1} of H. This energy generated in the core leaks continuously outwards in a very gentle manner and most of it is ultimately radiated into space.

In the middle of the last century, Helmholtz and Kelvin had shown that if the Sun's energy arose purely from its own gravitational contraction it would last only 2×10^7 yr; it was this which led Eddington to conclude in 1925 that the interior of the Sun must be a gigantic atomic reactor converting nuclear energy. The magnetic diffusion-time (Section 2.6.1) for the original magnetic flux that threaded the plasma cloud from which the Sun contracted is of the same order as the Sun's age, so it is not known how much of this primordial magnetic field is still present in the solar interior.

The overall structure of the Sun is sketched in Figure 1.1. The interior is divided into three regions, namely the *core*, *intermediate* (or *radiative*) *zone* and *convection zone*, where different physical processes are dominant. Across them the temperature and density fall by $3\frac{1}{2}$ and $8\frac{1}{2}$ orders of magnitude to 6600 K and 4×10^{-4} kg m^{-3} at the visible surface. The *core* contains only half the mass of the Sun in only about one-fiftieth of its volume, but generates 99% of the energy. This energy is slowly transferred outwards across the intermediate zone by radiative diffusion, as the photons are absorbed and emitted many times. The solar interior is so opaque that,

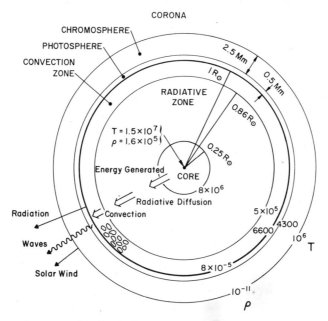

Fig. 1.1. The overall structure of the Sun, indicating the sizes of the various regions and their temperatures (in degrees K) and densities (in kg m^{-3}). The thicknesses of the photosphere and chromosphere are not to scale, and recent models place the base of the convection zone at about 0.7 $R_\odot$ rather than 0.86 $R_\odot$

whereas an unimpeded photon would take 2 s to reach the surface from the centre, there are so many collisions (absorptions and re-emissions) that photons in practice take 10^7 yr for the journey! The effect of these collisions is to increase the typical wavelength from that of high-energy gamma rays in the core to that of visible light at the solar surface.

Every Scot who heats his pan of porridge oats knows that, when the bottom of the pan is hot enough, the porridge starts to bubble away and exhibit a cellular circulation. Such convective motions also take place in the solar *convection zone* and for exactly the same reason: the temperature gradient is too great for the material to remain in static equilibrium and convective instability ensues (Sections 7.5.6, and 8.1). Convection transports energy because an individual blob of plasma carries heat as it rises and then gives up some of it before falling and picking up more. In fact, convection is the dominant means of energy transport in the convection zone. According to *dynamo theory* (Section 9.4), this zone (or its lower boundary) is also the region where the Sun's magnetic field is generated.

1.2.2. OUTER ATMOSPHERE

The visible solar atmosphere consists of three regions with different physical properties. The lowest is an extremely thin layer of plasma, called the *photosphere* (Section 1.3.2), which is relatively dense and opaque and emits most of the solar radiation. Above it lies the rarer and more transparent *chromosphere* (Section 1.3.3), while the *corona* (Section 1.3.4) extends from the top of a narrow *transition region* to the Earth and beyond. Hydrogen is almost wholly ionised in the upper chromosphere, but neutrals are important in the lower chromosphere and photosphere. The density (n) decreases rather rapidly with height above the solar surface; typical values are

$$n \approx \begin{cases} 10^{23} \text{ m}^{-3} & \text{in photosphere } (n_e \approx 10^{19}), \\ 10^{15} \text{ m}^{-3} & \text{in transition region,} \\ 10^{12} \text{ m}^{-3} & \text{at a height of 1 } R_\odot, \\ 10^{7} \text{ m}^{-3} & \text{at 1 AU,} \\ 10^{6} \text{ m}^{-3} & \text{in interstellar medium.} \end{cases}$$

By comparison the gas density at the Earth's surface is 10^{25} particles m^{-3}.

Before 1940 it was thought, quite naturally, that the temperature decreases as one goes away from the solar surface. But, since then, it has been realised that, after falling from about 6600 K (at the bottom of the photosphere) to a minimum value of about 4300 K (at the top of the photosphere), the temperature rises slowly through the lower chromosphere and then dramatically through the transition region to a few million degrees in the corona (Figure 1.2). Thereafter, the temperature falls slowly in the outer corona, which is expanding outwards as the *solar wind*, to a value of 10^5 K at 1 AU. The photospheric temperature of a few thousand degrees may be compared with the temperature of red-hot iron (1400 K) and that of the white-hot filament of an electric bulb (3900 K). The reason for the temperature rise above the photosphere has been one of the major problems in solar physics and is not yet fully answered: the low chromosphere is probably heated by sound waves that are generated in the

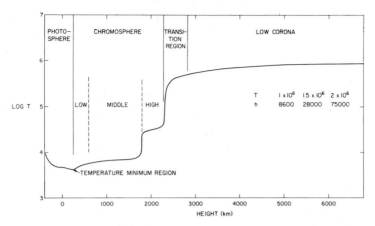

Fig. 1.2. An illustrative model for the variation of the temperature with height in the solar atmosphere (Athay, 1976).

noisy convection zone, propagate outwards and then dump their energy after steepening to form shocks (Section 6.3); higher levels may well be heated by several magnetic mechanisms (Section 6.4). The coronal temperature of a few million degrees is so high that H nuclei (protons) completely lose their planetary electrons and heavier nuclei may lose up to 15 of theirs.

Imagine what would happen to a hand if it were placed in a bottle of coronal plasma: its temperature would rise by only 1 degree, because the coronal density is so small that its heat content is tiny.

The bulk of the solar radiation comes from the *photosphere*, which emits a continuous spectrum with superimposed dark *absorption lines*. Light of all wavelengths is emitted by the photosphere and most of it goes straight through the overlying atmosphere into space. However, at certain specific wavelengths it is absorbed by particles in the overlying atmosphere; this effect is due to an increased opacity and it gives rise to the absorption lines. For example, the H Balmer-alpha line (Hα) is due to a H atom dropping from its third to its second quantum level. Such lines give us much information on temperature (from the intensity), magnetic field strength (from Zeeman splitting) and local line-of-sight plasma motion (from Doppler broadening).

Most lines are formed in the upper photosphere, but some, such as Hα, come from the chromosphere. Thus, when the Sun is observed through filters of different wavelengths, pictures can be obtained of the Sun's structure at a variety of levels (Figures 1.3(a)–(d)). For example, the *lower chromosphere* is shown up by using an Hα filter, which is most important for following the evolution of *active regions* (Section 1.4.1) and *prominences* (Section 1.4.3) and for observing the low-atmospheric part of a *solar flare* (Section 1.4.4). Just at the start of an eclipse you can see light that has originally come up from the photosphere and is then scattered towards you at the chromospheric level as well as the intrinsic chromospheric emission. This can be a very colourful effect, with reds from Hα and greens from Hβ, and it led Young in 1870 to give the chromosphere its name (from the Greek word for 'colour'). The chromosphere has sometimes been modelled as a static plane-parallel region, but in reality it is a constantly seething mass of plasma with a far-from-uniform structure; it has been

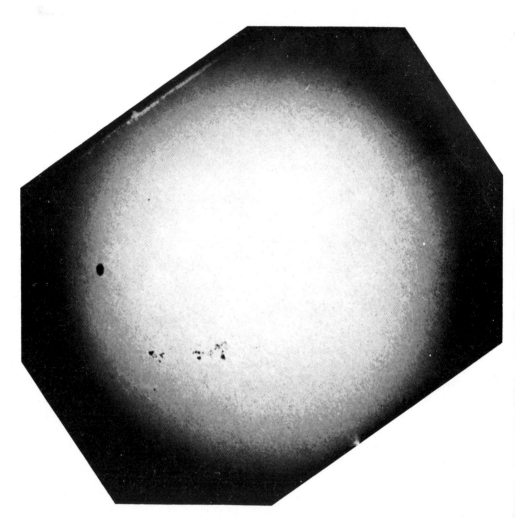

Fig. 1.3 (a).

Fig. 1.3 (b).

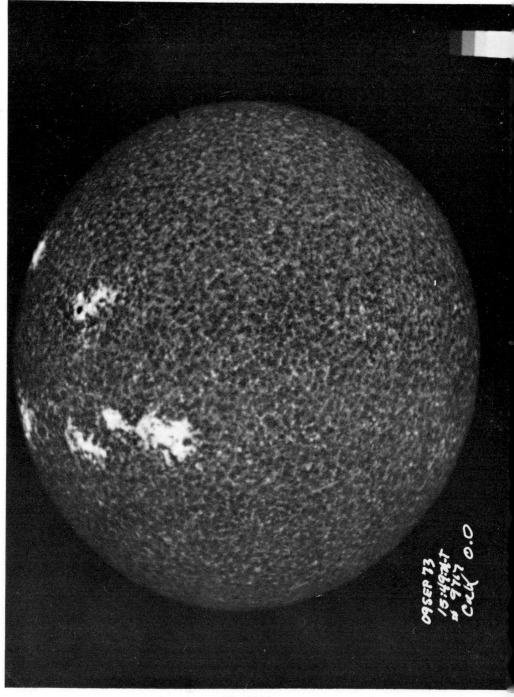

Fig. 1.3 (c).

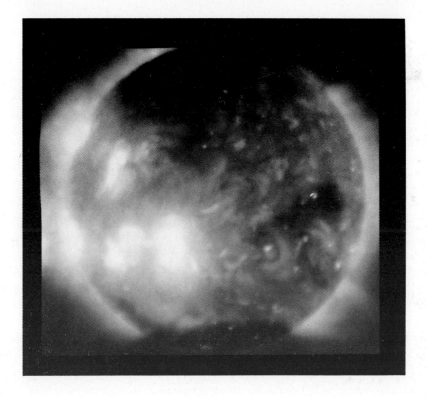

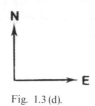

Fig. 1.3 (d).

Fig. 1.3(a)–(d). The appearance of the Sun on 9 September 1973, at different levels in the atmosphere: (a) photosphere (white light), (b) lower chromosphere (Hα), (c) middle chromosphere (Ca K), which reflects the magnetic flux pattern, (d) corona (soft X-ray). ((a)–(c) are from Sacramento Peak Observatory, © AURA Inc; (d) is courtesy of D. Webb, American Science and Engineering.)

described rather picturesquely as a burning prairie or as the spray of the photosphere thrown up from below.

The *corona* (from the Latin for 'crown') is observed at eclipses as a faint halo of very low density and high temperature, about as bright as the full Moon. (Its existence has been known for many centuries. The Egyptians used to worship the Sun, and on the back of the golden chair of Tutankamen is an emblem that is very reminiscent of the shape of the corona. The Chinese thought an eclipse was the effect of a dragon swallowing up the Sun, and so they used to employ an astronomer to shoot the beast!) This corona is due to the scattering of photospheric light coming up from

Fig. 1.4. White-light eclipse photographs of the corona taken during the eclipses of (a) 12 November 1966 and (b) 7 March, 1970, showing (1) prominence, (2) streamer, (3) coronal hole (courtesy G. Newkirk, High Altitude Observatory). Superimposed on the 1970 eclipse is a soft X-ray photograph of the inner corona from Skylab (courtesy A. Krieger, American Science and Engineering).

below, both off electrons (the K-corona) and off dust (the F-corona). Within 2.3 solar radii, the K-corona is dominant and its intensity is proportional to the electron density, so it appears brighter where more plasma is situated. The overall shape of the corona varies with the solar cycle: near sunspot maximum, bright features called *streamers* extend out in all directions; near sunspot minimum, streamers are present only in the equatorial region and *polar plumes* are seen to fan out from the poles. An example of an eclipse photograph is reproduced in Figure 1.4(a). It was taken near sunspot minimum using a radially graded filter that is 10 000 times more sensitive near its edge than near its centre: much structure can be seen, including that due to *closed field lines* low down, and *open field lines* further out, stretched radially by the outward-flowing *solar wind*. Another eclipse photograph, taken near sunspot maximum is shown in Figure 1.4(b)

1.3. The Quiet Sun

1.3.1. THE INTERIOR

1.3.1A. The Core

In the core, He nuclei are being built up from H nuclei, mainly by the proton–proton cycle but partly by the C–N cycle. At the end of these cycles, which last about 10^7 yr, groups of four protons (^{1}H) have been converted into He nuclei (^{4}He) according to the reaction

$$4\,^1\text{H} \rightarrow\,^4\text{He} + 2e^+ + 2\nu + 26.7 \text{ MeV},$$

and other nuclei have just acted as catalysts. Since each proton has a charge of $+1$ and each He nucleus has a charge of $+2$, the net charge is $+4$ on both sides of this relation. However, each He nucleus is smaller in mass by 3% than the original protons and this mass defect appears as energy; according to Albert Einstein's equivalence of mass and energy ($E = mc^2$), each kilogram of mass is equivalent to 9×10^{16} J. In the above reaction, the energy is released in the form of two high-frequency γ-rays (26.2 MeV) and two neutrinos (0.5 MeV), denoted by ν. The neutrinos are so tiny that they escape unimpeded from the core through the rest of the solar interior and so are our only direct diagnostic of core conditions. They are extremely difficult to capture; the number of those detectable on Earth depends critically on the core temperature, and, using the temperature from a solar model, it was expected that the capture-rate would be 6 SNU (solar neutrino units). However, only 1.6 SNU have been detected (Davis, 1972), which suggests that the solar model is in error, with a central temperature that is too high. Several ways of reducing the model temperature or neutrino flux have been proposed, such as: a significant mixing between the core and intermediate zone, driven by a build-up of ^{3}He; the presence of large magnetic fields or high rotation in the core: a non-uniform composition; the existence of a finite mass for the electron neutrinos. (See Roxburgh (1981) for a comparison of these suggestions.)

1.3.1B. A Model

A model for the interior of the Sun may be computed by assuming that the pressure ($p(r)$), density ($\rho(r)$) and temperature ($T(r)$) are functions of radius (r) alone and that the interior is in hydrostatic and thermal equilibrium. The set of equations is discussed in detail in Chapter 2. It includes: a *gas law* (Equation (2.26))

$$p = \frac{k_B}{m} \rho T; \tag{1.1}$$

a *force balance* (Equation (2.21))

$$\frac{dp}{dr} = -\rho g, \tag{1.2}$$

where the gravitational acceleration (Equation (2.22b)) is

$$g(r) = \frac{MG}{r^2}$$

and $M(r)$ is the mass inside a sphere of radius r, given by

$$\frac{dM}{dr} = 4\pi r^2 \rho;$$

an *energy balance* below the convective zone (Equations (2.32) and (2.35a))

$$\frac{1}{r^2}\frac{d}{dr}\left(r^2 \kappa_r \frac{dT}{dr}\right) + \rho\varepsilon = 0, \tag{1.3a}$$

where ε is the nuclear energy generation-rate and

$$\kappa_r = \frac{16\sigma_s T^3}{3\rho\tilde{\kappa}}$$

in terms of the *opacity*($\tilde{\kappa}$). Within the convective zone, the temperature gradient is assumed to take the adiabatic value (see Equation (1.7))

$$\frac{dT}{dr} = -\frac{\gamma-1}{\gamma}\frac{gm}{k_B}. \tag{1.3b}$$

The above Equations (1.1) to (1.3) are solved subject to the boundary conditions

$$M = 0, \quad \frac{dT}{dr} = 0 \quad \text{at } r = 0$$

and

$$M = M_\odot, \quad L = L_\odot \quad \text{at } r = R_s,$$

where R_s is the radius of the surface (at which p and T become small), and the *luminosity* ($L(r)$) is given by

$$\frac{dL}{dr} = 4\pi r^2 \rho \varepsilon.$$

The resulting entropy at the surface depends on the details of the turbulence in the convective zone. In qualitative models, it is a function of the ratio of the *mixing-length* (l) (the mean-free-path of a turbulent eddy) to the local scale-height (Λ) (Section 3.1). This ratio is adjusted until the radius (R_s) of the model Sun equals the observed value ($R_\odot$), while the composition is obtained by integrating over 4.5×10^9 yr from a star of initially uniform composition. The results for a standard solar model are shown in Figure 1.5, in which $X(H)$ and $X(^4He)$ are the fractions of the mass in the form of

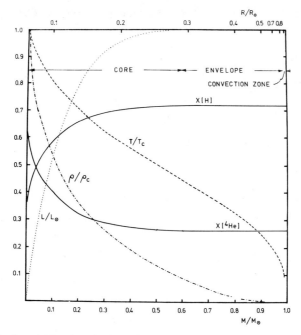

Fig. 1.5. A standard model for the solar interior, with the density (ρ), temperature (T), luminosity (L) and mass fractions (X) plotted as functions of radius (r) (Durrant and Roxburgh, 1977).

H and He, respectively; the central temperature and density are $T_c = 1.6 \times 10^7$ K and $\rho_c = 1.6 \times 10^5$ kg m^{-3}.

The core and envelope extend to radii of 0.25 to 0.3 $R_\odot$ and 0.8 to 0.9 $R_\odot$, respectively (typically 0.29 $R_\odot$ and 0.86 $R_\odot$), although more recent models give a radius of about 0.7 $R_\odot$ for the base of the convection zone.

More than a little apprehension with models of the above type has been expressed by Roxburgh (1976). He points out several ways in which recent observations do not support the model: the observed neutrino flux suggests a central temperature of 1.2×10^7 K rather than 1.6×10^7 K; geological evidence coupled with theoretical climate modelling tentatively suggests that the luminosity ($L_\odot$) has stayed constant to within 5% over the past 4×10^9 yr, rather than increasing monotonically from $0.65 \, L_\odot$ to $L_\odot$; the low surface Li abundance indicates that surface material has been mixed down to a temperature of 3×10^6 K (rather than 1.5×10^6 K) by, for example, convective overshooting; some mechanism, such as a convective core or a central instability, is making the Sun oscillate in its normal modes (Section 4.9.1), with long periods (up to 2 h 40 min) and a velocity amplitude of 2 m s^{-1} (Hill and Stebbins, 1975), whereas the model predicts stability to such oscillations.

1.3.1C. Convection Zone

As one moves outward in the solar envelope of the above model, beyond about 0.8 $R_\odot$, the opacity ($\tilde{\kappa}$) starts to increase rapidly, because new ions begin to be present as electrons recombine with other particles, and so photons can be absorbed more easily; this decreases the radiative conductivity (κ_r) and so, by Equation (1.3a), it increases the magnitude of the temperature gradient. Eventually, at about 0.86 $R_\odot$ (or possibly sooner) the temperature ($\approx 10^6$ K) is decreasing so rapidly that *convective instability* (Sections 7.5.6 and 8.1) sets in, and beyond this radius there lies the region of convective turbulence known as the *convection zone*. When one reaches the low photosphere, some radiation can escape directly from the Sun and so the opacity ($\tilde{\kappa}$) becomes smaller and the material returns to convective stability again.

The onset of instability when the vertical temperature gradient ($-dT/dr$) is too large may be explained by the following *parcel argument* (Figure 1.6). Consider a vertically stratified plasma in hydrostatic equilibrium with pressure $p(r)$, temperature $T(r)$ and density $\rho(r)$. Suppose an elementary parcel of the material is displaced vertically so slowly that it remains in *horizontal pressure equilibrium* with its surroundings. Then the parcel will feel a buoyancy force and so continue to rise if

$$\delta \rho_i < \delta \rho, \tag{1.4}$$

where $\delta \rho_i$ and $\delta \rho$ are the density changes inside the parcel and in the ambient medium. By the perfect gas law (Equation (1.1)), the changes in pressure, density and temperature are related by

$$\frac{\delta p_i}{p} = \frac{\delta \rho_i}{\rho} + \frac{\delta T_i}{T} \quad \text{and} \quad \frac{\delta p}{p} = \frac{\delta \rho}{\rho} + \frac{\delta T}{T}. \tag{1.5}$$

But horizontal pressure equilibrium means $\delta p_i = \delta p$ and so Equations (1.4) and (1.5) imply

$$-\delta T_i < -\delta T.$$

In other words, the parcel is unstable and will continue rising if

$$-\frac{dT}{dr} > -\frac{dT_i}{dr}, \tag{1.6}$$

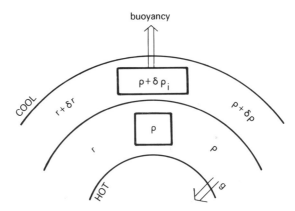

Fig. 1.6. The displacement of a parcel from r to $r + \delta r$, during which the density decreases by $-\delta \rho_i$ inside the parcel and $-\delta \rho$ outside the parcel.

so that the ambient temperature is falling with height faster than the temperature (T_i) within the parcel. Now, the parcel properties are governed by

$$p_i = \frac{k_B}{m} \rho_i T_i, \qquad \frac{dp_i}{dr} = -\rho_i g,$$

and, if the motion is so rapid (i.e., adiabatic) that there is no heat exchange with the surroundings,

$$\frac{p_i}{\rho_i^\gamma} = \text{constant}.$$

These may be combined to give

$$\frac{dT_i}{dr} = \frac{\gamma - 1}{\gamma} \frac{T_i}{p_i} \frac{dp_i}{dr} = -\frac{(\gamma - 1)}{\gamma} \frac{gm}{k_B}, \qquad (1.7)$$

which is known as the *adiabatic temperature gradient*. The criterion (1.6) for convective instability therefore finally becomes

$$-\frac{dT}{dr} > \frac{\gamma - 1}{\gamma} \frac{gm}{k_B}, \qquad (1.8)$$

and is known as the *Schwarzschild condition*.

Modelling the turbulent convection in the highly compressible convective zone is a formidable task. Numerical models normally consider an isolated layer; they also make the *Boussinesq assumption* (Section 8.1.2) and adopt a large Prandtl number, all of which are inappropriate. As an alternative, an order-of-magnitude treatment uses *mixing-length* theory, where collisions between turbulent eddies are described by a single mixing-length (of order the scale-height); the results of such a model (Spruit, 1974) which is made to fit onto the Harvard–Smithsonian Reference Atmosphere (Section 1.3.2) at its upper boundary are shown in Figure 1.7. However, it may well be that large-scale cells extending over several scale-heights carry most of the energy. From surface observations the convection appears to be dominated by cells of four fairly discrete scales, corresponding to *granulation, mesogranulation, supergranulation* and *giant cells*. It has been suggested that the first three types are driven by the ionisation of H and He and that they have scales comparable with the depths at which these processes take place. Hydrogen becomes highly ionised within about 1000 km below the photosphere and He becomes 90% singly- and doubly-ionised at depths of 5–10 000 km and 30 000 km, respectively, corresponding to the horizontal sizes of mesogranules (November *et al.*, 1981) and supergranules (Simon and Leighton, 1964). When these elements become ionised, they can absorb energy by more degrees of freedom, namely by ionisation and excitation; this decreases the value of γ (Equation (2.29c)) and so lowers the adiabatic temperature gradient (Equation (1.7)) which makes convection easier.

The existence of *giant cells* is much less well-established. They are believed to have a typical dimension of 300 000 km (comparable with the depth of the convection zone) and possibly a surface velocity of only 0.03 to 0.1 km s^{-1} and a lifetime of 14 mo; they may produce a large-scale organisation of weak surface magnetic fields and of filament distribution. A much stronger global motion is that of the solar rotation,

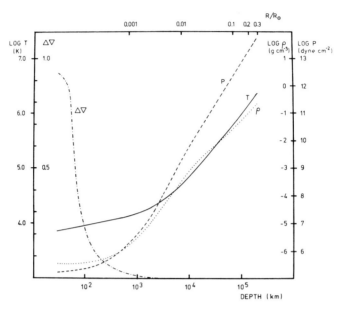

Fig. 1.7. Spruit's (1974) model of the convection zone, in which temperature (T), pressure (p) and density (ρ) are plotted as functions of distance (r) from the solar centre (Durrant and Roxburgh, 1977).

but the Sun does not rotate like a solid body. Rather, its atmosphere exhibits *differential rotation*, with the equatorial regions rotating once every 26 days, giving an equatorial speed of 2.0 km s^{-1}, while polar regions take nearly 37 days. Sunspots rotate somewhat more rapidly and, since they are presumably anchored somewhere below the photosphere, this implies that the angular velocity increases slightly with depth. However, *coronal holes* have very little differential rotation. Differential rotation is probably caused by the interaction of rotation and convection, but the difficulty in modelling it arises from the problems of treating compressible convection adequately, of reproducing a negligible variation with latitude of surface heat flux and of producing the right dynamo action to explain the solar cycle. Associated with the differential rotation there may exist an extremely slow *meridional flow*. Recently Duvall (1979) has detected such a polewards flow of 20 m s^{-1}, which supports the differential rotation model of, for example, Gilman (1976, 1977): he models numerically the nonlinear convection in a rotating spherical shell (with eddy diffusivities) and finds that giant convection cells can transport angular momentum towards the equator and so generate differential rotation, while the buoyancy forces drive a weak meridional flow.

In summary, the theory of the solar interior, which is largely outside the scope of this book, has the following main aims: to construct a model of the interior which gives a neutrino flux and Li abundance in agreement with observations; to explain the properties of the Sun's oscillations; to model the highly compressible convection zone and account for the three scales of cell size; and to understand differential rotation.

1.3.2. The photosphere

1.3.2A. Motions

The photosphere is the Sun's extremely thin visible surface layer and was named after the Greek word for 'light'. It may be defined either as the region (about 550 km thick) from $\tau_{5000} = 1$ up to the temperature minimum, or as the region (only 100 km thick and centred about $\tau_{5000} = 1$) from which most of the Sun's light escapes. However, the photosphere is neither uniformly bright nor perfectly still. In a high-resolution photograph, it appears covered with irregularly shaped 'cobble stones', and a film shows these bright granules to be in continual motion (Figure 1.8). Such a granular structure, or *granulation*, covers the whole Sun at the photospheric level and, at any one time, there are about a million granules present, which represent the tops of the convective cells that are overshooting the upper convection zone. The centre of a granule appears brighter than its boundary because it is composed of hot, rising (0.4 km s^{-1}) and horizontally outflowing plasma (0.25 km s^{-1}) rather than cool falling material. The typical diameter of a cell is 1–2″ (700 to 1500 km) and the mean distance between cell centres is about 1800 km; its mean lifetime is about 8 min but individual granules may live for 15 min. (So-called *exploding granules* are brighter than normal and form rings that expand at 1.5 to 2.0 km s^{-1} and fragment over about 10 min.) For a granule of size 800 km and a speed of 0.4 km s^{-1}, the turnover-time is of order $800/0.4$ s ≈ 30 min. This means that there are not many turnovers during the lifetime of a cell and the convection is therefore non-stationary, which is much harder to treat than stationary convection. High-resolution observations of the temperature-minimum region by Brueckner (1980) show that most of the EUV emission comes from small *grains* covering 10% of the solar surface; they have a size of 1500 km, a lifetime of 1 to 4 min or more, and a brightness temperature of 4300 to 4800 K for the quiet Sun and 4300 to 5000 K for plages.

In addition to the small-scale granular velocities, other motions include *oscillations* and large-scale velocity patterns known as *mesogranulation* and *supergranulation*. The up-and-down *oscillations* are best observed as Doppler shifts over a scale of about 10 000 km near the centre of the disc. Their properties are described in Section 4.9.1. The mean period is 5 min in the photosphere and decreases slightly with height. The velocity amplitude increases from about 0.15 km s^{-1} in the low photosphere to typically 0.5 km s^{-1} in the low chromosphere. These 5 min oscillations appear to be unrelated to granules; they are probably a result of acoustic waves trapped in a layer below the temperature minimum and driven by the Lighthill mechanism; in the temperature-minimum region itself they are evanescent but at chromospheric levels they may become travelling waves (Section 4.9). The properties of *mesogranulation* are much less well-known; mesogranules are typically 5000 to 10 000 km in size and possess speeds (both vertical and horizontal) of about 60 m s^{-1}.

Supergranulation was first studied in detail by Leighton *et al.* (1962). It comprises the tops of large convection cells, in which material rises at the centre at 0.1 km s^{-1}, moves horizontally outwards at typically 0.3 to 0.4 km s^{-1} and then descends at the edges of the cell at 0.1 to 0.2 km s^{-1}. (Such motions are faster than the winds (0.1 km s^{-1}) found in the most severe hurricanes on Earth.) Supergranule cells are irregular in shape and have diameters ranging from 20 000 to 54 000 km, with a mean value of

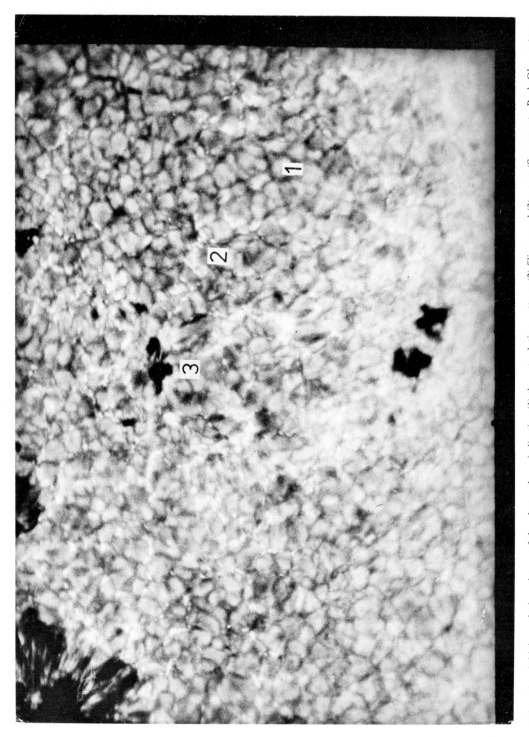

Fig. 1.8. A high-resolution picture of the photosphere, indicating (1) the granulation pattern, (2) filigree and (3) pores (Sacramento Peak Observatory, courtesy J. P. Mehltretter, © AURA Inc.).

32 000 km; they are about 10% larger than normal near active regions. In the photosphere they exhibit little if any brightness variation, but show up best near the limb as a pattern of horizontal motions. Individual supergranules last for 1 to 2 days, which is comparable with the turnover-time. Their boundaries are very prominent in the chromosphere (Section 1.3.3) and are regions where magnetic flux is concentrated. At the photosphere, supergranule boundaries show up brightly as a *photospheric network* in some Fraunhofer lines and as bright regions called *photospheric faculae* near the limb. They also appear in extremely high-resolution photographs away from active regions as *facular points*; near sunspots these are clustered more densely as a *crinkle* pattern, known as *filigree*, with a width of only 0.25" (200 km). Individual crinkles are up to 2.5" (1800 km) long and appear to be jostled around by granules. Within the supergranulation boundaries one finds downflows of 1 to 2 km s^{-1} (possibly falling to 0.1 km s^{-1} in the upper photosphere and then increasing to 3 to 4 km s^{-1} in the chromosphere and transition region), but these are concentrated in points (especially at supergranule junctions) rather than being present uniformly along the whole boundary.

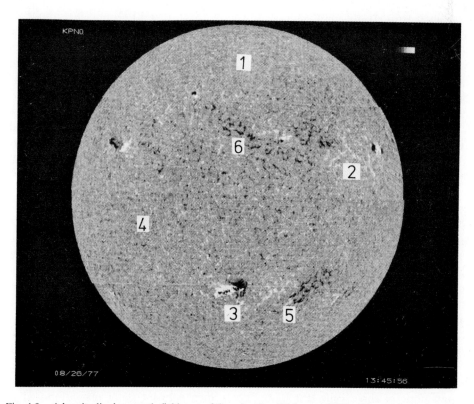

Fig. 1.9. A longitudinal magnetic field map of the solar disc (26 August 1977), with positive and negative polarities showing up bright and dark, respectively. The number (1) indicates polar field, (2) a large-scale unipolar field, (3) an active region, (4) an ephemeral region, (5) a remnant active region, (6) network field. The two sunspot bands north and south of the equator can clearly be seen (courtesy J. W. Harvey, Kitt Peak National Observatory).

1.3.2B. Magnetic Field

The photospheric *magnetic field* (e.g., Beckers, 1976b; Howard, 1977) is far from being the uniform dipole configuration that is sometimes assumed for other stars. Rather, it consists of small *magnetic elements* that are shuffled around and evolve rather rapidly (Figure 1.9). However, these small-scale magnetic structures are organised into large-scale patterns of several types, as follows.

(1) *Sunspots.* These represent exceptionally large concentrations of magnetic flux and are described in Section 1.4.2. They may act as a plentiful source of *Alfvén waves* (Section 4.3), which could heat the overlying atmosphere (Section 6.4); the Alfvén flux has been estimated as at most 2×10^7 W m^{-2}.

(2) *Plage regions.* The part of an active region outside sunspots is known as a *plage* and contains a *mean* magnetic field of a few hundred gauss (Figure 1.10(a)). However, the flux is concentrated into magnetic elements (known as *magnetic knots* or *micropores*) with an extremely high field strength of 1 to 2 kG and typical fluxes of 8×10^{10} Wb (8×10^{18} Mx); these elements are located at the edges of supergranule cells between granules and coincide with the bright facular and filigree network. They are associated with strong photospheric downflow (Section 1.4.1 D) and last about an hour. The effect of granular motions appears tentatively to be to give the magnetic elements random horizontal motions of 1 km s^{-1} over times of 500 s. This in turn could generate an Alfvén wave flux ($\rho v^2 v_A$) of roughly 2×10^6 W m^{-2}, which, when averaged over a supergranule cell, amounts to about 10^5 W m^{-2}, since the magnetic elements occupy about 5% of the area.

(3) *Large-scale unipolar areas.* These extend over several 100 000 km in both latitude and longitude. They contain elements of predominantly one polarity and are remarkably long-lived, with lifetimes of a year or more. Unipolar regions possibly rotate faster than the photospheric plasma and show less differential rotation than normal, which may imply that they are anchored deep inside the Sun.

Above some large unipolar regions one may find *coronal holes* (Section 1.3.4 C), and the boundaries to the east of unipolar regions tend to be more active than those to the west. The underlying cause of a unipolar region is possibly the slow diffusion of an active region or it may be associated with a giant convection cell. In particular, the polar field is believed to be composed mainly of the trailing parts of active regions that were earlier formed at lower latitudes and whose flux migrated polewards. The total flux above 60° latitude is typically 10^{13} Wb (10^{21} Mx); it is contained in small elements of strong field and gives an average field strength of about 1 G. Near the north and south poles, the net polarities are usually opposite; they reverse a year or two after the sunspot maximum, but near the time of reversal the two polarities can be the same for a few weeks or months.

(4) *Supergranulation fields.* Most (probably 90%) of the photospheric magnetic flux outside active regions is concentrated at supergranule boundaries, particularly at junctions of three cells (Figure 1.10(b)). Such *network fields* are composed of *magnetic knots* or *intense flux tubes*, which, like the filigree elements, tend to be located between granules. They possess extremely high field strengths of 1 to 2 kG, fluxes of about 3×10^9 Wb (3×10^{17} Mx) and characteristic sizes of 100 to 300 km. In old remnant active regions at the sunspot latitudes the network pattern shows up

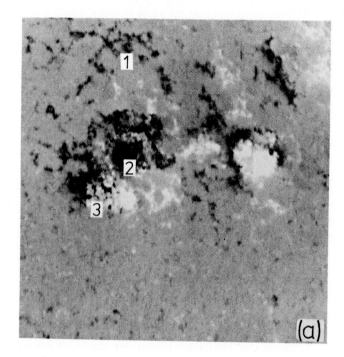

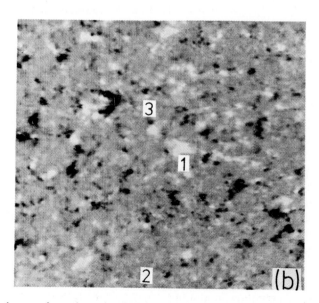

Fig. 1.10. (a) A close-up of an active region (27 July 1979) showing: (1) network field, (2) sunspot, (3) emerging flux region. (b) A high-resolution magnetic field map of a 200 arc sec square near disc centre (15 June 1979) showing: (1) network fields, (2) inner network fields and (3) ephemeral active region (courtesy J. W. Harvey, Kitt Peak National Observatory).

particularly well. In the quiet Sun the network is much more fragmented than in plage regions and it contains many fewer magnetic knots. The network fields are concentrated partly by supergranular and granular flow and partly by an instability (Section 8.7). Magnetic elements are also found *within* supergranule cells, where they are known as *inner network fields*; they are of mixed polarity with fluxes of about 10^9 Wb and lifetimes of 30 min. Over most of a cell the field is believed to form a horizontal canopy at (or just above) the temperature minimum. The height at which the network field has spread out to fill the available space is probably about 750 km for active regions and 1600 km for quiet regions (Spruit, 1981c).

(5) *Ephemeral regions*. These are tiny bipolar regions, which are visible on high-resolution photospheric magnetograms as *newly emerging regions* of magnetic flux and show up in the corona as *X-ray bright points* (Section 1.3.4 B). The average speed of emergence is 1.8 to 2.6 km s^{-1} and the average amount of flux is 2×10^{11} Wb (2×10^{19} Mx). They last for only about half a day, and the longer-lived features are found to contain more magnetic flux. As a function of size or lifetime, ephemeral regions appear at first sight to form a broad spectrum of activity which is continuous with active regions (Section 1.4.1). Their variation with the solar cycle has been studied by Martin and Harvey (1979). But *active regions are quite distinct from bright points*: they only appear within certain latitudes and are significantly hotter; also, active regions are probably much more deeply rooted, as indicated by the fact that they emerge at supergranule boundaries, whereas bright points have a random emergence in supergranule cells; active regions live much longer and show a preference for an east–west orientation, unlike ephemeral regions.

The magnetic field influences the plasma structure when the ratio β of plasma to magnetic pressure is less than or of order unity (Section 2.5). At the photospheric level this is only the case inside sunspots (Section 1.4.2) and at supergranule boundaries. In the chromosphere and corona it is certainly true for active regions (Section 1.4.1), but otherwise $\beta \approx 1$ (taking, e.g., $p = 0.016$ N m^{-2} and $B = 2$ G).

1.3.2C. A Model

Useful models of the solar photosphere and chromosphere are the *Harvard–Smithsonian Reference Atmosphere* (Gingerich *et al.*, 1971), and more recently the VAL model of Vernazza *et al.* (1981), which assume that thermodynamic properties vary only with height and are static. Of course, the solar atmosphere is in reality highly inhomogeneous, especially above the photosphere, and also rather turbulent, but nevertheless a model is important as a standard; its zero level is taken as the point where the optical depth at a wavelength of 5000 Å in the photosphere is equal to unity. The variation with height of several atmospheric properties is shown in Figures 1.11 and 2.1.

N_{HII} and N_{HI} refer to the number densities of ionised H and neutral H, respectively. In the model the photosphere is defined as the region of thickness 550 km from $\tau_{5000} = 1$ to the temperature minimum, which is 4170 K (although the VAL model gives a higher value of 4300 K). It can be seen that the density and pressure fall off by two-and-a-half orders of magnitude across the photosphere and a further four orders of magnitude across the next 2000 km. Below the photosphere and above

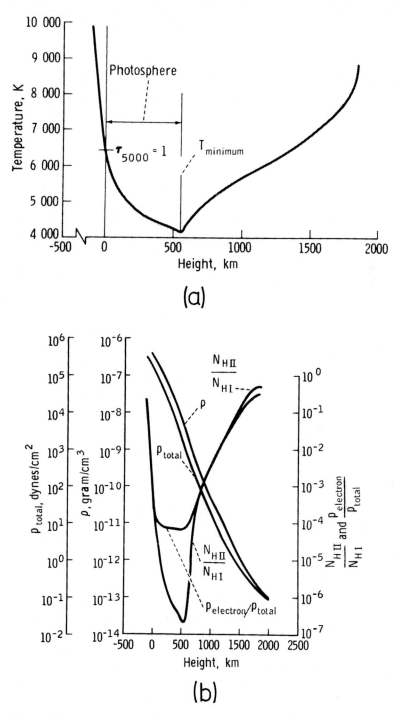

Fig. 1.11. The variation with height in the Harvard-Smithsonian Reference Atmosphere of (a) temperature (T) and (b) pressure (p), density (ρ) and H ionisation (N_{HII}/N_{HI}) (courtesy E. G. Gibson).

2000 km the H is almost fully ionised, but in the photosphere proper the inclusion of metals such as Mg, Fe and Si reduces the proportion of total material that is ionised to 10^{-4}.

In summary, the main aim of magnetohydrodynamic theory applied to the 'quiet' photospheric level has been to understand the concentration of intense flux tubes at supergranule boundaries and to model the waves and flows that are found within them (Section 8.7).

1.3.3. THE CHROMOSPHERE

The temperature rises monotonically through the chromosphere, in the manner shown in Figure 1.2 and described in detail by Athay (1976). The temperature increase is very rapid indeed at the boundary between the middle and high chromosphere and at the *transition region* between the chromosphere and corona. However, the heights (and therefore the pressures) at which those extreme temperature gradients occur may vary considerably from one feature to another, as indicated in Figure 1.12.

The chromosphere itself is highly non-uniform, as indicated in Figures 1.3(b) and 1.3(c). In particular, the middle chromospheric emission in the Ca II K line reveals the *network* of supergranulation boundaries very clearly as an irregular bright pattern. In the Hα wings the network is visible as a dark pattern, whereas in the Hα core it shows up bright. At the limb (Figure 1.13) one sees the chromosphere as a mass of plasma jets known as *spicules* (e.g., Beckers, 1972); they are ejected up from the high chromospheric part of supergranule boundaries (probably along magnetic field lines) and reach speeds of 20 to 30 km s^{-1} and heights of about 11 000 km before fading, although most show no signs of falling. One surprising feature is that, after an initial acceleration, the velocity of a spicule remains fairly constant over a long distance, despite the strong gravitational deceleration that is present. Spicules have typical lifetimes of 5 to 10 min, diameters 500 to 1200 km, maximum lengths 10 000 to 20 000 km, temperatures 1 to 2 × 10^4 K, electron densities 3 × 10^{16}

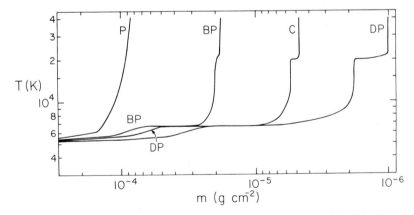

Fig. 1.12. The temperature versus height (mass column density) for several models of the chromosphere. C refers to the average quiet Sun, P to a plage, BP to a bright point and DP to a dark point in the models of Basri *et al.* (1979) (Habbal and Rosner, 1979).

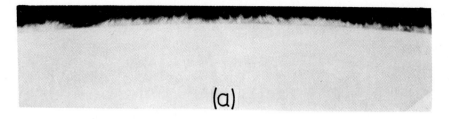

Fig. 1.13. Spicules as seen (a) at the limb in Hα and (b) near the limb in the wing of Hα outlining the network (© AURA Inc., Sacramento Peak observatory).

to 3×10^{17} m^{-3} and substantial rotational velocities. There are roughly 30 spicules to each supergranule cell at any given time, and their average inclination to the vertical is 20°. The region between spicules must be much hotter than the spicules themselves, possibly possessing coronal temperatures. On the disc in Hα, one sees elongated dark (and bright) features with dimensions $1'' \times 10''$ (700 × 7000 km) called *fine dark* (and bright) *mottles* (or *fibrils*), which are often grouped into clusters at supergranule junctions. Spicules cover 1% of the disc and are predominantly vertical and located near the network, whereas fibrils cover 50% of the disc and are mainly horizontal.

The cause of spicules has not yet been adequately established, but it is probably the interaction of plasma with the intense fields at supergranule boundaries. One possibility is that they are driven by a *resonance* between plasma motions along the intense flux tubes and granular or supergranular *buffeting* on the sides of the tubes (Section 8.7.3). Another is that they may be a result of the lack of thermal stability in the temperature profile along field lines (Section 6.5.1).

In polar regions, where the magnetic field is presumably open, large jets of chromo-

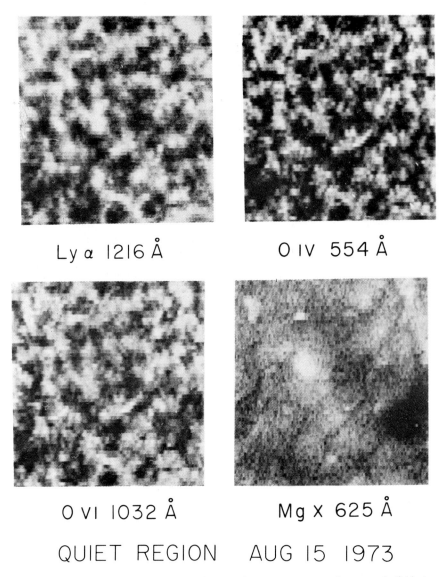

Fig. 1.14. The thickening of the network with increasing temperature as the magnetic field at supergranule boundaries spreads out. This is shown in the lines of Ly α ($\approx 10^4$ K), O IV (2×10^5 K), O VI (3×10^5 K) and Mg X (1.5×10^6 K) (courtesy R. Levine, Center for Astrophysics, Harvard).

spheric material are found. These resemble small surges and are called *macrospicules*. After rising with velocities of 10 to 150 km s^{-1} (apparently along the magnetic field) to a height of 5" to 50" (4000 to 40 000 km), they fall back or fade away. They last for 8 to 45 min and have diameters of 5" to 15" (4000 to 11 000 km) and densities of 10^{16} m^{-3}.

The narrow *transition region* separating the chromosphere from the corona is observed mainly in EUV emission lines and is sometimes taken to include the high chromosphere and its boundary with the middle chromosphere, indicated in Figure 1.2. Rather than being a static horizontal layer, it is probably composed of many dynamic *thin sheaths* around the cool spicules which are continually intruding into coronal plasma. In addition to spicules, Brueckner (1980) finds that the transition region has supersonic *jets* and small explosions, while high-speed *spikes* extend above macrospicules. Typical transition-region models give a thickness of only 30 km for the rapid temperature rise from 3×10^4 K to 3×10^5 K and 2500 km for the slower rise from 2×10^5 K to 10^6 K (Jordan, 1977). The average height at 10^5 K is 1700 ± 800 km for quiet regions. (It appears much higher for coronal hole regions such as near the poles, but this may just be due to the presence of macrospicules.) As one goes higher in the solar atmosphere, the magnetic field above supergranule boundaries continues to spread out and this causes the network to thicken with increasing temperature; eventually at coronal heights it ceases to exist (Figure 1.14).

In summary, two of the goals of chromospheric theory are to understand the generation and behaviour of spicules and to establish a model for the structure of the magnetic field and plasma above a supergranule cell (Section 6.2.2). Another is to understand the source of heating.

1.3.4. THE CORONA

1.3.4A. At Eclipses

In *white light* the corona cannot normally be seen through the dazzling light of the photosphere; it was originally observed only at the times of *eclipse*. On the average, there is only one total eclipse per year lasting about 3 min and the area covered by the Moon's shadow is only 100 km in diameter. The paucity of such opportunities to observe the corona prompted Lyot to create artificial eclipses by means of a *coronagraph*. This is a telescope containing an occulting disc to eliminate the glare of the photosphere, which is about a million times brighter than the corona. The apparent size of the disc needs to be somewhat greater than that of the Moon, but many weeks of observation have now been made with coronagraphs on board satellites such as Skylab and SMM, where the stray light is much reduced. This complements the many years of observation with ground-based K-coronameters.

In the quiet inner corona, the average electron density is several times 10^{14} m^{-3}, but this is enhanced by factors of 5 to 20 in many of the structures visible in Figure 1.4. The density rapidly falls off with distance from the solar surface: it is a few times 10^{12} m^{-3} at $1 R_\odot$ above the surface, 10^{11} m^{-3} at $4 R_\odot$ and less than 10^{10} m^{-3} at $10 R_\odot$. *Coronal streamers* are roughly radial structures extending from heights of 0.5 to 1 $R_\odot$ up to 10 $R_\odot$, with a density enhancement of 3 to 10. In particular, *helmet streamers* lie above prominences (Section 1.4.3) and active-region streamers above active regions (Section 1.4.1). A streamer consists of a round base (or arcade) of closed field lines surmounted by a blade of open field lines; from the end it looks like a helmet and from the side like a fan. It has been suggested that streamers form when the plasma in a closed-field region becomes hot enough to burst open the field and form a neutral current sheet. *Polar plumes* are ray-like structures near the poles, especially noticeable at times of sunspot minimum; they last for only about 15 h and presumably outline the local magnetic field. Plumes are also seen in coronal holes.

1.3.4B. In X-rays

In *soft X-rays*, the corona emits thermally and so may be viewed directly, since the contribution from the lower atmosphere is negligible. The disadvantage of observing in soft X-rays is that such wavelengths are normally absorbed by the Earth's atmosphere. But the resulting images from satellites such as Skylab are truly spectacular, and show the corona in a completely new guise (Figure 1.15). There are regions of two distinct types. Those in which the magnetic field is predominantly open appear relatively dark and are known as *coronal holes*; here the plasma is flowing outwards to give the *solar wind*. Those in which the magnetic

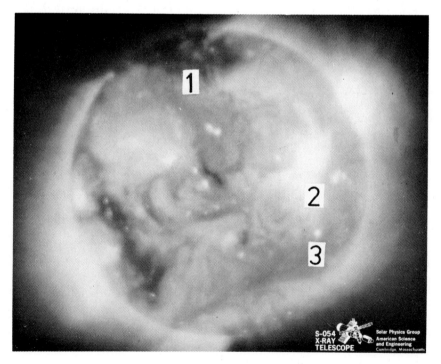

Fig. 1.15. A soft X-ray image of the corona (28 May 1973), showing: (1) coronal hole, (2) active region, (3) X-ray bright point (courtesy A. S. Krieger, American Science and Engineering).

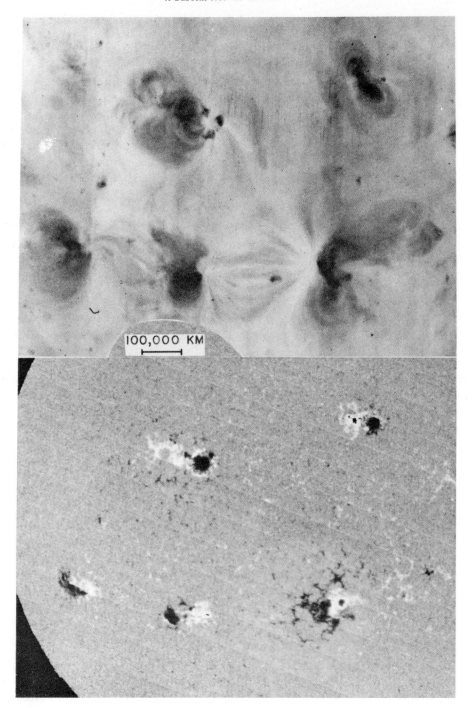

Fig. 1.16. Comparison between a (negative) picture in the EUV line of Fe XV (above) and a photospheric magnetogram, showing several active regions and their interconnecting loops (courtesy N. Sheeley, Naval Research Laboratory and Kitt Peak Observatory).

field is mainly closed consist of myriads of *coronal loops*. Also, small intense features called *X-ray bright points* are scattered over the whole disc.

Coronal loops may be split into five different types whose X-ray properties are summarised in Table 6.2; they are also seen in EUV light and in the red and green coronal lines of highly ionised Fe. Preliminary models for their thermal structure are outlined in Section 6.5. *Interconnecting loops* join different active regions and seem to form either when two loops stretch from separate active regions and reconnect or when one loop reconnects with some newly emerging flux (Figure 1.16). They may be up to 700 000 km long and in soft X-rays have a temperature of typically 2 to 3×10^6 K and a density of 7×10^{14} m^{-3}. Their ends are rooted in islands of strong magnetic field near the edges of active regions. A single loop lasts about a day, but a whole loop system may endure for many rotations. Loops that connect fully developed active regions have an intermittent visibility, but those that join an active region to an old remnant change little in shape and brightness for up to 12 days. Interconnecting loops sometimes brighten suddenly, for example, from a temperature of 2.1×10^6 K and a density of 7×10^{14} m^{-3} to values of 3.1×10^6 K and 1.3×10^{15} m^{-3}, by comparison with a quiet coronal temperature of 1.6×10^6 K. Such brightenings may be associated with a twisting of the foot points (Svestka, 1977). *Quiet-region loops* do not connect active regions, and in soft X-rays they are somewhat cooler, with a temperature of 1.5 to 2.1×10^6 K; their densities range much more extensively, from 2×10^{14} m^{-3} to 10^{15} m^{-3}. The other types of coronal loop are located in active regions and are described in Section 1.4.1C.

X-ray bright points have typical diameters of 22 000 km and bright cores of 4000 to 7000 km (Golub *et al.*, 1974, 1979). About 1500 of them appear per day. Their mean lifetime is 8 h; less than 15% live longer than one day and only a very few survive longer than two days. They appear to consist of two types: one has a lifetime of two days or less and is distributed uniformly over the solar disc, while the other is longer-lived and has a distribution similar to active regions, occurring within $\pm 30°$ of the equator. Bright points are the coronal manifestation of the tiny bipolar areas of emerging flux known as ephemeral regions (Section 1.3.2B) and appear to consist of several loops, typically 12 000 km long and 2500 km wide. Most of the magnetic flux emerging through the solar surface shows up as bright points rather than as active regions: the proportion of flux emerging as bright points was 40% in 1970 (sunspot maximum (see Section 1.4.2E)), 80% in 1973, 95% in 1976 (sunspot minimum) and 70% in 1978. However, its true importance is not clear, since it is not known how much of the bright point flux simply disappears below the photosphere again. A surprising result is that the *total* amount of flux emerging from 1970 to 1978 appears to be constant in time to within a factor two. In other words, there are more bright points during sunspot minimum than at sunspot maximum. This would imply that the *solar cycle* (see Section 1.4.2E) is an oscillation in the wavelength of the emerging flux rather than in the amount of flux!

1.3.4C. Solar Wind

Coronal holes are extended regions with a factor of three) than the typical background coron perature

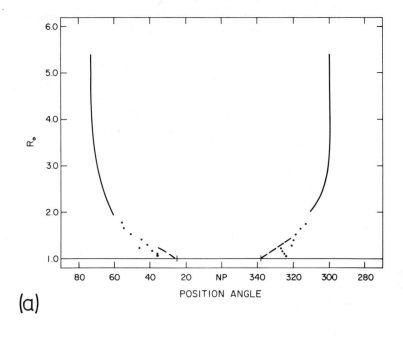

(a)

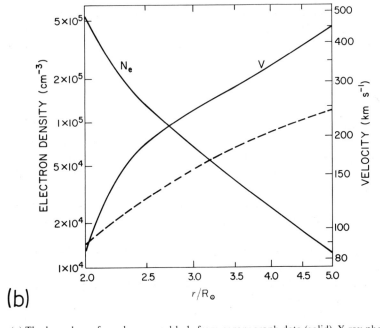

(b)

Fig. 1.17. (a) The boundary of a polar coronal hole from coronagraph data (solid), X-ray photographs (dashed) and K-coronameter data (dotted). (b) The variation of the electron density (n_e) with distance from the solar centre in units of solar radius. Also shown is the inferred velocity v (assuming a particle flux at 1 AU of 3×10^8 cm^{-2} s^{-1}) and the velocity (dashed) of a radially flowing, isothermal (2×10^6 K) solar wind, for which the speed of sound is about 150 km s^{-1} (Munro and Jackson, 1977).

too (1.4 to 1.8×10^6 K at 2 $R_\odot$). Their properties, especially those derived from Skylab observations, have been summarised by Bohlin (1976) and Zirker (1977). As an example, the geometry of a particular *polar hole* and the variation of its density with height are shown in Figure 1.17. Coronal holes possess an open diverging magnetic structure and lie above some of the more extensive large-scale unipolar regions in the photosphere. In the photosphere and low chromosphere, the plasma properties of these open-field regions are largely indistinguishable from those of their surroundings (although they do show up slightly in He 10 830Å). But at transition zone temperatures they appear quite different: the pressure is generally believed to be two or three times smaller and the temperature gradient five times smaller than in the normal quiet Sun (although it is possible that the pressure is the same); this makes the transition zone five times thicker and the conductive flux (60 W m^{-2}) an order of magnitude smaller than normal. (The dominant energy loss from coronal holes is convective transport by the solar wind (600 W m^{-2}) rather than the downward conductive flux; near the base of a hole the steady outflow is probably about 16 km s^{-1}.) During the Skylab period, 20% of the Sun was occupied by coronal holes, with 15% being in the form of polar holes. Coronal holes endure for several rotations and so are among the longest-lived of solar features. In addition to the longer-lasting *polar holes* (which disappeared near sunspot maximum), nine other holes were studied on Skylab; three lasted ten rotations or more and another four had lifetimes in excess of five rotations. Rather surprisingly, one coronal hole appeared to rotate more like a solid body than differentially, with only a 3% variation in rotational velocity from pole to equator (Figure 1.15); this is similar to the differential rotation of *unipolar regions* (Section 1.3.2B) and may imply that the magnetic field of coronal holes is anchored deep inside the Sun. The boundary of a coronal hole is occupied by an *arcade of coronal loops*, as indicated in Figure 1.18, and whether or not a coronal hole forms above a unipolar region seems to depend on the region's width. If the region is narrow, the bounding arcades are not very high and can be surmounted by a single streamer (Figure 1.18(b)). But, when the unipolar region is wider than 3 to 4×10^5 km, field lines from the centre of the region reach such a height that they can be pulled out by the solar wind. In practice, the emergence of an active region may change the large-scale coronal magnetic configuration and, in particular, it may stimulate the formation of a coronal hole, which often joins with the polar hole of the same polarity.

In open-field regions, the solar corona is not in hydrostatic equilibrium, but is continuously expanding outwards as the *solar wind* (e.g., Hundhausen, 1972). Most of it probably escapes along open field lines from coronal holes, especially the two polar coronal holes that are normally present, but small, open regions above active regions may also exist. The two polar holes with their (usually) oppositely directed magnetic fields are, at small distances from the solar surface, separated by the closed magnetic configuration typical of active regions. But beyond one or two solar radii above the solar surface they come into contact at a *neutral (heliomagnetic) current sheet*. In an idealised solar atmosphere, this current sheet would lie along the magnetic equator, but the presence of large-scale photospheric fields, such as the unipolar areas (Section 1.3.2), causes the sheet to be *warped*, like the brim of a sombrero (Figure 12.7). It is also inclined by about 7° to the Earth's orbit. As the Sun rotates,

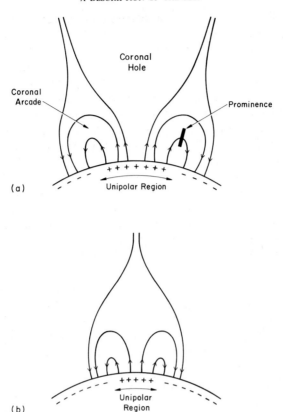

Fig. 1.18. The magnetic configuration above (a) a wide unipolar region and (b) a narrow unipolar region.

an observer above the solar equator sees a sequence of alternating polarities (typically four in number), successively from one side or the other of the current sheet. The current sheet itself is called a *sector boundary* as it moves past the observer and the whole pattern is known as *sector structure*. Near sunspot maximum the current sheet tends to be highly distorted, but near sunspot minimum the warping is slight.

The flow speed increases monotonically from very low values in the inner corona and eventually becomes super-Alfvénic beyond the Alfvén radius (r_A) (Section 12.3.3). Within this radius the magnetic field dominates the plasma dynamics: beyond it the magnetic field is carried by the plasma and the rotation of the Sun means that the magnetic field takes on a spiral structure, while the plasma flows roughly radially, like water from a rotating garden hose. Typical values of solar wind properties at 1 AU are given in Table 1.1, while some models for the solar wind are described in Chapter 12.

The solar-wind plasma does not quite flow radially from the Sun; its velocity is inclined at about 1.5° to the radius vector. This means that angular momentum is being transferred from the Sun to the solar wind and the Sun is being braked in the process; this effect is sufficient to slow the Sun down significantly over its lifetime. Near sunspot maximum the solar wind speed and density, as measured near the

TABLE 1.1
Properties of the solar wind at 1 AU

	Minimum	Average	Maximum
Velocity	200 km s^{-1}	400 km s^{-1}	900 km s^{-1}
Density	4×10^5 m^{-3}	6.5×10^6 m^{-3}	10^8 m^{-3}
Electron temperature	5×10^3 K	2×10^5 K	10^6 K
Proton temperature	3×10^3 K	5×10^4 K	10^6 K
Magnetic field	2×10^{-6} G	6×10^{-5} G	8×10^{-4} G
Alfvén speed	30 km s^{-1}	60 km s^{-1}	150 km s^{-1}

Earth, each seem to be about 30% lower than near minimum, while the magnetic field strength remains constant. Whereas sunlight takes only 8 min to reach the Earth's orbit, the solar wind plasma travels the distance in about 5 days. Beyond the Earth, the solar wind is believed to extend out to a boundary with the interstellar medium at 50 or 100 AU, where a shock slows the flow from supersonic to subsonic.

The solar wind is far from uniform. Near the Earth it consists of a series of *high-speed streams* in which the speed may increase from 400 km s^{-1} up to 800 km s^{-1} in two days and thereafter decline more slowly. At the same time the proton temperature increases to a maximum, while the electron temperature remains constant. The density and magnetic field rise sharply in the leading edge of the stream and possess low values within the bulk of the stream, so that the mass flux remains the same (10^{12} m^{-2} s^{-1}) to within a factor of two for all wind conditions. High-speed streams appear to come from coronal holes and are the most uniform parts of the solar wind, while slow streams may originate in the open fields above active regions. Complicated interactions between streams of different speeds are found, with compression regions and shock waves being formed. High-speed streams rotate with the Sun with the equatorial rotation period of about 27 days, and their rotation past the Earth is highly correlated with the recurrence of geomagnetic storms. The most common pattern (equivalent to the sector structure) is a set of four streams, two of high speed and two of low speed, with a unipolar magnetic field in each. Streams last from 1 to 18 rotations and are especially persistent during the declining phase of the solar cycle. On a small scale the solar wind is highly irregular. Magnetic-field fluctuations are continually present and consist mainly of *Alfvén waves*, propagating outwards and possibly generated by supergranulation. Also *tangential* and *rotational discontinuities* are very common features and *fast magnetoacoustic shocks* are often found.

Since coronal structures are dominated by the magnetic field, the corona provides magnetohydrodynamic theory with a wealth of problems. Amongst them are the the following: to model the coupling between the solar wind and the solar magnetic field, as exhibited by streamers and coronal holes (Section 12.4); to explain why only certain field lines show up as coronal loops and to model their structure (Section 6.5): to understand how the corona is heated and how bright points are produced (Section 6.4).

1.4. Transient Features

When viewed at low resolution in white light, the Sun appears a rather dull creature, but a closer look at the photosphere and the overlying atmosphere reveals a tantalising structure, which changes dynamically in a rich variety of ways. For example, an Hα photograph such as Figure 1.19 shows up many features superimposed on the quiet atmosphere. *Active regions* appear as bright *plages* of emission in the equatorial belt within $\pm 30°$ of the equator; they represent moderate concentrations of magnetic flux with mean fields of 100 G or so. Within a mature active region one finds dark regions of intense magnetic field called *sunspots*, and near sunspots there is occasionally a brilliant region of intense emission, called a *solar flare*, which represents the violent instability of part of an active-region magnetic field and the resulting energy release. Furthermore, around and far away from active regions there are thin, dark ribbons called *filaments* (or *prominences*). It is interesting to compare the appearance of all these features at different levels in the atmosphere, as for instance in Figure 1.3; in white light, sunspots represent the dominant departure from uniformity, whereas in soft X-rays the active regions as a whole are most prominent (see also Figure 1.15). Furthermore, the eclipse photograph in Figure 1.4 shows clearly the streamers that lie above prominences and active regions, while a

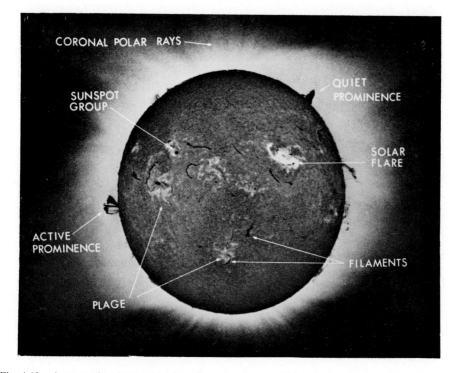

Fig. 1.19. A composite photograph of the Sun, showing several forms of activity. It includes the solar disk in Hα (July 16, 1959), prominences at the limb (1928) and the corona during the 1963 eclipse (courtesy S. Martin, Lockheed Solar Observatory).

magnetic field map, such as Figure 1.9, is dominated by the intense concentrations of flux in sunspots and, to a lesser extent, in active regions.

All the above forms of activity owe their existence to the *magnetic field*. Rather than being distinct, they simply represent different ways in which the solar plasma is responding to the underlying magnetic field development. They evolve on a variety of time-scales. The distribution of sunspots varies with an 11-yr periodicity known as the *solar cycle*. Prominences, the most stable of all surface features, may endure for 200 days, whereas a large sunspot group may last half that time and a solar flare is usually over in an hour or so.

1.4.1. ACTIVE REGIONS

1.4.1A. Development

The life of an active region has been summarised clearly by Svestka (1976a) and current theories of its structure are presented in Orrall (1981). When new magnetic flux pops up from below the photosphere as an *emerging flux region*, the atmosphere is heated and produces an X-ray bright point. Most bright points fade away in less than a day, but sometimes magnetic flux in the equatorial belt continues to emerge and a bright point grows into an active region. Emerging flux regions sometimes appear within or near an existing active region and sometimes they form the first stage of a new region; they are bipolar and first give rise to a small Hα plage connected by a few small, dark filaments. After a day or so, the bipolar region typically consists of a pair of sunspots joined by a system of dark loops called an *arch-filament system*. The loops are fairly low-lying, with a length of up to 30 000 km and a height possibly less than 5000 km. Their summits rise at up to 10 km s^{-1}, while plasma falls down near both ends with speeds up to 50 km s^{-1}. These magnetic flux loops expand outwards into the corona; in Hα the individual filaments fade after about 20 min and are replaced by a new set of loops coming up from below. Since the Sun rotates from east (i.e., left) to west (viewed from the Earth) the western part of an active region is known as the *preceding* part. It usually contains a large, regular sunspot called the *p-spot* or *leader spot*; the main spot in the easterly, following part is called the *f-spot* or follower. After about 3 or 4 days a well-developed active region has been formed with a bright Hα plage maybe 200 000 km across: its lower level shows up as a sunspot group surrounded by photospheric faculae, while its upper portion appears as an X-ray enhancement. The region continues to grow and reaches its maximum activity in 10 or 15 days. In a typical development, almost all the sunspots except the preceding one have disappeared after one rotation.

The decay of an active region is much slower than its rise and is marked by the slow dispersal of magnetic flux. The plage continues to increase in size and the flare incidence greatly decreases unless a revival of activity is stimulated by newly emerging flux. An *active-region filament* (Section 1.4.3), which may have formed typically when the region was quite young, often becomes more prominent and increases in size; it occupies the magnetic inversion line between opposite polarities. After two rotations, the Hα plage continues to decrease in brightness and disappears altogether after about four rotations. In X-rays, the compact active-region core has decayed away to leave a simple, diffuse loop structure joining opposite polarities; the *remnant* of an active region is often marked by a cool *filament* embedded in an *arcade* of hot loops. The filament is subsequently stretched out by differential rotation to become a huge *quiescent filament*, which migrates towards one of the poles (Section 1.4.3). The quiescent filament may erupt sometime during its life and give a coronal brightening; occasionally this is accompanied by a *two-ribbon* Hα flare (Figure 10.1).

New active regions have a tendency to develop near existing or remnant active regions and so form *complexes of activity*; they occupy preferred longitudes, which may persist for several years and rotate with a period of about 27 days. An example of the appearance of an active region in Hα and X-rays is given in Figure 1.20. In Hα a chromospheric plage can be seen; after two solar rotations it has dispersed and become a *remnant-active region*, but meanwhile a new active region has been

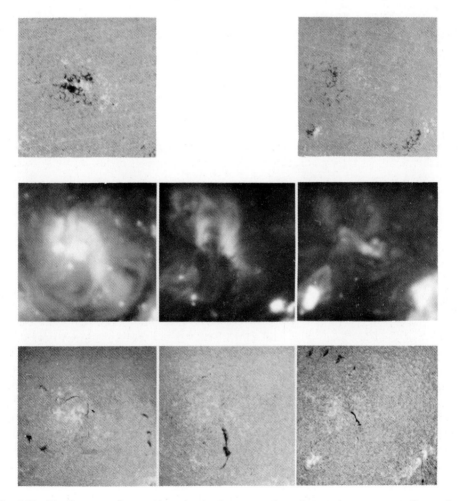

Fig. 1.20. Development of an active region in three successive solar rotations, corresponding to the three vertical columns. The top row of frames shows the magnetic field (Kitt Peak); the middle row gives the soft X-ray structure; the bottom row is Hα (Svestka, 1976a).

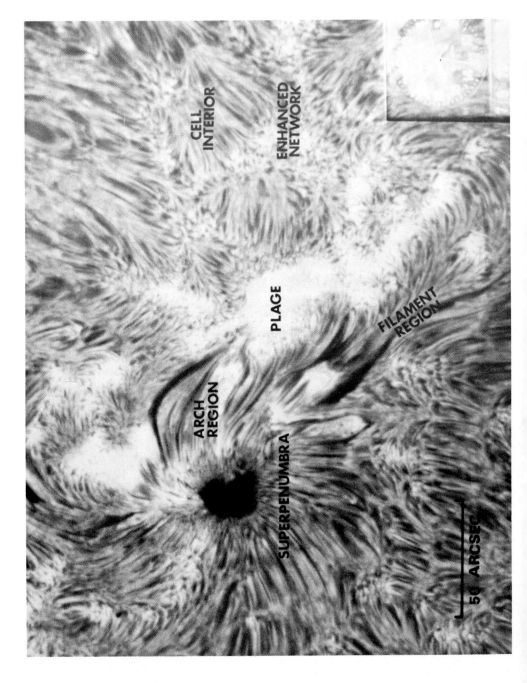

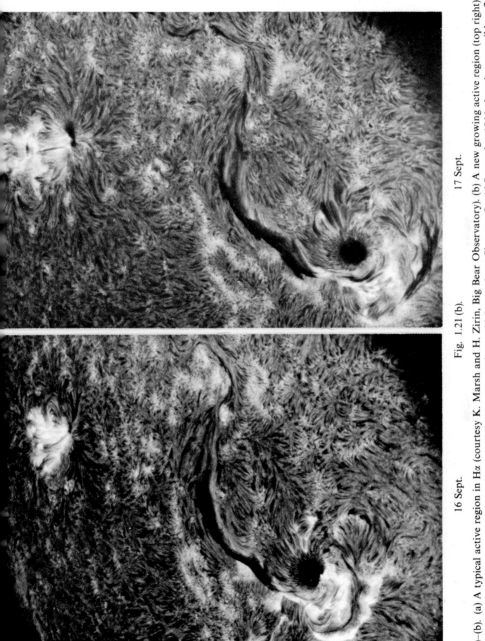

Figs. 1.21 (a)–(b). (a) A typical active region in Hα (courtesy K. Marsh and H. Zirin, Big Bear Observatory). (b) A new growing active region (top right) together with an older active region (bottom left) containing a single large sunspot above which stretches a filament. On 16 September 1966, there is a two-ribbon flare to the bottom left of the large spot (courtesy S. Martin, Lockheed Solar Observatory).

born in the bottom left corner. The regions of enhanced magnetic flux occupy much the same area as the plage but the soft X-ray emission is more diffuse.

1.4.1B. Structure

Most active regions are bipolar with the flux well-ordered into two islands of opposite polarity, but occasionally a magnetically complex region forms as new flux emerges with a different orientation or as a new region appears within an existing one. The total flux in a medium size region is 10^{14} Wb (10^{22} Mx). In the photosphere the most intense concentrations of magnetic flux show up as *sunspots*. They form during the emergence of flux and decay away during the slow dispersal of an active region, but the region may remain active with an enhanced magnetic field for weeks or months after the sunspots have disappeared. A typical well-developed active region (Figure 1.21) has a single sunspot; its preceding flux is concentrated there and its following flux is much more diffuse. The long, thin, dark streaks are called *fibrils* and probably follow magnetic field lines, some connecting opposite polarities. Individual fibrils have widths 700 to 2200 km, an average length of 11 000 km and lifetimes of 10 to 20 min, although their overall pattern remains constant for hours. The arch-filament systems present during the emergence of new flux are subsequently replaced by fibrils called *field transition arches*, which continue to join opposite polarity areas. The pattern surrounding large, old sunspots is composed of *superpenumbral fibrils*. The extremities of an active region are sometimes marked by a radial or slightly spiral pattern of fibrils. The distinction between thick fibrils and active-region filaments is that the latter always lie along a magnetic inversion line. Flare-like brightenings can be seen in X-ray bright points or young active regions, but major *solar flares* usually occur near the peak of an active region's development, when sunspots are present and when the magnetic structure is most complex.

The corona above an active region possesses a density and temperature that are enhanced by about 10 and at least 2 times, respectively, which makes it visible in white light at the limb during eclipses (Figure 1.4) and also in EUV, X-ray and radio wavelengths against the disc. This active-region corona is referred to as a *coronal condensation or enhancement*. Its hot core has densities up to 10^{16} m^{-3} and temperatures in excess of 3×10^6 K and is surrounded by a longer-lived permanent condensation with densities of a few times 10^{15} m^{-3} and temperatures 1.5 to 2.5×10^6 K. Its evolution in soft X-rays has been described by Howard and Svestka (1977). Overlying an active region one finds an *active-region streamer* (Figure 1.4) extending outwards for 3 to 4 $R_\odot$ as a series of fans or rays; the most conspicuous ones narrow to a throat or neck at 2 to 3 $R_\odot$ above the base and then diverge slightly.

1.4.1C. Loops

At high temperatures, the active region is seen to consist of *active-region loops* (Figures 6.8 and 1.22). In soft X-rays, only a few are distinguishable and the centre of the region appears as a bright core; the loops have a temperature of 2.5×10^6 K with a very small variation of only 0.3×10^6 K for a wide range of density, from 5×10^{14} m^{-3} to 5×10^{15} m^{-3}. The loop lengths lie typically between 10^4 km and 10^5 km; the shorter ones appear brighter in X-rays. In transition-region lines, formed over a small range of temperature, a whole jungle of loops is visible and it appears that the cool loops represent the cores of hotter and thicker loops. A startling revelation by Jordan (1975) and Foukal (1975) is that extremely cool loops, with temperatures at least an order of magnitude lower than the surrounding corona, exist right up to a height of more than 50 000 km above the photosphere. (This is in stark contrast to early theories of atmospheric structure that model a more or less horizontal transition region at only 2000 to 5000 km but it supports the earlier evidence from optical coronal lines for cool plasma up in the corona.) Jordan found that the pressure and temperature in the *core* of a loop are always lower than in a surrounding *sheath*, but they can be either larger or smaller than the ambient coronal values. One huge loop reached a height of 150 000 km and, at an altitude of 30 000 km, possessed a core pressure of $2\,p_e$ and a sheath pressure of $4\,p_e$, in terms of the external coronal pressure (p_e).

The most prominent active-region loops at transition-region temperatures connect sunspots (Foukal, 1975) and so may be called *sunspot loops*. Most are about 100 000 km long and 10 000 km wide and have *cool cores* with a temperature lower than 2×10^5 K, by comparison with the ambient coronal temperature of probably 2×10^6 K. The core density is less certain but is probably the same as the surrounding density, so that the core pressure is about a tenth of the coronal pressure. Also, there may be a sheath around the core with a plasma density three or four times bigger than

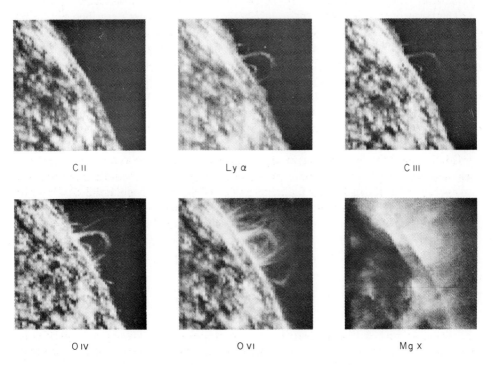

Fig. 1.22. An active-region sunspot loop about 60 Mm high in the EUV lines of C II (3×10^4 K), Ly α ($\simeq 10^4$ K), C III (7×10^4 K), O VI (1.8×10^5 K), Mg X (1.5×10^6 K). Hotter emission comes from progressively thicker co-axial shells (courtesy P. Foukal, Harvard).

the ambient value. The low core-temperature accounts for its visibility in EUV lines, whereas the sheath density-enhancement shows up the loop in X-rays against the background emission. The pressure and energy balance are steady over several hours, much larger than the free-fall time, and their intensities are not directly related to umbral area: large loops can be rooted in insignificant spots and some large spots may have no bright loops at all. The main loop in Figure 1.22 has a density of 7×10^{14} m^{-3} and a temperature of 1.3×10^5 K, 4×10^5 K and 8×10^5 K at distances of 2500, 5000 and 7500 km from the axis, respectively.

1.4.1D. Internal Motions

The magnetic field of an active region probably evolves slowly through a series of essentially stationary states, mainly force-free (Section 3.5). As far as movement normal to the magnetic field is concerned, the plasma is completely dominated by the field, since β (the ratio of plasma to magnetic pressure) is much less than unity (Section 2.5). But, along the field, the plasma is observed to be in continual motion rather than in a static state. A cursory glance at ground-based films shows a sunspot region to be truly active and this has been confirmed by observations from space. It is now known that *active-region plasma is dynamic*, with continual activity in the form of a wide range of flows (Priest, 1981c). The physical properties of some of the coronal flows are not yet known, while the modelling of dynamic loops (Section 6.5.2) has barely begun.

The types of flow revealed by *ground-based observations* are described below and indicated schematically in Figure 1.23(a), in which the solid curves represent the magnetic field lines of a typical active region; it possesses a single preceding sunspot and more diffuse following flux (Figure 1.20). The dashed lines are drawn to indicate rough levels in the atmosphere, even though the notion of a plane-parallel atmosphere is no longer completely relevant; for example, as described above, transition region material shows up as loop-structures rather than plane sheets and may be spread over a wide range of heights by spicular motions. Possible causes of the flow are outlined in Section 6.5.2.

(1) *Evershed flow* in a sunspot penumbra is described in Section 1.4.2D. It is present as an outflow of 6 to 7 km s^{-1} along dark filaments at the photosphere (with an average over light and dark filaments of 0.5 to 1 km s^{-1}) and a rather faster inflow at chromospheric levels.

(2) *Downflow* is found within the *intense fields* of supergranule boundaries (Section 1.3.2), which show up particularly well at the trailing edge of an active region. The speed is 1 to 2 km s^{-1} at the photosphere and 3 to 4 km s^{-1} at the chromosphere.

(3) *Surges* tend to be located especially at the leading edge of an active region. They are streams of plasma that are ejected upwards along slightly curved paths at typically 20 to 30 km s^{-1}, though occasionally a surge may move at 100 to 200 km s^{-1}, reach a height of up to 200 000 km and last 10 to 20 min. The surge plasma either fades from view or returns back along the same path. It originates from a small flare-like brightening close to a sunspot. At the feet of the small surges that occur close to sunspots in evolving or developing active regions, one finds bright dots in the wings of Hα known as *Ellerman bombs*. They are probably located

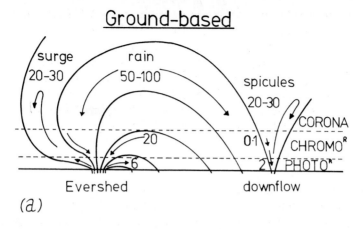

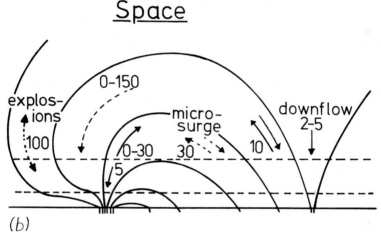

Fig. 1.23. Several types of active-region flow from (a) ground-based and (b) space observations. The schematic active region has preceding magnetic flux (left) concentrated as a sunspot and following flux (right) more diffuse. Heavy-headed arrows indicate the flow directions and the numbers give the typical speeds in km s^{-1}. In (b) transient flows have dashed arrows, while large-scale steady flows are indicated by solid arrows.

at the sites of tiny satellite spots and last for typically 20 min (though sometimes a few hours); they appear rapidly over 2 to 3 min and disappear just as quickly.

(4) *Spicules* are ejected at supergranule boundaries possibly from the low chromosphere or below although they are seen only in the high chromosphere. They reach a speed of 20 to 30 km s^{-1} and a height of typically 11 000 km before fading from view or occasionally falling back (Section 1.3.3).

(5) *Coronal rain* is cool plasma flowing down along curved paths at the free-fall speed of 50 to 100 km s^{-1}.

Observations from space on OSO-8 and the HRTS rockets have shown the existence of flows in the transition region (Figure 1.23(b)), which may be classified as *transient* or *steady*, although their cause remains a mystery.

(1) *Transient, small-scale, fast flows* have lifetimes of a minute or less and dimensions of only 1" to 2". Over a sunspot these flows are often downward at 0 to 150 km s^{-1}, but it is not known whether they are related to coronal rain. Over a plage, EUV *microsurges* (or *bursts*) are seen, in which the intensity increases by a factor of 10 to 100 and up-and-down flows of 30 km s^{-1} are generated; they may represent surges or coronal rain (Athay *et al.*, 1980). Occasionally, tiny *explosions* are found, with upward and downward motions of 100 km s^{-1}, but any connection with microsurges, X-ray bright points or magnetic reconnections has not yet been established.

(2) *Steady, large-scale, slow flows* last for hours or more. Over sunspots they are generally downward at 4 to 6 km s^{-1}, but they may sometimes be upward at less than 30 km s^{-1} (Lites *et al.*, 1976), while over the network they are downward at 2 to 5 km s^{-1}. In plage regions both upflows and downflows are found at typically 10 km s^{-1}. At present, the conditions which give rise to an upward rather than a downward motion are unknown; for instance, there appears to be no simple relation to the magnetic field strength or configuration.

In summary, active regions owe their very existence to the magnetic field, and so magnetohydrodynamic theory is uniquely qualified to explain their structure. In particular, the main aims are to understand the structure of the active-region loops of which the region is composed (Section 6.5) and to model the rich variety of motions that are present (Section 6.5.2).

1.4.2. SUNSPOTS

1.4.2A. Development

The most intense phase of an active region is characterised by the presence in the photosphere of sunspots, which are cooler than their surroundings and represent exceptionally strong concentrations of magnetic flux (e.g., Bray and Loughhead, 1964). They are observed to form in the following manner. Magnetic flux first appears in the upwelling at the centre of a supergranulation cell and is seen in Hα as an *arch-filament system* (Section 1.4.1A). The footpoints migrate to the cell boundary during the next 4 or 5 h and flux tends to be concentrated most at a junction of three cells where a *pore* eventually appears over about 45 min. Pores are darker than the surrounding photosphere and have no penumbra; they have diameters of 1" to 5" (700 to 4000 km), about 50% of the photospheric brightness and field strengths in excess of 1500 G. Often they last only hours or days, but sometimes one develops into a small sunspot. During the growth-phase of the sunspot, say between 3 and 10 days, more and more magnetic flux is added to it. This is evidenced by the approach to the spot of *moving magnetic features* (or *magnetic knots*) with speeds of 0.25 to 1.0 km s^{-1}; they have the same polarity as the spot and many appear as pores visible in white light. Typically, the spots are formed in pairs in this way, both spots remaining between supergranules but gradually moving apart to a maximum separation of about 150 000 km, five times a supergranule diameter.

Most sunspots disappear within a few days of forming, but some large ones (usually leaders) last much longer, slowly decaying over a few months. For these an annular cell or *moat* develops, with a typical diameter of 40 000 to 60 000 km. It is swept clear of magnetic flux except for tiny *moving magnetic features* that may have either polarity; these migrate outwards at constant speeds up to 2 km s^{-1}, the net outwards flux transport being 6×10^9 to 8×10^{11} Wb h^{-1} and of the same polarity as the sunspot. Eventually, a sunspot either decreases in size or breaks up. Often a *light-bridge* of normal photospheric intensity forms across the umbra, especially just before a spot divides.

Fifty-three percent of sunspot groups are bipolar (β-type), with the spots concentrated at the preceding and following sides of a group and having opposite polarity.

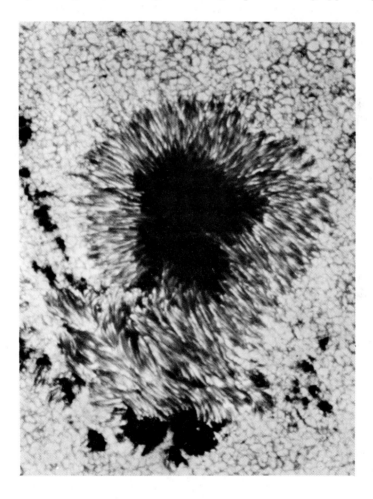

TACHE SOLAIRE

Fig. 1.24. A large sunspot showing fine penumbral structure (Muller, 1973).

Forty-six percent of spots are unipolar (α-type) and only 1% are complex in their polarity (γ-type). Examples are given in Figures 1.24 and 1.27.

1.4.2B. Umbra

The central dark area of a sunspot is known as the *umbra*, with a typical diameter of 10 000 to 20 000 km, about 0.4 of the total spot diameter. In it the magnetic field strength and temperature are fairly uniform and the intensity in visible light is only 5 to 15% of the photospheric value.

The magnetic field in the centre of an umbra is vertical and has a strength of about 2000 to 3000 G, though it may reach as much as 4000 G. It decreases gradually to 1000 to 1500 G near the penumbral-photospheric boundary. The flux of a large spot is typically 10^{13} Wb (10^{21} Mx) and that of a large sunspot group may be 2×10^{14} Wb.

As a large spot develops, its maximum field strength B_{max} increases over a few days, remains constant and then decreases as the spot decays, as shown in Figure 1.25. Strangely, however, the spot area decreases with time while the maximum field is constant.

The effective temperature of a spot umbra is typically only 3700 K, some 2100 K cooler than that of the surrounding photosphere. The radiative energy flux from a sunspot is therefore only 1.2×10^7 W m^{-2} (1.2×10^{10} erg cm^{-2} s^{-1}), or one-fifth of the normal photospheric value (6×10^7 W m^{-2}). Many models for the variation of the umbral temperature with height have been proposed; for example, Mattig (1958) suggested a surface value of 3300 K, increasing to 5250 K at a depth of 300 km.

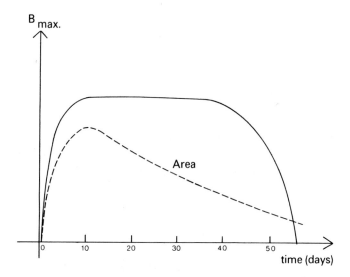

Fig. 1.25. A typical variation with time of the maximum magnetic field strength (solid) and area (dashed) of a large sunspot. The field strength peaks at 3 kG, while the maximum area is 4×10^{-4} of the Sun's hemisphere (after Cowling, 1946).

Within an umbra at high resolution one may find 20 or so *umbral dots*, with a diameter of only 150 to 200 km and a normal photospheric brightness; they have depths of about 100 km and are moving upwards at 0.5 km s^{-1} with a lifetime of 1500 s and a temperature of 5700 K. It appears that the magnetic field in umbral dots is the same as in the umbra. At lower resolution, umbral dots are described as *umbral granulation*. Umbral granules resemble ordinary photospheric granules, but they are fainter, more closely packed and have substantially longer lifetimes. At the photosphere, there are also *umbral oscillations* with periods 145 to 185 s, vertical motions of 0.2 km s^{-1} and a horizontal size of up to 2000 km. They are also present in the chromosphere (Hα), but there the period is shorter and the velocity amplitude much larger (1 to 6 km s^{-1}).

At the chromosphere there appear (in the Calcium K line) *umbral flashes*, which last only 50 s (with a rapid increase and slow decrease in brightness) and tend to repeat every 145 s. They have diameters of 2000 km and field strengths of 2000 G and may be caused by upward-propagating magneto-acoustic waves. The material is seen to move up at 6 km s^{-1} and towards the penumbra at 40 km s^{-1}, although the exciter may move up at 90 km s^{-1}.

1.4.2C. Penumbra

The umbra is surrounded by the *penumbra*, which consists of light and dark radial filaments that are typically 5000 to 7000 km long and 300 to 400 km in width. Individual penumbral filaments endure typically $\frac{1}{2}$ to 6 h, by comparison with a lifetime of days or months for the sunspot as a whole. The intensity of a bright filament is typically 95% of the surrounding photosphere, while that of a dark filament is only

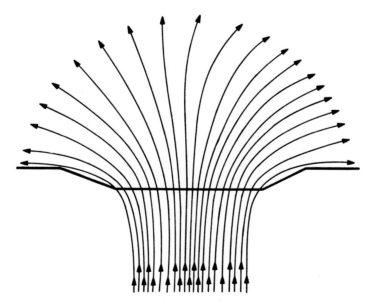

Fig. 1.26. A sketch of the magnetic field lines in a vertical plane through an idealised sunspot, regarded as a single flux tube. The thick line marks the photospheric level and the Wilson depression (Parker, 1979b).

60%. At high resolution the bright filaments seem to consist of bright grains aligned on a dark background (Muller, 1973). A grain may form anywhere in the penumbra and then move slowly towards the umbra; its lifetime is 40 min to 3 h and its temperature is 6300 K by comparison with 5700 K for the dark background.

Running penumbral waves start at the umbral boundary of a regular spot and propagate outwards at about 10 to 20 km s^{-1}. Their velocity amplitude is 1 km s^{-1} and the period is typically 260 to 280 s.

Measurements of the transverse field imply that the field lines fan out as one moves outwards from the centre of the spot, as indicated in Figure 1.26. Nearly all the magnetic flux from a sunspot probably returns to the photosphere through the many small *magnetic knots* that are located at supergranule boundaries (Section 1.3.2).

1.4.2D. Motion

At the beginning of this century, Evershed (1909) discovered the existence of radial motions at the photosphere, outwards from the umbra. In well-established spots the flow is continuous, with speeds of 6 to 7 km s^{-1} along the dark penumbral filaments; this is comparable with the sound speed but less than the local Alfvén speed. The radial *outflow* appears to reach a maximum near the outer boundary of the penumbra and the possible existence of azimuthal and vertical components is somewhat controversial. (A *slower inflow* is found along the bright penumbral filaments, which are probably moving upwards relative to the dark filaments; the tiny elongated grains (of which a bright filament is composed) move inwards from rest at the outer boundary to speeds of 0.5 km s^{-1} at the umbral boundary.) Higher up in the atmosphere the *Evershed outflow* becomes slower and eventually reverses its direction at chromospheric levels. This *Evershed inflow* also extends over a fairly large region surrounding the sunspot. It is observed along *superpenumbral fibrils*, which are loops about 5000 km high: the flow speed is typically 20 km s^{-1}, but may reach as much as 50 km s^{-1}.

Sunspots can exhibit *proper motion* relative to the surrounding photosphere in several ways. During the first few days of its life, a *p*-spot moves westwards and then, as the region decays, it slowly returns to roughly its original longitude. Occasionally, spots may divide or merge and a single spot or a bipolar pair may rotate; such motions may precede the occurrence of a large *flare*. Large flares may also be produced when the spots have a δ-configuration (with umbrae of opposite polarity inside the same penumbra) or an A-configuration (with spots of opposite polarity usually close together); in both these cases the *magnetic gradient* is exceptionally high.

The complexity of an active region may be enhanced by the presence of isolated areas of one polarity, completely surrounded by the opposite polarity; they are called *parasites* or *inclusions* and are preferred locations for flares, surges and Ellerman bombs. In particular, *magnetic satellites* often surround a large spot of opposite polarity and sometimes show up as *satellite spots*.

The appearance of a sunspot changes as it passes from the east to the west limb of the Sun, the east side of the penumbra being thinner than the west side when the spot is located near the west limb and vice versa, as shown in Figure 1.27. This is known

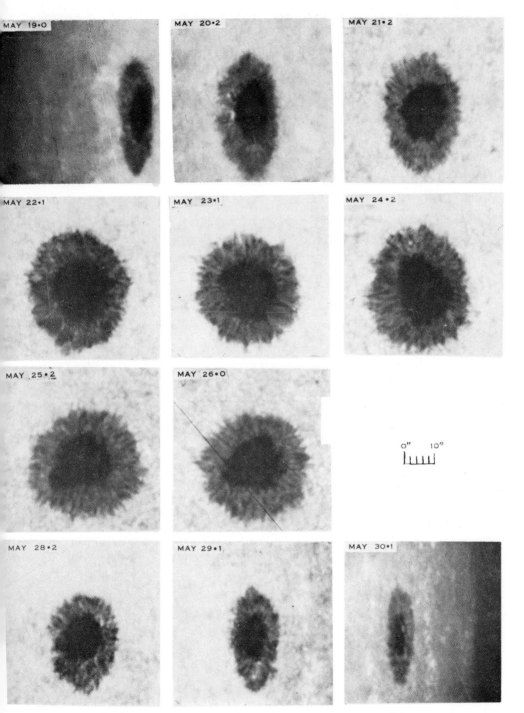

Fig. 1.27. The Wilson effect in a sunspot, showing the changing appearance of the spot as it passes from the east to the west limb of the Sun (Bray and Loughhead, 1964).

52   CHAPTER 1

as the *Wilson effect* (Wilson, 1774) and implies that the sunspot is a saucer-like depression of about 500 to 700 km below the photosphere. The effect is caused by the fact that the sunspot is more transparent than the surrounding photosphere (because of its lower temperature and density) and so the observed light comes from a greater depth.

1.4.2E. Solar Cycle

Traditionally, the state of the solar cycle is measured by counting the number of sunspots (f) and sunspot groups (g) visible on the disc and combining them as the

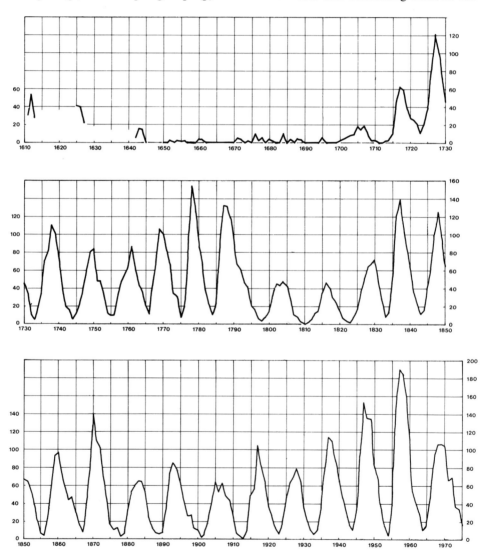

Fig. 1.28.   The variation of the sunspot number with time (courtesy J. Eddy).

Wolf number

$$R = K(10g + f),$$

where K is an observer correction factor (typically 0.6). The resulting sunspot number has a cyclic behaviour, as indicated in Figure 1.28.

Although the periodicity was discovered only in 1843 (by Schwabe) and the Wolf number was introduced only in 1848, this index of solar activity has been evaluated for earlier times by searching previous records. In particular, auroral records have been used to extrapolate the level of activity back for 2000 yr before the time ($\approx$ A.D. 1700) when sunspot records cease to be reliable. With the exception of granulation, almost all the observable features of the solar atmosphere owe their existence to the magnetic field, of which a sunspot is just the most intense example. It is not surprising, therefore, that all these features, such as prominences and flares (but excluding the network, fibrils and spicules), also vary in phase with the sunspot cycle; they are not purely local in character, but are constrained by some kind of global magnetic oscillation in the solar interior.

The *sunspot cycle* is often referred to loosely as the *11-yr cycle*, but there is a considerable variation in the period and in this century the average has been nearer $10\frac{1}{2}$ yr. For example, between 1750 and 1958 the average time between maxima was 10.9 yr, with a range from 7.3 to 17.1 yr; whereas the average period between minima was 11.1 yr, with a variation between 9.0 and 13.6 yr. Also, a wide range in the maximum and minimum number of sunspots is found: the average maximum is 108.2 sunspots, with a variation from 48.7 to 201.3, while the average minimum is 5.1, ranging between 0 and 11.2. Another feature of the cycle is that (except for the low maxima) the rise from minimum to maximum is shorter than the decline to minimum again: the average rise-time is 4.5 yr, while the decay time is 6.5 yr, with the asymmetry increasing with the height of the maximum. However, for a period of 70 yr from 1645, known as the *Maunder minimum*, the Sun appears to have been largely without sunspots at all! Thus, as pointed out by Eddy (1976), the solar cycle is not as regular as we once thought.

The short-term variation in sunspot number is considerable and appears largely random, although a 27-day variation is present; this corresponds to the rotation period and is caused partly by the persistence of large spots for more than one rotation and partly by preferred longitudes for their appearance. A tendency can be seen in Figure 1.28 for very low or high peaks in the solar cycle to be present every seven or eight cycles, which suggests that a longer cycle may be operating on an 80-yr period. Even longer periods of several hundred years have also been suggested.

Some intriguing observational facts about bipolar sunspot groups are that, throughout an 11-yr cycle, the polarity of all leading spots in the northern hemisphere is the same and reverses its sense at the start of a new cycle; also, leading spots in the southern hemisphere have the opposite polarity to those in the north (Figure 1.29). These *laws of sunspot polarity* were put forward by Hale and Nicholson in 1925 and are found to be obeyed by about 97% of sunspot groups. (Indeed, those freaks with inverted polarity are often associated with high flare activity.) Another feature indicated in Figure 1.29 is that the magnetic axis of a bipolar group is inclined (by typically 10°) in such a way that the leading spot tends to be closer to the equator.

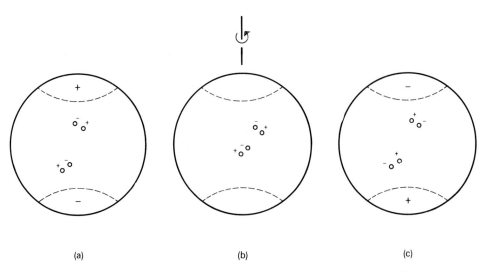

Fig. 1.29. The polarity of sunspots and polar regions for (a) the start of one cycle, (b) the maximum of that cycle and (c) the start of the next cycle.

It is also found that the net polar flux in one hemisphere has the same polarity as the leader spots there, and the polar field reverses sign at solar maximum or a year or two afterwards, although the reversals at the two poles may occur at different times. It appears that a polar reversal is caused by a migration (over about a year) of flux (with the polarity of trailing spots) from 40° latitude to the pole and this is accompanied by a migration of filaments, which occupy the boundary between large-scale unipolar regions of opposite polarity. At sunspot minimum, the polar fields are strong and make the polar coronal holes relatively large and symmetric, but near sunspot maximum the polar holes contract and disappear as the polar fields weaken (e.g., Sheeley, 1980). Two sunspot cycles are needed to return the Sun to the same magnetic state; the resulting 22-yr periodicity is known as the *Hale cycle*.

Most sunspots are confined to belts between the equator and latitudes $\pm 35°$. At any one time there may be a considerable spread of latitude, but, in 1859, Carrington discovered that the average latitude depends on the phase of the cycle: this remarkable latitude drift (Sporer's Law) towards the equator as the cycle proceeds, is indicated in Maunder's 'butterfly diagram' (Figure 1.30). At the beginning of a new solar cycle, the average latitude is 28° from the equator; after 6 yr it is 12° and after 11 yr it has fallen to 7°. Consecutive cycles may overlap by typically 2 yr, in the sense that near sunspot minimum the last spots of one cycle will be present at low latitudes, while the new spots of the next cycle are appearing at higher latitudes and with the opposite polarity.

The above general behaviour of the Sun's most intense magnetic fields shows in practice a considerable variability, which cannot at present be predicted in advance. For example, as the number of spots climbs from a sunspot minimum, it is uncertain just how high the maximum will reach, and at the end of one cycle there is an air

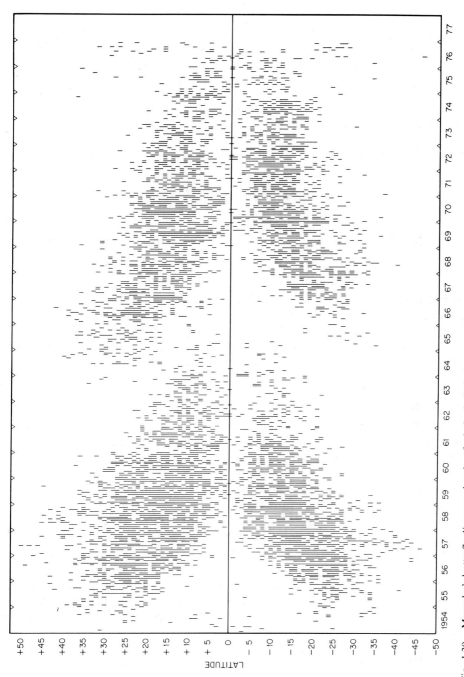

Fig. 1.30. Maunder's butterfly diagram, showing the latitude drift of sunspot occurrence as a function of time for two sunspot cycles (courtesy R. Howard).

of expectancy as one awaits the birth of the first sunspots at high latitudes, heralding the dawn of a new cycle.

Sunspots provide the most fundamental challenges to magnetohydrodynamic theory, and a lively debate on the following basic questions is raging. How are sunspots cooled (Section 8.3) and formed (Section 8.6.1), and what is their structure (Section 8.4)? Is the solar cycle caused by a dynamo or by an oscillation in the primordial field (Section 9.4.4)? How are the various umbral motions produced (Section 4.9.4.)? What causes the penumbral filamentation and Evershed flow (Section 8.5 and Section 6.5.2)?

1.4.3. PROMINENCES

1.4.3A. Introduction

Prominences are amazing objects. They are located in the corona but possess temperatures a hundred times lower and densities a hundred or a thousand times greater than coronal values. In eclipse or coronagraph pictures, these cool, dense features appear bright at the limb (Figures 1.4 and 1.19), but in Hα-photographs of the disc they show up as thin, dark, meandering ribbons called *filaments* (Figures 1.3, 1.19 and 1.31).

Prominences have been classified morphologically in several different ways, but there appear to be two basic types; their characteristics have been well-described by Tandberg-Hanssen (1974) and Jensen *et al.* (1979).

(1) *A quiescent prominence* is an exceedingly stable structure and may last for many months. It begins life as a relatively small *active-region* (or *plage*) *filament*, which is located either along the magnetic inversion line between the two main polarity regions of an active region or at the edge of an active region where it meets a surrounding region of opposite polarity. Sometimes it may enter a sunspot from one side. As the active region disperses, the prominence grows thicker and longer to become a *quiescent filament*. It may continue growing for many months up to 10^6 km (600 000 mi!) in length, and in the process it migrates slowly towards the nearest pole. Typical values for the properties of a quiescent prominence are:

density (n_e): 10^{17} m^{-3},
temperature (T_e): 7000 K,
magnetic field (B): 5–10 G,
length: 200 000 km,
height: 50 000 km,
width: 6000 km.

(2) *Active prominences* are located in active regions and are usually associated with solar flares. They are dynamic structures with violent motions and have lifetimes of only minutes or hours. There are various types, such as *surges*, *sprays* (which may well be erupting plage filaments) and *loop prominences*: both their magnetic field (about 100 G) and average temperature are much higher than for quiescent prominences. A description of active prominences can be found in Section 1.4.4; for the remainder of this section, our remarks concern mainly their quiescent cousins.

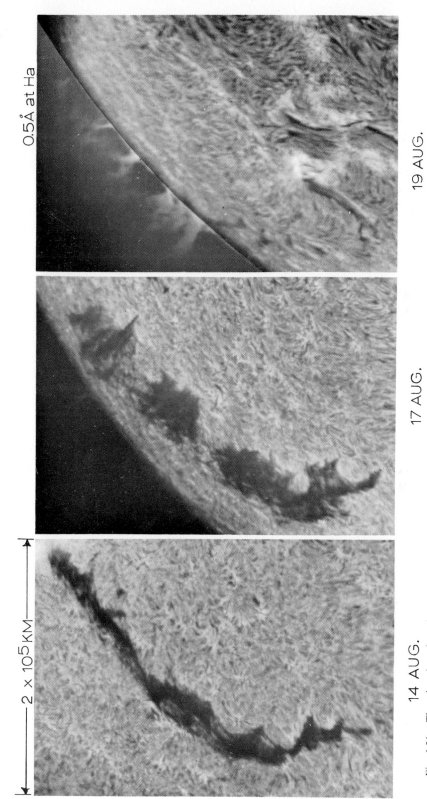

Fig. 1.31. The migration of a quiescent prominence in 1966 to the limb due to solar rotation, as viewed in Hα (courtesy S. Martin, Lockheed Solar Observatory).

Quiescent prominences have captivated the solar observer for centuries. One was observed at an eclipse in the Middle Ages (1239) and described as a 'burning hole'; another in 1733 was called a 'red flame'. However, by the beginning of the nineteenth century, their existence had been forgotten. At the 1842 eclipse they were rediscovered, but the observers were so surprised that they did not give a reliable description: some even thought they were 'mountains' on the Sun. In 1860 they were photographed, and in 1868 spectroscopic techniques were introduced which led to the discovery of He. Babcock found in the 1920s that, when viewed on the disc, a filament always lies along a so-called *polarity* (or *magnetic*)-*inversion line*, where the line-of-sight magnetic field reverses its sign. In the next decade, Lyot's invention of the *coronagraph* enabled him to observe prominences at the limb without waiting for an eclipse; then, in 1957, Kippenhahn and Schlüter put forward their classic model for the support of the dense prominence material against gravity by a magnetic field (Section 11.2.1).

1.4.3B. Properties

A quiescent prominence is a huge, almost vertical sheet of dense, cool plasma surrounded by a hotter and rarer coronal environment. Its *density* ranges between 10^{16} m^{-3} and 10^{17} m^{-3}, with the ratio of protons to neutral H atoms lying between about 1 and 10. The central *temperature* lies between 5000 K and 8000 K. The *dimensions* may have the following range: length 60 000 km to 600 000 km, height 15 000 to 100 000 km, thickness 4000 to 15 000 km. An active-region prominence is typically a factor of three or four smaller than its mature quiescent form; its temperature is much the same, but its density is rather larger ($\gtrsim 10^{17}$ m^{-3}) and its height is at most 20 000 km.

The *magnetic field* in quiescent prominences observed at the limb has a line-of-sight component (from the Zeeman effect) that varies from no observable field to 30–40 G. Tandberg-Hanssen (1974) finds a mean value of 7.3 G, with about half the observations in the range 3 to 8 G. Harvey (1969) gives a mean value of 6.6 G for 1967, while Rust (1967) gives 5 G for 1965 and finds that the magnetic field increases by roughly 50% over the height of the prominence. The average angle between the direction of the magnetic field and the long axis of a prominence is about 15°. (Similar results have been given more recently by Leroy (1979) using the Hanle effect.) Active-region prominences have higher field strengths, often from 20 to 70 G, though possibly as much as 100 to 200 G; it appears that for this type the magnetic field may be aligned approximately with the filament, whereas for quiescent ones the field runs more across the filament.

1.4.3C. Development

Prominences change slowly in overall shape with a lifetime between 1 and 300 days. Low-latitude quiescent prominences possess a mean lifetime of about 2 rotations (50 days), whereas those at high latitudes have an average duration of 5.1 rotations (140 days). The formation of a filament within an active region takes typically a few hours or a day. A *quiescent* filament may appear more slowly (over the course of a day or so), either between two nearby active regions, or at the boundary of an active region, or in a remnant active region (Martin, 1973); but always the birth takes place at a polarity inversion line. Another condition for the formation of many filaments within an active region or between adjacent active regions is that the Hα fibrils first align themselves end-to-end along a path called a *filament channel*, which eventually becomes a filament. Once the filament is formed the fibrils on either side are found to be directed roughly parallel to the filament. This alignment of fibrils suggests that the magnetic field is directed approximately along the filament. Eventually, a quiescent prominence disappears by either slowly dispersing and breaking up, or erupting, or flowing down to the chromosphere.

As a prominence migrates towards a pole, it is stretched more and more by the action of differential rotation into an east–west direction, while its width and height remain relatively constant. Prominences tend to occupy *zones* about 10° polewards of the sunspot latitudes, which move towards the equator as the solar cycle progresses. They are also located in *polar zones*, where they are oriented nearly parallel to the equator and sometimes form a *polar crown* around a polar cap at about latitude 70°. The polar zones form about 3 yr after spot maximum and migrate towards the poles; at about the next maximum they reach the poles and are accompanied by the reversal in polarity of the polar magnetic fields.

1.4.3D. Structure

Often a prominence reaches downwards towards the chromosphere in a series of regularly spaced *feet*, which resemble great tree trunks. These feet are located at supergranule boundaries and are joined by huge arches (Figure 1.32b). Within a prominence there is much *fine structure* in the form of vertical *threads* of length 5000 km and diameter about 300 km or less (Figure 1.32b); material continually streams slowly down these threads and down the arches into the chromosphere at speeds of only 1 km s^{-1}, which is much less than the free-fall speed. The resulting loss of mass is immense and would drain the prominence in a day or so if it were not being replenished somehow. Hardly any motion along the axis of a quiet quiescent filament is observed, unless it interacts with a sunspot, but active-region filaments often show matter flowing along the axis into a sunspot.

In eclipse and coronagraph pictures, one finds a region of reduced density, known as a *coronal cavity*, surrounding a prominence; it can also be seen in soft X-ray photographs as a region of reduced intensity, but the mass defect is insufficient by an order of magnitude to account for the prominence mass. A hot, closed coronal arcade or arch is present above and around the cavity, and above that there lies a *helmet streamer*; it may live for several months and has a broad base (up to a solar radius in diameter) with a pointed top at 1 or 2 $R_\odot$ above the limb. At transition-region temperatures the large-scale region around a prominence is found to possess upflows at 6 to 10 km s^{-1}. The *prominence-corona interface* between the cool threads and the hot corona is extremely thin (at most only a few hundred kilometres) and so possesses an exceptionally high temperature gradient; the electron pressure there is about 2×10^{-3} N m^{-2} (2×10^{-2} dyne cm^{-2}), which is about a quarter of the quiet-Sun transition region pressure but about the same as the prominence

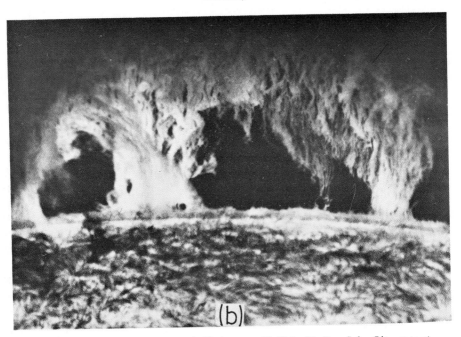

Fig. 1.32. Examples of quiescent prominences seen at the limb. (a) Three quiescent prominences observed in the lines of Ly α ($\simeq 10^4$ K), Mg x (1.5×10^6 K) and O vi (3×10^5 K) by Skylab (courtesy E. Schmahl, Harvard).

(b) A prominence 40 000 mi high seen in Hα (courtesy H. Zirin, Big Bear Solar Observatory).

pressure. The EUV data from Skylab shows that, with 5 arc-second resolution, the prominence size is identical in lines formed between 10^4 and 3×10^5 K; above this temperature the prominence begins to broaden and merge with the surrounding corona (Schmahl, 1979). Also, the data are consistent with a model of typically four vertical threads (and eight sandwiching interfaces) located across the width of a prominence; the conductive flux in each interface is estimated to be 140 W m^{-2} (1.4×10^5 erg cm^{-2} s^{-1}) and is sufficient to balance the radiative loss from the interface and possibly the whole prominence.

1.4.3 E. Eruption

Active-region and quiescent prominences can become *activated* and exhibit several types of large-scale motion. For example, a prominence may become larger and darker (when viewed on the disc as a filament) or brighter (when viewed at the limb). At the same time, there may be an increase in turbulent (or helical) motion or flow along the filament. This type of activation sometimes fades away after an hour or so and sometimes it leads to an *eruption*, as described below. In other cases, prominence material may drain away from the summit along a curved arc at speeds of 100 km s^{-1}. A third type of activation is called a *winking filament*, when the prominence performs a damped oscillation for 2 to 5 oscillations with a period of 6 to 40 min; it is initiated by the passage of a shock wave from a distant large flare.

At some stage in its life, an active-region filament or quiescent filament may become completely unstable and erupt, especially once it exceeds about 50 000 km in height. It ascends as an *erupting prominence* (Figure 1.33) and eventually disappears; some of the material escapes from the Sun altogether while some descends to the chromosphere along helical arches.

In two-thirds of cases, the prominence reforms in the same place and with much the same shape over the course of 1 to 7 days. The eruption of an old *quiescent prominence* is referred to as a *disparition brusque* (sudden disappearance); it starts with a slow rising motion at a few km s^{-1} and may take several hours. It is accompanied by an X-ray brightening (e.g. Figure 1.34) and occasionally by the appearance of Hα flare ribbons. Usually the cause of the eruption is a mystery, but sometimes it may be initiated by a disturbance from an emerging flux region or a flare. The eruption of an *active-region filament* is much more rapid and takes about $\frac{1}{2}$ h or less; while it is still ascending at high speed, a two-ribbon Hα flare begins (Section 1.4.4) and, in this case, the filament generally reforms after only a few hours.

1.4.3F. Coronal Transients

Erupting prominences are accompanied by *coronal transients* (e.g., MacQueen, 1980), which have been observed in the outer corona by Skylab (Figure 1.35). They represent outward-moving loops or clouds; the material probably originates in the low corona above the prominence, rather than from the interior of the prominence. The loops are oriented in planes inclined at less than 20° to the original filament and the thickness of the legs of the loop *increases linearly with height*. During the Skylab period,

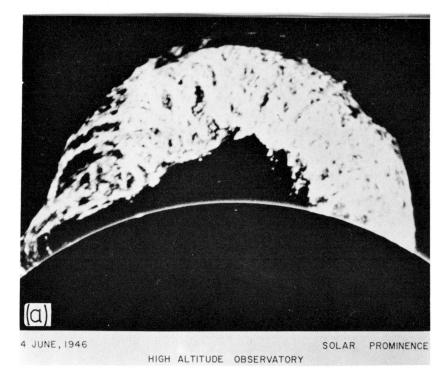

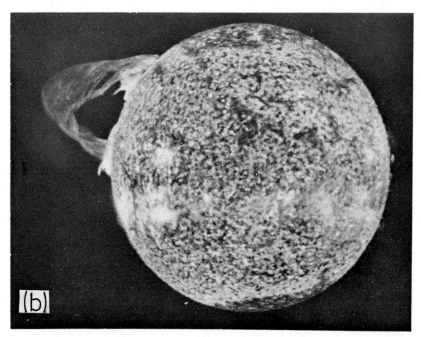

Fig. 1.33. Examples of erupting prominences: (a) 4 June 1946 in Hα (courtesy G. Newkirk, High Altitude Observatory); (b) 19 December 1973 in He II 304 (courtesy R. Tousey, Naval Research Laboratory).

SEPTEMBER 1, 1973

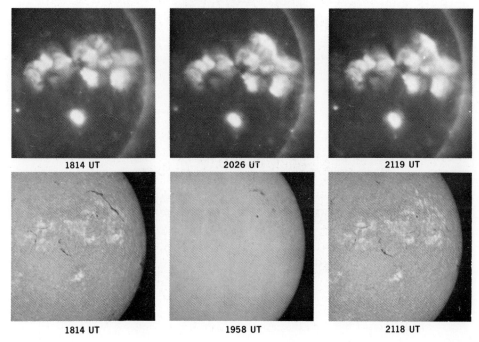

Fig. 1.34. The eruption of a quiescent filament from a remnant active region. The upper frames are in soft X-ray and two of the lower ones are in Hα. The Hα filament became active between 1800 and 1930 UT and the northern part disappeared between 1946 and 1958 UT when it was moving upwards at 150 km s^{-1}. The eruption produced an X-ray increase but only a slight Hα brightening (courtesy D. Rust, American Science and Engineering).

which coincided with the declining phase of the solar cycle, 110 coronal transients were observed, at a rate of roughly 1 per day.

At least 70% of the transients are found to be associated with erupting filaments and most of the remainder occur with sprays and large flares; it is possible that all transients accompany erupting filaments, only some of which give flares. The importance of coronal transients can be gauged from the fact that the kinetic energy of the transient mass may be twice the radiative energy from a major flare. Also, transients cause modifications in the large-scale structure of the corona and a depletion of coronal material. The transient mass is typically ten times the prominence mass and transients account for 5% of the total mass loss from the Sun.

Most of the acceleration of a transient occurs below $2 R_\odot$ so that between 2 and $6 R_\odot$ its speed is either constant or slightly increasing. The speeds have a wide range, from less than 100 km s^{-1} up to 1200 km s^{-1}, with an average value within the field of view of the Skylab instrument (2 to $6 R_\odot$) of 470 km s^{-1}. However, transients associated with large flares escape much more rapidly than those associated with erupting quiescent prominences; the average values are 775 km s^{-1} and 330 km s^{-1}, respectively. Most of the events moving faster than 400 km s^{-1} were also associated with type II or type IV radio bursts, the former being generated by shock waves.

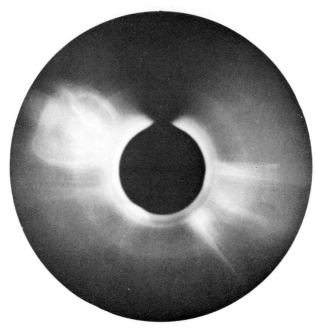

Fig. 1.35. A coronal transient associated with a prominence eruption (10 June 1973), observed by the High Altitude Observatory coronagraph on board Skylab (courtesy R. MacQueen).

Assuming synchrotron emission, the radio data imply a rather large magnetic field of 4 to 5 G at 3 to 5 $R_\odot$, which in turn means a value for β of only 0.1, so that the driving force must be magnetic.

From a theoretical viewpoint there are many questions that need answering. How does a prominence form (Section 11.1)? What is the magnetic configuration and how is the plasma supported against gravity (Sections 11.2, 11.3)? What causes the fine structure? Why does a prominence erupt (Section 10.3)? What is the driving mechanism for a coronal transient (Section 11.4)?

1.4.4. SOLAR FLARES

The solar flare is a truly remarkable and beautiful phenomenon. It varies from being a simple, localised brightening to a bewilderingly complex event, the most violent in the solar system. It may be naively defined as a *rapid brightening* in $H\alpha$, but simultaneously can have manifestations right across the electromagnetic spectrum and may eject high-energy particles and blobs of plasma into the solar wind. The $H\alpha$ brightening is, in fact, believed to be only a secondary response to the conversion of magnetic energy into heat and particle energy much nearer up to the coronal part of *active-region loops*. This section gives a description of the basic features of a flare and a brief summary of some recent observations. Some of the theories are outlined in Chapter 10.

1.4.4A. Basic Description

The optical flare is most often observed in Hα, formed in the low chromosphere. It has two basic stages. During the *flash phase*, which lasts typically 5 min (but sometimes an hour), the intensity and area of the emission rapidly increase in value. Then, in the *main phase*, the intensity slowly declines over about an hour (though sometimes as much as a day).

However, the flare is probably initiated and the energy released in a high-temperature region above the cool Hα flare. This overlying region of coronal loops may be heated to tens of millions of degrees; it exhibits variations co-incident with the flash and main phases, but also shows two more distinct phases. As seen in Figure 1.36, the soft X-ray emission (< 10 keV) possesses a *preflare phase* for minutes (or possibly longer) before flare onset, because of an enhanced thermal emission from the coronal plasma. Also, for 100 to 1000 s at the start of the flare, an *impulsive phase* is sometimes present, as indicated by the appearance of a microwave burst and a hard X-ray burst (> 30 keV) caused by highly accelerated electrons. After the impulsive phase, some

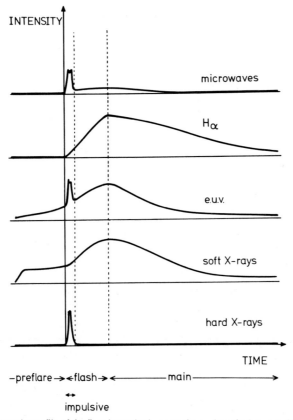

Fig. 1.36. A schematic profile of the flare intensity in several wavelengths (see, e.g., Kane, 1974; Lin, 1974). There is a great variation in the duration and complexity of the various phases. In a large event the preflare phase lasts typically 10 min, the impulsive phase a minute, the flash phase 5 min, and the main phase an hour.

of the largest events show a distinct second hard X-ray component due to a second phase of particle acceleration. For other events, the impulsive phase may take place after the Hα intensity has started increasing or it may be absent. In the latter case, where there is little particle acceleration, the events are known as *thermal flares*. They tend to occur in less complex regions and have a slower rise to flare maximum.

The disturbance of the solar atmosphere by a large flare may be manifested in many ways. Except when the flare occurs well within a closed field region, the hard X-ray emission is accompanied by type III radio bursts, which indicate the presence in the corona of streams of *high-energy electrons*, accelerated up to about a third the speed of light. Type II radio emissions and a Moreton wave both suggest the outwards propagation from the flare site (at more than 1000 km s^{-1}) of a fast *magnetohydrodynamic shock wave*. A *coronal transient* indicates the propagation into the solar wind of plasma contained in a huge magnetic loop system; such a *mass ejection* moves faster than the ambient solar wind and a shock wave is formed at its leading edge. Also, long-lasting type IV bursts show the presence of an *energetic plasma cloud* above the flare site. On the limb one may observe a *loop prominence*, which roughly outlines the loops of a potential magnetic field; down the legs of the loop drains an immense amount of material – far exceeding, in fact, the amount of matter in the surrounding corona!

The energy released in a flare varies from, say, 10^{22} J in a sub-flare to 3×10^{25} J (3×10^{32} erg) in the largest of events. The division between various types of energy in the latter case has been estimated as follows:

Electromagnetic radiation up to X-rays	10^{25} J
Interplanetary blast wave	10^{25} J
Fast electrons (hard X-rays)	5×10^{24} J
Subrelativistic nuclei	2×10^{24} J
Relativistic nuclei	3×10^{24} J
Total energy output	3×10^{25} J

It should be noted that other estimates suggest the hard X-ray contribution is instead possibly as much as 2×10^{25} J.

Since flares invariably occur in active regions and the other sources of energy seem inadequate, it has usually been assumed that it is the magnetic field which supplies the energy for a flare. 3×10^{25} J of energy would be released if, for instance, the whole of a 500 G magnetic field in a cube of side 30 000 km (or of a 100 G field in a cube of side 90 000 km) were annihilated; it also corresponds to the energy released if the field in a cube of side 40 000 km falls from 500 to 400 G.

There are several properties of the flare which are of particular importance when it comes to constructing a theoretical model. For example, active regions which are *complex* and *rapidly evolving* are the most likely flare producers; by contrast, simple bipolar regions show little activity. Also, there are no large-scale changes in the photosphere beneath a flare; variations in the magnetic flux through sunspots during all but the largest flares are usually no different from normal evolutionary changes. In any case, one would not expect the changes to be very great, since the flare energy is only a small fraction of the energy (10^{27} J) of a large sunspot group. Other interesting properties of some flares are those of 'homology' and 'sympathy'.

In the first case, a flare may repeatedly occur in the same place and with very similar characteristics. In the second, a flare may be triggered by the occurrence of another flare, even though the two are widely separated.

Flares have been categorised in many different ways, but two particular types (the *simple-loop* (or *compact*) flare and the *two-ribbon* flare) may be particularly significant. The former is small and consists of a loop (or collection of loops) which simply brightens and fades, without moving or changing its shape. It may occur in a large-scale unipolar region or near a simple sunspot and is sometimes accompanied by a surge, when a stream of plasma (with an average density of 10^{16} m^{-3}) may be ejected for up to 500 s. The two-ribbon flare is much larger and takes place near the active-region filament which often snakes its way through a complex active region. During the flash phase two ribbons of Hα emission form, one on each side of the filament (or filament channel), and throughout the main phase the ribbons move apart at 2 to 10 km s^{-1}. Frequently, they are seen to be connected by an arcade of so-called '*post*'-*flare loops*. Occasionally, the filament remains intact, though slightly disturbed, but usually it rises and disappears completely. Such an eruption of the filament begins slowly in the preflare phase, typically 10 min (but up to an hour) before flare onset, and continues at the flash phase with a much more rapid acceleration than before. A two-ribbon flare occasionally appears in a region completely devoid of sunspots when a quiescent prominence erupts. It is most important that we try to understand such 'pure' events as the spotless flare and the simple-loop flare, without the additional complication of complex sunspot fields. There are many differences between simple-loop and two-ribbon events. Simple-loop flares tend to have at most a single hard X-ray spike lasting about a minute, whereas two-ribbon flares may have multiple spikes. In soft X-rays simple-loop flares are characterised by small volumes, low heights, large energy densities and short time-scales, while two-ribbon flares have the opposite properties and often produce coronal transients. Furthermore, the energy release may be confined to the impulsive phase of a simple-loop event, but it continues throughout the main phase of a two-ribbon event.

1.4.4B. Ground-Based Observations

The main advances in flare observations over the past ten years have come from an improvement in the resolution (up to 1 arc sec) of ground-based photography and magnetic field measurements, and a high-resolution (up to 2 to 5 arc sec) study of the EUV and X-ray emitting parts of the flare. This has led to a much more detailed and broader description of the flare phenomenon.

Very high quality Hα photographs of flares have been produced by the Big Bear Solar Observatory, and these have enabled Zirin (1974) to give a detailed description of the development of the Hα flare. He stresses that the majority of flares arise after the emergence of new magnetic flux from below the photosphere, as sometimes evidenced by the appearance of small satellite sunspots close to large sunspots of the opposite polarity (Rust, 1968; Martres *et al.*, 1968). Flares are especially likely when the satellite is of 'following' polarity and appears just in front of the main spot with 'preceding' polarity. The most flare-productive active regions have steep field gradients (i.e., high electric currents) and also polarity inversions which result

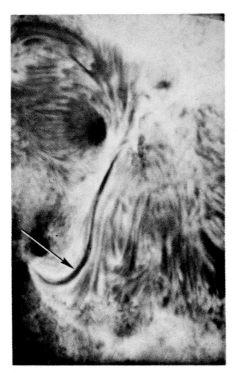 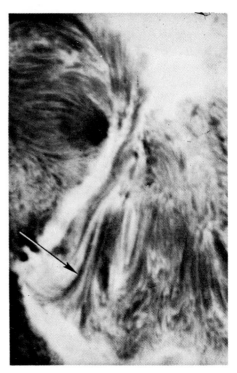

Before   **After**

Fig. 1.37. An example of Hα structure which is suggestive of reconnection during a flare. Arrows point to the lower end of a filament and show how it changed from being part of the whole filament *before* the flare to being part of an arch-filament system (which indicates emerging flux) *after* the flare (courtesy D. Rust, taken at Sacramento Peak Observatory).

from the arrival of these satellites. The presence of emerging flux in sheared fields is found to be particularly conducive to flaring (Neidig, 1978).

During the course of the flare, changes in fibril direction may be seen which suggest a reconnection of the magnetic field. An example which seems to indicate reconnection is shown in Figure 1.37. In some cases, fibril changes suggest a relaxation of the field to a lower energy state (Neidig, 1978, 1979).

Zirin also finds that great flares may begin at several points and then spread over a large area, giving the impression that the initial flare energy does not simply spread out but stimulates an extra energy release over a large region. Sometimes the initial brightening occurs simultaneously at points situated at opposite ends of a line of force. Furthermore, prominence eruptions are often associated with the flares which are slow *thermal* (no impulsive phase) and sometimes very large. There is a tendency for slowly rising flares to be thermal and fast-rising ones to produce non-thermal particles; also *knots* or *kernels* of Hα emission appear during the impulsive phase simultaneously with the hard X-ray and microwave bursts.

Magnetic field measurements have been important in aiding our understanding of the flare process and in supporting the inferences from Hα structure. Flares tend to occur near the '$H_{\|} = 0$ line' or 'polarity inversion line', along which the longitudinal magnetic field component changes sign and the field is largely horizontal. At Sacramento Peak Observatory, Rust (1972) showed that the impulsive-phase Hα kernels are located close to that part of the neutral line which skirts a satellite spot. He also demonstrated that, except possibly in the largest flares, flare-related photospheric magnetic field changes are confined to the satellite regions; they show an increase in strength throughout the preflare phase (half an hour) followed by a decrease during the flare itself. Moreover, it is those satellites which are evolving that tend to give rise to flares; but there is no obvious feature of the field morphology which indicates whether the resulting flare will be small or large. When observing a two-ribbon flare, one's attention may tend to be focussed on the filament and the ribbons, but it is the local emergence of new flux which seems to initiate the activation or eruption of the filament (Rust, 1976) and the brightest flare knots appear near the points of emerging flux (Rust and Bridges, 1975).

During June, 1972, a Campaign for Integrated Observations of Solar Flares (CINOF) was organized to investigate subflares. They were found to be very common and occurred at bright crossing points at the borders of supergranules, usually close to (and often either side of) the polarity inversion. All the main subflares which they studied were associated with surges or filament activations, which sometimes appeared before the events. In one case, a number of small flares seemed to indicate a long term build-up before a major event. Also, the initiation of two small flares after new flux emergence was observed.

By contrast with June 1972, and its subflares, August of the same year presented the opportunity to study several great flares, which were produced by one unusually complex sunspot group. The group possessed inverted polarity and extremely high field gradients, with a continual invasion of one polarity into the other and a distortion of the polarity-inversion line. Zirin and Tanaka (1973) have described its evolution. They see, in one impulsive flare, small (1″), short-duration (5 s) emission knots at the foot-points of magnetic flux loops; the knots occur at the same time as a hard X-ray burst and may be caused by the dumping of high energy ($ > 50$ keV) electrons in the low chromosphere. In another two-ribbon flare most of the Hα emission is produced in small kernels and there is a substantial velocity shear (6 km s^{-1}) across the polarity-inversion line. Rust and Bar (1973) discussed the loop prominences which join the two ribbons during the main phase of the same flare. The observed loops are situated progressively higher in the atmosphere and gradually become aligned more normal to the filament channel (indicative of a potential field). They suggested that particle acceleration continues through the main phase and that the fast particles bombard the chromosphere at the bright outer edges of the flare ribbons, whereas the dense, loop-prominence material falls onto the inner edges. Tanaka and Nakagawa (1973) demonstrated further that the observed proper motions of the sunspots before the flare could store enough energy in a force-free configuration to supply the flare.

Photospheric velocity measurements before and during flares have not been well-studied in the past. For the August 7th (1972) flare, Rust (1973) observed upward motions of 1 to 2 km s^{-1} in the half-hour before the flare, and downward motions of 5 km s^{-1} during and after the flare. The upward motions are said by Harvey (1974) to be a fairly general preflare property; maybe they are linked with emerging flux. Harvey also stressed that flares occur when there are shearing motions, in particular when the velocity and magnetic inversion lines cross (Martres *et al.*, 1971); whereas Levine and Nakagawa (1974) found flare activity to be located where the rate of strain is a local maximum and the vertical magnetic field or its gradient is zero.

1.4.4C. Space Observations

Perhaps the most striking of the recent advances in solar flare observations has been of the structure and time-evolution of the *high-energy flare* plasma. The EUV and X-ray emitting parts have been studied in great detail with instruments aboard the Apollo Telescope Mount, as summarised in Sturrock (1980) and Svestka (1981), where detailed references may be found. Most of these observations, however, have been of small events. They indicate the appearance of a high-temperature (up to 3×10^7 K) *kernel* or *flare core* during the impulsive phase and the start of the flash phase. It may be situated at some point (often the top) of a loop structure, which presumably outlines a magnetic flux tube in the upper chromosphere or lower corona; also, the loop may brighten gradually 10 min before the flare. The loop is typically 4000 to 13 000 km high and the kernel has a plasma density of at least 10^{17} m^{-3} (possibly as much as 3×10^{18} m^{-3}); its dimension may be as small as 4000 km or 1500 km and its temperature as large as 2×10^7 K. As the flare progresses, the EUV and X-ray emission may spread along the loop through the kernel and jump to other nearby loops; in one example, an arcade of loops brightens sequentially as a disturbance propagates along at about 200 km s^{-1}.

Studies have recently been made of the association between filament activation and the preflare soft X-ray enhancement which starts an average of 2 minutes before the Hα emission. Preflare filament activation is accompanied by a slight soft X-ray enhancement, with X-ray increases synchronised to filament expansion. This preflare soft X-ray increase is localised near a slowly erupting filament. Most filament disappearances outside active regions exhibit X-ray brightening but no Hα emission.

Although most of the Skylab observations were of subflares, one two-ribbon event (29 July 1973) is of particular interest (Figure 1.38). The preflare X-ray increase comes from a linear structure overlying a filament channel and is associated with the activation of a filament. At flare onset a high-temperature flare kernel is located at the intersection of the linear structure with a loop. During the early phase of the flare, the filament erupts, being most distorted near the loop, which then becomes the dominant part of the flare.

During the main phase, the two Hα ribbons are found to be connected by an arcade of 'post'-flare loops, which rises and becomes located more nearly perpendicular to the filament channel as the flare proceeds. The lowest loops show up as a cool *loop-prominence system* and above them are situated much hotter soft X-ray loops. These hot loops are typically 100 000 km high, with a bright summit 30 000 km wide and legs that are 10 000 km thick; even after a time of 30 h the X-ray emission is still significantly enhanced. The rise-speed is initially 10 to 20 km s^{-1}, falling to 1.5 km s^{-1} after 5 h and 0.5 km s^{-1} after 10 h. The density appears initially to be fairly uniform

3B FLARE OF JULY 29, 1973

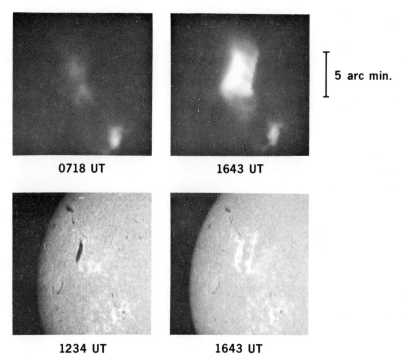

Fig. 1.38. Hα and X-ray pictures before and after the start of a two-ribbon flare. Beforehand there is an Hα filament and a faint X-ray cloud. Afterwards one sees the Hα ribbons joined by a system of X-ray loops, which have a temperature of 10^7 K and are brightest at their tops (courtesy D. Rust, American Science and Engineering).

along the loop, with a value of 7×10^{15} m^{-3} after 3 h, whereas after 12 h the summit and base values are 5×10^{15} m^{-3} and 6×10^{15} m^{-3}. However, the temperature is systematically higher inside the summit region than in the (relatively isothermal) legs; summit and leg values are 5×10^6 K and 3.5×10^6 K, respectively, after 3 h, and 4.4×10^6 K and 3.1×10^6 K after 12 h. The variations in X-ray intensity are mainly due to changes in density rather than temperature, and clear evidence has been presented for continued heating of the summit late in the main phase.

Analysis of Solar Maximum Mission data is revealing many exciting features (e.g., in Jordan, 1981). Images of the hard X-ray source for the first time show that the emission comes from loop footpoints in the impulsive phase and from the loop summit later on. It seems that the hard X-ray burst and Hα knots during the impulsive phase are caused by thick-target electrons. Simultaneously, a UV burst is produced as chromospheric material is being heated through 10^5 K up to 10^7 K. The particles may well be accelerated at the footpoints, and the hard X-ray burst sometimes possesses structure with a time-scale of $\frac{1}{10}$ s. Also, some asymmetric loops have been observed with hard X-rays produced at the wide end and thermal conductive heating at the other. Finally, much direct evidence has been produced for chromospheric evaporation.

Some fundamental theoretical questions are posed by solar flares (Chapter 10). What is the preflare magnetic configuration and why does it become unstable? How is the magnetic energy converted into heat, particle energy and mass motion? Can the flaring of a single loop be modelled? What is the explanation for the eruption of a filament and the subsequent "post-flare" loop phenomenon?

CHAPTER 2

THE BASIC EQUATIONS OF MAGNETOHYDRODYNAMICS

In the magnetohydrodynamic approximation, the behaviour of a continuous plasma is governed by a simplified form of Maxwell's equations (Section 2.1.1), together with Ohm's Law (Section 2.1.2), a gas law (Section 2.2.3) and equations of mass continuity (Section 2.2.1), motion (Section 2.2.2) and energy (Section 2.3). For simplicity, we shall use these equations as a basis, although an alternative would be to start with the Boltzmann equations for electrons and protons (e.g., Boyd and Sanderson, 1969). (Then the zeroth, first and second moments would give the equations of continuity, momentum and energy for each species, while the fluid equation of motion and the generalised Ohm's Law (Section 2.1.3) would follow from the sum and difference of the momentum equations for electrons and protons.) It is useful to eliminate the electromagnetic fields between Maxwell's equations (and Ohm's Law) and so reduce them to a single equation, known as the *induction equation* (Section 2.1.4); it relates the *plasma velocity* (**v**) to the *magnetic induction* (**B**) (often loosely referred to as the magnetic field in an astrophysical context).

2.1. Electromagnetic Equations

2.1.1. MAXWELL'S EQUATIONS

We begin with Maxwell's equations in m k s units,

$$\nabla \times \mathbf{B} = \mu \mathbf{j} + \frac{1}{c^2} \frac{\partial \mathbf{E}}{\partial t}, \tag{2.1}$$

$$\nabla \cdot \mathbf{B} = 0, \tag{2.2}$$

$$\nabla \times \mathbf{E} = -\frac{\partial \mathbf{B}}{\partial t}, \tag{2.3}$$

$$\nabla \cdot \mathbf{E} = \frac{\rho^*}{\varepsilon}, \tag{2.4}$$

where the constitutive relations $\mathbf{H} = \mathbf{B}/\mu$, $\mathbf{D} = \varepsilon \mathbf{E}$ have been used to eliminate the magnetic field (**H**) and electric displacement (**D**). (For the solar plasma, μ and ε are invariably approximated by their vacuum values, μ_0 and ε_0, respectively.) In these equations, **E** is the *electric field*, ρ^* the *charge density*, **j** the *current density*, $\mu_0 (= 4\pi \times 10^{-7} \text{ H m}^{-1})$ the *magnetic permeability* and $\varepsilon_0 (\approx 8.854 \times 10^{-12} \text{ F m}^{-1})$ the *permittivity of free space*, such that the *speed of light* in a vacuum is

$$c = (\mu_0 \varepsilon_0)^{-1/2} \approx 2.998 \times 10^8 \text{ m s}^{-1}.$$

E is measured in volts per metre (V m^{-1}), **B** in teslas (T) or webers m^{-2} (Wb m^{-2}), **j** in amps per square metre (A m^{-2}) and v in m s^{-1}. The relationship of the m k s system to the Gaussian system of units is given in Appendix 1, but note in particular that 1 G = 10^{-4} T. Magnetic induction values in formulae are usually measured in T, but in the text they are quoted in gauss, which are much more familiar to solar physicists. Furthermore, it is sometimes convenient to quote lengths in Mm, which is a useful measure for many solar features (1 Mm = 10^6 m); for example, an arc second is somewhat less than 1 Mm and the width of a filament is typically 5 Mm, while *granules* and *supergranules* have diameters of about 1 Mm and 30 Mm, respectively.

The first Maxwell equation shows that either currents or time-varying electric fields may produce magnetic fields, whereas the third and fourth equations imply that either electric charges or time-varying magnetic fields may give rise to electric fields. The second equation assumes that there are no magnetic poles and implies that a magnetic flux tube has a constant strength along its length (Section 2.9.2).

A fundamental supposition of magnetohydrodynamics is that the electromagnetic variations are non-relativistic or 'quasi-steady'. In other words,

$$V_0 \ll c, \tag{2.5}$$

where $V_0 = l_0/t_0$ is a characteristic electromagnetic (or plasma) speed, while l_0 and t_0 are a typical length and time. In addition, assume that

$$\frac{E_0}{l_0} \approx \frac{B_0}{t_0}, \tag{2.6}$$

where E_0 and B_0 are typical values of E and B, so that the two sides of Equation (2.3) are the same in order of magnitude. By comparing the sizes of the terms in Equation (2.1), it can be seen that the second term on the right-hand side (the displacement current) has magnitude

$$\frac{E_0}{c^2 t_0} \approx \frac{B_0 l_0}{c^2 t_0^2} = \frac{V_0^2}{c^2} \frac{B_0}{l_0}$$

$$\approx \frac{V_0^2}{c^2} |\text{curl } \mathbf{B}|,$$

which by Equation (2.5) is much smaller than the left-hand side of Equation (2.1).

Thus, one consequence of Equation (2.5) is that the term $c^{-2} \partial \mathbf{E}/\partial t$ may be neglected in Equation (2.1). Another is that the equation of charge continuity, which is obtained from the divergence of Equation (2.1), becomes $\nabla \cdot \mathbf{j} = 0$; this implies physically that local accumulations in time of charge are negligible and electric currents flow in closed circuits. A further consequence (upon using Equation (2.6)) is that the ratio of electrostatic to magnetic energy density, namely

$$\frac{\varepsilon_0 E_0^2}{B_0^2/\mu_0} \approx \frac{l_0^2}{t_0^2 c^2} = \frac{V_0^2}{c^2},$$

is much less than unity.

A characteristic of many processes in the solar atmosphere is that the plasma is, to a high degree of approximation, *electrically neutral* (otherwise, the resulting electric fields are enormous); in other words,

$$n_+ - n_- \ll n, \tag{2.7}$$

where n_+ and n_- are the *number densities* of positive and negative ions per unit volume and n is the total number density. The magnitude of the charge imbalance ($\rho^* = (n_+ - n_-)e$) is given from Equation (2.4) as

$$\rho^* \approx \frac{\varepsilon_0 E_0}{l_0},$$

or, after substituting for E_0 from Equation (2.6) and putting $t_0 = l_0/V_0$

$$\rho^* \approx \frac{\varepsilon_0 V_0 B_0}{l_0}.$$

The condition (2.7) for charge neutrality thus becomes

$$\frac{\varepsilon_0 V_0 B_0}{el_0} \ll n,$$

or, since $\varepsilon_0 \approx 8.8 \times 10^{-12}$ F m^{-1} and $e \approx 1.6 \times 10^{-19}$ C,

$$6 \times 10^7 \frac{V_0 B_0}{l_0} \ll n, \tag{2.8}$$

with B_0 in tesla.

Conditions (2.5) and (2.8) are well satisfied for many applications in solar physics. For example, in the photosphere near a *sunspot* one may observe motions with speeds $V_0 \approx 10^4$ m s^{-1}, which is a factor of 3×10^4 smaller than c. Typical values for magnetic field strength, length-scale and number density there are $B_0 \approx 0.1$ T (10^3 G), $l_0 \approx 10^5$ m, $n \approx 10^{20}$ m^{-3}, so that $6 \times 10^7 V_0 B_0/l_0 \simeq 6 \times 10^5$ and Equation (2.8) is easily satisfied.

In practice, a local charge imbalance produces an electrical field with a spatial range of the *Debye length* ($\lambda_D = (kT/(4\pi ne^2))^{1/2}$), which is a measure of the distance over which n_- can deviate appreciably from n_+ (e.g., Boyd and Sanderson, 1969). In fact, a *plasma* may be defined as an ionised gas for which λ_D is much smaller than other scales of interest.

2.1.2. Ohm's Law

Plasma moving at a non-relativistic speed in the presence of a magnetic field is subject to an electric field ($\mathbf{v} \times \mathbf{B}$) in addition to the electric field ($\mathbf{E}$) which would act on material at rest. Ohm's Law asserts that the current density is proportional to the *total* electric field (in a frame of reference moving with the plasma), and it may be written

$$\mathbf{j} = \sigma(\mathbf{E} + \mathbf{v} \times \mathbf{B}), \tag{2.9}$$

where σ is the *electric conductivity*, measured in mho m^{-1}.

2.1.3. GENERALISED OHM'S LAW

A generalisation of Ohm's Law may be more relevant in some regions of the photosphere and corona. It arises from a 'three-fluid' model for electrons, protons and neutral atoms (with number densities n_e, n_e and n_a, respectively) and may be written

$$n_e e \left(\mathbf{E}_0 + \frac{\nabla p_e}{n_e e} \right) = \frac{m_e}{e} \left(\frac{\partial \mathbf{j}}{\partial t} + \nabla \cdot (\mathbf{vj} + \mathbf{jv}) \right) + \left(\frac{1}{\Omega \tau_{ei}} + \frac{1}{\Omega \tau_{en}} \right) B\mathbf{j} + \mathbf{j} \times \mathbf{B} +$$

$$+ \frac{f^2 \Omega \tau_{in}}{B} \{ \nabla p_e \times \mathbf{B} - (\mathbf{j} \times \mathbf{B}) \times \mathbf{B} \}, \qquad (2.9a)$$

where $\mathbf{v}$ is the velocity of the centre of mass, $\mathbf{E}_0 \equiv \mathbf{E} + \mathbf{v} \times \mathbf{B}$ is the *total electric field*, $\Omega = eB/m_e$ is the electron gyration frequency, $f = n_a(n_a + n_e)^{-1}$ is the fraction of ions not ionised, τ_{ei} is the *electron-ion collision interval*, while τ_{en} and τ_{in} are the collision intervals for neutrals with electrons and ions (Cowling, 1976). In the derivation of this equation, terms of order m_e/m_i by comparison with unity have been neglected. The first term on the right-hand side, namely $(m_e/e) \partial \mathbf{j}/\partial t$, is due to electron inertia and is important only for time-scales as small as the collision times τ_{ei} or τ_{en}.

In many applications, both the electron inertia and electron pressure gradient terms are negligible. The generalised Ohm's Law (2.9a) then reduces to

$$\sigma \mathbf{E}_0 = \mathbf{j} + \frac{\sigma}{n_e e} \mathbf{j} \times \mathbf{B} - \frac{\sigma}{n_e e} \frac{f^2 \Omega \tau_{in}}{B} (\mathbf{j} \times \mathbf{B}) \times \mathbf{B}, \qquad (2.10)$$

where

$$\sigma = \frac{n_e e^2 m_e^{-1}}{\tau_{ei}^{-1} + \tau_{en}^{-1}} \qquad (2.11)$$

is the electrical conductivity in the absence of a magnetic field.

When the current is parallel to $\mathbf{B}$, Equation (2.10) reduces considerably to $\sigma \mathbf{E}_0 = \mathbf{j}$ with conductivity σ. On the other hand, when the current is normal to the magnetic field it may be rewritten

$$\sigma_3 \mathbf{E}_0 = \mathbf{j} + \frac{\sigma_3}{n_e e} \mathbf{j} \times \mathbf{B}, \qquad (2.12)$$

where

$$\sigma_3 = \frac{\sigma}{1 + f^2 B \sigma \Omega \tau_{in}/(n_e e)}$$

is the *Cowling conductivity*. It may be interpreted as the conductivity normal to the magnetic field (defined as the ratio of j to the electric field component $(\mathbf{E}_0 \cdot \mathbf{j}/j)$ in the direction of $\mathbf{j}$). Equation (2.12) may be solved for $\mathbf{j}$ to give

$$\mathbf{j} = \sigma_1 \mathbf{E}_0 + \sigma_2 \mathbf{B} \times \mathbf{E}_0/B,$$

where

$$\sigma_1 = \frac{\sigma_3}{1 + (\sigma_3 B/(n_e e))^2}$$

is the *direct conductivity* and

$$\sigma_2 = \frac{\sigma_3 B}{n_e e} \sigma_1$$

is the *Hall conductivity*.

Also, from Equation (2.12) the rate of energy dissipation ($\mathbf{E}_0 \cdot \mathbf{j}$) may be written $\mathbf{E}_0 \cdot \mathbf{j} = j^2/\sigma_3$, this is larger than the dissipation (j^2/σ) in the absence of neutrals, because collisions of neutrals with charged particles make σ_3 smaller than σ. In the *photosphere*, for example, σ_3 may be much smaller than σ, and so there may be a significant diffusion of charged particles relative to neutrals (known as *ambipolar diffusion*).

When the gas is fully ionised so that $n_a = 0$, the conductivities reduce to

$$\sigma_1 = \frac{\sigma}{1 + \Omega^2 \tau_{ei}^2}, \qquad \sigma_2 = \Omega \tau_{ei} \sigma_1, \qquad \sigma_3 = \sigma = \frac{n_e e^2 \tau_{ei}}{m_e}.$$

By comparison with Ohm's Law ($j = \sigma E$) in an unmagnetised medium, it can be seen that the effect of the magnetic field is to reduce the current flow parallel to $\mathbf{E}_0$ and to give rise to an extra (Hall) current normal to $\mathbf{E}_0$ and $\mathbf{B}$, the total dissipation (j^2/σ) being unaltered. The Hall term is produced by the drifting of charged particles across the magnetic field. It dominates when $\Omega \tau_{ei} \gg 1$, so that the electrons spiral freely between collisions, and it is much less important than the direct conduction term when $\Omega \tau_{ei} \ll 1$. (However, it is only in extremely thin regions of high current concentration that Ohm's Law departs at all from $\mathbf{E} + \mathbf{v} \times \mathbf{B} = \mathbf{0}$, as stressed in Sections 2.1.4 and 2.6.)

2.1.4. Induction Equation

It is convenient to eliminate $\mathbf{E}$ and $\mathbf{j}$ between Equations (2.1), (2.3) and the simple form (2.9) of Ohm's Law, to give

$$\frac{\partial \mathbf{B}}{\partial t} = -\nabla \times (-\mathbf{v} \times \mathbf{B} + \mathbf{j}/\sigma)$$

$$= \nabla \times (\mathbf{v} \times \mathbf{B}) - \nabla \times (\eta \nabla \times \mathbf{B}),$$

where $\eta = 1/(\mu\sigma)$, is the *magnetic diffusivity* (which must not be confused with the *electrical resistivity* (σ^{-1})). After using Equation (2.2) and the vector identity

$$\nabla \times (\nabla \times \mathbf{B}) = \nabla(\nabla \cdot \mathbf{B}) - (\nabla \cdot \nabla)\mathbf{B},$$

one finds (for uniform η)

$$\frac{\partial \mathbf{B}}{\partial t} = \nabla \times (\mathbf{v} \times \mathbf{B}) + \eta \nabla^2 \mathbf{B}, \tag{2.13}$$

known as the *induction equation*. When $\mathbf{v}$ is prescribed, it can be used to determine $\mathbf{B}$, subject to the condition

$$\text{div } \mathbf{B} = 0. \tag{2.14}$$

The resulting current density and electric field follow from Ampere's Law,

$$\mathbf{j} = \mathbf{V} \times \mathbf{B}/\mu, \qquad (2.15)$$

and Ohm's Law

$$\mathbf{E} = -\mathbf{v} \times \mathbf{B} + \mathbf{j}/\sigma, \qquad (2.16)$$

while the charge density can be determined, if need be, from Equation (2.4). Physical consequences of the induction equation are discussed further in Section 2.6.

The magnetic field is here regarded as primary and the electric current and electric field as secondary. Note, in particular, that Equation (2.15) rather than Equation (2.16) determines the current. This is because, in many solar applications, the length-scales are so large that the current term in Equation (2.16) is completely negligible; thus, save in regions of high current concentration such as *current sheets*, the presence of motion implies that $\mathbf{E} \approx -\mathbf{v} \times \mathbf{B}$. This is sometimes referred to as the infinite conductivity limit, but it would be better called the 'large length-scale' limit, since what makes the last term so small in astrophysical plasmas (compared with most terrestrial plasmas) is the relatively large length-scale rather than the conductivity (which varies relatively little).

When there are no motions, the magnetic field is determined from *magnetohydrostatic* considerations (Chapter 3), and the resulting electric current and electric field are given by Equations (2.15) and (2.16); such electric fields are, in practice, extremely small due to the large length-scale (l_0) for magnetic field variations. For example, a typical active-region magnetic field (B_0) of 100 G and a plasma speed (V_0) of 10^3 m s^{-1} give an electric field

$$E_0 \approx V_0 B_0 \approx 10 \text{ V m}^{-1}.$$

But, if there are no motions and the length-scale l_0 is, say, 10 Mm (10^7 m), the current density and electric field become only

$$j_0 \approx \frac{B_0}{\mu l_0} \approx 8 \times 10^{-4} \text{ A m}^{-2}$$

and (for $\sigma = 10^3$ mho m^{-1})

$$E_0 \approx \frac{j_0}{\sigma} \approx 8 \times 10^{-7} \text{ V m}^{-1}!$$

2.1.5. Electrical Conductivity

For a *fully-ionised*, collision-dominated, plasma Equation (2.11) gives

$$\sigma = \frac{n_e e^2 \tau_{ei}}{m_e},$$

where, according to Spitzer (1962), the *effective electron collision-time* is

$$\tau_{ei} = 0.266 \times 10^6 \frac{T^{3/2}}{n_e \ln \Lambda} \text{ s},$$

TABLE 2.1

The variation of ln Λ with density $n(\text{m}^{-3})$ and temperature $T(\text{K})$

Temperature	Density					
	10^{12}	10^{15}	10^{18}	10^{21}	10^{24}	10^{27}
10^4	16.3	12.8	9.43	5.97		
10^5	19.7	16.3	12.8	9.43	5.97	
10^6	22.8	19.3	15.9	12.4	8.96	5.54
10^7	25.1	21.6	18.1	14.7	11.2	7.85

with n measured in m^{-3}. The *Coulomb logarithm* (ln Λ) is generally between 5 and 20 in value and has a weak dependence on temperature and density, as indicated in Table 2.1. The electrical conductivity therefore becomes

$$\sigma = 1.53 \times 10^{-2} \frac{T^{3/2}}{\ln \Lambda} \text{ mho m}^{-1}, \tag{2.17}$$

while the corresponding expression for magnetic diffusivity is

$$\eta = \frac{m_e}{\mu n_e e^2 \tau_{ei}} = 5.2 \times 10^7 \ln \Lambda \ T^{-3/2} \text{ m}^2 \text{ s}^{-1}. \tag{2.18}$$

In particular, typical values for η in the solar chromosphere and corona are $8 \times 10^8 T^{-3/2}$ and $10^9 T^{-3/2}$ m² s⁻¹, respectively, assuming a fully-ionised plasma.

When the hydrogen plasma is only *partially ionised* with a neutral density n_n, say, Equation (2.11) implies that the above expressions for η should be multiplied by $(1 + \tau_{ei}/\tau_{en})$, where

$$\frac{\tau_{ei}}{\tau_{en}} \approx 5.2 \times 10^{-11} \frac{n_n}{n_e} \frac{T^2}{\ln \Lambda}.$$

Thus, for example, near the Sun's temperature-minimum region we have τ_{ei}/τ_{en} of order 0.001 (n_n/n_e); here less than 10^{-6} of the H is ionised, so that most of the electrons arise from metal ionisation. (See Figure 1.11(b), which shows that for the Harvard-Smithsonian Reference Atmosphere the fraction of H ionised falls from 3×10^{-4} at $\tau_{5000} = 1$ to a minimum of 2×10^{-7} at an altitude of 500 km and then rises through 10^{-3} at 1000 km to 1 at 2000 km.)

For a turbulent plasma, the collision time and corresponding electrical conductivity can often be much smaller than the Spitzer values. For example, low-frequency ion-sound turbulence has an *anomalous collision-time*

$$\tau^* \approx \omega_{pe}^{-1} \frac{n k_B T}{W}$$

and corresponding *anomalous conductivity*

$$\sigma^* \equiv \frac{n_e e^2 \tau^*}{m_e},$$

where W is the *turbulent energy density* and ω_{pe} the *electron plasma frequency*. After putting, for instance, $W \approx 0.01\, nk_B T$, these give numerically

$$\tau^* \approx 1.8 n_e^{-1/2}\, \text{s},$$
$$\sigma^* \approx 5.0 \times 10^{-8} n_e^{1/2}\, \text{mho m}^{-1}.$$

Such turbulence, however, occurs only for extremely high current densities (j_0) and correspondingly small magnetic field scale-lengths (l_0). The condition for its onset is (when $T_e \gg T_i$) that the *electron conduction speed* (u) exceeds the *thermal speed*, namely,

$$u > \left(\frac{k_B T_e}{m_e}\right)^{1/2}.$$

But $u \equiv j_0/(n_e e)$ and $j_0 \approx B_0/(\mu l_0)$, so this condition may be rewritten as

$$l_0 < \frac{B_0}{\mu n_e e}\left(\frac{m_e}{k_B T_e}\right)^{1/2}$$

or

$$l_0 < 1.3 \times 10^{21} \frac{B_0}{n_e T_e^{1/2}}\, \text{m},$$

with B_0 in tesla. For example, conditions characteristic of the corona ($T_e = 10^6$ K, $n_e = 10^{15}$ m^{-3}, $B_0 = 10^{-3}$ T) yield length-scales of less than 2 m!

If the plasma is known to be in a fluid turbulent state, such as, for instance, in the convection zone, it may be useful to introduce *eddy* transport coefficients. Thus, if the small-scale eddies have a speed v and a scale-length d, an *eddy magnetic diffusivity* may be defined as $\tilde{\eta} = vd$. For *granulation* (where $v \lesssim 1$ km s^{-1} and $d \approx 10^3$ km) it has a typical value of at most 10^9 m^2 s^{-1}, whereas for supergranules (where $v \approx 0.2$ km s^{-1} and $d \approx 3 \times 10^4$ km) it is somewhat larger (6×10^9 m^2 s^{-1}). By comparison, it is possible to infer values of 2×10^7 m^2 s^{-1} and 10^9 m^2 s^{-1} from the observed decay of sunspots (Section 8.6.2) and the dispersal of active regions.

2.2. Plasma Equations

2.2.1. Mass Continuity

The behaviour of the magnetic field, which is described by the induction equation, is coupled to that of the plasma by the presence of the velocity term in Equation (2.13). The plasma motion is in turn governed by equations of continuity, motion and energy. Consider first the equation of mass conservation, which may be written

$$\frac{D\rho}{Dt} + \rho \nabla \cdot \mathbf{v} = 0 \tag{2.19}$$

or

$$\frac{\partial \rho}{\partial t} + \nabla \cdot (\rho \mathbf{v}) = 0, \tag{2.20}$$

where

$$\frac{D}{Dt} \equiv \frac{\partial}{\partial t} + \mathbf{v} \cdot \mathbf{V}$$

is the material derivative for time variations following the motion. Equation (2.20) expresses the fact that the density at a point *increases* ($\partial \rho / \partial t > 0$) if mass flows *into* the surrounding region ($\mathbf{V} \cdot (\rho \mathbf{v}) < 0$), whereas it decreases when there is a divergence rather than a convergence of the mass flux.

2.2.2. Equation of Motion

Under conditions of electrical neutrality, the equation of motion may be written

$$\rho \frac{D\mathbf{v}}{Dt} = -\nabla p + \mathbf{j} \times \mathbf{B} + \mathbf{F}, \tag{2.21}$$

where ρ is the mass density, p is the plasma pressure (assumed scalar) and the material is, in general, subject to a plasma pressure gradient ∇p, a *Lorentz force* $\mathbf{j} \times \mathbf{B}$ per unit volume (discussed in Section 2.7), and a force $\mathbf{F} = \mathbf{F}_g + \mathbf{F}_v$, which represents the effects of *gravity* ($\mathbf{F}_g$) and *viscosity* ($\mathbf{F}_v$).

The gravitational force is

$$\mathbf{F}_g = -\rho g(r)\hat{\mathbf{r}}, \tag{2.22a}$$

where the unit vector ($\hat{\mathbf{r}}$) acts radially outwards from the centre of the Sun and the local gravitational acceleration ($g(r)$) may be written

$$g(r) = \frac{M(r)G}{r^2}, \tag{2.22b}$$

in terms of the mass ($M(r)$) of the Sun inside a radius r and the gravitational constant (G). At the surface $r = R_\odot (= 6.96 \times 10^8 \text{ m})$, $M = M_\odot$ ($= 1.991 \times 10^{30}$ kg) and the gravitational acceleration is $g_\odot = 274$ m s^{-2}.

The viscous force

$$\mathbf{F}_v = \rho v(\nabla^2 \mathbf{v} + \tfrac{1}{3}\nabla(\nabla \cdot \mathbf{v})) \tag{2.23a}$$

simplifies to

$$\mathbf{F}_v = \rho v \nabla^2 \mathbf{v} \tag{2.23b}$$

for incompressible flow, where v is the *coefficient of kinematic viscosity* (assumed uniform). For a fully-ionised H plasma, Spitzer (1962) gives

$$\rho v = 2.21 \times 10^{-16} \frac{T^{5/2}}{\ln \Lambda} \text{ kg m}^{-1} \text{ s}^{-1}.$$

In a frame of reference that is rotating with instantaneous *angular velocity* $\mathbf{\Omega}$ relative to an inertial frame, the equation of motion (2.21) at a distance $\mathbf{r}$ from the rotation axis is modified to

$$\rho \left(\frac{D\mathbf{v}}{Dt} + 2\mathbf{\Omega} \times \mathbf{v} \right) = -\nabla p + \mathbf{j} \times \mathbf{B} + \mathbf{F} + \rho \mathbf{r} \times \frac{d\mathbf{\Omega}}{dt} + \tfrac{1}{2}\rho \nabla |\mathbf{\Omega} \times \mathbf{r}|^2, \tag{2.24}$$

under the assumption that $|\mathbf{\Omega} \times \mathbf{r} + \mathbf{v}| \ll c$. In particular, when $\mathbf{\Omega}$ and ρ are both constant and $\boldsymbol{\omega} = \operatorname{curl} \mathbf{v}$ is the *relative vorticity*, the curl of Equation (2.24) gives

$$\rho \frac{D\boldsymbol{\omega}}{Dt} = \rho[(\boldsymbol{\omega} + 2\mathbf{\Omega}) \cdot \nabla]\mathbf{v} + \operatorname{curl}(\mathbf{j} \times \mathbf{B}) + \operatorname{curl} \mathbf{F} \qquad (2.25a)$$

or

$$\rho \frac{\partial \boldsymbol{\omega}}{\partial t} - \rho \nabla \times (\mathbf{v} \times \boldsymbol{\omega}) = 2\rho(\mathbf{\Omega} \cdot \nabla)\mathbf{v} + (\mathbf{B} \cdot \nabla)\mathbf{j} - (\mathbf{j} \cdot \nabla)\mathbf{B} + \nabla \times \mathbf{F}. \qquad (2.25b)$$

2.2.3. Perfect Gas Law

The gas pressure is determined by an equation of state, which is taken for simplicity as the perfect gas law

$$p = \frac{\tilde{R}}{\tilde{\mu}} \rho T, \qquad (2.26)$$

where $\tilde{R}$ is the *gas constant*, $\tilde{\mu}$ is the *mean atomic weight* (the average mass per particle in units of m_p). The tildes have been placed on $\tilde{\mu}$ and $\tilde{R}$ to distinguish them from the magnetic permeability μ (Section 2.1.1) and radial distance R in cylindrical polars. Also, the reader should beware that $\tilde{\mu}$ is sometimes incorporated into $\tilde{R}$ in Equation (2.26).

Often the *mean particle mass* (m) is used is place of $\tilde{\mu} = m/m_p$. Also, the *Boltzmann constant* (k_B) can be introduced, so that $\tilde{R}$ is replaced by $\tilde{R} = k_B/m_p$ and Equation (2.26) becomes

$$p = \frac{k_B}{m} \rho T. \qquad (2.26a)$$

In some cases it is more convenient to write the equations in terms of the *total number* (n) of particles per unit volume rather than ρ. Then the gas law becomes

$$p = nk_B T, \qquad (2.27)$$

and ρ is replaced by $\rho = \tilde{\mu} m_p n = mn$ in the equation of motion.

For a *fully-ionised H plasma*, there are two particles (a proton and an electron, denoted by subscripts p and e, respectively) for every proton; this implies that $\tilde{\mu} = 0.5$ and, in terms of the electron number density (n_e), we have

$$n \equiv n_p + n_e = 2n_e, \qquad \rho \equiv n_p m_p + n_e m_e \approx n_e m_p,$$

assuming electrical neutrality. In the solar atmosphere the presence of extra elements makes $\tilde{\mu} \approx 0.6$ (and $n \approx 1.9 n_e$), except close to the photosphere (where H and He are not fully-ionised) and in the core (where the composition is different), as indicated in Figure 2.1(a).

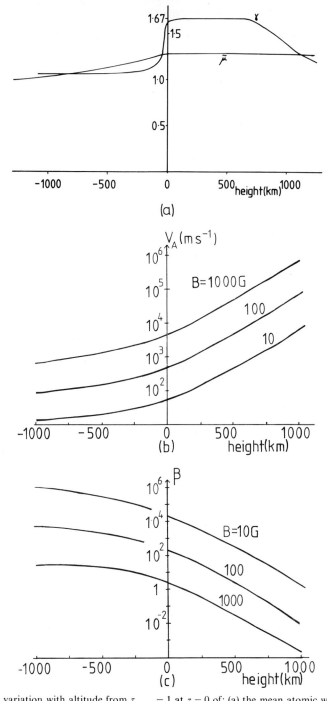

Fig. 2.1. The variation with altitude from $\tau_{5000} = 1$ at $z = 0$ of: (a) the mean atomic weight ($\tilde{\mu} = m/m_p$) and the ratio of specific heats (γ); (b) the Alfvén speed (v_A); (c) the plasma beta (β). For $z > 0$ the values are taken from the Harvard–Smithsonian Reference Atmosphere (Gingerich et al., 1971) and for $z < 0$, they come from Spruit's (1974) convection-zone model. Above 2000 km H is fully ionised and $\gamma \approx 5/3$. (Courtesy of A. Webb and P. Cargill.)

2.3. Energy Equations

2.3.1. DIFFERENT FORMS OF THE HEAT EQUATION

The last fundamental equation is the heat equation

$$\rho T \frac{Ds}{Dt} = -\mathscr{L}, \qquad (2.28a)$$

where the *energy loss function* $\mathscr{L}$ is the net effect of all the sinks and sources of energy and s is the *entropy* per unit mass of the plasma. The heat equation simply states that the rate of increase of heat for a unit volume as it moves in space is due to the net effect of the energy sinks and sources, which are described below. When the energy losses and gains balance, so that $\mathscr{L} \equiv 0$, the entropy is conserved.

The heat equation may be written in many alternative ways. For example, in terms of the *internal energy* (e) per unit mass, it becomes

$$\rho \left(\frac{De}{Dt} + p \frac{D}{Dt}\left(\frac{1}{\rho}\right) \right) = -\mathscr{L} \qquad (2.28b)$$

or

$$\rho \frac{De}{Dt} - \frac{p D\rho}{\rho\, Dt} = -\mathscr{L}. \qquad (2.28c)$$

For an ideal polytropic gas the internal energy is $e = c_v T$, where c_v is the *specific heat at constant volume*. Furthermore, the *specific heat at constant pressure* (c_p) and *ratio of specific heats* (γ) are defined by

$$c_p = c_v + \frac{k_B}{m}, \qquad \gamma = \frac{c_p}{c_v},$$

which together imply that

$$c_p = \frac{\gamma}{\gamma-1}\frac{k_B}{m}, \qquad c_v = \frac{1}{\gamma-1}\frac{k_B}{m}, \qquad (2.29a)$$

and

$$e = \frac{p}{(\gamma-1)\rho}. \qquad (2.29b)$$

The ratio of specific heats may also be written as

$$\gamma = \frac{\tilde{N}+2}{\tilde{N}}, \qquad (2.29c)$$

in terms of $\tilde{N}$, the number of degrees of freedom in the plasma. For fully-ionised H $\tilde{N} = 3$ and so $\gamma = 5/3$, but, in general, γ lies between 1 and 5/3. In a model solar atmosphere such as the Harvard–Smithsonian Reference Atmosphere, the effect of partial ionisation is to make γ fall typically from a maximum of 5/3 at $\tau_{5000} = 1$ to a minimum of 1.1 at an altitude of a few thousand km and then to rise rapidly to 5/3 again, as shown in Figure 2.1(a).

The energy Equation (2.28) may be rewritten in several other ways as follows. After substituting for e in (2.28c) and using the perfect gas law (2.26) for T, the terms on the left-hand side may be combined to give the concise form

$$\frac{\rho^\gamma}{\gamma - 1}\frac{D}{Dt}\left(\frac{p}{\rho^\gamma}\right) = -\mathscr{L}, \tag{2.28d}$$

or

$$\rho c_v T \frac{D}{Dt}\log\frac{p}{\rho^\gamma} = -\mathscr{L}, \tag{2.28e}$$

or

$$\frac{Dp}{Dt} - \frac{\gamma p}{\rho}\frac{D\rho}{Dt} = -(\gamma - 1)\mathscr{L}, \tag{2.28f}$$

or

$$\rho\frac{D}{Dt}(c_p T) - \frac{Dp}{Dt} = -\mathscr{L}, \tag{2.28g}$$

where $c_p T = \gamma p/((\gamma - 1)\rho)$ is the *enthalpy* per unit mass. Alternatively, Equation (2.19) may be used to substitute for $D\rho/Dt$ in Equation (2.28c) so that

$$\rho\frac{De}{Dt} + p\nabla\cdot\mathbf{v} = -\mathscr{L}, \tag{2.28h}$$

where the internal energy e is given by Equation (2.29b).

When the *pressure remains constant*, a convenient version of the energy equation is

$$\rho c_p \frac{DT}{Dt} = -\mathscr{L}. \tag{2.30a}$$

When the plasma is *thermally isolated* from its surroundings, in the sense that there is no exchange of heat ($\mathscr{L} = 0$), the change of state is said to be *adiabatic* and Equation (2.28d) shows that, following the motion, $p/\rho^\gamma = $ constant, for each plasma element. (Sometimes 'thermally isolated' refers instead to a state of vanishing heat flux across the boundaries.) This in turn means that the entropy of each element remains constant, since a comparison of Equations (2.28a) and (2.28e) implies in general

$$s = c_v \log(p/\rho^\gamma) + \text{constant}. \tag{2.31}$$

Formally, $\mathscr{L}$ is negligible in Equation (2.28) when the time-scale for changes in p, ρ and T is much smaller than the time-scales for radiation, conduction and heating; this is often valid for rapid changes associated with wave motions or instabilities.

2.3.2. Thermal conduction

Now consider in turn the various terms combined in the energy loss function ($\mathscr{L}$), which, in general, may be written as the rate of energy loss minus the rate of energy gain, namely,

$$\mathscr{L} = \nabla\cdot\mathbf{q} + L_r - j^2/\sigma - H, \tag{2.32}$$

where $\mathbf{q}$ is the *heat flux* due to particle conduction; L_r is the net radiation; j^2/σ is the *ohmic dissipation*; H represents the sum of all the other heating sources.

The heat flux vector may be written

$$\mathbf{q} = -\kappa \nabla T, \tag{2.33}$$

where κ is the *thermal conduction tensor*; in this case the divergence of the heat flux may be split into two parts,

$$\nabla_\parallel \cdot (\kappa_\parallel \nabla_\parallel T) + \nabla_\perp \cdot (\kappa_\perp \nabla_\perp T),$$

where subscripts $\parallel$ and $\perp$ refer to values along and across the magnetic field. Conduction along the magnetic field is primarily by electrons, and, for a fully-ionised H plasma, Spitzer (1962) gives

$$\kappa_\parallel = 1.8 \times 10^{-10} \frac{T^{5/2}}{\ln \Lambda} \text{ W m}^{-1} \text{ K}^{-1}. \tag{2.34}$$

Typical values are $4 \times 10^{-11} T^{5/2}, 10^{-11} T^{5/2}, 9 \times 10^{-12} T^{5/2}$ for photospheric, chromospheric and coronal regions, respectively. Conduction across the magnetic field is mainly by the protons in a fully-ionised H plasma, and it depends on the product

$$\Omega_i \tau_{ii} = 1.63 \times 10^{15} \frac{B T^{3/2}}{\ln \Lambda\, n}$$

of the *ion gyro-frequency* Ω_i and the *ion-ion collision time* τ_{ii} (which is $(m_i/m_e)^{1/2}$ times the electron collision time τ_{ei}). In most solar applications, the ions spiral many times between collisions, so that $\Omega_i \tau_{ii} \gg 1$ and Spitzer gives

$$\frac{\kappa_\perp}{\kappa_\parallel} = 2 \times 10^{-31} \frac{n^2}{T^3 B^2},$$

with B in tesla. More general expressions for transport coefficients are given by Braginsky (1965) and Burgers (1969), while the conditions for anomalous conduction are discussed by e.g. Somov (1978).

In a magnetic field that is strong enough to make $\kappa_\perp \ll \kappa_\parallel$, conduction is mainly along the field and the heat conduction term may be approximated by $\nabla \cdot (\kappa_\parallel \nabla_\parallel T)$, or, in terms of the distance s along a particular magnetic field line,

$$\frac{d}{ds}\left(\kappa_\parallel \frac{dT}{ds}\right) - \frac{\kappa_\parallel}{B}\frac{dB}{ds}\frac{dT}{ds};$$

the second term arises because the magnetic field, in general, varies with s. An alternative expression is simply

$$\frac{1}{A}\frac{d}{ds}\left(\kappa_\parallel \frac{dT}{ds} A\right),$$

where $A(s)$ is the cross-sectional area of a flux tube, related to the magnetic field strength by $(d/ds)(BA) = 0$, as shown in Section 2.9.2. In the particular case of radial symmetry, the conduction term reduces to

$$\frac{1}{r^2}\frac{d}{dr}\left(\kappa_\parallel \frac{dT}{dr} r^2\right).$$

2.3.3. Radiation

In the *solar interior*, energy transport by radiation (or convection) dominates particle conduction and the net radiation may be written as the divergence of a radiative flux ($\mathbf{q}_r$), namely

$$L_r = \nabla \cdot \mathbf{q}_r, \tag{2.35a}$$

where $\mathbf{q}_r = -\kappa_r \nabla T$ and $\kappa_r = 16\sigma_s T^3/(3\tilde{\kappa}\rho)$ is the *coefficient of radiative conductivity*; here σ_s is the Stefan–Boltzmann constant, $\tilde{\kappa}$ the *opacity* or *mass absorption coefficient* and $\tilde{\kappa}\rho$ is the *absorption coefficient*. When κ_r is locally uniform, the radiative loss reduces to the simple form

$$L_r = -\kappa_r \nabla^2 T. \tag{2.35b}$$

Furthermore, it is sometimes convenient to introduce the *thermal diffusivity* κ defined by

$$\kappa = \frac{\kappa_r}{\rho c_p},$$

so that at constant pressure the energy Equation (2.30a) reduces to

$$\frac{DT}{Dt} = \kappa \nabla^2 T. \tag{2.30b}$$

When waves or instabilities are being modelled (Chapters 4 and 7) on a time-scale τ, say, it is customary to assume perturbations that are adiabatic. However, in the photosphere and chromosphere the effects of *radiative damping* must often be included (e.g., Stix, 1970). It takes place on the *radiative relaxation time-scale* τ_R, so that when $\tau_R > \tau$ the plasma variations are approximately adiabatic; but when $\tau_R < \tau$ they are no longer adiabatic and, after a plasma element is displaced, it comes into thermal equilibrium with its surroundings before the restoring forces have time to act. For long wavelengths (λ) such that $\lambda \gg (\tilde{\kappa}\rho)^{-1}$, the plasma is optically thick to the disturbance and the expressions (2.35a) or (2.35b) may be used for L_r. For $\lambda \ll (\tilde{\kappa}\rho)^{-1}$, however, the plasma is effectively optically thin, and it has been shown by Spiegel (1957) that

$$\tau_R = \frac{c_v}{16\sigma_s \tilde{\kappa} T^3}.$$

In this case, if equilibrium values and small disturbances are denoted by subscripts 0 and 1, respectively, Equation (2.28c) becomes (with $\mathscr{L} = L_r$)

$$\frac{\partial T_1}{\partial t} - (\gamma - 1)\frac{T_0}{\rho_0}\frac{\partial \rho_1}{\partial t} = -\frac{T_1}{\tau_R}.$$

Thus, at constant density, the temperature evolution is governed by *Newton's Law of Cooling*, since $\partial T_1/\partial t$ becomes proportional to T_1.

For the *optically thin* part of the atmosphere ($T \gtrsim 2 \times 10^4$ K in the chromosphere and corona), the *radiative loss* (L_r) is no longer coupled to the radiation field by the radiation transfer equations and it takes the form

$$L_r = n_e n_H Q(T), \tag{2.35c}$$

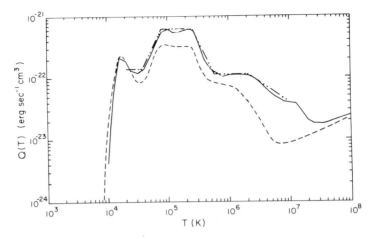

Fig. 2.2. The radiative loss function ($Q(T)$) derived by McWhirter et al. (1975) [- - -] and Raymond and Smith (1977) [——], together with an analytic fit of Rosner et al. (1978) [··—··]. (1 erg cm³ = 10^{-13} W m³.)

where n_e is the electron density (sometimes written as N) and n_H the density of H atoms or protons. (When the plasma is fully ionised, $n_H = n_e$.) The temperature variation (Q(T)) has been evaluated by a number of authors (Cox and Tucker, 1969; Tucker and Koren, 1971; McWhirter et al., 1975; Raymond and Smith, 1977) and is graphed in Figure 2.2. It is accurate only to within about a factor of two, and so the detailed variations should not be taken too seriously; the most important features are the presence of a maximum around 10^5 K and a minimum around 10^7 K. An analytic approximation is

$$Q(T) = \chi T^\alpha \text{ W m}^3, \qquad (2.35d)$$

with the temperature variation of the piecewise constants $\chi(T)$ and $\alpha(T)$ given in Table 2.2. (For lower temperatures than 2×10^4 K, Peres et al. (1982) have suggested that the *energetics* of the chromosphere may still be modelled to within a factor 2 by a net radiation of the form (2.35c), with $\alpha = 6.15$ and $\chi = 4.93 \times 10^{-62}$ for

TABLE 2.2
The variation with temperature $T(K)$ of α and χ (Equation (2.35d)) according to Rosner et al. (1978)

Range of T	α	χ
$10^{4.3} - 10^{4.6}$	0	$10^{-34.85}$
$10^{4.6} - 10^{4.9}$	2	10^{-44}
$10^{4.9} - 10^{5.4}$	0	$10^{-34.2}$
$10^{5.4} - 10^{5.75}$	-2	$10^{-23.4}$
$10^{5.75} - 10^{6.3}$	0	$10^{-34.94}$
$10^{6.3} - 10^7$	$-2/3$	$10^{-30.73}$

8×10^3 K $< T < 2 \times 10^4$ K, while $\alpha = 11.7$ and $\chi = 1.26 \times 10^{-83}$ for 4400 K $< T <$ < 8000 K. For temperatures above 10^6 K a good approximation is $\alpha = -0.5$, $\chi = 1.0 \times 10^{-32}$ for 10^6 K $< T < 10^{7.6}$ K and $\alpha = 0.5$, $\chi = 2.5 \times 10^{-40}$ for $10^{7.6}$ K $< T$, the latter contribution being due to bremsstrahlung.)

Less accurate forms that suffice for many purposes are

$$Q(T) = 10^{-34.44} \text{ W m}^3, 2 \times 10^4 \text{ K} < T < 2 \times 10^6 \text{ K},$$

or

$$Q(T) = 10^{-32.09} T^{-1/2} \text{ W m}^3, 2 \times 10^4 \text{ K} < T < 10^7 \text{ K},$$

or

$$Q(T) = 10^{-31.66} T^{-1/2} \text{ W m}^3, 10^5 \text{ K} < T < 10^7 \text{ K};$$

the departures from the full form are less than factors of 1.7, 3.7 and 1.4, respectively. (Occasionally, L_r is written proportional to $\rho^2 T^\alpha$ (e.g., Chapter 11), in which case the constant of proportionality in Equation (2.35d) is smaller by a factor 4 m^2 and is denoted by $\tilde{\chi}$.)

2.3.4. Heating

The heating term in Equation (2.32) may be written

$$H = \rho\varepsilon + H_v + H_w,$$

where ε is the *nuclear energy* generation rate (per unit mass) in the interior, H_v is the *viscous dissipation rate* (important for strong flows) and H_w is the *wave* (or other) *heating* term (for the outer atmosphere). The viscous heating is

$$H_v = \rho\nu(\tfrac{1}{2}e_{ij}e_{ij} - \tfrac{2}{3}(\nabla\cdot\mathbf{v})^2), \quad (2.36a)$$

where $e_{ij} = (\partial v_i/\partial x_j) + (\partial v_j/\partial x_i)$ is the rate of strain tensor. The wave heating contribution to H is not well-known. It is often assumed to be either uniform or proportional to density

$$H_w = \text{constant} \times n. \quad (2.36b)$$

A discussion of heating by acoustic waves, Alfvén waves or current dissipation can be found in Sections 6.3 and 6.4. In order of magnitude, wave-damping makes $H_w \approx F/\lambda$, where F is the wave flux at the base and λ is the *damping length-scale*. According to Rosner *et al.* (1978),

$$\lambda \approx \begin{cases} 7.5pT^{-2} \text{ m} & \text{for acoustic waves,} \\ 1.4 \times 10^4 T^{1/2} \text{ m} & \text{for acoustic shocks,} \\ R_c & \text{for Alfvén waves in a loop of radius } R_c. \end{cases}$$

For many applications, other sources and sinks of energy should be included on the right-hand side of Equation (2.28). For example, small-scale spicular motions may affect the large-scale coronal energy balance (Pneuman and Kopp, 1977); cool material is shot up into the corona (where it is heated) and falls back down, thus providing an additional heat sink for the lower corona and a source for the chromosphere. Furthermore, Raadu and Kuperus (1973) point out that quiescent prominences exhibit a continuous small-scale downflow of plasma at much less

2.3.5. ENERGETICS

The various equations we have set up in this chapter imply several relationships between the different types of energy such as heat, electrical energy and mechanical energy. First of all, the heat Equation (2.28) states that an

increase in entropy is due to *heat flow – radiation + heat sources.*

Also, the divergence of the *Poynting flux* ($\mathbf{E} \times \mathbf{H}$) may be written

$$\nabla \cdot (\mathbf{E} \times \mathbf{H}) = - \mathbf{E} \cdot \nabla \times \mathbf{H} + \mathbf{H} \cdot \nabla \times \mathbf{E}$$

and transformed, using Equation (2.15) for $\nabla \times \mathbf{H}$ and (2.3) for $\nabla \times \mathbf{E}$, to

$$- \nabla \cdot (\mathbf{E} \times \mathbf{H}) = \mathbf{E} \cdot \mathbf{j} + \frac{\partial}{\partial t}\left(\frac{B^2}{2\mu}\right). \tag{2.37}$$

Its physical interpretation is that an

inflow of electromagnetic energy $\mathbf{E} \times \mathbf{H}$ produces *electrical energy* $\mathbf{E} \cdot \mathbf{j}$ *for the plasma* + *a rise in magnetic energy* $B^2/(2\mu)$.

In turn, the electrical energy given to the plasma by the electromagnetic field may be rewritten, after substituting for $\mathbf{E}$ from Equation (2.16), as

$$\mathbf{E} \cdot \mathbf{j} = j^2/\sigma + \mathbf{v} \cdot \mathbf{j} \times \mathbf{B}; \tag{2.38}$$

this means that the

electrical energy appears as *heat by ohmic dissipation* + *work done by* $\mathbf{j} \times \mathbf{B}$ *force.*

Furthermore, the scalar product of $\mathbf{v}$ with the equation of motion (2.21) gives the so-called *mechanical energy equation*

$$\rho \frac{D}{Dt}\left(\frac{1}{2}v^2\right) = - \mathbf{v} \cdot \nabla p + \mathbf{v} \cdot \mathbf{j} \times \mathbf{B} + \mathbf{v} \cdot \mathbf{F}, \tag{2.39}$$

which implies that an

increase in speed is due to the *work done by* $-\nabla p, \mathbf{j} \times \mathbf{B}$ *and* $\mathbf{F}$.

Finally, Equations (2.38) and (2.39) may be combined with Equation (2.28h) to give another alternative to Equation (2.28), namely,

$$\rho \frac{D}{Dt}(e + \tfrac{1}{2}v^2) = - (\mathscr{L} + j^2/\sigma) + \mathbf{E} \cdot \mathbf{j} - \nabla \cdot (p\mathbf{v}) + \mathbf{v} \cdot \mathbf{F}, \tag{2.40a}$$

or, after using Equation (2.19),

$$\frac{\partial}{\partial t}(\rho e + \tfrac{1}{2}\rho v^2) + \nabla \cdot [(\rho e + \tfrac{1}{2}\rho v^2)\mathbf{v}] = -(\mathscr{L} + j^2/\sigma) + \mathbf{E} \cdot \mathbf{j} - \nabla \cdot (p\mathbf{v}) + \mathbf{v} \cdot \mathbf{F},$$

(2.40b)

where $-(\mathscr{L} + j^2/\sigma) = -\nabla \cdot \mathbf{q} - L_r + H$. Equation (2.40) expresses the fact that *the gain in material energy* (internal plus kinetic) is due to *heat flow, radiation, viscous dissipation, heat sources, electrical energy* and the *work done* by the *pressure* (and other forces **F**) in propulsion and compression. In the steady state it reduces to

$$\nabla \cdot [(\gamma p/(\gamma - 1) + \tfrac{1}{2}\rho v^2)\mathbf{v}] = -(\mathscr{L} + j^2/\sigma) + \mathbf{E} \cdot \mathbf{j} + \mathbf{v} \cdot \mathbf{F},$$

where $\gamma p/[(\gamma - 1)\rho]$ is the enthalpy per unit mass. Furthermore, in Equation (2.40) the sum of the terms H_v, $-\nabla \cdot (p\mathbf{v})$ and $\mathbf{v} \cdot \mathbf{F}_v$ may be written concisely as $-\partial/\partial x_j(p_{ij}v_i)$ in terms of the *stress tensor* p_{ij}.

2.4. Summary of Equations

The fundamental magnetohydrodynamic equations that are employed throughout this book are

$$\frac{\partial \mathbf{B}}{\partial t} = \nabla \times (\mathbf{v} \times \mathbf{B}) + \eta \nabla^2 \mathbf{B}, \tag{2.13}$$

$$\frac{D\rho}{Dt} + \rho \nabla \cdot \mathbf{v} = 0, \tag{2.19}$$

$$\rho \frac{D\mathbf{v}}{Dt} = -\nabla p + \mathbf{j} \times \mathbf{B} + \mathbf{F}, \tag{2.21}$$

$$p = \frac{k_B}{m}\rho T \left(= \frac{\tilde{R}}{\tilde{\mu}}\rho T \right), \tag{2.26}$$

$$\frac{\rho^\gamma}{\gamma - 1} \frac{D}{Dt}\left(\frac{p}{\rho^\gamma}\right) = -\nabla \cdot \mathbf{q} - L_r + j^2/\sigma + H, \tag{2.28}$$

where $\mathbf{F} = \mathbf{F}_g + \mathbf{F}_v$ is given by Equations (2.22) and (2.23), **q** is defined by Equation (2.33), L_r follows from Equation (2.35) and $H = \rho\varepsilon + H_v + H_w$ is given by Equation (2.36). These equations are, in general, coupled and serve to determine **v**, **B**, p, ρ and T. In addition, **j** and **E** are given explicitly by

$$\mathbf{j} = \nabla \times \mathbf{B}/\mu, \tag{2.15}$$

$$\mathbf{E} = -\mathbf{v} \times \mathbf{B} + \mathbf{j}/\sigma, \tag{2.16}$$

while **B** is subject to the condition

$$\nabla \cdot \mathbf{B} = 0. \tag{2.2}$$

Sometimes it is more convenient to deal in terms of the total number density (n) rather than ρ, in which case one replaces the density by $\rho = mn$ (for fully-ionised H) in the above equations.

2.4.1. ASSUMPTIONS

The assumptions made in deriving the above set of basic equations are as follows:

(1) The plasma is treated as a *continuum*. This is valid provided the length-scale for variations greatly exceeds typical internal plasma lengths such as the ion gyroradius.
(2) The plasma is assumed to be in thermodynamic equilibrium with distribution functions close to Maxwellian. This holds for time-scales much larger than the collision times and length-scales much longer than the mean free paths.
(3) The coefficients η and μ have been supposed uniform. Also most of the plasma properties are assumed *isotropic*. The exception is the coefficient of thermal conductivity, whose values along and normal to the magnetic field direction may differ greatly. A more comprehensive theory using tensor transport coefficients has been developed, but, for many applications, it is unnecessary.
(4) The equations are written for an *inertial* frame. The extra terms that arise for a frame rotating with the Sun may be important for large-scale processes.
(5) *Relativistic* effects are neglected, since the flow speed, sound speed and Alfvén speed are all assumed to be much smaller than the speed of light.
(6) The simple form of *Ohm's Law*, (2.16), is adopted for most applications, rather than its more general version (2.10).
(7) The plasma is treated as a *single fluid*, although two- or three-fluid models may be more relevant for the coolest or rarest parts of the solar atmosphere.

2.4.2. REDUCED FORMS OF THE EQUATIONS

Often, simplified forms of the basic equations are considered. For instance, approximations to the induction equation are discussed in Section 2.6. Also, if the temperature is assumed uniform and time-independent, the energy equation (2.28) is not needed. If the density changes of a moving plasma element are negligible ($D\rho/Dt = 0$), Equation (2.19) reduces to $\mathbf{V} \cdot \mathbf{v} = 0$; so that one needs to solve only the two coupled Equations (2.13) and (2.19) for $\mathbf{v}$ and $\mathbf{B}$, subject to the conditions that div $\mathbf{B}$ and div $\mathbf{v}$ vanish. This assumption of *incompressibility* is often made for simplicity in magnetohydrodynamics when one is more interested in other effects than compressibility and wishes to make analytic progress. But such a neglect of density variations is valid only when the *sound speed* $c_s = (\gamma p/\rho)^{1/2}$ is much larger than typical plasma speeds. Since plasma speeds are often of the order or less than the Alfvén speed $v_A = B/(\mu\rho)^{1/2}$, the condition for incompressibility becomes $c_s \gg v_A$ so that $\beta \equiv (2\mu p/B^2) \gg 1$. (Formally, the incompressible limit may be obtained by letting γ tend to infinity.) When performing a stability analysis, it is often found that the most unstable mode has an incompressible velocity perturbation, so that the density and pressure remain constant during a perturbation and no work is wasted in compressing the plasma unnecessarily.

These equations can be simplified further by prescribing $\mathbf{v}$ in Equation (2.13) at the outset and neglecting Equation (2.19), so that only **B** remains to be determined. With this so-called *kinematic* approach, frequently used in dynamo theory, only the effect of motions on the magnetic field is considered, while the influence of the magnetic field on the motions is neglected. This is valid provided the magnetic field is so small that $\beta \gg 1$ and pressure gradients dominate the Lorentz force.

Other reduced forms are obtained when $v \ll v_A$ and $v \ll c_s$, so that the inertial term in Equation (2.21) is negligible, and the magnetohydrostatic force balance results (Chapter 3). In particular, if also $\beta \ll 1$, the magnetic field is *force-free* (Section 3.5) and is determined by $\mathbf{j} \times \mathbf{B} = 0$, a particular case of which is the *potential field* (Section 3.4), with $\mathbf{j} = 0$. In these cases, the plasma motion along each field line is governed by the component of

$$\rho \frac{D\mathbf{v}}{Dt} = -\nabla p + \mathbf{F}$$

along the magnetic field; in particular, the equations for flow along a prescribed moving field line have been derived by Pneuman (1981). When the flow speed remains much smaller than the sound speed, this implies an evolution of the plasma through a series of equilibria parallel to the field. If the slow evolution is due to magnetic diffusion, **v** is determined by Equation (2.13), as in Section 3.5.5.

The Russian group of Syrovatsky and his coworkers (e.g., Somov and Syrovatsky, 1972; Ivanov and Platov, 1977) has used a strong-field approximation for a cold, perfectly conducting plasma ($c_s \ll v \ll v_A$). Here, the equations for the velocity (in two dimensions) reduce to

$$\frac{\partial A}{\partial t} + \mathbf{v} \cdot \nabla A = 0, \qquad \frac{\partial \mathbf{v}}{\partial t} \cdot \nabla A = 0,$$

where $A(x, y, t)$ is the z-component of the magnetic vector potential, which evolves through a series of prescribed potential solutions.

Often, when the energetics of a process are not the prime consideration, the energy equation is approximated either by $T = $ constant or, more generally, by the *polytropic approximation* $p/\rho^\alpha = $ constant, where α is a constant (distinct from γ). The latter approximation is simply meant to model temperature variations in a rough manner, but it may be derived from the full energy equation (2.28) when the only contribution to $\mathscr{L}$ is the conduction term and the conductive flux $\kappa \nabla T$ is proportional to the work done (vp) by the pressure.

2.5. Dimensionless Parameters

In terms of a typical plasma speed (V_0) and length-scale (l_0), the magnitude of the convective term in Equation (2.13) divided by that of the diffusive term is a dimensionless parameter

$$R_m = \frac{l_0 V_0}{\eta}, \tag{2.41}$$

known as the *magnetic Reynolds number*. (When V_0 is set equal to the Alfvén speed this becomes the Lundquist number.) It is a measure of the strength of the coupling between the flow and the magnetic field. In the laboratory, usually one finds $R_m \ll 1$

and the coupling is weak, whereas in the solar atmosphere generally $R_m \gg 1$ and the coupling is strong.

The *Reynolds number*

$$\mathrm{Re} = \frac{l_0 V_0}{\nu} \tag{2.42}$$

gives the ratio of the size of the inertial term to the viscous term in the equation of motion.

The *Mach number*

$$M = \frac{V_0}{c_s} \tag{2.43}$$

measures the flow speed (V_0) relative to the sound speed

$$c_s = \left(\frac{\gamma p_0}{\rho_0}\right)^{1/2}.$$

The *Alfvén Mach number*

$$M_A = \frac{V_0}{v_A} \tag{2.44}$$

gives the size of the flow speed in terms of the Alfvén speed

$$v_A = \frac{B_0}{(\mu \rho_0)^{1/2}},$$

where B_0 and ρ_0 are a typical magnetic field strength and plasma density.

The *plasma beta*

$$\beta = \frac{2\mu p_0}{B_0^2} \tag{2.45}$$

is the plasma pressure (p_0) divided by the magnetic pressure, so that when $\beta \ll 1$ we have a so-called low-β plasma.

The *Rossby number*

$$\mathrm{Ro} = \frac{V_0}{l_0 \Omega} \tag{2.46}$$

is the ratio of inertial terms to Coriolis terms in the equation of motion (2.24).

Certain other dimensionless numbers, not independent of the above ones, are sometimes met. For instance, the *magnetic Prandtl number* (sometimes the term is used for the ratio η/κ)

$$P_m = \frac{R_m}{\mathrm{Re}} = \frac{\nu}{\eta}$$

compares viscous and magnetic diffusion, while the *Ekman number*

$$E = \frac{\mathrm{Ro}}{\mathrm{Re}} = \frac{\nu}{l_0^2 \Omega}$$

is the ratio of the viscous force to the Coriolis force; furthermore, the *Taylor number* is

$$\mathcal{T} = E^{-2}$$

and measures the strength of the rotation.

For magnetoconvection the following numbers are also important (Section 8.1.2):

the *Rayleigh number*

$$\mathrm{Ra} = \frac{\alpha g \Delta T d^3}{\kappa \nu}$$

measures the importance of the buoyancy force relative to the stabilizing effects of nonmagnetic diffusion and is much larger than unity in the convection zone;

the *Chandrasekhar number* is

$$Q = \mathrm{Ha}^2,$$

where

$$\mathrm{Ha} = \frac{B_0 d}{(\mu \rho \eta \nu)^{1/2}}$$

is the *Hartmann number* representing the ratio of magnetic to viscous diffusion forces;

the *Prandtl number*

$$\mathrm{Pr} = \frac{\nu}{\kappa}$$

is a measure of the ratio of viscous to thermal diffusion and is generally very much less than unity (or of order unity) for the Sun, depending on whether molecular (or eddy) values are adopted; the *Nusselt number*

$$\mathrm{Nu} = 1 + \frac{F_c}{\kappa \Delta T/d},$$

where F_c and $\kappa \Delta T/d$ are the convected and conducted heat fluxes, is a measure of the vigour of convection.

In the solar atmosphere (for $\tilde{\mu} = 0.6$ and $\gamma = 5/3$),

$$R_m = 1.9 \times 10^{-8} l_0 V_0 T_0^{3/2}/\ln \Lambda, \tag{2.47}$$

$$c_s = 152 T_0^{1/2} \text{ m s}^{-1}, \tag{2.48a}$$

$$v_A = 2.8 \times 10^{16} B_0 n_0^{-1/2} \text{ m s}^{-1}, \tag{2.48b}$$

$$\beta = 3.5 \times 10^{-29} n_0 T_0 B_0^{-2}, \tag{2.49}$$

where l_0, V_0, T_0, n_0, B_0 are in m k s units. For example, above a sunspot, where typically $l_0 \approx 10^7$ m, $V_0 \approx 10^3$ m s^{-1}, $T_0 \approx 10^4$ K, $n_0 \approx 10^{20}$ m^{-3}, $B_0 \approx 0.1$ T(10^3 G), the relevant values are $R_m \approx 3 \times 10^7$, $c_s \approx 2 \times 10^4$ m s^{-1}, (M ≈ 0.05), $v_A \approx 3 \times 10^5$ m s^{-1} (M$_A \approx 4 \times 10^{-3}$), $\beta \approx 3 \times 10^{-3}$; also, with $\nu \approx 10^{-2}$ m^2 s^{-1} and $\Omega \approx 10^{-6}$ s^{-1}, we have Re $\approx 10^{12}$ and Ro ≈ 100. The way in which the plasma beta rapidly decreases with altitude, while the Alfvén speed increases, is indicated in Figures 2.1b and 2.1c, where they are plotted for the Harvard–Smithsonian Reference Atmosphere with a range of values of magnetic field. Whereas the plasma mostly dominates the magnetic field in the photosphere ($\beta > 1$), the opposite is true in the corona ($\beta < 1$).

2.6. Consequences of the Induction Equation

The induction equation

$$\frac{\partial \mathbf{B}}{\partial t} = \nabla \times (\mathbf{v} \times \mathbf{B}) + \eta \nabla^2 \mathbf{B} \tag{2.13}$$

has been derived in Section 2.1.4, and the magnetic Reynolds number

$$R_m = \frac{l_0 V_0}{\eta} \tag{2.41}$$

has been defined in Section 2.5 as the ratio of the convective to diffusive terms. The induction equation determines the behaviour of the magnetic field once **v** is known, and this behaviour depends crucially on whether R_m is small or large.

By using various vector identities, Equation (2.13) may be rewritten (for incompressible flow) as

$$\frac{D\mathbf{B}}{Dt} = (\mathbf{B} \cdot \nabla)\mathbf{v} + \eta \nabla^2 \mathbf{B},$$

which is similar in form to the vorticity equation (2.25)

$$\frac{D\boldsymbol{\omega}}{Dt} = (\boldsymbol{\omega} \cdot \nabla)\mathbf{v} + \nu \nabla^2 \boldsymbol{\omega}$$

for an incompressible fluid with no magnetic field (in an inertial frame). This is the basis of the so-called *vorticity-magnetic field analogy*, which implies that magnetic field lines respond in a similar way to the classical behaviour of vortex lines. For instance, field lines are, in general, partly transported with the flow and partly diffuse through it. Also, they may be stretched and increase in strength due to motions along their length. However, it should be noted that the analogy holds only for incompressible flow, and it is in any case not exact since the extra relation $\boldsymbol{\omega} = \text{curl } \mathbf{v}$ does not hold true for the magnetic field.

2.6.1. DIFFUSIVE LIMIT

If $R_m \ll 1$, the induction equation (2.13) reduces to a simple diffusion equation

$$\frac{\partial \mathbf{B}}{\partial t} = \eta \nabla^2 \mathbf{B}, \tag{2.50}$$

which implies that field variations on a length-scale l_0 are destroyed over a *diffusion time-scale*

$$\tau_d = \frac{l_0^2}{\eta}.$$

With the fully-ionised value for η (Equation (2.18)), this becomes numerically

$$\tau_d = 1.9 \times 10^{-8} l_0^2 T^{3/2}/\ln \Lambda \text{ s}. \tag{2.51}$$

Thus, for instance, with $T = 10^6$ K, $n = 10^{15}$ m^{-3} and a length-scale $l_0 = 10$ Mm (10^7 m), τ_d is equal to 10^{14} s, whereas a length-scale of 1 m gives $\tau_d \approx 1$ s. Since *solar flares* represent a release of magnetic energy over a time-scale of 100 or 1000 s, it seems that a length-scale as small as 100 or 1000 m is required (Chapter 10).

The smaller the length-scale, the faster the field diffuses away. As an example, consider the diffusion of a unidirectional magnetic field $\mathbf{B} = B(x, t)\hat{\mathbf{y}}$ with the initial step-function profile

$$B(x, 0) = \begin{cases} +B_0, & x > 0, \\ -B_0, & x < 0, \end{cases} \tag{2.52}$$

shown in Figure 2.3, where B_0 is a constant. Suppose the plasma in this initially infinitesimally thin *current sheet* remains at rest and that the magnetic field remains unidirectional. Then Equation (2.50) becomes

$$\frac{\partial B}{\partial t} = \eta \frac{\partial^2 B}{\partial x^2},$$

which has the following solution, subject to the initial conditions (2.52):

$$B(x, t) = B_0 \text{ erf}(\xi),$$

where $\xi = x/(4\eta t)^{1/2}$ and

$$\text{erf}(\xi) = \frac{2}{\pi^{1/2}} \int_0^\xi e^{-u^2} du.$$

Note that erf(ξ) approaches ± 1 as ξ tends to $\pm \infty$, so that the initial conditions are indeed satisfied. For $|x| \ll (4\eta t)^{1/2}$, $B(x, t) \approx B_0 x(\pi \eta t)^{-1/2}$ and the profile at a given time is linear, whereas for $|x| \gg (4\eta t)^{1/2}$, $|B(x, t)| \approx B_0$ and the profile is undisturbed from its initial state. The profiles are shown schematically in Figure 2.3 (a) at two times t_1 and $t_2 (0 < t_1 < t_2)$ and the corresponding field lines are sketched in Figure 2.3 (b).

The region to which the current (of density $j_z = \mu^{-1} dB/dx$) is concentrated is known as a *current sheet*. Its width is roughly $l = 4(\eta t)^{1/2}$, which increases with time at a continuously decreasing rate ($dl/dt = 2(\eta/t)^{1/2}$). Notice that the field strength at large distances ($x \gg (4\eta t)^{1/2}$) remains constant in time, whereas that at small distances decreases monotonically. The field lines in the sheet are not moving outwards, since those at large distances are unaffected. Rather, the field in the sheet is diffusing away, and one says that it is being *annihilated*. There is nothing mysterious about this effect; it is simply that magnetic energy is being converted into heat by ohmic dissipation. An interesting point is that, although the current density ($j_z = \mu^{-1} dB/dx$) at all points is changing in time, the total current in the sheet

$$J = \int_{-\infty}^{\infty} j_z dx = [B/\mu]_{-\infty}^{\infty} = 2B_0/\mu$$

remains constant.

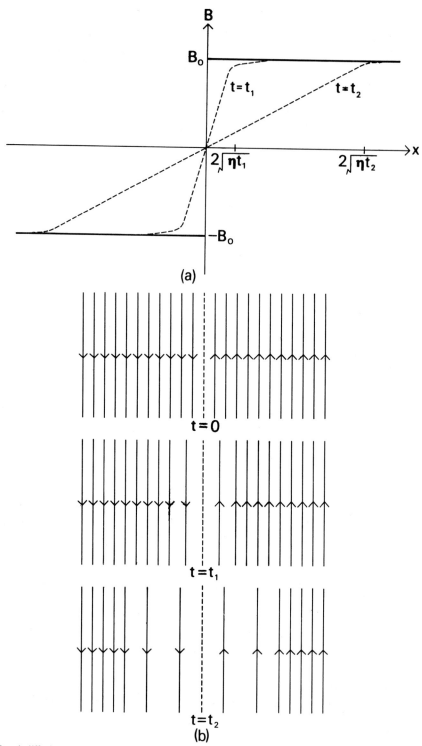

Fig. 2.3. A diffusing current sheet: (a) the variation with time of the magnetic field strength; (b) a sketch of the magnetic field lines at three times.

In practice, the above simple one-dimensional magnetic-field diffusion would be modified in several ways. For instance, the decrease of magnetic field strength near the neutral line leads to inward magnetic pressure gradients, which drive a flow towards the neutral line from the sides and outwards along it. This requires the inclusion of the convection term in the induction equation and gives rise to a coupling with the equation of motion (see Section 10.1 and Cheng, 1979; Kirkland and Sonnerup, 1979; Forbes *et al.*, 1982). Also, the one-dimensional nature of the current sheet may be disrupted by the occurrence of tearing-mode instabilities (Section 7.5.5).

2.6.2. Perfectly Conducting Limit

When $R_m \gg 1$, Equation (2.13) becomes approximately

$$\frac{\partial \mathbf{B}}{\partial t} = \nabla \times (\mathbf{v} \times \mathbf{B}), \tag{2.53a}$$

while Ohm's Law (2.16) reduces to

$$\mathbf{E} + \mathbf{v} \times \mathbf{B} = \mathbf{0}. \tag{2.53b}$$

It can be seen that Equation (2.53a) may be obtained by taking the curl of Equation (2.53b) and using Equation (2.13). Also, it should be stressed that, although Equation (2.53b) implies that the total electric field vanishes, currents are still present and are given by Equation (2.15).

In this large magnetic Reynolds number limit, the *frozen-flux theorem* of Alfvén holds: *in a perfectly conducting plasma, magnetic field lines behave as if they move with the plasma*. (By the vorticity-magnetic field analogy this is directly comparable with the classical *vorticity theorem of Helmholtz and Kelvin*.) An intuitive proof is as follows.

Consider a closed curve (C) bounding a surface (S) which is moving with the plasma, as indicated in Figure 2.4(a). During a time δt, an element ($\delta \mathbf{s}$) of C sweeps out an element of area $\mathbf{v}\, \delta t \times \delta \mathbf{s}$. A magnetic flux $\mathbf{B} \cdot (\mathbf{v}\, \delta t \times \delta \mathbf{s})$ passes through this area, and the properties of a triple scalar product may be used to write it as

$$-\delta t \mathbf{v} \times \mathbf{B} \cdot \delta \mathbf{s}. \tag{2.54}$$

The magnetic flux through S is $F = \iint \mathbf{B} \cdot d\mathbf{S}$, and its rate of change (DF/Dt) may be decomposed into two parts

$$\frac{DF}{Dt} = \int_S \int \frac{\partial \mathbf{B}}{\partial t} \cdot d\mathbf{S} - \oint_C \mathbf{v} \times \mathbf{B} \cdot d\mathbf{s};$$

the first is due to changes in the magnetic field with time, whereas the second is caused by the motion of the boundary and arises from the integral of Equation (2.54). After invoking Stokes' theorem, the second integral may be rewritten as

$$-\int_S \int \nabla \times (\mathbf{v} \times \mathbf{B}) \cdot d\mathbf{S},$$

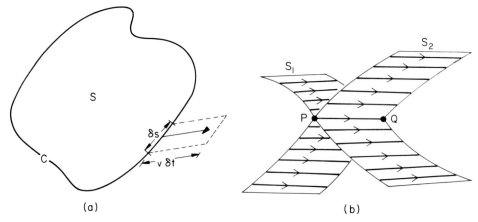

Fig. 2.4. (a) A surface S bounded by a curve C that moves with the plasma. (b) Parts of two magnetic flux surfaces S_1 and S_2 that intersect in the magnetic field line PQ.

so that the rate of change of flux becomes

$$\frac{DF}{Dt} = \int\int_S \left(\frac{\partial \mathbf{B}}{\partial t} - \nabla \times (\mathbf{v} \times \mathbf{B}) \right) \cdot d\mathbf{S},$$

which vanishes by Equation (2.53a). In other words, the magnetic flux through a surface moving with the plasma is conserved.

This result can now be used to show that two plasma elements (P and Q) that are initially located on the same magnetic field line will remain there for all time. Suppose the field line is defined initially by the intersection of two flux surfaces (S_1 and S_2), each of which is made up of field lines (Figure 2.4(b)). Then the flux through each surface is zero initially, and, in view of the above result, it remains zero as the surfaces move with the plasma. S_1 and S_2 therefore *remain* magnetic flux surfaces, and their intersection, namely the curve through P and Q, must remain a magnetic field line for all time. (Of course, the question as to whether the initial field line through P and Q is the same as the later one is simply a matter of labelling, so we shall assume it is the same.) This is Alfvén's result that field changes at a particular point are the same as if magnetic field lines move with the plasma. One refers to field lines being *frozen into the plasma*; plasma can move freely *along* field lines, but, in motion perpendicular to them, either the field lines are dragged with the plasma or the field lines push the plasma.

An alternative form for the induction equation (2.53a) is

$$\frac{D\mathbf{B}}{Dt} = (\mathbf{B} \cdot \nabla)\mathbf{v} - \mathbf{B}(\nabla \cdot \mathbf{v}).$$

This implies that changes in the magnetic field following the motion ($D\mathbf{B}/Dt$) are produced when a flux tube is stretched, sheared or expanded: according to the first term on the right-hand side, an accelerating motion along the field causes the field strength to increase, whereas a shearing motion normal to the field makes the field change direction by increasing the field component along the flow direction; the second term on the right-hand side implies that an expansion ($\nabla \cdot \mathbf{v} > 0$) causes the field strength to decrease, whereas a compression ($\nabla \cdot \mathbf{v} < 0$) makes it increase. By combining the above form of the induction equation with the continuity equation,

$$\frac{D\rho}{Dt} = -\rho \nabla \cdot \mathbf{v},$$

one can obtain

$$\frac{D}{Dt}\left(\frac{\mathbf{B}}{\rho}\right) = \left(\frac{\mathbf{B}}{\rho} \cdot \nabla\right)\mathbf{v}.$$

This in turn may be integrated (Roberts, 1967, p. 46) to give

$$\frac{\mathbf{B}}{\rho} = \left(\frac{\mathbf{B}_0}{\rho_0} \cdot \nabla_0\right)\mathbf{r}, \tag{2.55}$$

where $\mathbf{r}_0$ is the Lagrangian position vector, ∇_0 is the corresponding gradient operator, and $\mathbf{r} \equiv \mathbf{r}(\mathbf{r}_0, t)$, $\rho \equiv \rho(\mathbf{r}, t)$, $\rho_0 \equiv \rho_0(\mathbf{r}_0, t)$. The result shows how the vector $\mathbf{B}/\rho$ is changed from its initial value $\mathbf{B}_0/\rho_0$ by a displacement $\mathbf{r}$.

The above discussion shows that the value of the magnetic Reynolds number is crucial in determining the behaviour of the field. When R_m is much less than unity, the magnetic field slips through the plasma, whereas, when R_m has a value much larger than unity the field is frozen into the plasma. For some very small-scale phenomena in the solar atmosphere, such as current sheets a kilometre or less in thickness, R_m may be of order unity, but in most cases it is much larger. For instance, consider typical *sunspot motions* at a speed $V_0 \approx 10^3$ m s^{-1} and on a length-scale $l_0 \approx 10^7$ m. They make R_m of order 10^6 for a magnetic diffusivity $\eta \approx 10^4$ m^2 s^{-1}, and so the magnetic field is tied very closely to the plasma. According to Equation (2.51), the field would diffuse away over 300 yr, whereas sunspot fields are observed, in practice, to decay over only 100 days. Clearly, this cannot be caused by the above classical ohmic dissipation and would require an eddy diffusivity (Section 2.1.5) that is a factor of 10^3 larger than the classical value.

2.7. The Lorentz Force

Consider the physical effect of the $\mathbf{j} \times \mathbf{B}$ force, which was introduced in Equation (2.21). One point to notice is that it is directed *across* the magnetic field, so that any motion or density variation *along* field lines must be produced by other forces, such as gravity or pressure gradients. Another point is that the Lorentz force may be decomposed into a *magnetic tension* force and a *magnetic pressure* force. Thus, after substituting for $\mathbf{j}$ from Equation (2.15),

$$\mathbf{j} \times \mathbf{B} = (\nabla \times \mathbf{B}) \times \mathbf{B}/\mu,$$

which reduces, by means of a vector identity for the triple vector product, to

$$\mathbf{j} \times \mathbf{B} = (\mathbf{B} \cdot \nabla)\mathbf{B}/\mu - \nabla(B^2/(2\mu)). \tag{2.56}$$

The first term on the right-hand side of Equation (2.56) is non-zero if $\mathbf{B}$ varies along the direction of $\mathbf{B}$; it represents the effect of a tension parallel to $\mathbf{B}$ of magnitude

B^2/μ per unit area, which has a resultant effect when the field is curved. Putting $\mathbf{B} = B\hat{\mathbf{s}}$, in terms of the unit vector ($\hat{\mathbf{s}}$) along the field, we see that the tension term can be decomposed into

$$\frac{B}{\mu}\frac{d}{ds}(B\hat{\mathbf{s}}) = \frac{B}{\mu}\frac{dB}{ds}\hat{\mathbf{s}} + \frac{B^2}{\mu}\frac{d\hat{\mathbf{s}}}{ds}$$

$$= \frac{d}{ds}\left(\frac{B^2}{2\mu}\right)\hat{\mathbf{s}} + \frac{B^2}{\mu}\frac{\hat{\mathbf{n}}}{R_c},$$
(2.57)

where $\hat{\mathbf{n}}$ is the *principle normal* to the magnetic field line and R_c is its *radius of curvature*. Thus, the smaller the radius of curvature, the larger the tension force becomes.

The second term in Equation (2.56) represents a scalar pressure force of magnitude $B^2/(2\mu)$ per unit area, the same in all directions. Its component parallel to the magnetic field cancels with the corresponding tension component in Equation (2.57), as it must, since the $\mathbf{j} \times \mathbf{B}$ force is normal to $\mathbf{B}$.

The Lorentz force therefore has two effects. It acts both to shorten magnetic field lines through the tension force and also to compress plasma through the pressure term. It may be decomposed into a compressive force purely normal to $\mathbf{B}$ of amount $B^2/(2\mu)$ together with a stretching force along $\mathbf{B}$, also of amount $B^2/(2\mu)$. The former gives a resultant effect when $\mathbf{B}$ is changing in a direction normal to $\mathbf{B}$, while the latter produces a resultant force (normal to $\mathbf{B}$) when the field is curved.

A few simple examples may provide some physical insight. In constructing these, several forms for $\mathbf{B}$ that satisfy $\nabla \cdot \mathbf{B} = 0$ are taken, and, after calculating the resulting current (from $\mathbf{j} = \nabla \times \mathbf{B}/\mu$), it is shown how the Lorentz force may be interpreted in terms of magnetic pressure and tension.

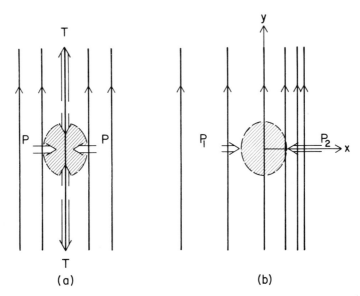

Fig. 2.5. The magnetic pressure (P) and tension (T) forces due to: (a) a uniform field; (b) a unidirectional field whose strength increases along the x-axis.

Consider first the forces exerted by a *uniform* field ($B_0\hat{\mathbf{y}}$) on an element of plasma shown in Figure 2.5(a). Since the current density vanishes, there is no resultant Lorentz force. Equal magnetic pressure forces (P) act from each direction, while equal and opposite tension forces (T) act along the field lines, so that the net effect is that the plasma element remains in equilibrium; the presence of neighbouring elements means that it is not stretched out along the field.

Next, suppose there is a field

$$\mathbf{B} = B_0 \, e^x \hat{\mathbf{y}}, \tag{2.58}$$

with corresponding current density

$$\mathbf{j} = \frac{dB_y}{dx}\frac{\hat{\mathbf{z}}}{\mu} = \frac{B_0 \, e^x}{\mu}\hat{\mathbf{z}}$$

and Lorentz force

$$\mathbf{j} \times \mathbf{B} = -\frac{B_0^2 \, e^{2x}}{\mu}\hat{\mathbf{x}},$$

which is directed along the negative x-axis. In the y-direction there is equilibrium, with the tension and pressure forces being just the same as before (Figure 2.5 (b)). But in the x-direction the magnetic pressure (P_1) on the left of the element is less than the magnetic pressure (P_2) on the right, since the field strength increases from left to right. The resultant magnetic pressure force acts in such a direction as to push the plasma to the left. As expected, the Lorentz force due to the magnetic field (2.58) comes entirely from the magnetic pressure force ($-\nabla(B^2/(2\mu))$), since the (resultant) magnetic tension force (($\mathbf{B}\cdot\nabla)\mathbf{B}/\mu$) vanishes.

Consider now the field

$$\mathbf{B} = -y\hat{\mathbf{x}} + \hat{\mathbf{y}}, \tag{2.59}$$

which arises from a current

$$\mathbf{j} = -\frac{dB_x}{dy}\frac{\hat{\mathbf{z}}}{\mu} = \frac{\hat{\mathbf{z}}}{\mu}$$

and produces a force on the x-axis of amount

$$(\mathbf{j} \times \mathbf{B})_{y=0} = -\frac{\hat{\mathbf{x}}}{\mu}. \tag{2.60}$$

The equation of the field lines is given by

$$\frac{dy}{dx} = \frac{B_y}{B_x} = -\frac{1}{y},$$

which has as solution the family of parabolas $x = -\frac{1}{2}y^2 + \text{constant}$, shown in Figure 2.6. In this case, on $y = 0$ the magnetic pressure force vanishes and so the Lorentz force is provided wholly by the magnetic tension. The pressure forces there balance one another and there remains the effect of only the tension forces (T) pulling along the curved field lines. As every mischievous child with a catapult

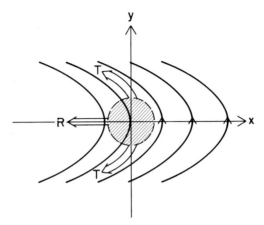

Fig. 2.6. The resultant magnetic force (R) due to a symmetrically curved field (Equation (2.59)).

knows, the effect of the tension is to produce a resultant force directed along the negative x-axis, in agreement with Equation (2.60).

As another example, suppose

$$\mathbf{B}_0 = y\hat{\mathbf{x}} + x\hat{\mathbf{y}}, \tag{2.61}$$

for which the current density and Lorentz force vanish. The origin is a so-called *X-type magnetic neutral point*, and the field lines are shown in Figure 2.7. They are the

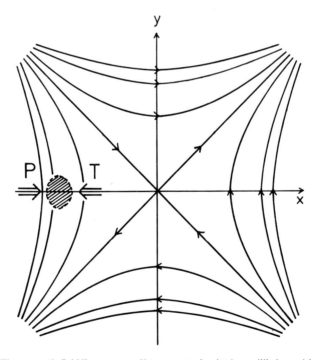

Fig. 2.7. The magnetic field lines near an X-type neutral point in equilibrium with no current.

solutions of
$$\frac{dy}{dx} = \frac{B_y}{B_x} = \frac{x}{y},$$
namely the hyperbolas $y^2 - x^2 =$ constant. In sketching the field lines, account has been taken of the fact that the field strength increases with distance from the origin so that the hyperbolas must be situated successively closer to one another. Any element of plasma, such as the one shown near the x-axis, experiences a resultant magnetic tension (T) due to the outwardly curving field lines. It acts outwards from the origin and is exactly balanced by a magnetic pressure force (P), which acts inwards because the magnetic field strength weakens as the origin is approached.

Finally, consider a magnetic field of the form
$$\mathbf{B}_1 = y\hat{\mathbf{x}} + \alpha^2 x\hat{\mathbf{y}}, \tag{2.62}$$
where $\alpha^2 > 1$. The corresponding field lines are given by $y^2 - \alpha^2 x^2 =$ constant and are sketched in Figure 2.8. Again, this represents an X-type magnetic neutral point, but the field lines ($y = \pm \alpha x$) that pass through it are now no longer inclined at $\frac{1}{2}\pi$ to one another. On the x-axis, the field lines are more closely spaced than those in Figure 2.7, so that the magnetic pressure is stronger; but their curvature

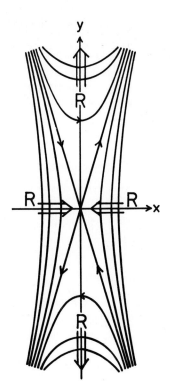

Fig. 2.8. The magnetic field lines near an X-type neutral point away from equilibrium with a uniform current.

is smaller than before, so that the magnetic tension has not increased as much as
the pressure. The dominance of the magnetic pressure therefore produces a resultant
force (R) acting inward. On the y-axis, the field lines have the same spacing as in
Figure 2.7, but they are more sharply curved so that the pressure force remains
the same while the tension force increases. The resultant force (R) therefore acts
outward as shown. These comments are borne out by evaluating the current density

$$j_{1z} = \frac{1}{\mu}\left(\frac{\partial B_{1y}}{\partial x} - \frac{\partial B_{1x}}{\partial y}\right) = \frac{\alpha^2 - 1}{\mu}$$

and the resulting Lorentz force

$$\mathbf{j}_1 \times \mathbf{B}_1 = -\frac{(\alpha^2 - 1)\alpha^2 x}{\mu}\hat{\mathbf{x}} + \frac{(\alpha^2 - 1)y}{\mu}\hat{\mathbf{y}}.$$

Dungey (1953) pointed out an interesting property of these fields, namely that, when $\mathbf{B}_0$ is perturbed to $\mathbf{B}_1$ (Equation (2.62)), the magnetic force is such as to *increase* the perturbation. Thus the *equilibrium X-type point* of Figure 2.7 is *unstable* (provided that conditions at distant boundaries allow such a displacement). Furthermore, as the instability proceeds, α increases and the limiting field lines through the origin close up; therefore, the current density and the ohmic heating (j^2/σ) also increase. This idea has been included by Syrovatsky (1966) in a solar flare model, and a similarity solution for the collapse has been presented by Imshennik and Syrovatsky (1967).

Much qualitative information can be gained from magnetic field configurations such as Figure 2.8. As well as seeing at a glance the magnetic field direction everywhere, some impression of relative field strengths at different points may be gained by comparing the spacing of field lines. Also, the directions of the magnetic pressure gradient and tension force can be estimated from spatial variations in field strength and field-line curvature.

2.8. Some Theorems

One very useful theorem (due to Alfvén) has already been mentioned in Section 2.6.2. In this section, several other theorems are outlined; they apply mainly to steady motions and follow very simply from either the equation of motion or the induction equation.

2.8.1. Cowling's Antidynamo Theorem

Steady plasma motions cannot maintain a magnetic field that is confined to a finite region of space and possesses axial symmetry.

For a proof, see Section 9.2.

2.8.2. Taylor–Proudman Theorem

Steady, slow motions of a perfectly conducting plasma permeated by a uniform magnetic field ($\mathbf{B}_0$) *must be two-dimensional, with no variation along* $\mathbf{B}_0$.

For steady flow with $\eta = 0$, the induction equation (2.13) reduces to

$$\nabla \times (\mathbf{v} \times \mathbf{B}) = \mathbf{0}, \tag{2.63}$$

where $\nabla \cdot \mathbf{B} = 0$ and, from Equation (2.20),

$$\nabla \cdot (\rho \mathbf{v}) = 0. \tag{2.64}$$

Consider small departures from a uniform plasma (of density ρ_0) at rest in a uniform field ($\mathbf{B}_0$). After putting

$$\rho = \rho_0 + \rho', \quad \mathbf{v} = \mathbf{v}', \quad \mathbf{B} = \mathbf{B}_0 + \mathbf{B}',$$

and linearising, Equation (2.64) becomes

$$\nabla \cdot (\rho_0 \mathbf{v}') = 0. \tag{2.65}$$

Also, Equation (2.63) gives $\nabla \times (\mathbf{v}' \times \mathbf{B}_0) = \mathbf{0}$, or, using Equation (2.65) and the uniformity of $\mathbf{B}_0$, $(\mathbf{B}_0 \cdot \nabla)\mathbf{v}' = \mathbf{0}$. This implies that $\mathbf{v}'$ has no variation in the direction of $\mathbf{B}_0$, as required. (The original, *non-magnetic* version of this theorem applies to slow, steady inviscid flow in a uniformly rotating fluid (with no magnetic field present); it states that such motions must be two-dimensional, with no variation along the axis of rotation.)

2.8.3. FERRARO'S LAW OF ISOROTATION

For a steady, axisymmetric flow and magnetic field, the angular speed (v_ϕ/R in cylindrical polars) is a constant along field lines.

In the special case when the plasma is perfectly conducting ($\eta = 0$) and the motion is around the z-axis with $\mathbf{v} = v_\phi(R, z)\hat{\boldsymbol{\phi}}$, the proof is particularly simple. Then the axisymmetric magnetic field

$$\mathbf{B} = B_R(R, z)\hat{\mathbf{R}} + B_z(R, z)\hat{\mathbf{z}}$$

reduces the steady induction equation

$$\nabla \times (\mathbf{v} \times \mathbf{B}) = \mathbf{0}$$

to

$$(\mathbf{B} \cdot \nabla)(v_\phi/R) = 0.$$

In other words, v_ϕ/R remains constant along magnetic field lines; this is a reasonable consequence of flux freezing, since, otherwise, the resulting differential rotation would generate a toroidal field component (B_ϕ).

2.8.4. THE VIRIAL THEOREM

Consider the equation of motion

$$\rho \frac{D\mathbf{v}}{Dt} = -\nabla p + \mathbf{j} \times \mathbf{B} + \rho \mathbf{g},$$

where the gravitational acceleration may be written in terms of a *gravitational potential* (Φ) such that $\mathbf{g} = \nabla\Phi$ and $\nabla^2\Phi = -4\pi G\rho$ for a self-gravitating plasma. This equation of motion holds locally everywhere in the plasma, and, after taking the scalar product with $\mathbf{r}$, it may be integrated over the volume occupied by the plasma to yield a relationship between the global mechanical energy contributions. The result is known as the *scalar virial theorem* and takes the form

$$\frac{1}{2}\frac{D^2 \mathscr{I}}{Dt^2} = 2\mathscr{T} + 3(\gamma - 1)\mathscr{U} + \mathscr{M} + \mathscr{W} + \mathscr{S}, \tag{2.66}$$

where $\mathscr{I} = \int \rho r^2 \, dV$ is the polar moment of inertia,

$\mathscr{T} = \int \frac{1}{2}\rho v^2 \, dV$ is the kinetic energy,

$\mathscr{U} = \int p/(\gamma - 1) \, dV$ is the internal energy,

$\mathscr{M} = \int B^2/(2\mu) \, dV$ is the magnetic energy,

$\mathscr{W} = -\frac{1}{2}G \int\int \rho(\mathbf{r})\rho(\mathbf{r}')|\mathbf{r} - \mathbf{r}'|^{-1} \, dV \, dV'$ is the gravitational potential energy,

$\mathscr{S} = \int_S (\mathbf{r} \cdot \mathbf{B})\mathbf{B}/\mu \cdot d\mathbf{S} - \int_S (p + B^2/(2\mu))\mathbf{r} \cdot d\mathbf{S}$ is the surface contribution.

In Equation (2.66) all the terms on the right-hand side are positive except for $\mathscr{W}$ (which is always negative) and $\mathscr{S}$ (which may sometimes be negative). Thus, there can be no equilibrium (i.e., $D^2\mathscr{I}/Dt^2 = 0$) or deceleration ($D^2\mathscr{I}/Dt^2 < 0$) without the effect of $\mathscr{W}$ and/or $\mathscr{S}$. (A more general *tensor virial theorem* is proved in Chandrasekhar (1961).)

2.9. Summary of Magnetic Flux Tube Behaviour

The two types of building block found in magnetic configurations are a *magnetic flux tube* and a *current sheet*. When discussing their characteristics, it is helpful to regard them simply as isolated entities, although it should be remembered that they are not really isolated at all and interact in an intimate way with the surrounding magnetic field. Their basic properties are described in detail at various points throughout the book and are summarised in this section and the next.

Perhaps the most important example of a flux tube is the *sunspot* (Sections 8.2–8.6), observed in the photosphere where a large magnetic tube breaks through the solar surface. Another possible example is an *erupting prominence* (Section 10.3). However, recently a renewed interest in the properties of flux tubes has been stimulated by two discoveries, namely the *intense magnetic fields* (Sections 8.7) present along supergranule boundaries and the myriads of coronal loops (Section 6.5) that fill the Sun's outer atmosphere.

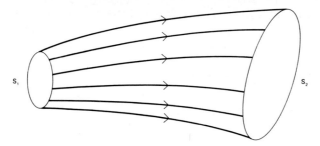

Fig. 2.9. A section of a magnetic flux tube.

2.9.1. Definitions

A *magnetic field* line is such that the tangent at any point is in the direction of **B**. Thus, in rectangular Cartesians, it is the solution of

$$\frac{dy}{dx} = \frac{B_y}{B_x}$$

in two dimensions or

$$\frac{dx}{B_x} = \frac{dy}{B_y} = \frac{dz}{B_z}$$

in three dimensions. For cylindrical or spherical polar coordinates the equations of field lines are given by solving

$$\frac{dR}{B_R} = \frac{R\,d\phi}{B_\phi} = \frac{dz}{B_z}, \qquad \frac{dr}{B_r} = \frac{r\,d\theta}{B_\theta} = \frac{r\sin\theta\,d\phi}{B_\phi},$$

respectively.

A *magnetic field* (or *flux*) *tube* is the volume enclosed by the set of field lines which intersect a simple closed curve (Figure 2.9). The *strength* (F) of such a flux tube may be defined as the amount of flux crossing a section S

$$F = \int_S \mathbf{B}\cdot d\mathbf{S}, \tag{2.67}$$

where d**S** is taken in the same sense as **B**, so that F is always positive.

2.9.2. General properties

(1) *The strength of a flux tube remains constant along its length.*
This is a consequence of the basic equation

$$\operatorname{div}\mathbf{B} = 0. \tag{2.68}$$

Integrate the flux (2.67) over the surface (S) enclosing the volume (V) between the two sections (S_1 and S_2). Then, since the contribution from the curved surface vanishes,

$$\int_S \mathbf{B}\cdot d\mathbf{S} = \int_{S_1} \mathbf{B}\cdot d\mathbf{S} + \int_{S_2} \mathbf{B}\cdot d\mathbf{S},$$

where d**S** is along the outward normal from V. But, by the divergence theorem,

$$\int_S \mathbf{B} \cdot d\mathbf{S} = \int \operatorname{div} \mathbf{B}\, dV,$$

and this vanishes by Equation (2.68), so that

$$\int_{S_1} \mathbf{B} \cdot d\mathbf{S} = -\int_{S_2} \mathbf{B} \cdot d\mathbf{S}.$$

In other words, the quantity

$$F \equiv \int_S \mathbf{B} \cdot d\mathbf{S},$$

which is known as the *strength* of a flux tube, remains constant along its length.

(2) *The mean field strength of a flux tube increases when it narrows and decreases when it widens.*

Equation (2.67) may be written in terms of the mean field ($\bar{B}$) across the tube as $F = \bar{B}A$, where A is the cross-sectional area. Thus, if the flux tube narrows as one moves along it, so A decreases and the mean field strength ($\bar{B}$) increases, and vice versa. Strong-field regions have the field lines close together, while weak-field regions have them further apart.

(3) *A compression of a flux tube increases B and ρ in the same proportion.*

Consider a cylindrical flux tube whose dimensions change from l_0 and L_0 to λl_0 and $\lambda^* L_0$, as shown in Figure 2.10. If the original uniform density and field strength are ρ_0 and B_0, and the field is frozen to the plasma, the conservation of matter gives

$$\rho \pi (\lambda l_0)^2 (\lambda^* L_0) = \rho_0 \pi l_0^2 L_0,$$

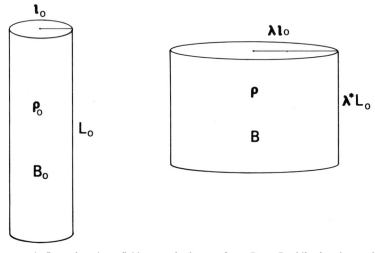

Fig. 2.10. A magnetic flux tube whose field strength changes from B_0 to B while the plasma density changes from ρ_0 to ρ and the dimensions alter by fractions λ and λ^* as shown.

so that the final density is

$$\rho = \frac{\rho_0}{\lambda^2 \lambda^*}. \tag{2.69}$$

Also, conservation of magnetic flux, namely

$$B\pi(\lambda l_0)^2 = B_0 \pi l_0^2,$$

implies a final magnetic field strength of

$$B = \frac{B_0}{\lambda^2}. \tag{2.70}$$

Thus, if the length of the tube remains constant ($\lambda^* = 1$), we see that $B/\rho = $ constant (see also Equation (2.55)); in other words a transverse compression ($\lambda < 1$) increases B and ρ by the same proportion, while a transverse expansion decreases them. This holds only if a constant length of plasma is being considered. Thus, for example, it would not apply to a *coronal loop* whose magnetic field strength was observed to increase, since plasma could flow in or out of the loop footpoints.

(4) *An extension of a flux tube without compression increases the field strength.*

If the plasma is not compressed so that the density is unchanged, Equation (2.69) implies that $\lambda^2 \lambda^* = 1$. Then Equation (2.70) gives $B = \lambda^* B_0$, and so an extension of the tube ($\lambda^* > 1$) produces a corresponding rise in field strength, while a shortening causes the field to weaken. An increase in the length of a flux tube may be produced by, for example, the shearing motions which arise in subphotospheric *convection*, or *differential rotation*.

(5) *For a cylindrical flux tube in magnetohydrostatic equilibrium, the plasma pressure $(p(R))$ and magnetic field components $(B_\phi(R), B_z(R))$ are related by*

$$0 = \frac{dp}{dR} + \frac{d}{dR}\left(\frac{B_\phi^2 + B_z^2}{2\mu}\right) + \frac{B_\phi^2}{\mu R},$$

where R is the radial distance from the axis. The second term represents the magnetic pressure force and acts outward if the magnetic pressure ($B^2/(2\mu)$) decreases with R. The third term gives the effect of magnetic tension and acts inward. The corresponding twist ($\Phi(R)$) of a field line about the axis in going from one end of the tube (of length L) to the other is given by

$$\Phi(R) = \frac{L B_\phi(R)}{r B_z(R)}.$$

Magnetostatic and force-free flux tubes are described in Section 3.3.

(6) *Cylindrical flux tubes in force-free equilibrium with free ends* have several interesting properties (Section 3.3.3), many of which have been demonstrated in masterly fashion by Parker (1979a).

Suppose the magnetic field ($B(a)$), at some variable distance $R = a$, is held fixed in value. Then the mean-square value of B_z across the tube can be shown to possess the value $B^2(a)$, and so it is *invariant with respect to twisting*. Also, as the tube is twisted so the mean value of B_z across the tube *decreases* (and is always less than $B(a)$), since an increasing part of the total field ($B(a)$) goes into B_ϕ.

Suppose a flux tube expands, with its axial and azimuthal magnetic flux held fixed. Then the *tube becomes increasingly more and more twisted*. (Ultimately, when the magnetic pressure of B_ϕ exceeds the magnetic tension of B_z, the tube will no longer be under tension and will *buckle*; that is, provided an instability does not set in first.)

Suppose just one section of a flux tube expands, with its axial flux and azimuthal torque held fixed. Then *coils of flux are transferred into the expanded portion*, which becomes more twisted than the remainder of the tube.

(7) *A twisted flux tube with free ends is unstable to the helical kink instability* (Section 7.5.3).

According to the Kruskal–Shafranov–Tayler criterion, all helical kink perturbations whose axial wavelength ($-2\pi/k$) is so long that $-kL \leqslant \Phi$ are unstable.

The effect of *line-tying* at the ends of a flux tube is stabilising: a uniform-twist force-free tube, for example, requires a twist (Φ) larger than 2.6π before it becomes kink unstable. A plasma pressure that increases away from the axis is a stabilising influence too.

(8) *The fundamental wave modes in a uniform plasma are modified by the geometry when they propagate along a flux tube.*

Torsional Alfvén waves propagate at the Alfvén speed (Section 4.3) and the *magnetoacoustic tube waves* possess an amplitude that is radially dependent. A *slow (sausage) surface wave* propagates at speeds less than the so-called *tube speed* c_T (Section 8.7.4); this is smaller than both the sound speed (c_s) and Alfvén speed (v_A) and may be written as

$$c_T^2 = \frac{c_s^2 v_A^2}{c_s^2 + v_A^2}.$$

There are also slow (kink) surface waves with speed $v_A/\sqrt{2}$, slow (kink and sausage) body waves at c_T, and fast (body or surface) waves at the external sound speed. At critical radii, where there are Alfvén or cusp resonances, waves may be absorbed and so heat the plasma (Section 6.4.3).

2.9.3. FLUX TUBES IN THE SOLAR ATMOSPHERE

(1) *Convection* can expel magnetic flux from a convecting eddy and concentrate it to form a vertical flux rope with a field strength that exceeds the photospheric equipartition value of typically a few hundred gauss (Section 8.1.3).

(2) A horizontal flux tube embedded in a gravitationally stratified medium is subject to a *magnetic buoyancy* force, which tends to make it rise. It can remain in equilibrium as an arch if the feet are anchored at points that are separated by less than a few scale-heights (Section 8.2.1). Also, magnetic buoyancy may destabilise an equilibrium magnetic field whose strength declines too rapidly with height (Section 8.2.2); the resulting speed of rise is estimated in Section 8.2.3.

(3) Sunspots may be cool because of either the *inhibition of convection* or *overstable Alfvén waves* or a *subphotospheric downdraft* (Section 8.3).

(4) A sunspot may consist of either a *single large flux tube* in equilibrium or a *cluster of small tubes* held together by magnetic buoyancy and a downdraft (Section 8.4.1).

(5) A model sunspot whose diameter decreases monotonically with depth is magnetohydrodynamically *stable* when its flux exceeds about 10^{11} Wb (10^{19} Mx). However, slender flux tubes with less flux than this are unstable to *fluting* (Section 8.4.2).

(6) A sunspot may be formed either because inhibition of convection in an intermediate strength field causes plasma to cool and fall, or because individual tubes are assembled by hydrodynamic attraction and aerodynamic drag (Section 8.6.1). The decay may be modelled in terms of an eddy diffusivity (Section 8.6.2).

(7) Intense flux tubes may be formed if a static photospheric tube of a few hundred gauss is subject to instability, which causes a downflow of plasma and a *compression* of the field to a new equilibrium at typically 1 to 2 kG (Section 8.7.2).

(8) A slender subphotospheric tube in thermal and hydrostatic equilibrium widens with height (Section 8.7.1). If it is cooler than its surroundings it becomes evacuated over a few scale-heights.

(9) The static equilibrium of a coronal loop is in general governed by

$$\mathbf{j} \times \mathbf{B} - \nabla p + \rho \mathbf{g} = \mathbf{0}.$$

In *active regions* the magnetic field is approximately *force-free*, so that **B** satisfies $\mathbf{j} \times \mathbf{B} = \mathbf{0}$, while the plasma structure along each field line follows from $-\nabla p + \rho \mathbf{g} = \mathbf{0}$ together with an energy equation (Section 6.5.1).

(10) *Siphon flows* may be driven along photospheric and coronal loops by an imposed pressure difference, while downflows and upflows may arise in a variety of ways (Section 6.5.2).

2.10. Summary of Current Sheet Behaviour

In most of the solar atmosphere, the length-scale (L) over which the magnetic field (B) varies is rather large, say typically 1000 to 10,000 km, and the corresponding current density

$$j \approx \frac{B}{\mu L} \tag{2.71}$$

is very small. However, it is believed that *current sheets* may exist, with thicknesses (l) much smaller than L and corresponding current densities much larger than the value (2.71). They are sometimes likely to be rather transient in nature, and the energy that is liberated within them may well heat the corona (Section 6.4.4). Also, current sheets may play an important role in the *solar flare* process (Chapter 10) and are possible sites for *prominence formation* (Section 11.1.3). In addition, they may be present on the surface of a flux tube such as a *sunspot* (Section 8.4) and on the edge of *helmet streamers* (Section 12.4.1) at the boundary between closed and open fields.

114 CHAPTER 2

A current sheet may be defined as a non-propagating boundary between two plasmas, with the magnetic field tangential to the boundary. In other words, it may be regarded loosely as a *tangential discontinuity* (Section 5.4.2) without any flow. The tangential field components are arbitrary in magnitude and direction, subject only to the condition that the total pressure be continuous (for transverse equilibrium):

$$p_2 + \frac{B_2^2}{2\mu} = p_1 + \frac{B_1^2}{2\mu}, \qquad (2.72)$$

where subscripts 1 and 2 denote conditions on the two sides of the current sheet. It may be noted that this holds even when the sheet is curved, since the magnetic tension does not give rise to an extra term.

Inside active regions the magnetic field is so strong that for many purposes the plasma pressures (p_1 and p_2) outside the current sheet may be neglected. Then Equation (2.72) implies that the magnetic field strength must be the same on both sides of the sheet, while the field directions may differ (Figure 2.11 (a)). Suppose the *x*-axis is taken normal to the plane of the sheet and the *y*-axis bisects the angle between the two field directions. Then, as one passes through the current sheet, B_y may remain constant, while B_z reverses its direction; at the centre of the sheet, the plasma pressure

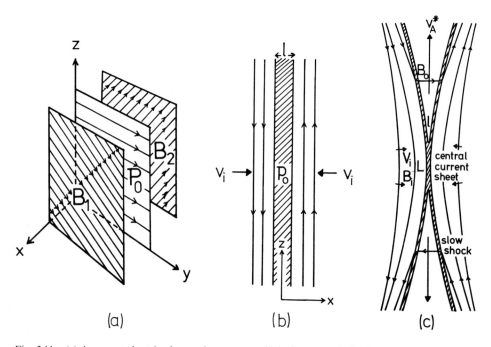

Fig. 2.11. (a) A current sheet in the *yz* plane across which the magnetic field rotates from B_1 to B_2. (b) A section across a neutral current sheet, in the centre of which the magnetic field vanishes and the plasma pressure is p_0. (c) The reconnection of magnetic field lines by their passage through a current sheet. The central sheet bifurcates into two pairs of slow shocks.

is enhanced by an amount

$$p_0 = \frac{B_{z1}^2}{2\mu} = \frac{B_{z2}^2}{2\mu}.$$

The particular case when B_y vanishes gives a *neutral* current sheet (or *neutral sheet*); the magnetic field vanishes completely in the centre of such a sheet and the field reverses its direction from one side of the sheet to the other (Figure 2.11 (b)). It is the neutral current sheet that is, for simplicity, most often analysed.

A current sheet is rather like a *shock wave* in the sense that it may be regarded as a discontinuity separating two regions where the equations of ideal magnetohydrodynamics hold; also its width and the details of its interior are determined by diffusive processes. However, the similarity ends there. Unlike shocks, current sheets do not propagate; they tend to diffuse away in time and jets of plasma are squirted from their ends at Alfvénic speeds. The ways in which such sheets may form and some elements of their behaviour are summarised below and detailed in other chapters.

2.10.1. Processes of Formation

There are three ways in which a current sheet may be formed.
(1) The region near an X-type neutral point can collapse (Section 2.7).
(2) When topologically separate parts of a magnetic configuration are pushed together, a current sheet can appear at the boundary between them (Section 6.4.4B). This may give rise to a *solar flare* when either a complex active-region field is evolving or new flux is emerging (Section 10.2.1).
(3) Current sheets may develop when a magnetohydrostatic equilibrium becomes *unstable* or even ceases to exist (i.e., non-equilibrium). In particular, as the photospheric footpoints of a complex coronal force-free field move, the coronal field cannot always adjust to a new force-free equilibrium; instead a current sheet forms and magnetic energy is rapidly released, a process known as *topological dissipation* (Section 6.4.4B).

2.10.2. Properties

(1) In the absence of flow, a current sheet *diffuses away* at a speed η/l, where η is the magnetic diffusivity. The magnetic field is *annihilated* and magnetic energy converted into heat by ohmic dissipation (Section 2.6.1).

(2) The region outside a current sheet is effectively frozen to the plasma. Plasma and magnetic flux may be brought towards the sheet from the sides at speed v_i, say. If v_i is less than η/l the sheet expands, whereas if v_i exceeds η/l the sheet will become thinner. When $v_i = \eta/l$ a steady state is maintained.

(3) The enhanced plasma pressure in the centre of the sheet expels material from the ends of the sheet at the Alfvén speed (v_A^*) based on the external magnetic field and internal density. Magnetic flux is ejected with the material, and so one effect of the sheet is to *reconnect the field lines* (Figure 2.11 (c)). The centre of the sheet is an X-type neutral point. In a steady flow the rate at which magnetic flux is trans-

ported remains constant (see Section 5.3); in other words, the rate $(v_i B_i)$ at which flux enters the sheet equals the rate $(v_A^* B_o)$ at which it leaves, where subscripts i and o denote input and output values, respectively. Thus, for sub-Alfvénic inflow $(v_i < v_A)$ the outflow field strength,

$$B_o = \frac{v_i}{v_A^*} B_i,$$

is smaller than the inflow field strength (B_i). An important effect of a current sheet is therefore to *convert magnetic energy* into *heat and flow energy*.

(4) Pairs of *slow-mode shock waves* propagate from the ends of the current sheet and remain as standing waves in a steady flow. The dimensions of the central current sheet are given by

$$l = \frac{\eta}{v_i}, \qquad L = \frac{\rho_o v_A^*}{\rho_i v_i} l;$$

the second equation is an expression of mass conservation. *Magnetic reconnection may occur for a wide range of inflow speed* (v_i); as v_i is varied, so the dimensions of the sheet respond. There is a *maximum inflow speed*, which lies typically between $0.01\, v_{Ae}$ and $0.1\, v_{Ae}$ and is weakly dependent on the magnetic Reynolds number (Section 10.1), where v_{Ae} is the external inflow Alfvén speed at large distances.

(5) Consider an equilibrium current sheet, which has a unidirectional field $(B(x)\hat{y})$ reversing at $x = 0$ and a uniform total pressure

$$p(x) + \frac{B(x)^2}{2\mu} = \text{constant}.$$

Such a sheet is subject to the *tearing-mode instability* on a time-scale of typically the geometric mean $(\tau_d \tau_A)^{1/2}$ of the diffusion time (τ_d) and the Alfvén travel time (τ_A) (Section 7.5.5). In the nonlinear development, one is likely to find (depending on the boundary conditions) either a state of quasi-steady magnetic reconnection or the creation of stationary loops which slowly diffuse away.

CHAPTER 3

MAGNETOHYDROSTATICS

3.1. Introduction

A comparison of the sizes of terms in the equation of motion (2.21)

$$\rho \frac{D\mathbf{v}}{Dt} = -\nabla p + \mathbf{j} \times \mathbf{B} + \rho \mathbf{g}$$

shows that the inertial terms on the left-hand side may be neglected when the flow speed is much smaller than both the sound speed $(\gamma p_0/\rho_0)^{1/2}$, the Alfvén speed $B_0/(\mu \rho_0)^{1/2}$ and the gravitational free-fall speed $(2gl_0)^{1/2}$ for a vertical scale-length l_0. The result is a magnetohydrostatic balance

$$0 = -\nabla p + \mathbf{j} \times \mathbf{B} + \rho \mathbf{g} \tag{3.1}$$

between the *pressure gradient*, the *Lorentz force* and the *gravitational force*. The object of the present chapter is to solve Equation (3.1) together with the subsidiary equations

$$\mathbf{j} = \operatorname{curl} \mathbf{B}/\mu, \tag{3.2}$$

$$\operatorname{div} \mathbf{B} = 0, \tag{3.3}$$

$$\rho = \frac{mp}{k_B T}, \tag{3.4}$$

and (in general) an energy equation for the temperature.

If gravity acts along the negative z-axis and s measures the distance along magnetic field lines inclined at an angle θ to the vertical, the component of Equation (3.1) parallel to $\mathbf{B}$ is

$$0 = -\frac{dp}{ds} - \rho g \cos \theta,$$

with no contribution from the Lorentz force. Since $\delta s \cos \theta = \delta z$ (Figure 3.1) this becomes

$$0 = -\frac{dp}{dz} - \rho g, \tag{3.5}$$

where p and ρ are functions of z along a particular field line. After substituting for ρ from Equation (3.4) in Equation (3.5) and integrating, we find

$$p = p_0 \exp - \int_0^z \frac{1}{\Lambda(z)} dz, \tag{3.6}$$

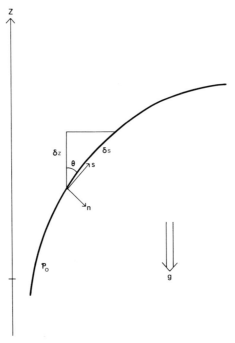

Fig. 3.1. A magnetic field line inclined at θ to the vertical z-axis. Distance s is measured along the field line and p_0 is the pressure at the reference height $z = 0$.

where p_0 is the base pressure (at $z = 0$) which may vary from one field line to another; also

$$\Lambda(z) = \frac{k_B T(z)}{mg} \left(= \frac{p}{\rho g} = \frac{\tilde{R} T(z)}{\tilde{\mu} g} \right) \qquad (3.7)$$

is the (*pressure*) *scale-height*, which represents the vertical distance over which the pressure falls by a factor e. In terms of density Equation (3.6) becomes

$$\frac{\rho}{\rho_0} = \frac{T_0}{T(z)} \exp - \int_0^z \frac{1}{\Lambda(z)} dz. \qquad (3.8)$$

Equation (3.6) shows that the pressure along a given magnetic field line decreases exponentially with height. The rate of decrease depends on the temperature structure as determined by the energy equation. It is here that the magnetic field enters implicitly, since the length of the field line depends on the magnetic structure and may influence both the conductive and heating terms in the energy balance. The corresponding density variation follows from Equation (3.8). When the temperature increases with height the density decreases faster than the pressure; but, when the temperature falls with height, the density may either increase or decrease locally depending on whether the factor T^{-1} or the exponential dominates in Equation (3.8).

For the particular case when the temperature is uniform along a field line (due to.

for instance, the dominance of thermal conduction), Λ is constant and Equation (3.6) reduces to

$$p = p_0 e^{-z/\Lambda}. \tag{3.9}$$

In the definition (3.7) of Λ the gravitational acceleration g varies with distance r from the solar centre like $g_\odot r_\odot^2/r^2$, where $\odot$ refers to values at the photosphere, namely $g_\odot \approx 274 \text{ m s}^{-2}$ and $r_\odot \approx 696 \text{ Mm}$ ($= 6.96 \times 10^8$ m). Thus, assuming $\tilde{\mu} = 0.6$, the scale-height can be written numerically in terms of the temperature (T) as

$$\Lambda = 50\, T(r/r_\odot)^2 \text{ m}. \tag{3.10}$$

For example, a temperature of 10 000 K gives a scale-height of 500 km (for $r \approx r_\odot$), while a temperature of 10^6 K makes the scale-height equal to $50\,(r/r_\odot)^2$ Mm ($= 5 \times 10^7 (r/r_\odot)^2$ m).

In many applications not all the terms in Equation (3.1) are equally important. For example, the force of gravity may be neglected by comparison with the pressure gradient when the height of a structure is much less than the scale-height. When, in addition, the ratio

$$\beta \equiv \frac{2\mu p_0}{B_0^2} \tag{3.11}$$

of plasma to magnetic pressure is much smaller than unity (see Equation (2.49) for its numerical value), any pressure gradient is dominated by the Lorentz force and Equation (3.1) reduces to

$$\mathbf{j} \times \mathbf{B} = \mathbf{0}. \tag{3.12}$$

Magnetic fields satisfying this condition are called *force-free* (or *Beltrami*). A significant Lorentz force is not allowed in this approximation, because a pressure gradient would not be strong enough to balance it. The particular case when $\mathbf{j} = \mathbf{0}$ is called *current-free* or *potential*.

Magnetohydrostatics is relevant to a variety of solar structures that appear to remain motionless for long periods of time. It has been applied, for example, to the overall structure of sunspots (Section 8.4) and of prominences (Section 11.2) and to the large-scale structure of the coronal magnetic field, which often appears stationary for times long compared with the Alfvén travel time. This chapter begins by discussing some simple solutions for unidirectional fields and cylindrically symmetric flux tubes and then proceeds to analyse current-free and force-free fields. The properties of a slender flux tube are discussed in Section 8.7 and a summary of flux tube behaviour in general is given in Section 2.9.

3.2. Plasma Structure in a Prescribed Magnetic Field

When $\beta \ll 1$, the magnetic structure is determined by the appropriate force-free solution to Equation (3.12), as described in Section 3.5. The plasma structure follows by solving Equation (3.5), together with an appropriate energy equation for the temperature and density. For example, a *coronal arcade* has been modelled by Priest and Smith (1979) using the force-free solution (3.44) sketched in Figure 3.8. The special case when the atmosphere is isothermal leads to a pressure

$$p = p_0(x, y) e^{-z/\Lambda}$$

along each field line (from Equation (3.6)). If the base pressure (p_0) is uniform, we have a simple plane-parallel atmosphere, with the magnetic field having no influence on the plasma structure at all. But, if the base of a particular flux loop has an enhanced pressure, it will possess a higher pressure than its surroundings throughout its length, as indicated in Figure 3.8.

Consider the case when β is no longer small, but the magnetic field is either vertical or horizontal. A *purely vertical field* (independent of y) of the form $\mathbf{B} = B(x)\hat{\mathbf{z}}$ gives Equation (3.5) as the z-component of Equation (3.1). This has solution (3.6), with the base pressure ($p_0(x)$) and temperature varying from one field line to another. The horizontal x-component is

$$0 = -\frac{\partial}{\partial x}\left(p + \frac{B^2}{2\mu}\right)$$

and has the solution

$$p + \frac{B^2}{2\mu} = f(z), \tag{3.13}$$

say, where $f(z)$ gives the vertical variation of the total pressure at some particular value of x. Then the derivative of Equation (3.13) with respect to z gives $\partial p/\partial z = df/dz$, which is independent of x; so, from Equation (3.5), the density is also independent of x. In other words, the density is unaffected by the presence of the magnetic field, a result which is relevant for sunspot theory (Section 8.4). In addition, Equation (3.13) implies that for a given z the pressure (and therefore the temperature) is strongest where the magnetic field is weakest (Figure 3.2).

When the magnetic field is *purely horizontal* (independent of y) and of the form $\mathbf{B} = B(z)\hat{\mathbf{x}}$, the horizontal component of Equation (3.1) implies that the pressure is a function of z alone, while the vertical component gives

$$0 = -\frac{d}{dz}\left(p + \frac{B^2}{2\mu}\right) - \rho g.$$

Since the field lines are straight, the only contribution from the Lorentz force is the magnetic pressure. In general ρ and T can vary along the field in such a way as to keep the pressure at that level constant. If $B(z)$ and T are prescribed, the above equation with $\rho = pm/(k_B T)$ can be integrated to give p. In particular, when T is uniform we find

$$p = \left[p_0 - \int_0^z e^{z/\Lambda}\, d/dz(B^2/2\mu)\, dz\right] e^{-z/\Lambda},$$

which shows how a magnetic field that decreases (or increases) with height raises (or lowers) the pressure over its hydrostatic value. When both the sound speed and

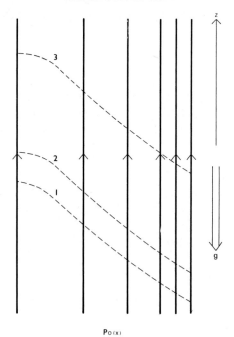

Fig. 3.2. For plasma situated in a vertical magnetic field, the isodensity contours are horizontal, while the constant-pressure contours (dashed) are inclined as shown, with the labels 1, 2, 3 indicating successively lower pressure values.

Alfvén speed are uniform the equilibrium solution is

$$p = p_0 e^{-z/\Lambda_B}, \qquad \rho = \rho_0 e^{-z/\Lambda_B}, \qquad B = B_0 e^{-z/2\Lambda_B}$$

with a modified scale-height

$$\Lambda_B = \frac{p_0 + B_0^2/(2\mu)}{\rho_0 g}.$$

Magnetic buoyancy instability in such a configuration is discussed in Section 8.2.2.

3.3. The Structure of Magnetic Flux Tubes (Cylindrically Symmetric)

Consider a cylindrically symmetric flux tube whose magnetic field components

$$(0, B_\phi(R), B_z(R)) \tag{3.14}$$

in cylindrical polar coordinates are functions of R alone. The field lines are then helical and lie on cylindrical surfaces, as indicated in Figure 3.3, while the electric current components are, from Equation (3.2),

$$\left(0, -\frac{1}{\mu}\frac{dB_z}{dR}, \frac{1}{\mu R}\frac{d}{dR}(RB_\phi)\right). \tag{3.15}$$

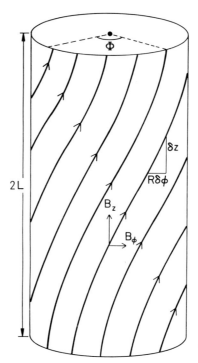

Fig. 3.3. The notation for a cylindrically symmetric flux tube of length 2L.

Under the neglect of gravity the force-balance equation then reduces to

$$\frac{dp}{dR} + \frac{d}{dR}\left(\frac{B_\phi^2 + B_z^2}{2\mu}\right) + \frac{B_\phi^2}{\mu R} = 0, \tag{3.16}$$

the second term representing the magnetic pressure and the third term the magnetic tension due to the azimuthal component (B_ϕ) that encircles the axis.

On each cylindrical surface the field lines have a constant inclination, but this may vary from one radius to another. The field lines are given by

$$\frac{R\,d\phi}{B_\phi} = \frac{dz}{B_z},$$

and the amount by which a given line is twisted in going from one end of the tube (length 2L) to the other is

$$\Phi = \int d\phi = \int_0^{2L} \frac{B_\phi}{RB_z}\,dz,$$

or

$$\Phi(R) = \frac{2LB_\phi(R)}{RB_z(R)}. \tag{3.17}$$

($4\pi L/\Phi$ is sometimes called the *pitch* of the field and gives the axial length of a field line that encircles the axis once.)

Equation (3.16) contains three dependent variables p, B_ϕ, B_z, so that any two can be prescribed and the third deduced. The most natural choice for many applications is to impose the longitudinal component (B_z) and the twist Φ (or equivalently B_ϕ) and deduce p. But sometimes one may want to fix B_z and p, say, from observations and determine B_ϕ. Several special cases arise as follows.

3.3.1. PURELY AXIAL FIELD

When no azimuthal component (B_ϕ) is present, Equation (3.16) reduces to

$$\frac{d}{dR}\left(p + \frac{B^2}{2\mu}\right) = 0,$$

with solution $p + B^2/(2\mu) = $ constant, so that the total pressure (gas plus magnetic) is conserved.

3.3.2. PURELY AZIMUTHAL FIELD

When the axial component vanishes, Equation (3.16) becomes

$$\frac{dp}{dR} + \frac{d}{dR}\left(\frac{B_\phi^2}{2\mu}\right) + \frac{B_\phi^2}{\mu R} = 0, \tag{3.18a}$$

where, according to Equation (3.2), B_ϕ is related to the current by

$$j_z = \frac{1}{\mu R}\frac{d}{dR}(RB_\phi). \tag{3.18b}$$

If, in particular, the current flows with uniform total value I within a cylinder of radius a, an integration of Equation (3.18b) yields

$$B_\phi = \begin{cases} \dfrac{\mu I R}{2\pi a^2}, & R < a, \\ \dfrac{\mu I}{2\pi R}, & R > a, \end{cases} \tag{3.19}$$

assuming B_ϕ to be finite and continuous. The corresponding plasma pressure results from integrating Equation (3.18a). Assuming that it takes the value p_∞ outside the current column, we find

$$p = \begin{cases} p_\infty + \tfrac{1}{4}\mu(I/(\pi a^2))^2(a^2 - R^2), & R < a, \\ p_\infty, & R > a. \end{cases}$$

The magnetic field lines are shown in Figure 3.4. Within the cylinder of radius a, B_ϕ increases linearly with R, while the gas pressure decreases, so that the outwards gas pressure is balanced by inwards magnetic pressure and tension forces. Outside the cylinder the pressure is uniform and the magnetic field is potential, so that the outwards magnetic pressure and inwards tension balance one another.

In the laboratory, a plasma configuration in which the current is axial and the

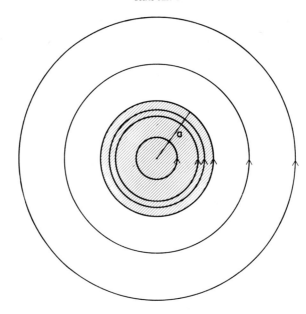

Fig. 3.4. The purely azimuthal magnetic field lines in a section across a column of uniform current and radius a.

magnetic field azimuthal is known as a *linear pinch*. A simple relation may be derived in this case between the current

$$I \equiv \int_0^{R_0} j_z 2\pi R \, dR$$

flowing through the plasma column (of radius R_0) and the number

$$N \equiv \int_0^{R_0} n 2\pi R \, dR$$

of particles per unit length of the column. Equation (3.18a) may first be multiplied by R^2 and integrated to give

$$\int_0^{R_0} R^2 \, dp = -\int_0^{R_0} RB_\phi/\mu \, d(RB_\phi).$$

Then, assuming that the plasma pressure vanishes at R_0 and the temperature ($T = p/(nk_B)$) is uniform across the column, an integration by parts of the left-hand side together with the use of Equation (3.18b) on the right-hand side yields the expression

$$I^2 = (8\pi/\mu)k_B T N,$$

known as *Bennett's relation*.

3.3.3. Force-Free Fields

3.3.3A. Linear Field

In the absence of pressure, Equation (3.16) reduces to

$$\frac{d}{dR}\left(\frac{B_\phi^2 + B_z^2}{2\mu}\right) + \frac{B_\phi^2}{\mu R} = 0. \tag{3.20}$$

Here, either B_ϕ or B_z may be prescribed and the other deduced. For the so-called 'constant-α' field (Section 3.5) one assumes that $\mu \mathbf{j} = \alpha \mathbf{B}$, where α is uniform. After using Equation (3.2), the ϕ-component of this becomes

$$-\frac{dB_z}{dR} = \alpha B_\phi. \tag{3.21}$$

Finally, an elimination of B_ϕ between Equations (3.20) and (3.21) yields Bessel's equation whose solution subject to $B_z = B_0$ and $dB_z/dR = 0$ at $R = 0$ is

$$B_\phi = B_0 J_1(\alpha R), \qquad B_z = B_0 J_0(\alpha R), \tag{3.22}$$

where J_0, J_1 are Bessel functions. This solution is due to Lundquist (1951), but it has the rather undesirable feature (for solar applications) that the axial component possesses field reversals at the zeros of J_0. Solutions in toroidal geometry have been presented by Miller and Turner (1981).

3.3.3B. Nonlinear Fields

An easy way to generate solutions to Equation (3.20) is to choose (Lüst and Schlüter, 1954)

$$B^2 = f(R), \tag{3.23}$$

and then Equation (3.20) gives

$$B_\phi^2 = -\tfrac{1}{2} R \frac{df}{dR} \tag{3.24}$$

and

$$B_z^2 = B^2 - B_\phi^2. \tag{3.25}$$

The restrictions that B_ϕ^2 and B_z^2 be positive imply that df/dR is negative and that f approaches zero slower than R^{-2} as $R \to \infty$. The limiting case $f = R^{-2}$ gives the purely azimuthal field $R^{-1}\hat{\phi}$.

Another simple example of a force-free field is the 'uniform-twist' field, for which Φ (given by Equation (3.17)) is constant and the field components are

$$B_\phi = \frac{B_0 \Phi R/(2L)}{1 + \Phi^2 R^2/(2L)^2}, \qquad B_z = \frac{B_0}{1 + \Phi^2 R^2/(2L)^2}. \tag{3.26}$$

They have the property that field lines at different radii are twisted through the same angle, as shown in Figure 3.5, so that the whole tube is twisted like a rigid body.

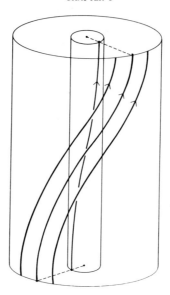

Fig. 3.5. Magnetic field lines at two radii for the uniform-twist field.

3.3.3C. *Effect of Twisting a Tube*

Parker (1977) considered the effect of twisting a force-free flux tube of finite radius (a), confined by a fixed plasma pressure ($B^2(a)/(2\mu)$). While the tube is being twisted up, the field is assumed to remain cylindrically symmetric, but the radius a is allowed to vary from an initial value ($a^{(0)}$). Then the mean-square axial field $\langle B_z^2 \rangle$ is unaffected by the twisting, since

$$\langle B_z^2 \rangle \equiv \frac{2}{a^2} \int_0^a R B_z^2 \, dR$$

$$= \frac{2}{a^2} \int_0^a Rf + \tfrac{1}{2}R^2 \frac{df}{dR} \, dR \quad \text{from Equation (3.25)}$$

$$= \frac{1}{a^2} \int_0^a \frac{d}{dR}(R^2 f) \, dR$$

$$= f(a),$$

or, by Equation (3.23), $\langle B_z^2 \rangle = B^2(a)$.

In particular, Parker considers the uniform-twist field (3.26) with the value of B_z on the axis $R = 0$ set equal to

$$B_0 = B^{(0)}(1 + \Phi^2 a^2/(2L)^2)^{1/2}.$$

This varies with twist (Φ) in such a way as to keep the total field equal to a constant ($B^{(0)}$) at the edge of the tube. $B^{(0)}$ is the initial uniform axial field strength when the tube is untwisted. The variation of the tube radius with twist is determined by the condition that the longitudinal magnetic flux through the tube be conserved, namely

$$2\pi \int_0^a RB_z \, dR = \pi a^{(0)2} B^{(0)}$$

or

$$(1 + \Phi^2 a^2/(2L)^2)^{1/2} \log_e (1 + \Phi^2 a^2/(2L)^2) = a^{(0)2} \Phi^2/(2L)^2.$$

It transpires from this equation that, as Φ increases, so Φa and a increase. At the same time $B_z(0)$ increases (since an outwards magnetic pressure is required to balance the inwards tension force produced by twisting), while $B_z(a)$ decreases (since an increasing proportion of $B(a)$ goes into the azimuthal component).

3.3.3D. *Effect of Expanding a Tube*

Parker (1974c) shows how the radial expansion of a tube makes B_ϕ/B_z increase. Since the twist $(2LB_\phi/(RB_z))$ remains constant while the mean value of R increases, it is clear that the mean value of B_ϕ/B_z must increase. To demonstrate the effect,

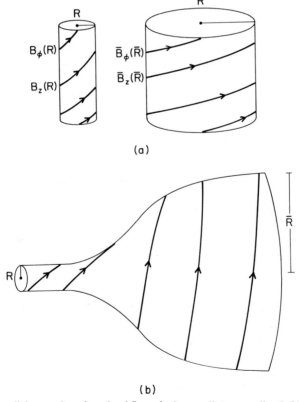

Fig. 3.6. (a) The radial expansion of a twisted flux tube from radius a to radius $\bar{a}$. (b) The concentration of azimuthal flux in the widest part of a flux tube.

suppose the confining pressure decreases on the surface of a cylindrical tube with radius a and field $(B_\phi(R), B_z(R))$, and so causes it to expand to a radius $\bar{a}$, in such a way that the radius R becomes $\bar{R}$ and the field becomes $(\bar{B}_\phi(\bar{R}), \bar{B}_z(\bar{R}))$, as indicated in Figure 3.6(a). Suppose the plasma pressure is negligible (or uniform), so that the initial and final fields are given by generating functions f and $\bar{f}$ according to Equation (3.23). Then the conservation of longitudinal and azimuthal flux initially through the annulus $(R, R + dR)$ and finally through $(\bar{R}, \bar{R} + d\bar{R})$ gives

$$B_z(R) R \, dR = \bar{B}_z(\bar{R}) \bar{R} \, d\bar{R}$$

and

$$B_\phi(R) \, dR = \bar{B}_\phi(\bar{R}) \, d\bar{R}.$$

After using Equations (3.24) and (3.25) to rewrite these in terms of f and $\bar{f}$, $\bar{f}$ may be eliminated to yield an equation for the mapping $R(\bar{R})$, namely

$$0 = \frac{du}{d\bar{u}} \left\{ \frac{d^2 u}{d\bar{u}^2} \left(\frac{d(uf)}{du} - \bar{u} \frac{df}{du} \right) + \frac{1}{2} \left(\frac{du}{d\bar{u}} \right)^2 \left(\frac{d^2(uf)}{du^2} - \frac{d^2 f}{du^2} \right) - \frac{df}{du} \frac{du}{d\bar{u}} \right\},$$

where $u = R^2$ and $\bar{u} = \bar{R}^2$. This may be solved once the initial generating function (f) is prescribed. For example, with $a^2 = \frac{1}{2}$ and

$$f(R)/(2\mu) = 1 - R^2, \quad 0 \leq R^2 \leq \tfrac{1}{2},$$

so that

$$\frac{B_z^2}{2\mu} = 1 - 2R^2, \quad \frac{B_\phi^2}{2\mu} = R^2,$$

Parker finds for a large expansion $(\bar{a} \gg \tfrac{1}{2})$

$$R^2 = \frac{\log_e(1 + \bar{R}^2)}{4 \log_e \bar{a}}$$

and

$$\frac{\bar{B}_\phi^2}{2\mu \bar{R}^2} = \frac{\bar{B}_z^2}{2\mu} = \frac{1}{16(1 + \bar{R}^2)^2 \log_e^2 \bar{a}}.$$

Thus, the initial field has a B_ϕ that increases with R and a B_z that is uniform near the axis and falls to zero at the surface ($R = a$), while the final field is mainly azimuthal ($\bar{B}_\phi \gg \bar{B}_z$) over most of the radius ($1 < \bar{R} < \bar{a}$). A similar increase in B_ϕ/B_z occurs if the flux tube is compressed in the axial direction, while the opposite effect may be produced by compressing the tube radially or stretching it axially. It may be noted that three types of field are invariant with respect to expansion, namely a purely axial field, a purely azimuthal field or a uniform-twist field (3.26).

3.3.3E. A Tube of Non-Uniform Radius

Parker (1974c) also considers a flux tube whose radius increases with axial distance (z) due to a fall in the confining pressure. He supposes the radius a is uniform for

$z < -h$ (with field $(B_\phi(R), B_z(R))$ and generating function f) and the radius $\bar{a}$ is uniform for $z > h$ with field $(\bar{B}_\phi(\bar{R}), \bar{B}_z(\bar{R}))$ and generating function $\bar{f}$, as shown in Figure 3.6(b). Then again the problem is to determine the relationship between the two fields. Conservation of longitudinal flux gives

$$B_z(R) R \, dR = \bar{B}_z(\bar{R}) \bar{R} \, d\bar{R},$$

as before, but conservation of azimuthal flux is no longer appropriate, since the azimuthal flux may be concentrated at certain locations along the tube. However, the azimuthal Maxwell stress is $B_\phi B_z / \mu$, and so conservation of its torque gives

$$R(B_\phi B_z) R \, dR = \bar{R}(\bar{B}_\phi \bar{B}_z) \bar{R} \, d\bar{R}.$$

The field components may be written in terms of f and $\bar{f}$, and then $\bar{f}$ is eliminated between these two equations. The result for an initial generating function $f/(2\mu) = 1 - R^2 (0 \leq R^2 \leq \frac{1}{2})$ and tube radius $a = 1/\sqrt{2}$ is

$$0 = \frac{du}{d\bar{u}} \left\{ (1 - 2u) \frac{d^2 u}{d\bar{u}^2} - \left(\frac{du}{d\bar{u}}\right)^2 + \frac{u}{\bar{u}} \right\}$$

for the mapping $R(\bar{R})$, where $u = R^2$ and $\bar{u} = \bar{R}^2$. We note that this equation is singular at the tube surface where B_z vanishes. If the tube is expanded greatly so that the new radius $\bar{a} \gg 1/\sqrt{2}$, the solution is

$$R = \frac{\bar{R} J_1(2\bar{R})}{\sqrt{2} \bar{a} J_1(2\bar{a})},$$

and the resulting field components at large z are given by

$$\frac{\bar{B}_z^2}{2\mu} = \frac{J_0^2(2\bar{R})}{2\bar{a}^2 J_1^2(2\bar{a})}, \qquad \frac{\bar{B}_\phi^2}{2\mu} = \frac{J_1^2(2\bar{R})}{2\bar{a}^2 J_1^2(2\bar{a})},$$

which are valid only out as far as $\bar{R} \simeq 1.2$ where $\bar{B}_z^2$ vanishes. Beyond this radius the mapping $\bar{R} = \bar{R}(R)$ is no longer single-valued and it is rather uncertain what happens, although Parker suggests that the field becomes purely azimuthal. Browning and Priest (1982, *GAFD*, in press) have included B_R in the calculation. The surface becomes cusp-like in shape, and they suggest that, when the internal gas pressure exceeds the external pressure, the flux tube bursts, with its outer layers being stripped off.

3.3.4. MAGNETOSTATIC FIELDS

One simple solution to the magnetostatic equation (3.16) possesses an axial component

$$B_z = \frac{B_0}{1 + R^2/a^2},$$

which decreases from a maximum on the axis over a length-scale a, while the azimuthal component

$$B_\phi = \frac{\Phi R B_z}{2L}$$

is chosen to make the twist (Φ) uniform. The pressure for this *uniform-twist* field is

$$p(R) = p_\infty + \frac{(\Phi^2 a^2/(2L)^2 - 1)B_0^2}{(1 + R^2/a^2)2\mu}.$$

For an untwisted tube this has a minimum on the axis and, as the twist increases, so $p^{(0)}$ increases in value. Ultimately, it becomes a maximum when Φ exceeds $2L/a$.

Another simple solution occurs when the axial field is uniform with $B_z = B_0$, while the twist,

$$\Phi(R) \equiv \frac{2LB_\phi}{RB_0} = \frac{\Phi_0}{1 + R^2/a^2},$$

decreases from a maximum Φ_0 on the axis. This *variable-twist field* has a pressure (from (3.16)) of

$$p(R) = p_\infty + \frac{\Phi^2 B_0^2 a^2}{8\mu L^2}.$$

It is uniform when the twist is zero, and its maximum, which always occurs on the axis, increases in value with increasing Φ_0. In future, there is a need to study in detail the effect of flux tube curvature by considering solutions in toroidal geometry with a varying cross-section.

3.4. Current-Free Fields

When the current density vanishes everywhere, Equation (3.2) gives curl $\mathbf{B} = \mathbf{0}$, and the curl of this, together with Equation (3.3), yields

$$\nabla^2 \mathbf{B} = \mathbf{0}. \tag{3.27}$$

Thus the magnetic field is *potential* and the whole body of potential theory may be applied. It is sometimes convenient to write

$$\mathbf{B} = \nabla \psi, \tag{3.28}$$

where ψ, the *scalar magnetic potential*, also satisfies Laplace's equation,

$$\nabla^2 \psi = 0. \tag{3.29}$$

Some standard results of potential theory are of particular interest to solar physics. For instance, the solution within a closed volume that takes prescribed values of ψ or $\partial\psi/\partial n (= B_n)$ on the boundary is unique, where n refers to a direction normal to the boundary. If B_n is prescribed on the surface, the corresponding potential field contains the smallest possible amount of magnetic energy $W = \int B^2/(2\mu) dV$ (*Minimum energy theorem*). Thus magnetic fields with non-zero currents but the same B_n on the boundary must contain more energy than the potential field. This result is true for a semi-infinite region such as the solar atmosphere provided there are no sources at infinity, so that the magnetic field at large distances L falls off faster than L^{-1}. It also holds if, on part of the boundary, B_n is not prescribed but instead the tangential field component vanishes.

Standard solutions of Equation (3.29) may be derived by using the method of separation of variables. For instance, in *spherical polar coordinates* (r, θ, ϕ), the general solution is

$$\psi = \sum_{l=0}^{\infty} \sum_{m=-l}^{l} (a_{lm} r^l + b_{lm} r^{-(l+1)}) P_l^m(\cos \theta) e^{im\phi},$$

in terms of the associated Legendre polynomial (P_l^m); the particular case when the potential is independent of ϕ reduces to

$$\psi = \sum_{l=0}^{\infty} (a_l r^l + b_l r^{-(l+1)}) P_l(\cos \theta),$$

where P_l is the Legendre polynomial. In *cylindrical polars* (R, ϕ, z), the general solution may be written

$$\psi = \sum_{n=-\infty}^{\infty} (c_n J_n(kR) + d_n Y_n(kR)) e^{in\phi \pm kz},$$

where J_n and Y_n are Bessel functions, or, when there is no z-dependence,

$$\psi = C \log R + \sum_{n=-\infty}^{\infty} (C_n R^n + D_n R^{-n}) e^{in\phi}.$$

In each of the above cases, the constants can be determined by applying boundary conditions.

Several computer codes have been developed to obtain potential magnetic configurations appropriate to the solar atmosphere. These employ the measured line-of-sight photospheric magnetic field as a boundary condition. The actual solar magnetic field above the photosphere is expected to be potential when the photospheric sources have remained stationary long enough that the field has reduced to its minimum-energy configuration. This is often not the case, either in *active regions* (where force-free and more general magnetostatic configurations may be present) or in *coronal streamers* (where current sheets may be located).

The first code for current-free coronal fields was developed by Schmidt (1964), who gave the magnetic field in rectangular coordinates from a distribution of monopoles on a region of the photospheric boundary. Its shortcomings are the built-in assumption that no field lines leave the region, the neglect of curvature and the use of the normal field component at the boundary, which together mean that it can be applied only to small regions near the disc centre. Semel (1967) allowed the boundary to be inclined at any angle to the line-of-sight and then Altschuler and Newkirk (1969) developed a program for modelling the field above the whole solar disc. It approximates the magnetic field by a finite series of Legendre polynomials, the coefficients of which are determined to obtain a fit at the photosphere. The effect of the solar wind in dragging out field lines is simulated by forcing the field to become radial at a distance of typically 2.5 $R_\odot$. The resulting magnetic configuration is sketched as a so-called 'hairy ball', which has been widely used to compare with eclipse photographs (e.g., Figure 3.7) or, for instance, radio burst locations.

An alternative code to the Altschuler-Newkirk one has been set up by Adams and

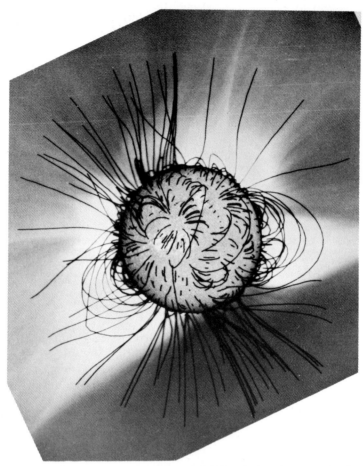

Fig. 3.7. A 'hairy ball', indicating potential magnetic field lines in the solar corona for 12 November 1966, calculated by the Altschuler–Newkirk code and superimposed on the eclipse photograph (courtesy G. Newkirk, High Altitude Observatory).

Pneuman (1976) employing a finite-difference method. Both methods yield similar solutions for the large-scale field at two solar radii. The Adams–Pneuman code possesses uniform accuracy at all heights, but, in order to treat small-scale surface features with the same accuracy, the Altschuler–Newkirk program needs to employ many more than the original nine polynomials. A much faster code has been developed by Riesebieter and Neubauer (1979), who use orthogonality relations of the spherical harmonics to determine recursion formulae for the harmonic coefficients. (The effect of a non-spherical source surface has been incorporated by Levine *et al.* (1982).) Limitations on these global calculations are the poor quality of data near the poles and the fact that variations on a time shorter than the solar rotation period cannot be studied.

Another ingenious code has been developed by Sakurai and Uchida (1977) for modelling the global field from several active regions; it models them by a series of discrete current loops and has the advantage of simplicity and of automatically giving flux balance. Also, Yeh and Pneuman (1977) have modelled the corona as a series of potential field regions with different pressures and separated by current sheets; this is particularly useful for representing a coronal streamer. Finally, Sakurai (1982a) has returned to the original philosophy of Schmidt (1964) and calculated the Green's functions for a spherical surface S (i.e., the field of a monopole on S, modified to make the component along the line-of-sight vanish everywhere else on S).

3.5. Force-Free Fields

The Lorentz force dominates the pressure gradient and gravitational force in Equation (3.1) when the plasma has a low beta ($\beta \ll 1$) and its vertical extent (H) is much smaller than Λ/β. Such conditions exist above an active region and, to the lowest order in β and $\beta H/\Lambda$, the force balance reduces to

$$\mathbf{j} \times \mathbf{B} = \mathbf{0}. \tag{3.30}$$

Thus, the electric current flows along magnetic field lines and, since it is given by Ampere's Law (3.2), we may write Equation (3.30) as

$$\text{curl } \mathbf{B} = \alpha \mathbf{B}, \tag{3.31}$$

where α is some function of position. The only restriction on α is that it remains constant along each field line. This follows by taking the divergence of Equation (3.31) so that the left-hand side vanishes identically and the right-hand side reduces (after using Equation (3.3)) to

$$(\mathbf{B} \cdot \nabla)\alpha = 0, \tag{3.32}$$

which implies that $\mathbf{B}$ lies on surfaces of constant α (as does also $\mathbf{j}$). If such a surface S is closed, it cannot in general be simply connected. For, otherwise, integrating along a magnetic field line C gives

$$\int_C \mathbf{B} \cdot \mathbf{ds} = \int_S \int \text{curl } \mathbf{B} \cdot \mathbf{dS}, \quad \text{by Stokes' theorem}$$

$$= \mu \int \int \mathbf{j} \cdot \mathbf{dS}, \quad \text{by Equation (3.2)}$$

$$= 0.$$

since $\mathbf{j}$ lies in the surface S. Thus, the simplest form for a constant-α closed surface is a torus on which the line of force spirals.

When α takes the same value on each field line, we have a so-called *linear* or *constant-α* field, for which the curl of Equation (3.31) reduces to

$$(\nabla^2 + \alpha^2)\mathbf{B} = \mathbf{0}, \tag{3.33}$$

so that the governing equation is linear; this is not true generally (see Equation (3.55) below, where the nonlinearity in Equation (3.30) is made explicit). Much attention has been paid to constant-α fields because of the difficulty that exists of finding more general solutions to Equation (3.30), despite its disarmingly simple form!

This section first presents some rather surprising results that hold for general force-free fields, and then it gives some simple analytical, constant-α solutions. Finally, the diffusion and evolution of force-free fields are discussed.

3.5.1. GENERAL THEOREMS

Roberts (1967) and Cowling (1976) refer to a number of interesting general results that lend an air of mystery to the structure of force-free fields; because of their inherent nonlinearity, such fields are not yet fully understood. The generalisation of the *minimum-energy theorem* for potential fields is as follows: suppose the normal magnetic field component is prescribed on the bounding surface S of a volume V, together with the correspondence between points of entrance and exit (i.e., fixing the footpoint positions and so the value of the vector potential $\mathbf{A}$ on S); then, if the field energy in V is stationary (i.e., an extremum), it must be force-free. A corollary of this is that, *if the flux and topological connections on S are given and the field possesses a minimum energy, then it is force-free*. But the converse is not true so that a force-free field does not necessarily produce a *minimum* energy.

From the *virial theorem* (2.66), the magnetic energy within a volume V may be written in terms of the magnetic field on the surface S as

$$\int_V B^2/(2\mu)\,dV = \int_S \left[(\mathbf{r}\cdot\mathbf{B})\mathbf{B} - \tfrac{1}{2}B^2\mathbf{r}\right]\cdot d\mathbf{S}/\mu.$$

When, in particular, V is the region above the plane $z = 0$, this becomes

$$\int_V B^2/(2\mu)\,dV = \int_S (xB_x + yB_y)B_z/\mu\,dx\,dy.$$

Another interesting theorem is that, if $\mathbf{j}\times\mathbf{B}$ vanishes everywhere within a volume (V) and on its surface (S), then the magnetic field is identically zero. (This follows from the virial theorem above by taking a new surface at large distances so that the right-hand side vanishes.) Thus a *non-trivial magnetic field* ($\mathbf{B}\not\equiv 0$) *that is force-free within V must be stressed somewhere on S*, since the Lorentz force cannot vanish everywhere on S. In other words, force-free fields are possible but they must be anchored down somewhere on the boundary. Therefore, attempts to build force-free fields from currents enclosed entirely within a volume are doomed to failure.

A similar theorem is that no magnetic field having a finite energy can be force-free everywhere. For, if a magnetic field falls off like r^{-2} (or faster) at large distances from the origin, the energy

$$W = \int B^2/(2\mu)\,dV$$

can be transformed to

$$W = \int \mathbf{r}\cdot\mathbf{j}\times\mathbf{B}/\mu\,dV,$$

which vanishes when the field is force-free everywhere. Thus a non-trivial field that is force-free everywhere must possess a singularity (which holds in particular for potential fields, of course).

The following result warns one against trying to construct fields in polar coordinates that are too simple: *an axisymmetric, force-free, poloidal magnetic field must be current-free*. A poloidal field possesses no azimuthal component and so has the form

$$\mathbf{B} = B_R \hat{\mathbf{R}} + B_z \hat{\mathbf{z}},$$

where axisymmetry implies that **B** is independent of azimuth ϕ. The current from Ampère's Law is then

$$\mathbf{j} = \frac{1}{\mu}\left(\frac{\partial B_R}{\partial z} - \frac{\partial B_z}{\partial R}\right)\hat{\boldsymbol{\phi}},$$

so that

$$\mathbf{j} \times \mathbf{B} = j(B_z \hat{\mathbf{R}} - B_R \hat{\mathbf{z}}).$$

If this is to vanish, then so must j and the theorem is established.

For a constant-α, force-free field Molodensky (1974) has obtained the lower limit

$$\delta W \geq \tfrac{1}{2}\mu^{-1}\left\{\int B_1^2\, dV - \left[\int \alpha^2 B_1^2\, dV\right]^{1/2}\left[\int r^2 B_1^2\, dV\right]^{1/2}\right\}$$

for the change in energy (δW) produced by a perturbation ($\mathbf{B}_1$) from a force-free field. If d is the maximum dimension of the finite volume, so that $r \leq d$, we have

$$\delta W > \tfrac{1}{2}\mu^{-1}\int B_1^2\, dV(1 - |\alpha|d),$$

which is positive provided $|\alpha| < d^{-1}$. Thus, the field is stable provided the electric current scale-length exceeds the size of the region. In later papers Molodensky (1975, 1976) strengthened this result considerably by showing that a constant-α, force-free field in a spherical region of diameter d is stable if $|\alpha| < q/d$ and that a non-constant-α field is stable if the maximum value of α is small enough. However, general conditions for *instability* have not yet been obtained.

Boström (1973) established the result that, in a plasma of finite electrical conductivity, a force-free solution of constant α within a *finite* region cannot be matched to a potential field outside the region. This limits the applicability of constant-α solutions.

Woltjer (1958) has shown that, *for a perfectly conducting plasma in a closed volume* (V_0), the integral

$$\int_{V_0} \mathbf{A} \cdot \mathbf{B}\, dV = K_0 \qquad (3.34)$$

is invariant and the *state of minimum magnetic energy is a linear (constant-α) force-free field*. The proof is as follows. For a perfectly conducting plasma,

$$\frac{\partial \mathbf{B}}{\partial t} = \nabla \times (\mathbf{v} \times \mathbf{B}),$$

or, in terms of the magnetic vector potential (**A**),

$$\frac{\partial \mathbf{A}}{\partial t} = \mathbf{v} \times \mathbf{B}, \tag{3.35}$$

where

$$\mathbf{B} = \nabla \times \mathbf{A}, \tag{3.36}$$

and the gauge has been chosen to make the scalar potential vanish. Thus

$$\frac{\partial}{\partial t} \int_{V_0} \mathbf{A} \cdot \mathbf{B} \, dV = \int_{V_0} \mathbf{A} \cdot \frac{\partial \mathbf{B}}{\partial t} \, dV + \int_{V_0} \frac{\partial \mathbf{A}}{\partial t} \cdot \mathbf{B} \, dV$$

$$= \int_{V_0} -\nabla \cdot \left[\mathbf{A} \times \frac{\partial \mathbf{A}}{\partial t} \right] + 2\frac{\partial \mathbf{A}}{\partial t} \cdot \nabla \times \mathbf{A} \, dV$$

after putting $\mathbf{B} = \nabla \times \mathbf{A}$ in the first integral and rewriting a triple scalar product. Now, the second term vanishes by Equation (3.35) and the first may be transformed by Gauss' Theorem to

$$\frac{\partial}{\partial t} \int_{V_0} \mathbf{A} \cdot \mathbf{B} \, dV = - \int_{S_0} \mathbf{A} \times \frac{\partial \mathbf{A}}{\partial t} \cdot d\mathbf{S}.$$

This vanishes for a closed volume on whose boundary **A** is constant and so Equation (3.34) follows. The next stage of the proof is to consider the magnetic energy

$$W = \int_{V_0} B^2/(2\mu) \, dV \tag{3.37}$$

contained within the volume and the effect on it of a small arbitrary change of **A** and **B** to $\mathbf{A} + \delta\mathbf{A}$ and $\mathbf{B} + \delta\mathbf{B}$, such that

$$\delta\mathbf{A} = 0 \quad \text{on } S \tag{3.38}$$

and

$$\delta\mathbf{B} = \text{curl } \delta\mathbf{A}. \tag{3.39}$$

Then, by linearising and subtracting α_0 times Equation (3.34), one finds

$$2\mu \, \delta W = \int_{V_0} 2\mathbf{B} \cdot \delta\mathbf{B} - \alpha_0(\delta\mathbf{A} \cdot \mathbf{B} + \mathbf{A} \cdot \delta\mathbf{B}) \, dV$$

or, after using Equation (3.39) and rewriting the triple scalar products,

$$2\mu \, \delta W = \int_{V_0} \nabla \cdot (-2\mathbf{B} \times \delta\mathbf{A} + \alpha_0 \mathbf{A} \times \delta\mathbf{A}) \, dV +$$

$$+ 2\int_{V_0} (\nabla \times \mathbf{B} - \alpha_0 \mathbf{B}) \cdot \delta\mathbf{A} \, dV.$$

The first integral may be reduced to a surface integral that vanishes by Equation (3.39), whereas the second integral shows that $\delta W = 0$ for all perturbations ($\delta \mathbf{A}$) if and only if

$$\nabla \times \mathbf{B} = \alpha_0 \mathbf{B}. \qquad (3.40)$$

Thus, if the energy is a minimum, the field must satisfy Equation (3.40) for some α_0, and if a field satisfies Equation (3.40), the energy must be an extremum (but not necessarily a minimum). The possible values of α_0 may be found in terms of the invariant K_0 and the amount of flux that is present. Also, it may be noted that the theorem holds as well if the magnetic field component normal to the boundary is prescribed (but not necessarily zero), so that again $\partial \mathbf{A}/\partial t$ and $\delta \mathbf{A}$ vanish.

Taylor (1974, 1976) has extended Woltjer's theorem for force-free fields in a laboratory torus, whose walls are rigid and perfectly conducting, so that the normal magnetic field component vanishes and $\mathbf{A}$ is constant on the surface. In such experiments an initial toroidal magnetic field ($\mathbf{B}_0$) is produced by external coils and then a toroidal current ($\mathbf{I}$) is induced. It pinches the plasma, and, after a violent dissipative phase, the magnetic field can relax to a stable state which depends on the value of the *pinch ratio*

$$\theta = \frac{\mu I}{2\pi a B_0}, \qquad (3.41)$$

where a is the minor radius of the torus and R the major radius. When $\theta \lesssim a/R$, one has the *tokamak regime*, whereas when $\theta \gtrsim 1.2$ the configuration is known as a *reversed-field pinch* and contains a reversed field in the outer regions. The stable field must possess a minimum energy subject to whatever constraints are imposed on the possible motion, but what are the relevant constraints? Taylor points out that in a *perfectly conducting plasma*,

$$\int_V \mathbf{A} \cdot \mathbf{B} \, dV = K \qquad (3.42)$$

is invariant for every infinitesimal flux tube; the minimum-energy field subject to the constraint that all the K's be invariant is then the force-free field $\nabla \times \mathbf{B} = \alpha(\mathbf{r})\mathbf{B}$, where α is constant on each field line. In other words, in an ideal plasma the final state can be any equilibrium. For an *imperfect plasma*, with small non-zero resistivity, Taylor suggests that the resulting changes in topology are accompanied by small changes in $\mathbf{B}$, so that $\mathbf{A} \cdot \mathbf{B}$ is redistributed amongst the field lines but its integral over all field lines is almost unchanged. In other words, *Taylor's hypothesis* is that $\int_{V_0} \mathbf{A} \cdot \mathbf{B} \, dV$ over the total volume is approximately invariant, so that by Woltjer's theorem the *minimum energy configuration is a linear force-free field*.

3.5.2. SIMPLE CONSTANT-α SOLUTIONS

The simplest solution has the form

$$\mathbf{B} = (0, B_y(x), B_z(x)),$$

in which case Equation (3.30) gives

$$\frac{d}{dx}(B_y^2 + B_z^2) = 0$$

and div **B** = 0 is satisfied identically. An integration with respect to x implies

$$B_y^2 + B_z^2 = B_0^2,$$

say, and the solution becomes

$$\mathbf{B} = (0, B_y, (B_0^2 - B_y^2)^{1/2}).$$

As one moves along the x-axis the magnetic field rotates, keeping a constant magnitude (B_0).

For the particular case of constant α, the z-component of Equation (3.31) is

$$\frac{dB_y}{dx} = \alpha(B_0^2 - B_y^2)^{1/2},$$

so that, with the origin chosen to be a zero of B_y, the solution is

$$B_y = B_0 \sin \alpha x, \qquad B_z = B_0 \cos \alpha x. \tag{3.43}$$

Another useful one-dimensional force-free field with constant field magnitude (although without a constant α) is $(0, \tanh x, \operatorname{sech} x)$ as used by Chiuderi and Van Hoven (1979). The cylindrical analogue to (3.43) is the Lundquist solution (3.22), but in spherical polars the only solution to div **B** = 0 and $\mathbf{j} \times \mathbf{B} = 0$ having the form $\mathbf{B}(r)$ is proportional to $r^{-2}\hat{\mathbf{r}}$.

There are several simple *two-dimensional* solutions of separable form. Suppose first that in rectangular cartesians

$$\begin{aligned} B_x &= A_1 \cos kx \, e^{-lz}, \\ B_y &= A_2 \cos kx \, e^{-lz}, \\ B_z &= B_0 \sin kx \, e^{-lz}. \end{aligned} \tag{3.44}$$

Then the y- and z-components of Equation (3.30) yield

$$A_1 = -(l/k)B_0, \qquad A_2 = -(1 - l^2/k^2)^{1/2} B_0.$$

The field is periodic in the x-direction, but a section of it may provide a simple model for a coronal arcade of lateral extent π/k, as shown in Figure 3.8. The inclination of the field lines to the x-direction is $\gamma = \tan^{-1}(k^2/l^2 - 1)^{1/2}$ at their summits. Thus, when $l = k$, γ vanishes and we have the potential field

$$B_x = -B_0 \cos kx \, e^{-kz}, \qquad B_z = B_0 \sin kx \, e^{-kz};$$

but, as l decreases from k to 0, so the shear angle γ increases from 0 to $\tfrac{1}{2}\pi$. This example of a constant-α solution has $\alpha = (k^2 - l^2)^{1/2}$.

The analogous solution in cylindrical polars (R, ϕ, z) is (Schatzman, 1965)

$$\begin{aligned} B_R &= (l/k) B_0 J_1(kR) \, e^{-lz}, \\ B_\phi &= (1 - l^2/k^2)^{1/2} B_0 J_1(kR) \, e^{-lz}, \\ B_z &= B_0 J_0(kR) \, e^{-lz}, \end{aligned} \tag{3.45}$$

in terms of Bessel functions J_0 and J_1. As one moves out in the radial direction, so the horizontal field component reverses in sign at each zero of $J_1(kR)$, but, if it is limited at the first zero (kR_0), the solution may be suitable for a simple model of a

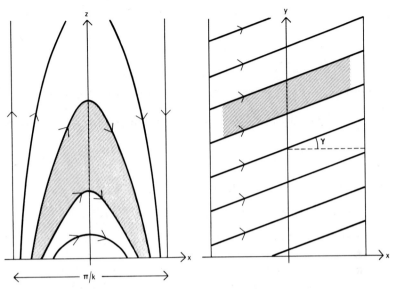

Fig. 3.8. Vertical and horizontal sections through a magnetic configuration described by Equation (3.44) with $B_0 < 0$. It may be used to model a coronal arcade. The shaded loop possesses a pressure that is enhanced at the base and therefore also at all heights (Section 3.2).

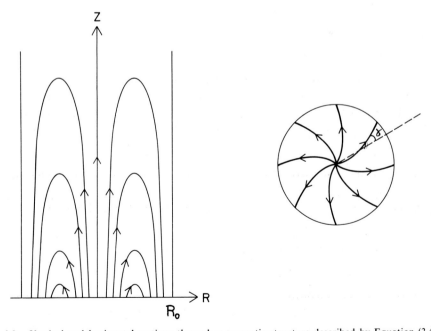

Fig. 3.9. Vertical and horizontal sections through a magnetic structure described by Equation (3.45). It may model the field above a sunspot.

sunspot field above the photosphere (Figure 3.9). The spiral angle
$$\gamma = \tan^{-1}\frac{B_\phi}{B_R} = \tan^{-1}\left(\frac{k^2}{l^2}-1\right)^{1/2}$$
is uniform throughout the structure. It vanishes for the potential field ($l = k$)
$$B_R = B_0 J_1(kR)\,e^{-lz}, \qquad B_z = B_0 J_0(kR)\,e^{-lz},$$
and, as l decreases from k to 0, so γ increases from 0 to $\tfrac{1}{2}\pi$ and the field winds up more and more.

Again, in spherical polar coordinates (r, θ, ϕ) one finds corresponding solutions of the form

$$\begin{aligned}
B_r &= C_n n(n+1) r^{-3/2} J_{n+\frac{1}{2}}(\alpha r) P_n(\cos\theta),\\
B_\theta &= C_n\bigl[-nr^{-3/2} J_{n+\frac{1}{2}}(\alpha r) + \alpha r^{-1/2} J_{n-\frac{1}{2}}(\alpha r)\bigr]\frac{dP_n(\cos\theta)}{d\theta},\\
B_\phi &= -C_n \alpha r^{-1/2} J_{n+\frac{1}{2}}(\alpha r)\frac{dP_n(\cos\theta)}{d\theta}.
\end{aligned} \qquad (3.46)$$

3.5.3. GENERAL CONSTANT-α SOLUTIONS

Solutions to
$$(\nabla^2 + \alpha^2)\mathbf{B} = \mathbf{0} \qquad (3.33)$$
are required that satisfy
$$\nabla\cdot\mathbf{B} = 0. \qquad (3.3)$$
In general, they may be written in the form
$$\mathbf{B} = \alpha\nabla\times(\psi\mathbf{a}) + \nabla\times(\nabla\times(\psi\mathbf{a})), \qquad (3.47)$$
where $\mathbf{a}$ is any constant vector and the scalar function ψ satisfies
$$(\nabla^2 + \alpha^2)\psi = 0. \qquad (3.48)$$

In the case when $\mathbf{a} = \hat{\mathbf{z}}$, the solution of Equation (3.48) in *rectangular Cartesians* for which each term decays away to zero as $z\to\infty$ is (Nakagawa and Raadu, 1972)

$$\psi = \int_0^\infty\!\!\int_0^\infty A(k_x, k_y)\,e^{i\mathbf{k}\cdot\mathbf{r}-lz}\,dk_x\,dk_y,$$

where $\mathbf{k} = k_x\hat{\mathbf{x}} + k_y\hat{\mathbf{y}}$, $l = (k^2-\alpha^2)^{1/2}$, and $A(k_x, k_y)$ are complex constants ($k\neq 0$). Thus the magnetic field components become

$$\begin{aligned}
B_x &= \int_0^\infty\!\!\int_0^\infty i(\alpha k_y - lk_x)A(k_x, k_y)\,e^{i\mathbf{k}\cdot\mathbf{r}-lz}\,dk_x\,dk_y,\\
B_y &= -\int_0^\infty\!\!\int_0^\infty i(\alpha k_x + lk_y)A(k_x, k_y)\,e^{i\mathbf{k}\cdot\mathbf{r}-lz}\,dk_x\,dk_y,\\
B_z &= \int_0^\infty\!\!\int_0^\infty k^2 A(k_x, k_y)\,e^{i\mathbf{k}\cdot\mathbf{r}-lz}\,dk_x\,dk_y.
\end{aligned} \qquad (3.49)$$

The particular solution $k_y = 0$ yields Equation (3.44). Nakagawa et al. (1973) have attempted to seek solutions of the above form for an active region, based on the observed line-of-sight field. Reasonable agreement with Hα structures is obtained, but a shortcoming of the method is that the fields are periodic in x and y (with field lines leaving the active region domain); also, the condition that l remain real implies an upper limit on α (in practice 0.07).

With $\mathbf{a} = \mathbf{r}$ the general solution of Equation (3.48) in *spherical polars* (r, θ, ϕ) is (Chandrasekhar and Kendall, 1957)

$$\psi = \sum_{n=0}^{\infty} \sum_{m=0}^{\infty} A_n^m r^{-1/2} J_{n+(1/2)}(\alpha r) P_n^m(\cos \theta) e^{im\phi},$$

in terms of Bessel functions (J_n) and associated Legendre functions $(P_n^m(\cos \theta))$. It leads to the following magnetic field components with complex constants C_n^m

$$B_r = \sum_{n=0}^{\infty} \sum_{m=0}^{\infty} C_n^m n(n+1) r^{-3/2} J_{n+(1/2)}(\alpha r) P_n^m e^{im\phi},$$

$$B_\theta = \sum_{n=0}^{\infty} \sum_{m=0}^{\infty} C_n^m \left[-nr^{-3/2} J_{n+(1/2)}(\alpha r) + \alpha r^{-1/2} J_{n-(1/2)}(\alpha r) \right] \frac{dP_n^m}{d\theta} e^{im\phi} +$$

$$+ C_n^m \alpha r^{-1/2} J_{n+(1/2)}(\alpha r) im P_n^m e^{im\phi} / \sin \theta,$$

$$B_\phi = \sum_{n=0}^{\infty} \sum_{m=0}^{\infty} C_n^m [nr^{-3/2} J_{n+(1/2)}(\alpha r) - \alpha r^{-1/2} J_{n-(1/2)}(\alpha r)] im P_n^m e^{im\phi} / \sin \theta -$$

$$- C_n^m \alpha r^{-1/2} J_{n+(1/2)}(\alpha r) \frac{dP_n^m}{d\theta} e^{im\phi}.$$

In *cylindrical polars* (R, ϕ, z) the general solution for which each term decays to zero as $z \to \infty$ and remains finite on the axis is

$$\psi = \sum_{n=0}^{\infty} e^{in\phi} \int_0^\infty C_n(k) J_n(kR) e^{-lz} \, dk, \tag{3.50}$$

where $k = (l^2 + \alpha^2)^{1/2}$. The corresponding axisymmetric solution (independent of ϕ) is

$$\psi = \sum_{n=0}^{\infty} \int_0^\infty A_n(k) \frac{\partial^n}{\partial l^n} [J_0(kR) e^{-lz}] \, dk,$$

with again $k = (l^2 + \alpha^2)^{1/2}$, and the resulting field components are

$$B_R = -\frac{\partial}{\partial z} \left(R \frac{\partial \psi}{\partial z} - z \frac{\partial \psi}{\partial R} \right),$$

$$B_\phi = \alpha \left(R \frac{\partial \psi}{\partial z} - z \frac{\partial \psi}{\partial R} \right),$$

$$B_z = \frac{1}{R} \frac{\partial}{\partial R} \left(R^2 \frac{\partial \psi}{\partial z} - Rz \frac{\partial \psi}{\partial R} \right).$$

The value $n = 0$ gives a solution with topological properties that are similar to those of Figure 3.9 except that the spiral angle (γ) is not uniform. These solutions have been compared by Nakagawa et al. (1971) with spiralling Hα fibrils near isolated sunspots. Also, Sheeley and Harvey (1975) have computed the force-free fields that arise from discrete flux sources.

An alternative prescription to Equation (3.47) for *axisymmetric solutions* is obtained by writing (Lüst and Schlüter, 1954)

$$\mathbf{B} = \nabla \times (\tilde{\psi}\hat{\phi}) + \alpha\tilde{\psi}\hat{\phi},$$

where

$$\left(\nabla^2 + \frac{2}{R}\frac{\partial}{\partial R} + \alpha^2\right)\tilde{\psi} = 0.$$

The solutions that behave like e^{-lz} and are finite on the axis take the form

$$B_R = -\frac{\partial \tilde{\psi}}{\partial z}, \qquad B_\phi = \alpha\tilde{\psi}, \qquad B_z = \frac{1}{R}\frac{\partial}{\partial R}(R\tilde{\psi}),$$

where

$$\tilde{\psi} = \sum_{n=0}^{\infty}\int_0^\infty A_n(k)\frac{\partial^n}{\partial l^n}(J_1(kR)\,e^{-lz})\,dk.$$

The value $n = 0$ gives our previous solution Equation (3.45).

More recently, Chiu and Hilton (1977) have pointed out that the above forms do not give the most general magnetic field that satisfies Equation (3.33) and decays to zero as z approaches infinity. For example, the expression (3.50) should be supplemented by the extra term

$$\sum_{n=0}^{\infty} e^{in\phi}\int_0^\alpha J_n(kR)\{D_n(k)\cos[(\alpha^2 - k^2)^{1/2}z] + E_n(k)\sin[(\alpha^2 - k^2)^{1/2}z]\}\,dk.$$

This has the surprising effect that, if only α and the value of B_z on the plane $z = 0$ are prescribed, the solution is not unique, by contrast with the special case of a potential field ($\alpha = 0$). In order to obtain a unique boundary-value problem, one needs to impose additional information about, for example, the tangential field component on the boundary.

Chiu and Hilton have also set up the solution in Cartesian geometry in terms of Green's functions. Their work has been extended by Barbosa (1978), who has obtained the constant-α, force-free field above the (photospheric) plane $z = 0$ (and below a plane $z = L$ at which the normal field component is prescribed to vanish) in the form

$$\mathbf{B}(x, y, z) = \int\int B_{z0}(x', y')\mathbf{G}_\alpha(x - x', y - y', z)\,dx'\,dy'.$$

Here B_{z0} is the normal field component on $z = 0$, $\mathbf{G}_\alpha$ is the *Green's function* and the integration is performed over the plane $z = 0$. However, constant-α fields are inadequate to model the global solar atmosphere completely, since at large distances they need to be contained by either a boundary or a uniform field.

3.5.4. NON-CONSTANT-α SOLUTIONS

When the assumption of constant α is dropped, it becomes much more difficult to solve

$$\mathbf{j} \times \mathbf{B} = 0, \tag{3.30}$$

where $\mathbf{j} = \text{curl } \mathbf{B}/\mu$ and

$$\text{div } \mathbf{B} = 0, \tag{3.51}$$

even when one seeks magnetic fields that are independent of one of the coordinates. For example, consider a field independent of y. It has rectangular Cartesian components that may be written

$$B_x \equiv \frac{\partial A}{\partial z}, \quad B_y, \quad B_z = -\frac{\partial A}{\partial x},$$

so as to satisfy Equation (3.51) automatically. The components of Equation (3.30) become

$$\nabla^2 A \frac{\partial A}{\partial x} + B_y \frac{\partial B_y}{\partial x} = 0, \tag{3.52}$$

$$\frac{\partial B_y}{\partial z}\frac{\partial A}{\partial x} - \frac{\partial B_y}{\partial x}\frac{\partial A}{\partial z} = 0, \tag{3.53}$$

$$\nabla^2 A \frac{\partial A}{\partial z} + B_y \frac{\partial B_y}{\partial z} = 0. \tag{3.54}$$

Equation (3.53) implies that B_y is a function of the flux function (A) alone and so it remains constant along a field line in the flux surface $A = \text{constant}$. Either Equation (3.52) or (3.54) then yields

$$\nabla^2 A + \frac{d}{dA}(\tfrac{1}{2}B_y^2) = 0, \tag{3.55}$$

which determines A and so (B_x, B_z) once the function $B_y(A)$ is prescribed. The particular cases $B_y = \text{constant}$ and $B_y = \text{constant} \times A$ give potential and constant-α fields, respectively. More generally, there is no need to introduce α, but it is of interest to notice that it equals $-dB_y/dA$ and so is a function of A alone (cf Equation (3.32)).

The difficulty in dealing with Equation (3.55) lies in the fact that it is in general a *nonlinear* equation, so that, for instance, the possibility of multiple solutions arises (see Section 10.3.1). One general result is that, *if* the solution of Equation (3.55) is unique, then

$$\frac{d^2}{dA^2}(\tfrac{1}{2}B_y^2) \leq 0$$

for all A (Courant and Hilbert, 1963).

A similar analysis applies in cylindrical polars (R, ϕ, z) for an axisymmetric field (independent of ϕ) with components

$$B_R \equiv -\frac{1}{R}\frac{\partial A}{\partial z}, \quad B_\phi \equiv \frac{b_\phi}{R}, \quad B_z = \frac{1}{R}\frac{\partial A}{\partial R}.$$

144   CHAPTER 3

In this case b_ϕ is a function of A alone, $\alpha = db_\phi/dA$ and Equation (3.30) reduces to

$$\Delta_1 A + \frac{d}{dA}(\tfrac{1}{2}b_\phi^2) = 0, \tag{3.56a}$$

where

$$\Delta_1 \equiv \frac{\partial^2}{\partial R^2} - \frac{1}{R}\frac{\partial}{\partial R} + \frac{\partial^2}{\partial z^2}. \tag{3.56b}$$

3.5.5. DIFFUSION

The slow resistive diffusion of a magnetic field through a series of force-free equilibria is governed by the equations

$$\frac{\partial \mathbf{B}}{\partial t} = \mathrm{curl}\,(\mathbf{v} \times \mathbf{B}) + \eta \nabla^2 \mathbf{B}, \tag{3.57}$$

$$\nabla \times \mathbf{B} = \alpha \mathbf{B}, \tag{3.58}$$

where

$$\nabla \cdot \mathbf{B} = 0 \quad \text{and} \quad (\mathbf{B} \cdot \nabla)\alpha = 0. \tag{3.59}$$

If the medium is stationary ($\mathbf{v} \equiv 0$) and $\alpha = $ constant, an initially force-free field diffuses in such a way as to remain force-free (Chandrasekhar and Kendall, 1957). For then Equations (3.58) and (3.59) imply

$$\nabla^2 \mathbf{B} = -\alpha^2 \mathbf{B},$$

and so Equation (3.57) reduces to

$$\frac{\partial \mathbf{B}}{\partial t} = -\eta \alpha^2 \mathbf{B},$$

with solution

$$\mathbf{B} = \mathbf{B}_0 \, e^{-\eta \alpha^2 t}, \qquad \mathbf{j} = \mathbf{j}_0 \, e^{-\eta \alpha^2 t}.$$

Thus, given an initial magnetic field ($\mathbf{B}_0$) and current ($\mathbf{j}_0$) that are parallel, the subsequent field and current remain parallel for all time. For a stationary medium the converse is also true, namely that, if a field remains force-free as it diffuses, then α must be constant (Jette, 1970).

Low (1973, 1974) sought non-constant-α solutions for a so-called '*passive*' medium whose plasma velocity is determined by Equation (3.57). He considered one-dimensional solutions similar to Equation (3.43) having the form

$$B_y = B_0 \cos \phi, \qquad B_z = B_0 \sin \phi,$$

where $\alpha = -\partial \phi/\partial x$.

The functions $\phi(x, t)$ and $v_x(x, t)$ are determined by the y- and z-components of Equation (3.57), which may be written

$$\frac{\partial \phi}{\partial t} - \eta \frac{\partial^2 \phi}{\partial x^2} + v_x \frac{\partial \phi}{\partial x} = 0, \qquad \eta \left(\frac{\partial \phi}{\partial x}\right)^2 + \frac{\partial v_x}{\partial x} = 0.$$

This pair of nonlinear equations possesses steady solutions that are linearly unstable and have a density singularity at the origin. There are also self-similar solutions based on the variable $xt^{-1/2}$ that represent both relaxing and steepening forms. One particular class of solutions has the interesting feature of evolving slowly at first and then rapidly developing an electric current singularity. Conditions for its occurrence have been given by Reid and Laing (1979), but, before the singularity is reached, the force-free assumption breaks down and dynamics need to be included. In a cylindrical geometry, Low has derived the corresponding solutions, while Reid and Laing (1979) find that a reversal in the axial field component can develop, and Boström (1973) has presented some solutions of separable form. It should be noted, however, that in many solar applications the resistive diffusion time is much longer than the time-scales of interest.

3.5.6. CORONAL EVOLUTION

A problem of great interest is the evolution of an active-region magnetic field through a series of force-free equilibria that are determined by the motion of the photospheric footpoints. In particular, one would like to know whether the equilibrium configuration always exists and whether it is stable as the footpoints on the photospheric boundary slowly evolve. An equilibrium that ceases to exist or becomes unstable when the photospheric shear reaches a critical value would allow the release of magnetic energy as the field relaxes back to one of lower energy (such as a potential field). This may aid our understanding of such large-scale disruptions as *eruptive prominences* and *two-ribbon flares*.

The first numerical attempt at following the evolution of force-free fields was by Sturrock and Woodbury (1967), who considered a line-dipole configuration representing a *magnetic arcade*, a single field line of which is sketched in Figure 3.10.

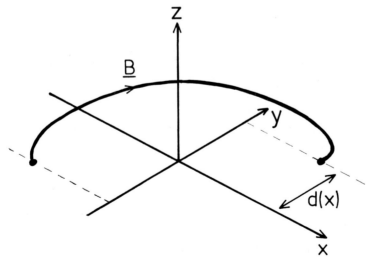

Fig. 3.10. Schematic representation of the photospheric displacement ($d(x)$) of the footpoints of a force-free magnetic field line. Initially, the field is potential and the field line lies in the $x-z$ plane with $d \equiv 0$.

This is similar to the constant-α structure of Figure 3.8 that evolves in response to a photospheric displacement in the y-direction proportional to x. Sturrock and Woodbury, however, found *non-constant-α* fields subject to an imposed displacement of
$$\tfrac{1}{2}L \quad \text{for } x > 0, \qquad -\tfrac{1}{2}L \quad \text{for } x < 0.$$
They characterised each field line by two scalars $a(x, z)$ and $y + c(x, z)$ such that
$$\mathbf{B} \equiv \nabla a \times \nabla(y + c),$$
for which the force-free condition ($\mathbf{j} \times \mathbf{B} = 0$) reduces to
$$\begin{aligned}
(1 + c_z^2)a_{xx} + (1 + c_x^2)a_{zz} - a_z c_z c_{xx} - a_x c_x c_{zz} \\
= 2c_x c_z a_{xz} - (a_x c_z + a_z c_x) c_{xz}, \\
a_z c_z a_{xx} + a_x c_x a_{zz} - a_z^2 c_{xx} - a_x^2 c_{zz} \\
= (a_x c_z + a_z c_x) a_{xz} - 2 a_x a_z c_{xz},
\end{aligned} \quad (3.60)$$
where subscripts denote differentiation. Since this pair of nonlinear equations is linear in the second derivatives, it is possible to solve it numerically by a relaxation method. As the shear increases, so the field lines become more distorted and rise higher in the atmosphere. Seen from above they are S-shaped, in contrast to the constant-α field lines of Figure 3.8, which are straight. There is no evidence for lack of equilibrium or multiple solutions and the energy in the force-free field remains less than that in the corresponding open field, since the latter is infinite.

Barnes and Sturrock (1972) repeated the above calculation for a structure with cylindrical symmetry similar to Figure 3.9, such that
$$\mathbf{B} \equiv \nabla a \times \nabla(\phi - c),$$
where a and c are functions of z and R. Again, a pair of equations for a and c similar to Equation (3.60) is solved subject to a prescribed amount of photospheric twist (c). As the field is twisted up, so the field lines expand outwards (Figure 3.11), and,

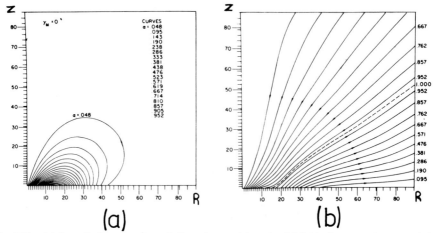

Fig. 3.11. (a) A vertical section through the axisymmetric potential field of a sunspot surrounded by a unipolar region of opposite polarity. As the spot rotates so a force-free field develops. (b) The open magnetic configuration that has the same normal photospheric component as the force-free field after a rotation of π (from Barnes and Sturrock, 1972).

when a twist of about π is exceeded, the energy in the force-free field exceeds that in the open field based on the same photospheric (but not outer) boundary conditions. Barnes and Sturrock suggest that, at this point, the force-free field erupts into the open configuration containing a current sheet, which is then susceptible to the tearing-mode instability (Section 7.5.5). They propose such a process for a solar flare mechanism, but they do not treat the boundary condition at large distances adequately, so the validity of their conclusion is uncertain (Priest, 1976).

Low and Nakagawa (1975) have also considered the evolution of a magnetic arcade, using a rather different formulation from that of Sturrock and Woodbury. They solved Equation (3.55) iteratively (Courant and Hilbert, 1963), subject to

$$A = \cos x \quad \text{on } z = 0, \quad \text{for } |x| < \tfrac{1}{2}\pi,$$

(so that the normal component (B_z) at the photosphere is fixed) and

$$A = 0 \quad \text{on } z = 10\,(|x| < \tfrac{1}{2}\pi) \quad \text{and} \quad |x| = \tfrac{1}{2}\pi\,(0 < z < 10).$$

The function $B_y(A)$ is prescribed, and then afterwards the corresponding photospheric displacement ($d(x)$) from potential is deduced (Figure 3.10). As the form of

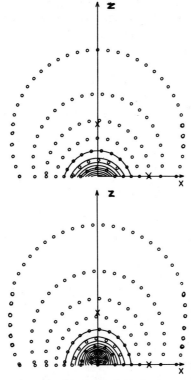

Fig. 3.12. A vertical section through a pair of force-free fields having the same values of B_z and B_y but different shear on the boundary $z = 0$. Open circles trace field lines and solid curves represent currrent-density lines (from Jockers, 1978).

B_y varies, so the configuration evolves through a series of equilibria. In particular, the cases

(I) $\quad B_y(A) = -2\varepsilon A(1 + 3A^2)^{-1/2}$

and

(II) $\quad B_y(A) = -\varepsilon A^2(1 + A^2)^{-1}$

were investigated. Case I causes the field lines to rise as the parameter ε increases in value, and the resulting footpoint displacement $d(x)$ increases with distance from the axis $x = 0$. By contrast, in case II the field lines descend as ε increases, and $d(x)$ possesses a maximum at a certain distance from the axis. Although the solution is in general not unique, Low and Nakagawa seek only the so-called *'maximal'* solution, whose A is greater than or equal to any other solution and which possesses no closed field lines above the photospheric base.

Jockers (1978) too has investigated the shearing of magnetic arcades numerically. His boundary conditions for solving Equation (3.55) are

$$A = \frac{1}{1 + x^2} \quad \text{on } z = 0,$$

and that the field be dipole-like as $(x^2 + y^2)$ approaches infinity. He treats the half-space $z \geq 0$ much more carefully than before by transferring it to the interior of a circle. The prescribed form for the axial field is (with $n \geq 4$)

$$B_y(A) = \lambda A^{(n+1)/2}, \quad A \geq 0,$$

and, for each value of the constant λ less than a value λ_{max}, Jockers found *two* solutions. An example is shown in Figure 3.12 (for $n = 8$ and $y = 0.77$). The one solution (B_y^I) contains only field lines with $0 \leq A \leq 1$ that end on the photospheric boundary, whereas the other solution (B_y^{II}) sometimes possesses a *magnetic island* with closed field lines ($A > 1$). However, two solutions correspond to different amounts of shear $d(x)$ (both of which, incidentally, possess a maximum). Thus, for a given shear, Jockers has only one solution; as the shear increases, λ increases to λ_{max}, and B_y^I is the solution; then λ decreases and B_y^{II} is the solution. The non-existence of solutions for $\lambda > \lambda_{max}$ is therefore no evidence for the onset of instability, since it is the shear ($d(x)$) that must be prescribed rather than $B_y(A)$. Jockers stressed that one needs to demonstrate the *existence of more than one configuration with the same shear* (and the same normal field (B_z)) in order to obtain a flare mechanism with magnetic energy released as one configuration evolves suddenly to another. He also considered the case

$$B_y^2(A) = c^2 A^{n+1} [2 + 2(n+1)(1-A) +$$
$$+ (n+1)(n+2)(1-A)^2](n+2)^{-1}(n+3)^{-1},$$

for A between 0 and 1. For given B_y, there are again two solutions, but this time neither of them contains a magnetic island, the second solution being stretched out vertically much more than the first.

The existence of solutions with magnetic islands has also been demonstrated

analytically by Low (1977), subject to the conditions
$$A = \log_e(1 + x^2) \quad \text{on } z = 0,$$
$$B_y(A) = \lambda e^{-A}.$$

The two solutions are
$$A = \log_e(1 + x^2 + z^2 + z(\lambda/\mu - 2)),$$

where $\mu = (2 \pm \sqrt{(4 - \lambda^2)})/\lambda$ and the maximum value of λ is 2. The field lines have a similar appearance to those in Figure 3.12, except that they are all circles or arcs of circles in the x–z plane.

For other solutions of relevance to the eruption of magnetic arcades to give two-ribbon flares, see Section 10.3.1.

3.6. Magnetohydrostatic Fields

In general, we saw in Section 3.1 that the component of the force-balance
$$\mathbf{0} = -\nabla p + \mathbf{j} \times \mathbf{B} - \rho g \hat{\mathbf{z}} \tag{3.1}$$
along the magnetic field has the solution
$$p = p_0(A) \exp - \int_0^z \frac{1}{\Lambda(z)} \, dz \tag{3.6}$$
along a single field line that is defined by the magnetic vector potential $\mathbf{A}$. When the vertical extent of the region under consideration is less than a scale-height, the exponential approximates to unity, so that the pressure becomes constant along a particular magnetic field line and Equation (3.1) reduces to
$$\mathbf{0} = -\nabla p + \mathbf{j} \times \mathbf{B}. \tag{3.61}$$
Taking the scalar product with $\mathbf{B}$ and $\mathbf{j}$ in turn, we have
$$\mathbf{B} \cdot \nabla p = \mathbf{j} \cdot \nabla p = 0, \tag{3.62}$$
so that the magnetic field and electric current lie in surfaces of constant pressure. In other words, p is constant along both magnetic field lines and electric current lines.

The particular cases of magnetic fields that are uniformly directed or have cylindrical symmetry have already been dealt with in Section 3.2 and 3.3. Moreover, many of the general methods for treating force-free fields (Section 3.5), can be extended easily to include the additional pressure gradient term in Equation (3.61). For instance, consider an axisymmetric magnetic field whose cylindrical polar components may be written
$$B_R \equiv -\frac{1}{R}\frac{\partial A}{\partial z}, \qquad B_\phi \equiv \frac{b_\phi}{R}, \qquad B_z = \frac{1}{R}\frac{\partial A}{\partial R},$$
so as to satisfy div $\mathbf{B} = 0$ identically. Then the electric current components (from

$\mathbf{j} = \operatorname{curl} \mathbf{B}/\mu$) are

$$j_R = -\frac{1}{R}\frac{\partial b_\phi}{\partial z}, \qquad j_\phi = -\frac{1}{R}\Delta_1 A, \qquad j_z = \frac{1}{R}\frac{\partial b_\phi}{\partial R},$$

with the operator Δ_1 defined as before by Equation (3.56b). For an axisymmetric pressure, we have $\partial p/\partial \phi = 0$ and the ϕ-component of Equation (3.61) becomes

$$\frac{\partial b_\phi}{\partial R}\frac{\partial A}{\partial z} - \frac{\partial b_\phi}{\partial z}\frac{\partial A}{\partial R} = 0,$$

which implies that $b_\phi = b_\phi(A)$. Also, Equation (3.62) gives

$$\frac{\partial p}{\partial R}\frac{\partial A}{\partial z} - \frac{\partial p}{\partial z}\frac{\partial A}{\partial R} = 0,$$

so that $p = p(A)$ is also a function of A alone. Then the R-component of Equation (3.61) gives

$$\Delta_1 A + \frac{d}{dA}(\tfrac{1}{2}b_\phi^2) = -\mu R^2 \frac{dp}{dA} \tag{3.63}$$

as the generalisation of Equation (3.56a). It determines $A(R, z)$ after $b_\phi(A)$ and $p(A)$ have been prescribed. Numerical methods for solving this have been described by, among others, Thomas (1976) for a laboratory application.

In spherical polar coordinates, Comfort et al. (1979) have sought analytical solutions to the magnetohydrostatic balance (3.1), using the method of separation of variables to construct several sets of solutions with axial symmetry. An energy equation is not satisfied even though the temperature is forced to have a certain behaviour; for instance, in one case T falls off like r^{-1}, and so the particular solutions are rather artificial.

Consider finally the case when the variables depend on x and z alone so that the magnetic field components may be written

$$B_x \equiv \frac{\partial A}{\partial z}, \qquad B_y(A), \qquad B_z \equiv -\frac{\partial A}{\partial x},$$

in terms of the flux function $(A(x, z))$, which defines a field line in the x–z plane or a flux surface in three dimensions. Then Equation (3.6) becomes

$$p(A, z) = p_0(A) \exp - \int_0^z \frac{mg}{k_B T(A, z)} dz, \tag{3.64}$$

and the component of Equation (3.1) normal to the magnetic field reduces to

$$\nabla^2 A + \frac{d}{dA}(\tfrac{1}{2}B_y^2(A)) = -\mu\frac{\partial}{\partial A}(p(A, z)) \tag{3.65}$$

as the extension of Equation (3.55). These equations are in general coupled to an energy equation through the appearance of $T(A, z)$ in Equation (3.64), but in special cases, the coupling is absent. When the temperature is uniform along each field line

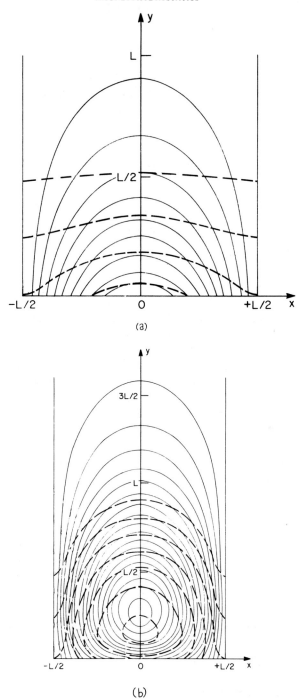

Fig. 3.13. Magnetohydrostatic arcade models showing magnetic field lines (solid) and isobars (dashed) for (a) $2\alpha\Lambda = 3$ and (b) $2\alpha\Lambda = 5.5$. In each case the arcade width is $L = (2\pi/3)\Lambda$, where Λ is the scale-height (after Zweibel and Hundhausen, 1982).

Equation (3.64) simplifies to

$$p(A, z) = p_0(A) \exp(-z/\Lambda(A)),$$

where the scale-height $\Lambda(A) = k_B T(A)/(mg)$ varies from one field line to another. Moreover, if the scale for z-variations is much less than the scale-height, this reduces still further to $p(A) = p_0(A)$, so that Equation (3.65) may in principle be solved once the total pressure $(p_0(A) + B_y^2(A)/(2\mu))$ is prescribed on each field line.

It is clearly possible to use standard techniques to seek both linear and nonlinear solutions to Equation (3.65) that will represent the structure of a *magnetic arcade*. To date, only one simple case has been considered, namely linear solutions for isothermal plasma ($\Lambda(A)$ = constant) in a shear-free arcade ($B_y(A) \equiv 0$). Zweibel and Hundhausen (1982) put

$$\mu p(A, z) = \tfrac{1}{2}\alpha^2 A^2 e^{-z/\Lambda} + \text{constant},$$

so that Equation (3.65) reduces to

$$\nabla^2 A = -\alpha^2 A e^{-z/\Lambda}.$$

The solution subject to

$$B_z = B_0 \sin \frac{\pi x}{L}$$

at the base is just

$$A = A_0 \cos \frac{\pi x}{L} J_n(2\alpha\Lambda e^{-(1/2)z/\Lambda}),$$

where

$$A_0 = \frac{LB_0}{\pi J_n(2\alpha\Lambda)} \quad \text{and} \quad n = \frac{2\pi\Lambda}{L}.$$

When $\alpha = 0$ the magnetic field is potential and the isobars horizontal. When $2\alpha\Lambda$ is smaller than the first maximum of the Bessel function J_n, both the field lines and isobars are inflated slightly (Figure 3.13(a)). When $2\alpha\Lambda$ lies between the first maximum and the first zero of J_n, the magnetic field includes a magnetic island, below which there lies a pressure maximum (Figure 3.13(b)).

It should be possible in future to extend these solutions to a cylindrical geometry (similar to Figure 3.9) so as to model the field above a single sunspot. The extension in spherical geometry to model the global solar atmosphere has already been presented by Hundhausen et al. (1981) and Uchida and Low (1981). It has the property that an enhanced pressure at the equator causes the field lines to distend outwards. Also, if a source surface is placed out at $r = 3R_\odot$, say, the model has more magnetic flux on the open polar field lines than has the potential field model of Section 3.4.

CHAPTER 4

WAVES

4.1. Introduction

4.1.1. Fundamental modes

In a gas such as the air we breathe, one is aware of the continual propagation of *sound waves* – especially when there are children in the vicinity, willing to act as a source! Sound waves owe their existence to the presence of a pressure *restoring force*: a local compression or rarefaction of the gas sets up a pressure gradient in opposition to the motion, which tries to restore the original equilibrium. If the gas is uniform, the waves propagate equally in all directions at the sound speed c_s (Sections 2.4.2, 2.5). They carry energy away from the source but, for the most part, they possess such a small amplitude that the ambient gas is disturbed only slightly. When its amplitude is large enough, however, a wave may steepen into a *shock wave* (see Chapter 5), as with the sonic boom from a supersonic aircraft.

In a plasma such as the solar atmosphere, there are typically four modes of wave motion, driven by different restoring forces. The magnetic tension and Coriolis forces can drive so-called *Alfvén waves* and *inertial waves*, respectively. The magnetic pressure, the plasma pressure and gravity can act separately and generate *compressional Alfvén waves, sound waves* and (*internal*) *gravity waves*, respectively; but, when acting together, these three forces produce only two *magnetoacoustic gravity* modes. In the absence of gravity, the two modes are referred to as *magnetoacoustic waves*, and when the magnetic field vanishes they are called *acoustic gravity waves*. The aim of this chapter is to describe the properties of each of the pure modes, before considering the most important couplings between them. Then, one application is given, namely to *5-min oscillations*.

Waves are always present on the Sun, because it is such a dynamic body, containing features that are continually in motion over a wide range of scales. For example, films of sunspots reveal so-called *running penumbral waves* that propagate outwards from the umbra and may well be fast magnetoacoustic gravity waves. Also, after the occurrence of a large flare, a *Moreton* (or *flare-induced coronal*) *wave* is sometimes emitted from the flare site and moves rapidly across the disc; it is probably a fast magnetoacoustic wave. Furthermore, the whole photosphere and chromosphere exhibit small-scale wave motions with a period of about 300 s outside sunspots (*5-min oscillations*); they are thought to be standing acoustic waves. Another important possible effect of wave propagation is the heating of the outer solar atmosphere (Chapter 6). The photosphere is in continual motion, as evidenced by granulation, 5-min oscillations and supergranulation. It therefore generates waves, which propagate upwards and may dump their energy in the chromosphere or corona, so helping to raise the temperature above that of the photosphere. Short-period acoustic waves are believed to heat the lower chromosphere, while magnetic waves or magnetic dissipation may heat the upper chromosphere and corona.

The basic theory of wave motion can be found in many textbooks on magnetohydrodynamics, such as that by Cowling (1976), while its application to the solar atmosphere has been summarised by Schatzman and Souffrin (1967), Stein and Leibacher (1974) (which we follow in Section 4.9) and Bray and Loughhead (1974).

All mathematical discussion of wave motion follows a standard pattern. One first of all considers an equilibrium situation and then perturbs it slightly to see whether the resulting disturbance propagates as a wave. The basic equations are linearised, and the perturbation quantities are assumed to vary like exp $i(\mathbf{k}\cdot\mathbf{r} - \omega t)$, the object being to find the *dispersion relation*, which expresses ω in terms of k.

4.1.2. Basic equations

The basic equations for our discussion of waves are the continuity of mass, momentum and energy, together with the induction equation, in the form

$$\frac{D\rho}{Dt} + \rho \mathbf{V}\cdot\mathbf{v} = 0, \tag{4.1}$$

$$\rho\frac{D\mathbf{v}}{Dt} = -\nabla p + (\nabla \times \mathbf{B})\times \mathbf{B}/\mu - \rho g\hat{\mathbf{z}} - 2\rho \mathbf{\Omega} \times \mathbf{v}, \tag{4.2}$$

$$\frac{D}{Dt}\left(\frac{p}{\rho^\gamma}\right) = 0, \tag{4.3}$$

$$\frac{\partial \mathbf{B}}{\partial t} = \nabla \times (\mathbf{v} \times \mathbf{B}), \tag{4.4}$$

$$\nabla \cdot \mathbf{B} = 0. \tag{4.5}$$

The resulting electric current and temperature follow from

$$\mathbf{j} = \operatorname{curl} \mathbf{B}/\mu, \qquad T = \frac{mp}{k_B \rho}.$$

The equations are written in a frame of reference rotating with the Sun at a (constant) angular velocity ($\mathbf{\Omega}$) relative to an inertial frame. The rotation has a negligible effect on Maxwell's equations provided the absolute speed $|\mathbf{\Omega} \times \mathbf{r} + \mathbf{v}|$ is much less than the speed of light. It gives rise to the *Coriolis force* ($-2\rho\mathbf{\Omega} \times \mathbf{v}$) in the equation of motion together with a *centrifugal force* ($\frac{1}{2}\rho\nabla|\mathbf{\Omega} \times \mathbf{r}|^2$), which has here been omitted since it may be combined with the gravitational term. The gravitational force is locally $-\rho g\hat{\mathbf{z}}$, with g assumed constant and the z-axis directed along the outward normal to the solar surface. For simplicity, the plasma has been assumed frozen to the magnetic field and thermally isolated from its surroundings, so that p/ρ^γ remains constant following the motion. From Equation (2.28d) it can be seen that this is only valid provided $\tau \ll p/\mathscr{L}$, where τ is the wave-period and the energy loss $\mathscr{L}$ (given by Equation (2.32)) may include thermal conduction, ohmic heating, radiative cooling (L_r) and (smaller-scale) wave heating. For example, p/L_r takes a value of about 1 s in the low photosphere and 1 h in the upper chromosphere, whereas $p/\mathscr{L}$ due to thermal conduction is about 500 s for a scale-length of 1 Mm (10^6 m).

Suppose a uniform equilibrium magnetic field ($\mathbf{B}_0$) permeates a vertically stratified stationary plasma, with a uniform temperature (T_0) and a density and pressure which behave like

$$\rho_0(z) = \text{constant} \times e^{-z/\Lambda}, \qquad p_0(z) = \text{constant} \times e^{-z/\Lambda}, \tag{4.6}$$

and satisfy

$$0 = -\frac{dp_0}{dz} - \rho_0 g.$$

Here

$$\Lambda = \frac{p_0}{\rho_0 g} \tag{4.7}$$

is the scale-height, typically 150 km in the photosphere and 100 Mm (10^8 m) in the corona (see Section 3.1). Consider small departures from the equilibrium

$$\rho = \rho_0 + \rho_1, \qquad \mathbf{v} = \mathbf{v}_1, \qquad p = p_0 + p_1, \qquad \mathbf{B} = \mathbf{B}_0 + \mathbf{B}_1,$$

and linearise the basic Equations (4.1) to (4.5) by neglecting squares and products of the small quantities (denoted by subscript 1). The result is

$$\frac{\partial \rho_1}{\partial t} + (\mathbf{v}_1 \cdot \nabla)\rho_0 + \rho_0(\nabla \cdot \mathbf{v}_1) = 0, \tag{4.8}$$

$$\rho_0 \frac{\partial \mathbf{v}_1}{\partial t} = -\nabla p_1 + (\nabla \times \mathbf{B}_1) \times \mathbf{B}_0/\mu - \rho_1 g \hat{\mathbf{z}} - 2\rho_0 \mathbf{\Omega} \times \mathbf{v}_1, \tag{4.9}$$

$$\frac{\partial p_1}{\partial t} + (\mathbf{v}_1 \cdot \nabla)p_0 - c_s^2 \left(\frac{\partial \rho_1}{\partial t} + (\mathbf{v}_1 \cdot \nabla)\rho_0\right) = 0, \tag{4.10}$$

$$\frac{\partial \mathbf{B}_1}{\partial t} = \nabla \times (\mathbf{v}_1 \times \mathbf{B}_0), \tag{4.11}$$

$$\nabla \cdot \mathbf{B}_1 = 0, \tag{4.12}$$

where

$$c_s^2 = \frac{\gamma p_0}{\rho_0} = \frac{\gamma k_B T_0}{m} \tag{4.13}$$

is the sound speed and $\hat{\mathbf{z}}$ is a unit vector in the z-direction. The set of Equation (4.8) to (4.12) may be reduced to a single equation by differentiating Equation (4.9) with respect to time and substituting for $\partial \rho_1/\partial t$, $\partial p_1/\partial t$ and $\partial \mathbf{B}_1/\partial t$ from Equations (4.8), (4.10) and (4.11), respectively. After some manipulation involving the use of Equation (4.6), a generalised wave equation may be derived for the disturbance velocity ($\mathbf{v}_1$):

$$\frac{\partial^2 \mathbf{v}_1}{\partial t^2} = c_s^2 \nabla(\nabla \cdot \mathbf{v}_1) - (\gamma - 1)g\hat{\mathbf{z}}(\nabla \cdot \mathbf{v}_1) - g\nabla v_{1z} - 2\mathbf{\Omega} \times \frac{\partial \mathbf{v}_1}{\partial t} +$$

$$+ [\nabla \times (\nabla \times (\mathbf{v}_1 \times \mathbf{B}_0))] \times \frac{\mathbf{B}_0}{\mu \rho_0}. \tag{4.14}$$

The next step is to seek *plane-wave* solutions of the form

$$\mathbf{v}_1(\mathbf{r}, t) = \mathbf{v}_1 \, e^{i(\mathbf{k} \cdot \mathbf{r} - \omega t)},$$

where $\mathbf{k}$ is the *wavenumber vector* and ω the *frequency*. The *period* of the wave is then just $2\pi/\omega$, while its *wavelength* (λ) is $2\pi/k$ and the *direction of propagation* of the wave is $\hat{\mathbf{k}}(\equiv \mathbf{k}/k)$. The effect of the plane-wave assumption is simply to replace $\partial/\partial t$ by $-i\omega$ and ∇ by $i\mathbf{k}$ in Equation (4.14). For the case of vanishing magnetic field ($\mathbf{B}_0 = 0$), Equation (4.14) reduces to

$$\omega^2 \mathbf{v}_1 = c_s^2 \mathbf{k}(\mathbf{k} \cdot \mathbf{v}_1) + i(\gamma - 1)g\hat{\mathbf{z}}(\mathbf{k} \cdot \mathbf{v}_1) + igkv_{1z} - 2i\omega \boldsymbol{\Omega} \times \mathbf{v}_1. \tag{4.15}$$

However, when the magnetic field is present, the last term in Equation (4.14) does not have a constant coefficient but is proportional to $e^{z/\Lambda}$ through the presence of ρ_0. Nevertheless, provided the wavelength ($2\pi/k$) of the perturbations is small compared with the scale-height $\Lambda(\equiv c_s^2/(\gamma g))$, ρ_0 may be regarded as locally constant, which can be justified formally by the *WKB approximation* (McLellan and Winterberg, 1968). The equation then becomes

$$\omega^2 \mathbf{v}_1 = c_s^2 \mathbf{k}(\mathbf{k} \cdot \mathbf{v}_1) + i(\gamma - 1)g\hat{\mathbf{z}}(\mathbf{k} \cdot \mathbf{v}_1) + igkv_{1z} - 2i\omega \boldsymbol{\Omega} \times \mathbf{v}_1 +$$
$$+ \left[\mathbf{k} \times (\mathbf{k} \times (\mathbf{v}_1 \times \mathbf{B}_0))\right] \times \frac{\mathbf{B}_0}{\mu \rho_0}. \tag{4.16}$$

Equations (4.15) and (4.16) will be employed in this chapter as a basis for discussion of the fundamental wave modes. The main object is to determine the dispersion relation $\omega = \omega(\mathbf{k})$, which gives the frequency as a function of the magnitude of wavenumber $\mathbf{k}$ and its inclination to the directions of gravity and the magnetic field. Since Equations (4.15) and (4.16) each represent a set of three homogeneous equations for the three velocity components of $\mathbf{v}_1$, the relation between its coefficients (and so between ω and $\mathbf{k}$) can be found in principle by setting the determinant of coefficients equal to zero. But, in practice, this is often carried out by various vector manipulations.

The velocity $\mathbf{v}_p = (\omega/k)\hat{\mathbf{k}}$ is known as the *phase velocity* of the wave. Its magnitude (ω/k) gives the speed of propagation in the direction $\hat{\mathbf{k}}$ for a wave specified by a single wavenumber. By contrast, a *packet* (or *group*) of waves possesses a range of wavenumbers and travels at the *group velocity* ($\mathbf{v}_g$) with cartesian components

$$v_{gx} = \frac{\partial \omega}{\partial k_x}, \quad v_{gy} = \frac{\partial \omega}{\partial k_y}, \quad v_{gz} = \frac{\partial \omega}{\partial k_z}. \tag{4.17}$$

It is this which gives the velocity at which energy is transmitted, and in general it differs both in magnitude and direction from the phase velocity. When the phase speed varies with wavelength, the wave is said to be *dispersive*. However, in the special case when ω is linearly proportional to k, the wave is said to be *non-dispersive* and the phase and group velocities are identical. In general, it transpires that the propagation is *anisotropic*, since the phase speed varies with the direction of propagation. There are three preferred directions, due to the magnetic field, gravity and rotation, This causes great complexity, and so in the next few section each of the modes will be isolated in turn to derive their basic characteristics.

4.2. Sound Waves

When $g = B_0 = \Omega = 0$, so that the only restoring force is the pressure gradient, Equation (4.15) for the disturbance velocity ($\mathbf{v}_1$) reduces to

$$\omega^2 \mathbf{v}_1 = c_s^2 \mathbf{k}(\mathbf{k} \cdot \mathbf{v}_1). \tag{4.18}$$

After taking the scalar product of this equation with $\mathbf{k}$ and assuming $\mathbf{k} \cdot \mathbf{v}_1$ does not vanish, we find

$$\omega^2 = k^2 c_s^2.$$

For the outgoing disturbance alone, this becomes the *dispersion relation for sound (or acoustic) waves*, namely

$$\boxed{\omega = kc_s.} \tag{4.19}$$

Sound waves therefore propagate equally in all directions at a phase speed

$$v_p \left(\equiv \frac{\omega}{k} \right) = c_s$$

and a group velocity

$$v_g \left(\equiv \frac{d\omega}{dk} \right) = c_s$$

in the direction $\mathbf{k}$. From the definition (4.13) for the sound speed, we find numerically, when $\gamma = 5/3$ and $m = 0.5 m_p$ ($\tilde{\mu} = 0.5$), $c_s \approx 166 T_0^{1/2}$ m s^{-1}, which varies typically from about 10 km s^{-1} in the solar photosphere to about 200 km s^{-1} in the corona. The requirement that $\mathbf{k} \cdot \mathbf{v}_1$ (and so $\nabla \cdot \mathbf{v}_1$) be non-vanishing implies that sound waves owe their existence to the compressibility of the plasma. Also, it is noteworthy that the waves are *longitudinal* in the sense that, according to Equation (4.18), the velocity perturbation ($\mathbf{v}_1$) is in the direction of propagation ($\mathbf{k}$).

4.3. Magnetic Waves

In Section 2.7 it was pointed out that the Lorentz force may be interpreted as the sum of a magnetic tension of amount B_0^2/μ and a magnetic pressure of amount $B_0^2/(2\mu)$ per unit area. Now, one of the effects of the tension ($\mathcal{T}$) in an elastic string (of mass density ρ_0 per unit length) is to permit transverse waves to propagate along the string with speed $(\mathcal{T}/\rho_0)^{1/2}$. So, by analogy, it is reasonable to expect the magnetic tension to produce transverse waves that propagate *along* the magnetic field $\mathbf{B}_0$ (Figure 4.1(a)) with speed $[(B_0^2/\mu)/\rho_0]^{1/2}$. This is known as the *Alfvén speed*

$$v_A = \frac{B_0}{(\mu \rho_0)^{1/2}},$$

and, according to Equation (2.48b), it may be written numerically $v_A = 2.8 \times 10^{16} (B_0/n_0^{1/2})$ m s^{-1}, with B_0 in tesla. Thus, the values $B_0 \approx 10$ G and $n_0 \approx 10^{16}$ m^{-3} (characteristic of the corona above an active region) give typically

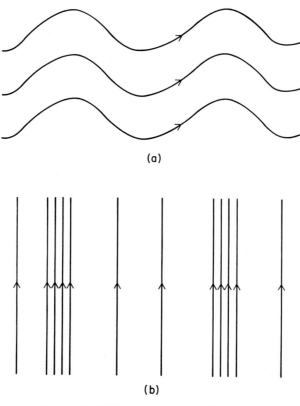

Fig. 4.1. (a) The ripples of magnetic field lines caused by an Alfvén wave propagating along the field. (b) The compression and rarefaction of magnetic field lines due to a compressional Alfvén wave propagating across the field.

300 km s^{-1}, whereas $B_0 \approx 10^3$ G and $n_0 \approx 10^{23}$ m^{-3} (characteristic of the photospheric network) give an Alfvén speed of only 10 km s^{-1}.

In Section 4.2 it was shown that the pressure of a gas obeying the adiabatic law, $p/\rho^\gamma =$ constant, produces (longitudinal) sound waves with phase speed $(\gamma p_0/\rho_0)^{1/2}$. Thus, by analogy, we might expect the magnetic pressure $p_m = B_0^2/(2\mu)$ to generate longitudinal magnetic waves propagating *across* the magnetic field (Figure 4.1(b)). If the magnetic field is frozen to the plasma (Section 2.6.2), the field strength and plasma density vary such that $B/\rho =$ constant, which means that, in terms of the magnetic pressure, $p_m/\rho^2 =$ constant. By comparing with the adiabatic law, this suggests that the effective value of γ for these magnetic waves is 2. Their wave speed must then be

$$\left(\frac{2p_m}{\rho_0}\right)^{1/2} = \frac{B_0}{(\mu\rho_0)^{1/2}},$$

which is again the Alfvén speed.

On the above intuitive grounds, a purely magnetic wave is expected to exist, driven by the $\mathbf{j} \times \mathbf{B}$ force, either along or across the field. The mathematical analysis below supports this; it also demonstrates that the two types of magnetic wave are distinct and may propagate obliquely to the magnetic field, not just along or across it.

When the magnetic field dominates the equilibrium, so that p_0 (and therefore c_s^2), Ω and g may all be set equal to zero, Equation (4.16) becomes

$$\omega^2 \mathbf{v}_1 = [\mathbf{k} \times (\mathbf{k} \times (\mathbf{v}_1 \times \hat{\mathbf{B}}_0))] \times \hat{\mathbf{B}}_0 v_A^2, \tag{4.20}$$

where $\hat{\mathbf{B}}_0$ is a unit vector in the direction of the magnetic field ($\mathbf{B}_0$). The vector products on the right-hand side may be expanded out to give

$$\omega^2 \mathbf{v}_1 / v_A^2 = (\mathbf{k} \cdot \hat{\mathbf{B}}_0)^2 \mathbf{v}_1 - (\mathbf{k} \cdot \mathbf{v}_1)(\mathbf{k} \cdot \hat{\mathbf{B}}_0) \hat{\mathbf{B}}_0 + [(\mathbf{k} \cdot \mathbf{v}_1) - (\mathbf{k} \cdot \hat{\mathbf{B}}_0)(\hat{\mathbf{B}}_0 \cdot \mathbf{v}_1)]\mathbf{k},$$

or, in terms of the angle (θ_B) which the direction of propagation ($\hat{\mathbf{k}}$) makes with the equilibrium magnetic field ($\mathbf{B}_0$),

$$\omega^2 \mathbf{v}_1 / v_A^2 = k^2 \cos^2 \theta_B \mathbf{v}_1 - (\mathbf{k} \cdot \mathbf{v}_1) k \cos \theta_B \hat{\mathbf{B}}_0 + \\ + [(\mathbf{k} \cdot \mathbf{v}_1) - k \cos \theta_B (\hat{\mathbf{B}}_0 \cdot \mathbf{v}_1)]\mathbf{k}. \tag{4.21}$$

Magnetic waves have several interesting properties. First, note from Equation (4.12) that $\mathbf{k} \cdot \mathbf{B}_1 = 0$, so that the magnetic field perturbation is normal to the direction of propagation. Next, the scalar product of Equation (4.21) with $\hat{\mathbf{B}}_0$ yields

$$\hat{\mathbf{B}}_0 \cdot \mathbf{v}_1 = 0, \tag{4.22}$$

so that the perturbed velocity is normal to the ambient magnetic field. This is not surprising, since only the Lorentz force ($\mathbf{j} \times \mathbf{B}$) is being allowed to drive motions and it is, of course, normal to $\mathbf{B}_0$ in the linear approximation. Finally, the scalar product of Equation (4.21) with $\mathbf{k}$ gives

$$(\omega^2 - k^2 v_A^2)(\mathbf{k} \cdot \mathbf{v}_1) = 0, \tag{4.23}$$

which possesses two distinct solutions that we discuss below.

4.3.1. SHEAR ALFVÉN WAVES

If the perturbation is *incompressible* ($\nabla \cdot \mathbf{v}_1 = 0$), so that

$$\mathbf{k} \cdot \mathbf{v}_1 = 0, \tag{4.24}$$

Equation (4.21) gives, after taking the positive square root,

$$\boxed{\omega = k v_A \cos \theta_B} \tag{4.25}$$

for *Alfvén waves*, sometimes known as *shear Alfvén waves*. (The positive square root gives waves propagating in the same direction as the magnetic field, whereas the negative root would correspond to waves in the opposite direction.) These waves have a phase speed $v_A \cos \theta_B$, which, for propagation along the magnetic field ($\theta_B = 0$), is just the Alfvén speed, in agreement with our intuitive discussion earlier. The variation of the phase speed with θ_B is most conveniently exhibited in a *polar diagram* (Figure 4.2), which takes the form of two circles of diameter v_A. In particular, it may be noted that the waves propagate fastest along the field but not at all in a direction normal ($\theta_B = \frac{1}{2}\pi$) to the field.

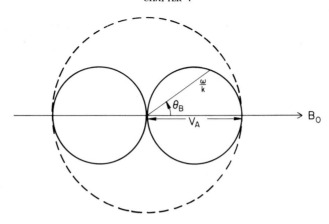

Fig. 4.2. A polar diagram for Alfvén waves (solid curve) and compressional Alfvén waves (dashed curve). The length of the radius vector at an angle of inclination θ_B to the equilibrium magnetic field ($\mathbf{B}_0$) is equal to the phase speed (ω/k) for waves propagating in that direction.

Taking the Z-axis along the magnetic field ($\mathbf{B}_0$), Equation (4.25) may be written alternatively as $\omega = k_z v_A$, and so differentiating with respect to k_z gives the group velocity (according to Equation (4.17)) as $\mathbf{v}_g = v_A \hat{\mathbf{B}}_0$. Energy is therefore carried at the Alfvén speed along the magnetic field, in spite of the fact that individual waves can travel at any inclination to the field (save $\tfrac{1}{2}\pi$).

The property (4.24) means that Alfvén waves are *transverse* in the sense that the velocity perturbation is normal to the propagation direction. Furthermore, with ρ_0 and p_0 uniform, Equations (4.8) and (4.10) imply that there are *no density or pressure changes* associated with the waves. Another characteristic can be shown by considering Equation (4.11), which for the plane-wave solution becomes

$$-\omega \mathbf{B}_1 = \mathbf{k} \times (\mathbf{v}_1 \times \mathbf{B}_0)$$

or

$$-\omega \mathbf{B}_1 = (\mathbf{k} \cdot \mathbf{B}_0)\mathbf{v}_1 - (\mathbf{k} \cdot \mathbf{v}_1)\mathbf{B}_0. \tag{4.26}$$

Since $\mathbf{k} \cdot \mathbf{v}_1$ vanishes (by Equation (4.24)), Equations (4.25) and (4.26) give

$$\mathbf{v}_1 = -\frac{\mathbf{B}_1}{(\mu \rho_0)^{1/2}}, \tag{4.27}$$

which implies that $\mathbf{B}_1$ and $\mathbf{v}_1$ are in the same direction, both lying in a plane parallel to the wave-front. (For propagation in the opposite direction to the magnetic field one would obtain $\mathbf{v}_1 = \mathbf{B}_1/(\mu\rho_0)^{1/2}$.) By Equation (4.22), this in turn means that

$$\mathbf{B}_0 \cdot \mathbf{B}_1 = 0, \tag{4.28}$$

so that the *magnetic field perturbation is normal to* $\mathbf{B}_0$. The directions of the perturbed quantities are indicated in Figure 4.3.

It is interesting to decompose the Lorentz force into the form

$$\mathbf{j}_1 \times \mathbf{B}_0 = (\mathbf{k} \times \mathbf{B}_1) \times \mathbf{B}_0/\mu$$

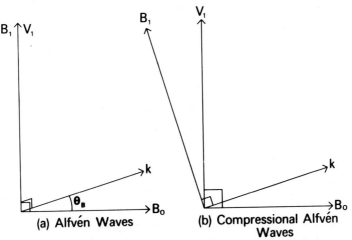

Fig. 4.3. The directions of the perturbed velocity ($\mathbf{v}_1$) and magnetic field ($\mathbf{B}_1$) relative to the equilibrium magnetic field ($\mathbf{B}_0$) and the wave propagation direction ($\mathbf{k}$). For (a) the vectors $\mathbf{v}_1$ and $\mathbf{B}_1$ are both normal to the plane of $\mathbf{k}$ and $\mathbf{B}_0$, whereas for (b) $\mathbf{v}_1$ and $\mathbf{B}_1$ lie in the same plane as $\mathbf{k}$ and $\mathbf{B}_0$.

or

$$\mathbf{j}_1 \times \mathbf{B}_0 = (\mathbf{k} \cdot \mathbf{B}_0)\mathbf{B}_1/\mu - (\mathbf{B}_0 \cdot \mathbf{B}_1)\mathbf{k}/\mu. \tag{4.29}$$

Here the first term on the right comes from the magnetic tension and the second term from the magnetic pressure. Thus, in view of Equation (4.28), the *driving force for Alfvén waves is the magnetic tension alone*. Also, the ratio of the magnetic energy to the kinetic energy in these waves is

$$\frac{B_1^2/(2\mu)}{\tfrac{1}{2}\rho_0 v_1^2},$$

which, by virtue of Equation (4.27) is unity, so that Alfvén waves involve an *equipartition between magnetic and kinetic energy*.

In a cylindrically symmetric geometry with an axial field, $B_0 \hat{\mathbf{z}}$, there exist waves which possess only an azimuthal component $\mathbf{B}_1 \sim \hat{\boldsymbol{\phi}} \cos k(v_A t \pm z)$ and are known as *torsional Alfvén waves*. Superposing two such waves propagating in opposite directions produces a *torsional oscillation* of a flux tube with $\mathbf{B}_1 \sim \hat{\boldsymbol{\phi}} 2 \cos kv_A t \cos kz$, as indicated in Figure 4.4.

When their amplitudes are no longer small, most disturbances cease to propagate as a wave with a constant profile, but instead distort in time. However, Alfvén waves are most unusual in that they can continue to propagate without distortion when their amplitudes are large. It can easily be verified that an Alfvénic disturbance with the properties

$$\mathbf{v}_1 = -\frac{\mathbf{B}_1}{(\mu\rho_0)^{1/2}}, \qquad |\mathbf{B}_0 + \mathbf{B}_1| = \text{constant},$$

satisfies the full equations, without the need to linearise. The latter property reduces to Equation (4.28) in the linear limit and it means that, as the wave passes by, the

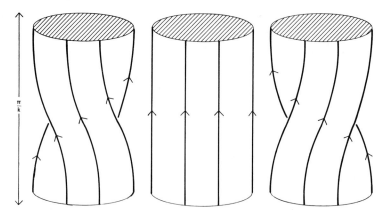

Fig. 4.4. Torsional oscillations of a magnetic flux tube of length π/k due to purely Alfvénic waves.

magnetic field vector just rotates, preserving a constant magnitude. An important consequence of the existence of finite-amplitude Alfvén waves is that they do not steepen and so tend to dissipate much less readily than other wave modes.

The time it takes a single Alfvén wave of wavelength λ to decay away due to ohmic diffusion is just the diffusion time λ^2/η (see Section 2.6.1), whereas most other finite-amplitude waves will decay much more rapidly since they steepen to form length-scales much shorter than λ. However, nonlinear interactions can transfer energy from Alfvén waves to acoustic waves, which then dissipate rapidly (Section 6.3.2). All of the above plasma properties are unaltered by the presence of a significant plasma pressure, provided the pressure variations (p_1) are adiabatic, so that Equation (4.24) implies the vanishing of ρ_1 (and hence of p_1).

4.3.2. COMPRESSIONAL ALFVÉN WAVES

The second solution to Equation (4.23) is

$$\boxed{\omega = kv_A} \qquad (4.30)$$

for *compressional Alfvén waves*. The phase speed is v_A, regardless of the angle of propagation, as indicated in Figure 4.2, and the group velocity is $\mathbf{v}_g = v_A \mathbf{k}$, so that the energy is propagated isotropically.

Equations (4.21) and (4.22) imply that the velocity perturbation ($\mathbf{v}_1$) lies in the $(\mathbf{k}, \mathbf{B}_0)$ plane in a direction normal to $\mathbf{B}_0$. It therefore possesses components both along and transverse to $\mathbf{k}$ in general, and it gives rise to both density and pressure changes. From Equation (4.26), the vector $\mathbf{B}_1$ lies in the plane of $\mathbf{v}_1$ and $\mathbf{B}_0$ but is normal to $\mathbf{k}$. Furthermore, the Lorentz force (4.29) is in the direction of $\mathbf{v}_1$, and in general it contains a contribution from both the magnetic tension and the magnetic pressure. In the particular case of propagation directly across the magnetic field ($\theta_B = \tfrac{1}{2}\pi$), Equation (4.21) shows that $\mathbf{v}_1$ is parallel to $\mathbf{k}$; the wave is thus longitudinal, and from Equation (4.29) only the magnetic pressure is playing a part, as anticipated in our previous intuitive discussion. By contrast, propagation along the field ($\theta_B = 0$)

makes the compressional Alfvén wave transverse and *identical* with an ordinary Alfvén wave: it is now driven wholly by the *magnetic tension* and produces no compression, in spite of its name!

4.4. Internal Gravity Waves

Consider a blob of plasma (Figure 4.5), which is displaced vertically a distance δz from equilibrium, and make the assumptions that

(i) it remains in pressure equilibrium with its surroundings,
(ii) the density changes inside the blob are adiabatic.

The first is valid if the motion is so slow that sound waves can traverse the system faster than the time-scale of interest, whereas the second holds provided the motion is so fast that the entropy is preserved.

At the original height (z), the pressure and density inside the blob are the same as those outside, namely p_0 and ρ_0. These satisfy

$$\frac{dp_0}{dz} = -\rho_0 g, \tag{4.31}$$

for an equilibrium balance between a pressure gradient and gravity. Outside the blob the pressure and density at height $z + \delta z$ are then $p_0 + \delta p_0$ and $\rho_0 + \delta \rho_0$, where, by Equation (4.31),

$$\delta p_0 = -\rho g\, \delta z, \qquad \delta \rho_0 = -\frac{d\rho_0}{dz}\, \delta z. \tag{4.32}$$

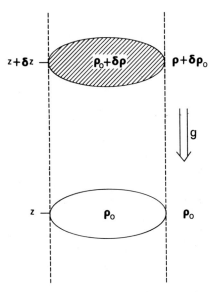

Fig. 4.5. An element of plasma moves vertically from a height z, where the external density is ρ_0, to a height $z + \delta z$, where the external density is $\rho_0 + \delta \rho_0$.

Inside the blob at height $z + \delta z$ the pressure and density are $p_0 + \delta p$ and $\rho_0 + \delta \rho$, say, where, by assumption (i),

$$\delta p = \delta p_0 = -\rho g\, \delta z. \tag{4.33}$$

Now, assumption (ii) means that, as the blob rises, its pressure and density obey $p/\rho^\gamma = $ constant, so that $\delta p = c_s^2 \delta \rho$, where the sound speed (c_s) is given by Equation (4.13). Substituting for δp from Equation (4.33) therefore gives the internal density change as

$$\delta \rho = -\frac{\rho_0 g \delta z}{c_s^2}. \tag{4.34}$$

Since the new density inside the blob differs from the ambient density at its new height, the blob experiences a buoyancy force of amount

$$g(\delta \rho_0 - \delta \rho) = -N^2 \rho_0\, \delta z, \tag{4.35}$$

say, where by Equations (4.32) and (4.34)

$$N^2 = -g\left(\frac{1}{\rho_0}\frac{d\rho_0}{dz} + \frac{g}{c_s^2}\right). \tag{4.36}$$

When it is real, N is known as the *Brunt–Väisälä* (or *Brunt*) *frequency*.

An alternative expression is obtained by using Equation (4.31) and the equation of state:

$$N^2 = \frac{g}{T_0}\left(\frac{dT_0}{dz} - \left(\frac{dT}{dz}\right)_{ad}\right), \tag{4.37}$$

where

$$\left(\frac{dT}{dz}\right)_{ad} = -(\gamma - 1)\frac{T_0 g}{c_s^2} \tag{4.38}$$

is the temperature gradient when the energy balance in the equilibrium structure is adiabatic. (It is obtained by eliminating p_0 and ρ_0 between the three equations $p_0 = k_B \rho_0 T_0/m$, $p_0/\rho_0^\gamma = $ constant and $dp_0/dz = -\rho_0 g$.) In general, N varies with height z but, in the particular case when the equilibrium temperature (T_0) is uniform, Equation (4.37) becomes

$$N^2 = \frac{(\gamma - 1)g^2}{c_s^2}. \tag{4.39}$$

Furthermore, in the presence of a horizontal magnetic field the Brunt–Väisälä frequency (4.36) is increased to

$$N^2 = -g\left(\frac{1}{\rho_0}\frac{d\rho_0}{dz} + \frac{g}{c_s^2 + v_A^2}\right),$$

or, in the case of a uniform temperature,

$$N^2 = \frac{g^2}{c_s^2}\left(\gamma - \frac{c_s^2}{c_s^2 + v_A^2}\right),$$

where v_A is the Alfvén speed (Chen and Lykoudis, 1972).

If the only resultant force acting on the plasma element is due to buoyancy (Equation (4.35)), the equation of motion becomes

$$\rho_0 \frac{d^2(\delta z)}{dt^2} = -N^2 \rho_0 \, \delta z. \tag{4.40}$$

Thus, the element executes simple harmonic motion with frequency

$$\omega = N, \tag{4.41}$$

provided

$$N^2 > 0, \tag{4.42}$$

so that the temperature decreases with height more slowly than adiabatic $(-dT_0/dz < -(dT/dz)_{ad})$.

The condition (4.42) is known as the *Schwarzschild criterion for convective stability*. If the temperature decreases with height faster than adiabatic, so that Equation (4.42) is violated, the solution to Equation (4.40) is exponentially growing and we have *convective instability* (see Section 7.5.6). The region of the solar interior where this is so is the *convection zone* (see Section 1.3.1C).

The above intuitive discussion leads us to expect the existence of gravity waves when $N^2 > 0$ due to the tendency for plasma to oscillate slowly with frequency N. Their dispersion relation can be determined by considering the basic equation (4.15) with $\Omega = 0$. Taking the scalar product with $\mathbf{k}$ and $\hat{\mathbf{z}}$ in turn and gathering together terms in v_{1z} and $\mathbf{k}\cdot\mathbf{v}_1$, we obtain

$$\begin{aligned}igk^2 v_{1z} &= (\omega^2 - c_s^2 k^2 - i(\gamma-1)gk_z)(\mathbf{k}\cdot\mathbf{v}_1), \\ (\omega^2 - igk_z)v_{1z} &= (c_s^2 k_z + i(\gamma-1)g)(\mathbf{k}\cdot\mathbf{v}_1).\end{aligned} \tag{4.43}$$

Then an elimination of $(\mathbf{k}\cdot\mathbf{v}_1)/v_{1z}$ between these two yields

$$(\omega^2 - igk_z)(\omega^2 - c_s^2 k^2 - i(\gamma-1)gk_z) = igk^2(c_s^2 k_z + i(\gamma-1)g). \tag{4.44}$$

The object is to seek waves with a frequency of the order of the Brunt–Väisälä frequency (N) and much slower than that of sound waves, so that

$$\omega \approx \frac{g}{c_s} \ll kc_s;$$

this implies that the wavelength is much smaller than a scale-height and Equation (4.44) reduces to

$$\omega^2 c_s^2 \approx (\gamma-1)g^2(1 - k_z^2/k^2).$$

In terms of N and the inclination ($\theta_g = \cos^{-1}(k_z/k)$) between the direction of propagation and the z-axis, this may be rewritten

$$\boxed{\omega = N \sin \theta_g} \tag{4.45}$$

for *(internal) gravity waves*. The word 'internal' is sometimes incorporated to distinguish them from *surface gravity waves*, which propagate along the surface between two fluids.

Several properties of this mode are of note. A typical value for N^{-1} is 50 s, so the gravity mode tends to be rather slow by comparison with the other waves (except for the inertial wave). Gravity waves do not propagate in the vertical direction ($\theta_g = 0$), since that would not allow a horizontal interaction with elements at the same height. Furthermore, Equation (4.45) implies that $\omega \leq N$, so that the waves cannot propagate faster than the Brunt–Väisälä frequency. For a given ω and N, it also means that they propagate along two cones centred about the z-axis with $\theta_g = \sin^{-1}(\omega/N)$. For the upward-propagating wave with

$$\omega = N(1 - k_z^2/k^2)^{1/2},$$

the z-component of the group velocity from Equation (4.17) is

$$v_{gz} = \frac{\partial \omega}{\partial k_z} = -\frac{\omega k_z}{k^2},$$

which is negative. Thus gravity waves have the unusual characteristic that a group of *upward* propagating waves carries energy *downward* and vice versa! In fact the group velocity is in a direction perpendicular to the surface of the cone with angle θ_g. (See for instance Lighthill (1967), who gives wave-vector surfaces for all the different modes.)

4.5. Inertial Waves

Consider now the effect of the Coriolis force alone, so that the linearised equation of motion (4.9) becomes

$$\frac{\partial \mathbf{v}_1}{\partial t} = -2\mathbf{\Omega} \times \mathbf{v}_1.$$

With the Z-axis along the axis of rotation, its components are

$$\frac{\partial v_{1X}}{\partial t} = 2\Omega v_{1Y}, \qquad \frac{\partial v_{1Y}}{\partial t} = -2\Omega v_{1X},$$

which possess a wave-like solution

$$v_{1X} = A \cos(kZ - 2\Omega t), \qquad v_{1Y} = A \sin(kZ - 2\Omega t).$$

This *inertial wave* propagates along the axis of rotation with frequency 2Ω, and the Coriolis force, acting perpendicular to the direction of motion, causes each plasma element to execute a circular orbit in the X–Y plane.

Waves driven by the Coriolis force may also propagate away from the rotation axis. With $g = B_0 = 0$, the basic Equation (4.15) becomes

$$\omega^2 \mathbf{v}_1 = c_s^2 \mathbf{k}(\mathbf{k} \cdot \mathbf{v}_1) - 2i\omega \mathbf{\Omega} \times \mathbf{v}_1. \tag{4.46}$$

Incompressible solutions may be sought such that

$$\mathbf{k} \cdot \mathbf{v}_1 = 0. \tag{4.47}$$

and so the vector product of Equation (4.46) with $\mathbf{k}$ gives

$$\omega \mathbf{k} \times \mathbf{v}_1 = 2i(\mathbf{k} \cdot \mathbf{\Omega}) \mathbf{v}_1.$$

Equating the magnitude of both sides then yields

$$\boxed{\omega = \pm \frac{2(\mathbf{k} \cdot \mathbf{\Omega})}{k}} \qquad (4.48)$$

as the dispersion relation for *inertial waves*. In terms of the angle θ_Ω between the rotation axis and the propagation direction, it may be written $\omega = \pm 2\Omega \cos\theta_\Omega$. As the wave propagates, so the velocity vector rotates about the direction of propagation, as shown in Figure 4.6. At a fixed location, $\mathbf{v}_1$ just performs a circular motion in a plane parallel to the wavefronts. In other words, the waves are *transverse* and *circularly polarised*. They also possess the rather unusual property that the vorticity ($\nabla \times \mathbf{v}_1$) is either parallel or antiparallel to the velocity $\mathbf{v}_1$, the value $\mathbf{v}_1 \cdot \nabla \times \mathbf{v}_1 = \mp k v_1^2$ being known as the *helicity* (see Section 9.4.2).

The group velocity from Equations (4.17) and (4.48) is

$$\mathbf{v}_g = \pm \frac{\mathbf{k} \times (2\mathbf{\Omega} \times \mathbf{k})}{k^3}.$$

It has a magnitude

$$v_g = \frac{2\Omega \sin\theta_\Omega}{k},$$

such that

$$(v_g^2 + v_p^2)^{1/2} = \frac{2\Omega}{k}$$

is independent of θ_Ω, where v_p is the phase speed (ω/k). Also, it may be noted that

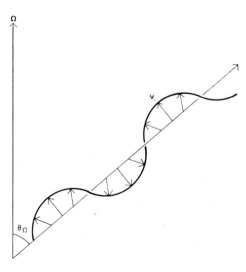

Fig. 4.6. The directions of the velocity vector ($\mathbf{v}_1$) at various locations in an inertial wave propagating along the direction $\mathbf{k}$ with frequency $2\Omega \cos\theta_\Omega$. The vector $\mathbf{v}_1$ is normal to $\mathbf{k}$ at each point, while, relative to an inertial frame, the $\mathbf{k}$-vector itself rotates with angular speed Ω about the rotation axis.

the group velocity is perpendicular to **k**, so that energy is transported in a direction normal to the phase velocity.

The effect of the Coriolis force on Alfvén waves can be seen from the basic Equation (4.16) with $g = 0$ and $\mathbf{k} \cdot \mathbf{v}_1 = 0$, namely

$$\omega^2 \mathbf{v}_1 = -2i\omega \boldsymbol{\Omega} \times \mathbf{v}_1 + [\mathbf{k} \times (\mathbf{k} \times (\mathbf{v}_1 \times \hat{\mathbf{B}}_0))] \times \hat{\mathbf{B}}_0 v_A^2.$$

After taking the vector product of this equation with **k**, it reduces to

$$\omega^2 \mp \omega_I \omega - \omega_A^2 = 0, \tag{4.49}$$

where

$$\omega_I = 2(\mathbf{k} \cdot \boldsymbol{\Omega})/k, \qquad \omega_A = \mathbf{k} \cdot \hat{\mathbf{B}}_0 v_A,$$

are the pure inertial and Alfvén wave frequencies.

When **k**, **Ω** and $\mathbf{B}_0$ are roughly parallel, the nature of the solutions depends on the ratio

$$\frac{\omega_A}{\omega_I} \approx \frac{kv_A}{2\Omega}.$$

On the Sun this ratio is generally large (typically $10 - 10^3$) and the solutions of Equation (4.49) approximate to $\omega^2 = \omega_A^2(1 \pm \omega_I/\omega_A)$, so that the Coriolis force produces a small *frequency-splitting* of the *Alfvén wave*. In the opposite extreme when $\omega_A/\omega_I \ll 1$, the solutions become

$$\omega^2 \approx \omega_I^2, \qquad \omega^2 \approx \frac{\omega_A^4}{\omega_I^2}, \tag{4.49'}$$

the former representing inertial waves and the latter so-called *hydromagnetic inertial waves*, which propagate much slower than Alfvén waves. A good summary of magnetohydrodynamics in a rotating frame of reference has been compiled by Acheson and Hide (1973), while general accounts of flow in a rotating system may be found in the books by Batchelor (1967) and Greenspan (1968).

4.6. Magnetoacoustic Waves

When both the magnetic force and the pressure gradient are important (but g and Ω vanish), Equation (4.16) gives a generalisation of Equation (4.21), namely,

$$\omega^2 \mathbf{v}_1 / v_A^2 = k^2 \cos^2 \theta_B \mathbf{v}_1 - (\mathbf{k} \cdot \mathbf{v}_1) k \cos \theta_B \hat{\mathbf{B}}_0 + \\ + [(1 + c_s^2/v_A^2)(\mathbf{k} \cdot \mathbf{v}_1) - k \cos \theta_B (\hat{\mathbf{B}}_0 \cdot \mathbf{v}_1)] \mathbf{k}.$$

Since $\mathbf{v}_1$ appears here in the combinations $\mathbf{k} \cdot \mathbf{v}_1$ and $\hat{\mathbf{B}}_0 \cdot \mathbf{v}_1$, we may proceed most easily by taking the scalar product with **k** and $\hat{\mathbf{B}}_0$ in turn to give

$$(-\omega^2 + k^2 c_S^2 + k^2 v_A^2)(\mathbf{k} \cdot \mathbf{v}_1) = k^3 v_A^2 \cos \theta_B (\hat{\mathbf{B}}_0 \cdot \mathbf{v}_1)$$

and

$$k \cos \theta_B c_S^2 (\mathbf{k} \cdot \mathbf{v}_1) = \omega^2 (\hat{\mathbf{B}}_0 \cdot \mathbf{v}_1). \tag{4.50}$$

If $\mathbf{k} \cdot \mathbf{v}_1$ vanishes, one finds the Alfvén wave solution (4.25) as before, but otherwise $(\mathbf{k} \cdot \mathbf{v}_1)/(\hat{\mathbf{B}}_0 \cdot \mathbf{v}_1)$ may be eliminated between the two equations above to yield the dispersion relation

$$\omega^4 - \omega^2 k^2 (c_S^2 + v_A^2) + c_S^2 v_A^2 k^4 \cos^2 \theta_B = 0 \quad (4.51)$$

for *magnetoacoustic (or magnetosonic) waves*. For outward-propagating disturbances ($\omega/k > 0$) there are two distinct solutions

$$\omega/k = [\tfrac{1}{2}(c_S^2 + v_A^2) \pm \tfrac{1}{2}\sqrt{(c_S^4 + v_A^4 - 2c_S^2 v_A^2 \cos 2\theta_B)}]^{1/2}. \quad (4.51')$$

The higher-frequency mode is known as the *fast magnetoacoustic wave* and the other one the *slow magnetoacoustic wave*. The Alfvén wave phase speed lies between that of the slow and fast waves, and so the Alfvén wave is sometimes referred to as the *intermediate mode*, a nomenclature that carries over into shock-wave terminology (Section 5.4.4).

The variation with propagation angle θ_B of the two phase speeds is sketched in Figure 4.7(a). For propagation along the field ($\theta_B = 0$), ω/k is either c_s or v_A, whereas for propagation across the field ($\theta_B = \tfrac{1}{2}\pi$) ω/k is $(c_s^2 + v_A^2)^{1/2}$ or 0. As θ_B tends to $\tfrac{1}{2}\pi$ for the slow wave, so ω/k approaches zero, but

$$\frac{\omega}{k_B} \equiv \frac{\omega}{k \cos \theta_B} \to c_T \equiv \frac{v_A c_s}{(v_A^2 + c_s^2)^{1/2}},$$

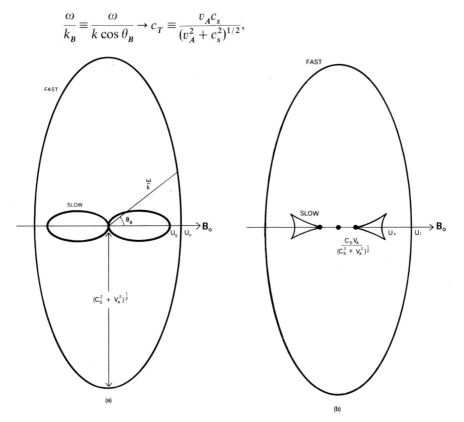

Fig. 4.7. Polar diagrams for fast and slow magnetoacoustic waves propagating at an angle θ_B to the equilibrium magnetic field. The speeds u_s and u_f are the slower and faster, respectively, of the Alfvén speed (v_A) and sound speed (c_s). (a) shows the phase velocities and (b) the group velocities.

where k_B is the wavenumber component along the field. Thus c_T is the component of the phase velocity *along* the field for propagation almost perperdicular to the field, so that the wavelength along the field is much longer than the wavelength across the field (see Sections 8.2.2 and 8.7.2). c_T also represents the *cusp speed* for the group velocity of the slow wave, as can be seen in Figure 4.7(b).

The two magnetoacoustic waves may be regarded as a sound wave, modified by the magnetic field, and a compressional Alfvén wave, modified by the plasma pressure, the modification being most marked for propagation away from the magnetic field direction. In the case of vanishing magnetic field ($v_A = 0$), the slow wave disappears and the fast wave becomes a sound wave. On the other hand, if the gas pressure vanishes ($c_s = 0$), the slow wave disappears and the fast wave becomes a compressional Alfvén wave.

When $\beta(\equiv 2\mu p_0/B_0^2$, the ratio of gas to magnetic pressure) is much larger than unity so that c_s^2/v_A^2 is also much larger than one, the fast and slow-wave dispersion relations reduce to

$$\frac{\omega}{k} \approx c_s, \qquad \frac{\omega}{k} \approx v_A \cos\theta_B,$$

respectively. For the latter case, Equation (4.50) becomes

$$\hat{\mathbf{k}} \cdot \hat{\mathbf{v}}_1 \approx \frac{v_A^2}{c_s^2} \cos\theta_B (\hat{\mathbf{B}}_0 \cdot \hat{\mathbf{v}}_1),$$

which is much less than unity, so the disturbance is nearly incompressible. This is a particular case of the more general result that, when $\beta \gg 1$, the plasma may be regarded as incompressible as far as magnetic field effects are concerned, since sound waves propagate almost instantaneously.

Polar diagrams for the group velocity are sketched in Figure 4.7(b). The slow-wave energy is seen to propagate in a narrow cone about the magnetic field direction, but the fast-wave energy propagates more isotropically. When $c_s \gg v_A$, the group velocities become $v_A \hat{\mathbf{B}}_0$ and $c_s \hat{\mathbf{k}}$ for the slow and fast waves, respectively.

An important limitation on magnetohydrodynamic theory is that $\omega < \Omega_i$, where $\Omega_i = eB/m_i$ is the *ion-gyration* (or *cyclotron*) frequency with which ions gyrate about the magnetic field. When this is violated the slow wave ceases to exist and the fast and Alfvén waves are modified in a way described by, for instance, Boyd and Sanderson (1969, p. 205). Numerically, we have for a H plasma $\Omega_i = 9.6 \times 10^7 B$, with B in tesla. Thus, for example, with a magnetic field (B) of 1 G, our analysis is valid provided the wave period ($2\pi/\omega$) exceeds 7×10^{-4} s.

4.7. Acoustic-Gravity Waves

When compressibility and buoyancy forces are present together, both the acoustic and gravity modes occur. They remain distinct modes but both are modified somewhat. Taking the scalar product of the basic equation (4.15) with $\mathbf{k}$ and $\hat{\mathbf{z}}$ in turn yields the pair of equations (4.43) and gives the dispersion relation

$$\boxed{\omega^2(\omega^2 - N_s^2) = (\omega^2 - N^2 \sin^2\theta_g') k'^2 c_s^2,} \tag{4.52}$$

where

$$N_s = \frac{\gamma g}{2c_s}\left(\equiv \frac{c_s}{2\Lambda}\right),$$

$$N = \frac{(\gamma-1)^{1/2}g}{c_s},$$

$$\sin^2\theta'_g = 1 - \frac{k'^2_z}{k'^2},$$

$$\mathbf{k}' = \mathbf{k} + i\frac{\gamma g}{2c_s^2}\hat{\mathbf{z}}.$$

N is the Brunt–Väisälä frequency and θ'_g is the angle between the wave vector $\mathbf{k}'$ and the vertical. N is always greater than or equal to N_s in value, equality occurring when $\gamma = 2$, but usually the difference between the two is rather small. When $\gamma = 5/3$, for instance, $N_s \approx 1.02\,N$. The change of wavenumber variable from k to k' means that the disturbances take the form

$$\mathbf{v}_1(\mathbf{r},t) = \mathbf{v}_1\, e^{(1/2)\gamma g z/c_s^2}\, e^{i(\mathbf{k}'\cdot\mathbf{r}-\omega t)},$$

and so they increase exponentially over a vertical distance $2c_s^2/(\gamma g)$, which is twice the scale-height (Λ).

In the limit $\omega \ll k'c_s$, the dispersion relation (4.52) reduces to $\omega = N\sin\theta_g$ for gravity waves, whereas when $\omega \gg N$ it becomes $\omega = k'c_s$ for pure acoustic waves. Vertical propagation ($\theta_g = 0$) gives $\omega^2 = N_s^2 + k'^2 c_s^2$, and so the wave exists ($\omega^2 > 0$) only if $\omega > N_s$.

Wave-like solutions propagating away from the vertical ($k'^2 > 0, \omega^2 > 0$) are allowed only in two frequency domains, namely $\omega < N\sin\theta_g$ and $\omega > N_s$. The higher-frequency mode is basically acoustic in nature, with a group speed $v_g < c_s$ and a phase speed $\omega/k > c_s$; the former approaches zero while the latter tends to infinity as the *lower-cut off* N_s is approached. The lower-frequency wave is basically a gravity mode; it has a phase speed $\omega/k < c_s$, which approaches zero at the *upper cut-off* $N\sin\theta_g$. For frequencies lying between $N\sin\theta_g$ and N_s the disturbance is *non-propagating* (or *evanescent*). Here k' is purely imaginary and so the disturbance grows or decays exponentially in space. In this band of frequencies, therefore, only standing waves can exist (Figure 4.8), but no energy can be propagated (since p_1 and v_1 are out of phase).

It is sometimes of interest to know whether disturbances of a given horizontal wavenumber (k_x) and frequency (ω) are vertically propagating or not. Since $k'^2 = k'^2_z + k_x^2$, Equation (4.52) may be recast in the form

$$k'^2_z \omega^2 c_s^2 = \omega^2(\omega^2 - N_s^2) - (\omega^2 - N^2)k_x^2 c_s^2.$$

Thus, for positive ω^2 and k_x^2, there are vertically propagating waves ($k'^2_z > 0$) provided

$$\omega^2(\omega^2 - N_s^2) > (\omega^2 - N^2)k_x^2 c_s^2.$$

This condition divides the $\omega - k_x$ plane into three domains, as indicated in Figure 4.9, sometimes referred to as a *diagnostic diagram*. For very small horizontal wave-

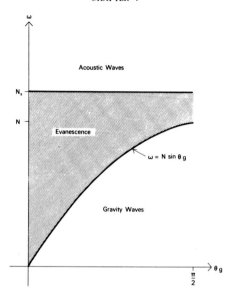

Fig. 4.8. The allowable domains for the propagation of acoustic-gravity waves of frequency ω at an angle θ_g to the vertical. In the shaded region, disturbances cannot propagate.

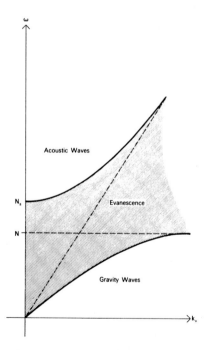

Fig. 4.9. A diagnostic diagram indicating the allowable regions for the vertical propagation of waves of frequency ω and horizontal wavenumber k_x. Disturbances in the shaded region are non-propagating (evanescent). The asymptotes $\omega = N$ and $\omega = k_x c_s$ are indicated by dashed lines.

numbers (k_x), it reduces to $\omega > N_s$ or $\omega < (N/N_s)k_x c_s$, whereas, for large values of k_x, it becomes $\omega < N$ or $\omega > k_x c_s$. For small wavenumbers, it can be seen that the acoustic waves are inhibited by gravity at frequencies less than N_s and the internal gravity waves are inhibited by compressibility at frequencies larger than about kc_s.

The extension of the theory to a non-isothermal atmosphere has been presented by Moore and Spiegel (1964) using the WKB approximation. Again no progressive waves are possible for frequencies lying between

$$N_s(z) = \frac{\gamma g}{2c_s(z)}$$

and $N(z)$ given by Equation (4.37).

The effect of the magnetic field on acoustic-gravity waves is to complicate the situation by introducing an extra restoring force and an extra preferred direction in addition to that of gravity. The Alfvén wave propagates unaltered but the magnetic field modifies the acoustic-gravity waves (or gravity modifies the magnetoacoustic waves) to give two *magnetoacoustic-gravity* (or 'magneto-atmospheric') modes. In particular, for an isothermal plasma in a uniform vertical magnetic field that is so strong that $v_A \gg c_s$ (e.g., Spruit, 1981a), the fast mode has

$$k_x^2 + k_y^2 + k_z^2 = 0; \qquad (4.53a)$$

for given real values of k_x and k_y, the fast mode has an imaginary k_z, and so it represents a wave that is propagating horizontally and is evanescent vertically. The slow mode is similar to an acoustic wave; it has

$$2ik_z = 1 - (1 - \omega^2/N_s^2)^{1/2} \qquad (4.53b)$$

and therefore is evanescent below the cut-off frequency (N_s).

4.8. Summary of Magnetoacoustic-Gravity Waves

In the absence of the Coriolis force, the basic equations for wave propagation in a plasma with a uniform magnetic field, a uniform temperature and a density proportional to $\exp(-z/\Lambda)$ allow the solution

$$\omega^2 = k^2 v_A^2 \cos^2 \theta_B \quad \text{(Alfvén wave)}$$

together with two magnetoacoustic-gravity modes.

When there is no magnetic field, Equation (4.15) gives the dispersion relation for the two latter modes as

$$\omega^4 - \omega^2(N_s^2 + k'^2 c_s^2) + N^2 \sin^2 \theta_g' k'^2 c_s^2 = 0 \quad \text{(acoustic-gravity waves)}, \qquad (4.54)$$

where

$$N_s = \frac{\gamma N}{2(\gamma - 1)^{1/2}},$$

$$N = (\gamma - 1)^{1/2} \frac{g}{c_s},$$

$$\sin^2 \theta'_g = 1 - \frac{k_z'^2}{k'^2},$$

$$k_z' = k_z + \frac{i}{2\Lambda},$$

$$k'^2 = k^2 + k_z'^2 - k_z^2,$$

$$\Lambda = \frac{c_s}{2N_s}\left(=\frac{c_s^2}{\gamma g}=\frac{p_0}{\rho_0 g}\right).$$

In the limit $\omega \gg N$, the effect of gravity can be neglected and Equation (4.54) reduces to

$$\omega^2 = k^2 c_s^2 \quad \text{(acoustic wave)}.$$

The alternative limit, namely

$$\omega \ll kc_s \quad \text{and} \quad N_s \ll kc_s \quad (\text{i.e., } k\Lambda \gg 1),$$

makes buoyancy dominate and gives rise to

$$\omega^2 = N^2 \sin^2 \theta_g \quad \text{((internal) gravity wave)}.$$

The presence of a magnetic field means that plane-wave solutions arise if $(k\Lambda)^{-1} \ll 1$, or, from the definition of Λ, if

$$N_s \ll kc_s. \tag{4.55}$$

This means that the wavelength is much shorter than a scale-height, which is a severe restriction near the photosphere where Λ may be as small as a few hundred kilometres (see Equation (3.10)). The magnetoacoustic-gravity dispersion relation can be obtained by eliminating $\mathbf{v}_1$ from Equation (4.16) and invoking Equation (4.55). It is

$$\omega^4 - \omega^2 k^2 (c_s^2 + v_A^2) + k^2 c_s^2 N^2 \sin^2 \theta_g + k^4 c_s^2 v_A^2 \cos^2 \theta_B = 0, \tag{4.56}$$

where θ_B is the angle between $\mathbf{k}$ and the magnetic field. There are several limiting cases. When N and v_A vanish, acoustic waves are recovered, whereas, when $v_A = 0$ and $\omega \ll kc_s$, the result is internal gravity waves. But it should be noted that, in the case of vanishing v_A, the full acoustic-gravity wave dispersion relation (4.54) is not recovered, since the coupling between acoustic and gravity waves is excluded by the condition (4.55).

When $N = c_s = 0$, Equation (4.56) becomes

$$\omega^2 = k^2 v_A^2 \quad \text{(compressional Alfvén wave)},$$

while the condition $N = 0$ alone gives

$$\omega^4 - \omega^2 k^2 (c_s^2 + v_A^2) + k^4 c_s^2 v_A^2 \cos^2 \theta_B = 0 \quad \text{(magnetoacoustic waves)}. \tag{4.57}$$

It should be noted that, if $c_s^2 \lesssim v_A^2$, the validity condition (4.55) reduces Equation (4.56) to (4.57), so that the interaction with gravity is excluded and the limit $c_s^2 \to 0$ produces just compressional Alfvén waves. However, if $c_s^2 \gg v_A^2$, the two magnetoacoustic-gravity solutions to Equation (4.56) become

$$\omega^2 = k^2 c_s^2 \quad \text{(acoustic wave)}$$

and

$$\omega^2 = N^2 \sin^2 \theta_g + k^2 v_A^2 \cos^2 \theta_B.$$

The latter mode reduces to an internal gravity wave at low frequencies (when $kv_A \ll N$) and a slow magnetoacoustic wave (see Equation (4.51)) at high frequencies (when $kv_A \gg N$); it is therefore known as a *magneto-gravity wave*.

In view of the condition (4.55), a coupling between the magnetic field and gravity is not allowed by the magnetoacoustic-gravity wave dispersion relation (4.56), save in the limit $c_s^2 \gg v_A^2$. This led Chen and Lykoudis (1972) to consider an alternative aproach in which the equilibrium magnetic field is horizontal and behaves like $\exp(-\tfrac{1}{2}z/\Lambda)$, so that the Alfvén speed is uniform. For this special case, they were able to derive a dispersion relation and construct diagnostic diagrams, while Nye and Thomas (1974) have used the approach to model running penumbral waves (Section 4.9.4). Otherwise, there do not appear to be any other vertically propagating plane-wave solutions that permit a coupling between magnetic and gravitational forces. One can, however, seek solutions in the form $v_1(z) \exp i(k_x x + k_y y - \omega t)$, with a variation that is sinusoidal in the horizontal but not the vertical direction. Equation (4.14) then reduces in general to a fourth-order ordinary differential equation for $v_1(z)$ that needs to be solved subject to relevant boundary conditions. This is the method adopted by, for example, Roberts and Webb (1978) in their consideration of intense photospheric fields at supergranulation boundaries, and by Nye and Thomas (1976a), who model *flare-induced coronal waves* in a uniform horizontal magnetic field. In both these special cases the differential equation for $v_{1z}(z)$ is only second order.

4.9. Five-Minute Oscillations

4.9.1. Observations

Leighton (1960) discovered that the quiet solar surface (photosphere and low chromosphere) is covered with regions oscillating up and down with a period of about 5 min. These regions are distinct from granules and consist of sinusoidal wave-packets, typically four or five cycles long (though sometimes up to nine). The wave packets are separated by low-amplitude noisy intervals, as shown in Figure 4.10, and it is noticeable that oscillations usually begin and end with small amplitudes. On average, the duration of a wave-packet is 23 min, though occasionally it may be as long as 50 min, but successive packets appear unrelated.

The oscillating regions are uniformly distributed over the solar disc, with about two-thirds of the surface exhibiting an oscillation at any instant. The velocity amplitude lies between about 0.1 and 1.6 km s^{-1} but it is typically 0.4 km s^{-1} in the photosphere. It increases slowly with height like $\rho^{-0.3}$, but the wave energy decreases. As the limb is approached, the velocity amplitude declines, indicating that the oscillations are largely vertical in direction. Shorter-period waves with periods as small as 15 s are also present in the photosphere with velocity amplitudes of 0.1 to 0.2 km s^{-1}.

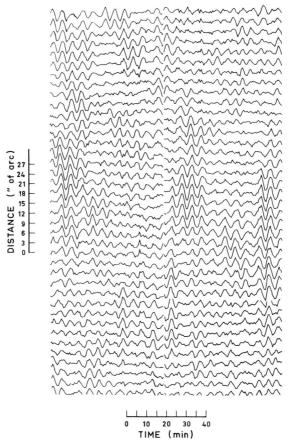

Fig. 4.10. The observed vertical velocity as a function of time at many photospheric locations, each separated by 3 arc sec (about 2200 km). The velocity scale is such that the distance between adjacent curves corresponds to 0.4 km s^{-1} (from Musman and Rust, 1970).

A wide range of frequencies is present in the oscillations (from 150 to 400 s), but the spectrum peaks at about 300 s. The higher frequencies down to 150 s are observable in the chromosphere and increase in importance with height. Lower frequencies (non-periodic) dominate in the photosphere and have been identified with photospheric granulation; their relative importance decreases with height.

The horizontal wavelength is typically 5000 to 10 000 km, which is distinctly larger than the width of a granule. It can clearly be seen in Figure 4.10 that many oscillations are in phase: an area the size of a supergranule cell appears to oscillate coherently with a superimposed amplitude modulation. The horizontal scale for amplitude coherence is between 5000 km and 10 000 km, whereas that for phase coherence is as much as 30 000 km.

Vertical phase speeds of 30 to 100 km s^{-1} (upwards) have been observed, as have horizontal phase speeds of the same magnitude. Oscillations in brightness (and so temperature) are observed; the brightness maxima lead the maxima in upward velocity by a phase of $\frac{1}{2}\pi$, which argues against the presence of progressive waves.

Furthermore, at least in the photosphere, there is no phase difference between velocity oscillations at different heights, which also implies that the waves are *standing* or *evanescent* rather than progressive. Deubner (1975) and Deubner et al. (1979) have observed oscillations with spatial and temporal resolution. These can be analysed in terms of two-dimensional power spectra in ω and k(horizontal), which show that the power is concentrated in well-defined ridges. Their widths result from a finite observation-time and imply that the duration of oscillations exceeds a day.

Athay (1981a) has summarised the observations of oscillations at chromospheric and transition region levels. In the chromosphere different regions of the solar surface are seen to oscillate with different periods, over a wide range from less than 30 s up to more than 400 s, but with a concentration of power near 300 s. Although the photospheric 5 min oscillations are clearly evanescent, chromospheric oscillations at 5 min or less appear to be propagating upwards at roughly the sound speed. In the transition region there are fluctuations of 2 to 3 km s^{-1} without any well-defined period. However, the mean time between intensity pulses is about 5 min, which may mean that the oscillations are still present but their periodicity has been destroyed during their propagation through the inhomogeneous chromosphere.

Global oscillations of the whole Sun depend on the spherical harmonic parameters (l, m) and the number (n) of nodes in the radial direction. For example, the zeroth-order $(l = n = 0)$ *radial pressure mode* has a period of about an hour, while the other pressure modes have shorter periods and most of the *gravity* modes have longer periods. Five-minute oscillations have been observed when looking at the whole Sun (Claverie et al., 1979), which implies that they are global, with l values between 0 and 4. This was repeated more accurately at the South Pole by Grec et al. (1980), who found 75 different pressure modes with periods between 3 and 10 min, and with amplitudes between 4 and 40 cm s^{-1}. They also observed a beating between modes and some modes that were rapidly excited and then decayed over the course of a couple of days (although the lifetime may be longer). No internal gravity modes have yet been identified.

4.9.2. Models

4.9.2A. Photospheric Ringing

The suggestion has been put forward by Schmidt and Zirker (1963), and developed by Meyer and Schmidt (1967) and Stix (1970), that the 5-min oscillation is a 'ringing' of the convectively stable photosphere, which is continually struck like a bell by granular blobs of plasma coming up from below. From Figure 4.8 it was seen that acoustic waves cannot propagate below the cut-off frequency (N_s) and that, as ω approaches N_s, so the group velocity tends to zero. The impulse from a granule therefore generates acoustic waves with frequency $\omega > N_s$, leaving behind a standing wave at a frequency N_s (see Section 8.7.4 and Roberts, 1981c).

Deubner (1973) has indeed observed high frequency (200 s) wave-trains initiated by bright granules, but one difficulty with the ringing hypothesis is that there is in general no observed coincidence between the appearance of a granule and the onset of oscillation. Furthermore, in the photosphere and low chromosphere N_s corresponds to too small a wave period, only 180 to 220 s.

4.9.2B. Wave Trapping

The more likely explanation of 5-min oscillations is that they represent waves trapped in a *resonant cavity* (Schatzman, 1956). Such a cavity is a layer of the solar atmosphere within which waves can propagate but which is bounded above and below by non-propagating (or evanescent) layers. The upper and lower boundaries of the cavity therefore reflect waves into the cavity and so give a standing (or stationary) wave inside the cavity.

At a constant temperature, a diagnostic diagram (Figure 4.9) shows whether a wave of a given frequency (ω) and horizontal wavenumber (k_x) is evanescent or propagating. The actual temperature structure of the solar atmosphere gives rise to four possible cavities whose ranges of ω and k_x are indicated by shaded regions in Figure 4.11. Cavities Ia and IIa contain *trapped acoustic waves*, while Ig and IIg contain *trapped gravity waves*. Ia and Ig give trapping around the temperature minimum region (4200 K), whereas IIa and IIg correspond to a region with a temperature of about 10^4 K, either the chromosphere or the upper convection zone. Initially, it was expected that 5-min oscillations would have a wave-length equal to the size of a granule, say 1800 km. This led Uchida (1965) and Thomas *et al.* (1971) to propose trapped gravity waves in cavity Ig as the explanation. But, after the discovery that the horizontal wavelength ($2\pi/k_x$) is instead about 5000 km, it became clear that the only cavity corresponding to the observed frequency ($\omega \approx 0.02 \text{ s}^{-1}$) and horizontal wavenumber ($k_x \approx 1 \text{ Mm}^{-1}$) is IIa. Trapped acoustic waves in the temperature minimum (Ia) possess a frequency that is too high, whereas trapped gravity waves (Thomas *et al.*, 1971) have too small a wavelength (Ig) or frequency

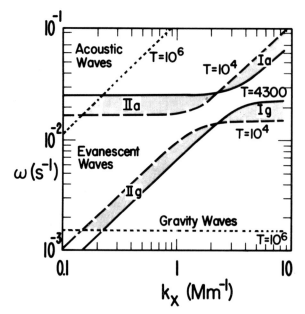

Fig. 4.11. Diagnostic diagrams for acoustic-gravity waves at the temperatures 4300 K (solid), 10^4 K (dashed) and 10^6 K (dotted). The shaded regions indicate ranges of frequency ω and horizontal wavenumber k_x for which the waves are trapped in the solar atmosphere (after Stein and Leibacher, 1974).

(IIg). (However, departures from adiabatic pressure variations due to radiation are important at photospheric levels (Souffrin, 1967). They modify the diagnostic diagrams somewhat, the main effect being to damp gravity waves within a few scale-heights of the photosphere, where the radiative damping times are short. Also, trapped gravity waves with a wavelength of 2 Mm (2000 km) or less would need high-resolution observations from space to be detected adequately.)

Since (IIa) lies *above* the dashed (10^4 K) curve, Figure 4.9 shows that acoustic waves with the corresponding ω and k_x will propagate at a temperature of 10^4 K. But, since IIa lies *below* the solid and dotted curves, the waves cannot propagate when they reach regions with temperatures of 4300 K or 10^6 K. There are two locations for a cavity of type IIa (Figure 4.12), either in the upper convection zone below the temperature minimum (Ulrich, 1970; Leibacher and Stein, 1971; Ando and Osaki, 1975; Graff, 1976) or in the chromosphere above the temperature minimum (Bahng and Schwarzschild, 1963). The lower of the two cavities contains acoustic waves that are reflected from above at some temperature T_1 less than 4300 K and from below at some temperature T_2 greater than 10^6 K; the values of T_1 and T_2 depend on ω and k_x and so on the position within the shaded region IIa. The upper (chromospheric) cavity is bounded above by T_2 and below by T_1. For 5-min oscillations observed in the photosphere, it is the lower cavity which is relevant. It gives reflection at the temperature T_1 when the frequency is N_s, which corresponds to periods between 220 and 420 s. Above this cavity the disturbance becomes evanescent and so decays away exponentially with height. Acoustic wave flux that *leaks* (or '*tunnels*') through the (evanescent) temperature-minimum region into the upper cavity may produce a heating of the upper chromosphere and corona as they

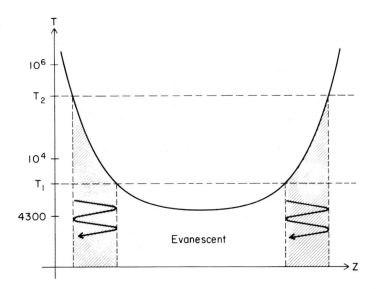

Fig. 4.12. A sketch of the temperature as a function of height in the solar atmosphere, indicating the location of two cavities (shaded), one in the chromosphere and the other in or below the photosphere. Acoustic waves may propagate within a cavity but not below or above it, where they become evanescent (non-propagating). Similar cavities exist which can trap gravity waves.

dissipate. However, such heating is now thought to be insignificant and shorter period waves are needed to heat the lower chromosphere (Section 6.3).

Ando and Osaki (1975) have given a full numerical solution for the vertical propagation of non-adiabatic acoustic waves in a realistic model envelope and atmosphere. The resulting $\omega - k$ diagram is in impressive agreement with the one obtained by Deubner *et al.* (1979) from observations. Their work has been complemented by that of Chiuderi and Giovanardi (1979), who are able to obtain an *analytical solution* by treating adiabatic variations in a particular temperature profile. The perturbed vertical velocity component v satisfies

$$\frac{\gamma \tilde{R} T(z)}{\tilde{\mu}(z)} \frac{\partial^2 v}{\partial z^2} - \gamma g \frac{\partial v}{\partial z} - \frac{\partial^2 v}{\partial t^2} = 0, \tag{4.58}$$

where Chiuderi and Giovanardi approximate the Harvard–Smithsonian Reference Atmosphere by a parabolic profile

$$T(z)/\tilde{\mu}(z) = A^2(1 + z^2/l^2),$$

with a minimum at height $z = 0$. The velocity perturbation is assumed to be oscillatory with frequency ω and amplitude $\phi(z)$, so that

$$v(z, t) = \phi(z) e^{i\omega t},$$

and Equation (4.58) reduces to a hypergeometric equation for $\phi(z)$, namely

$$\gamma \tilde{R} A^2 (1 + z^2/l^2) \frac{d^2 \phi}{dz^2} - \gamma g \frac{d\phi}{dz} + \omega^2 \phi = 0.$$

At large distances the asymptotic solution represents progressive waves if $\omega^2 > \gamma \tilde{R} A^2/(4l^2)$ and stationary waves if $\omega^2 < \gamma \tilde{R} A^2/(4l^2)$. The response of the atmosphere to a periodic disturbance in deep layers is simulated by imposing an oscillating boundary with the boundary condition $\phi = v_0$ at depth $z = z_0$, together with a radiation condition as $z \to +\infty$. The dominant feature of the results for v at a certain height as a function of ω is the presence of a very sharp maximum for frequencies corresponding to wave-periods between 180 and 300 s, where the velocity can reach a thousand times its general value. This seems to suggest that the 5-min oscillation may be interpreted as a resonant phenomenon, but the analysis refers only to the atmosphere and the resonance may be an artificial result of the lower boundary condition. By contrast, Christensen–Dalsgaard (1980) has obtained analytical Lamb-like solutions for oscillations in the envelope below the atmosphere.

4.9.3. WAVE GENERATION

Given that 5-min oscillations in the photosphere are probably acoustic modes trapped in a layer below the temperature minimum, the question arises as to how the acoustic waves are generated. Several methods of generation have been suggested, the first being the most likely in practice.

In the convection zone below the photosphere the buoyancy force drives turbulent motions. These in turn produce pressure perturbations, which can propagate away only as sound waves, since gravity waves do not exist there. Sound is generated mainly by quadrupole emission (Lighthill, 1953) and most efficiently at a wavelength comparable to the turbulence length-scale (L). For a sound speed (c_s) of 20 km s^{-1}, characteristic of the low photosphere, one would need $L \approx 6000$ km to obtain a period (L/c_s) of 300 s. Goldreich and Keeley (1977) suggest a *resonance* between one mode of pulsation and a convective eddy with the same time-scale. Application of this idea to the convection zone gives amplitudes consistent with those of the 5-min oscillations.

A second method is that of *penetrative convection* (Moore, 1967). Photospheric granulation is a manifestation of convection cells at the top of the convection zone, and the suggestion is that, when such convective elements penetrate into the (convectively) stable photosphere, they dump momentum and heat which generate new motions. However, the lack of correlation between granulation and oscillations argues against this mechanism.

Another possible driving mechanism for the 5-min oscillation is *thermal overstability* in the upper part of the convective zone (Ulrich, 1970), although the simulation of Ando and Osaki (1975) shows that it is a small effect. Overstability is a state of growing oscillation (Section 7.1), which needs a destabilising force, a restoring force and a dissipative process. During the oscillation, the dissipation reduces the size of the destabilising force, and so the restoring force is able to drive the amplitude of the oscillation larger and larger. For thermal overstability the plasma is destabilised by buoyancy, while the restoring force is due to pressure gradients and the dissipative process is radiation. Inside active regions one possible source for waves that heat the chromosphere and corona is *magneto-gravity overstability* in sunspots (Danielson, 1961; Roberts, 1976). In this case too, buoyancy is the destabilising force, but the restoring force is due to the magnetic field and dissipation is provided by thermal conduction.

Variations in opacity during an oscillation have a destabilising effect in some layers of the Sun, by causing the matter to gain heat when compressed and lose it when expanded, just like a classical Carnot engine. This so-called *kappa mechanism* was found by Ando and Osaki (1975) to render 5-min oscillations unstable. However, later calculations include a perturbation in the convective flux and cast doubt on the kappa mechanism, so that Lighthill's mechanism now appears the most viable.

4.9.4. Strong Magnetic Field Regions

Oscillations have been observed in regions where the magnetic field is strong, such as in sunspot umbrae (Beckers and Tallant, 1969), plages and supergranulation boundaries, but they are not yet well understood theoretically. For example, it is unclear why the velocity amplitude in regions of moderate magnetic field strength (greater than 80 G) is lower than in quiet regions by a factor of about 25%. The wave periods are unchanged at photospheric levels, but, in the chromosphere, the power is concentrated in the band 100 to 400 s, with a maximum at 330 s for plages (active regions) by comparison with a band of 200 to 400 s and a maximum of 300 s in the quiet sun. Within a sunspot umbra one finds instead '*three-minute oscillations*' (Beckers and Schultz, 1972); these are vertical motions with a small correlation length (2″ to 3″) and a period between 150 and 200 s, increasing back to 300 s or more in the penumbra. Umbral oscillations with periods 110 to 120 s and 300 to 470 s in the photosphere have also been reported, while the periods become 145 to 185 s in the chromosphere. Slow magnetoacoustic gravity waves of frequency ω (Equation (4.53b)) can be trapped near the temperature minimum between regions in which the cut-off frequency $N_s (\sim T^{-1/2})$ has fallen below ω. However, umbral oscillations may instead represent trapped fast modes (Equation (4.53a)) according to Scheuer and Thomas (1981) or trapped tube modes (Hollweg and Roberts, 1981).

Other wave-like phemomena in sunspots include *running penumbral waves* (Zirin and Stein, 1972; Giovanelli, 1972; Moore and Tang, 1975). These are observed in Hα films of penumbra as regular outgoing ripples with a period between 150 and 290 s, a horizontal wavelength of 2300 to 3800 km and a horizontal phase speed between 10 and 20 km s^{-1}. The wave motion is predominantly up-and-down and it may well be excited by the shorter-period umbral oscillations. It is almost always present in sizeable sunspots with a regular stable structure but occurs only rarely in complex active sunspots. Nye and Thomas (1974, 1976b) have given a model for running penumbral waves, which they suggest are gravity-modified fast magnetoacoustic waves, trapped in the photosphere and evanescent in the chromosphere. The trapping is due to the increase of sound speed below the region and the increase of Alfvén speed above. Also, inside umbrae one sometimes finds chromospheric *umbral flashes* (Beckers and Tallant, 1969), which are small upward moving elements lasting only 50 s (Section 1.4.2B). They are often repetitive but are uncorrelated with photospheric oscillations. Antia and Chitre (1979) have suggested they are overstable magnetoacoustic modes, and have investigated magnetoacoustic-gravity modes in a polytropic stratified atmosphere in the presence of a uniform vertical magnetic field (similar to Section 4.10.3). The resulting sixth-order differential equation for the perturbations leads to complex eigenfrequencies ω when the relevant boundary conditions are imposed.

4.9.5. The Future

A most exciting new topic is solar seismology: the use of global pressure and gravity modes to probe the solar interior and to determine the depth of the convection zone, the variation of rotation with depth and the presence of giant cells and large-scale magnetic fields. Also, there is a need for measurements of local magnetic field variations and for a continued development of oscillation theory in regions where the magnetic field is important.

4.10. Waves in a Strongly Inhomogeneous Medium

Most of the analysis of the present chapter has rested on the assumption that the ambient medium in which the waves are propagating is *uniform*. This is valid if the wavelength (λ) is much smaller than the length-scale (l_0) for variations in the medium. As such waves propagate, so their properties change slowly; for example, acoustic waves may steepen into shocks when they travel into a region of decreasing density (Sections 5.1.1 and 6.3.1).

The mathematical advantage of considering a *uniform* medium is that the partial differential Equations (4.8) to (4.12) for the perturbed quantities reduce to *algebraic* equations, (4.15) or (4.16), for the dispersion relation $\omega = \omega(k_x, k_y, k_z)$. But when $\lambda \gtrsim l_0$ the inhomogeneous nature of the medium determines the structure of the disturbance, which can no longer be assumed to be sinusoidal. If, for example, the medium is structured in the z-direction, the perturbation equations reduce to *ordinary differential equations* in z; their solution subject to certain boundary conditions determines the z-structure of the disturbance as well as the dispersion relation $\omega = \omega(k_x, k_y)$ for sinusoidal variations in the other two directions. However, one complication is that there may exist a *continuous spectrum* of modes in addition to a *discrete spectrum*; this can occur when the differential equation is singular in the region under consideration.

This topic of 'long-wavelength disturbances in the *inhomogeneous* solar atmosphere' is only in its infancy, but it is most important and is likely to receive much attention in future.

The main agents for creating inhomogeneity on the Sun are gravity and the magnetic field. Gravity causes the pressure to increase inwards towards the solar centre, while the magnetic field and its associated Lorentz force often cause the plasma pressure to increase in a direction normal to the magnetic field and away from regions of magnetic flux concentration. The stratification introduces several new effects:

(i) *amplification* – the amplitude of the wave may increase (or decrease) as it propagates;
(ii) *evanescence* – a time-oscillating disturbance may have a wave-like character in one region where it oscillates spatially, but then it may become evanescent and decay exponentially in another region;
(iii) *surface modes* – a discontinuity in the basic state may give rise to extra surface waves that decay away from the interface, while the 'body' waves that were present in a uniform medium are modified somewhat.

In order to give the reader a taste of this fascinating topic, the equations for three examples are presented briefly below, namely a magnetic interface, a flux tube with radial structure and a plane-parallel atmosphere in a uniform magnetic field. The possible relevance to magnetic heating is mentioned in Section 6.4.3.

4.10.1. Surface Waves on a Magnetic Interface

Consider a unidirectional magnetic field ($\mathbf{B} = B_0(x)\hat{\mathbf{z}}$) containing plasma at rest with equilibrium properties $p_0(x)$, $\rho_0(x)$, $T_0(x)$, such that

$$p_0(x) + \frac{B_0(x)^2}{2\mu} = \text{constant}.$$

A small perturbation velocity

$$v_{1x} = v_{1x}(x)\, e^{i(\omega t + k_y y + k_z z)},$$

with wavenumbers k_z and k_y in the z- and y-directions, can be shown (e.g., Wentzel,

1979; Roberts, 1981a) to satisfy

$$\frac{d}{dx}\left[\frac{\varepsilon(x)}{k_y^2 + m_0(x)^2}\frac{dv_{1x}}{dx}\right] = \varepsilon(x)v_{1x}, \qquad (4.59)$$

where

$$\varepsilon(x) = \rho_0(x)\omega^2 - k_z^2 B_0(x)^2/\mu \equiv \rho_0(x)(\omega^2 - k_z^2 v_A(x)^2),$$

$$m_0(x)^2 = \frac{(k_z^2 c_{s0}^2 - \omega^2)(k_z^2 v_A^2 - \omega^2)}{(c_{s0}^2 + v_A^2)(k_z^2 c_T^2 - \omega^2)},$$

$$v_A(x)^2 = B_0(x)^2/(\mu\rho_0(x)),$$

$$c_{s0}(x)^2 = \gamma p_0(x)/\rho_0(x),$$

$$c_T(x)^2 = \frac{c_{s0}(x)^2 v_A(x)^2}{c_{s0}(x)^2 + v_A(x)^2}.$$

An alternative way of writing (4.59) is

$$\frac{d}{dx}\left(\frac{(\omega^2 - \omega_A^2)(\omega^2 - \omega_T^2)}{(\omega^2 - \omega_+^2)(\omega^2 - \omega_-^2)}(c_{s0}^2 + v_A^2)\rho_0\frac{dv_{1x}}{dx}\right) = -(\omega^2 - \omega_A^2)\rho_0 v_{1x}, \qquad (4.59')$$

where $\omega_A = k_z v_A$ is the Alfvén frequency and $\omega_T = k_z c_T$ is the cusp (or tube) frequency, while ω_+ and ω_- are the frequencies for fast and slow waves given by Equation (4.51') with $k_x = 0$. Thus, the presence of the inhomogeneous magnetic field, varying with x, means that, in general, the perturbation velocity (v_{1x}) is not sinusoidal in x but satisfies Equation (4.59). The function m_0^2 may be positive or negative, since it changes sign at the places where the phase speed ω/k_z (assumed real) is equal to the sound speed (c_{s0}), the Alfvén speed (v_A) or the "cusp (or tube) speed" (c_T). The significance of c_T is that it represents the phase velocity of slow magnetoacoustic waves along the field in the limit of propagation normal to the field (Section 4.6).

In two cases the fundamental equation takes on a simpler form. When the basic state is uniform, ε and m_0 are constants and we may recover the previous results of Sections 4.3.1 and 4.6: as one solution, (4.58) has $\varepsilon = 0$ for Alfvén waves, while for another solution v_{1x} is proportional to $e^{ik_x x}$ and

$$k_x^2 + k_y^2 + m_0^2 = 0,$$

which is just the dispersion relation (4.51) for magnetoacoustic waves. The second simple case is that of transverse propagation (with ω^2 not close to $k_z^2 c_T^2$), so that $k_y^2 \gg m_0^2(x)$ and Equation (4.59) simplifies to

$$\frac{d}{dx}\left[\frac{\varepsilon(x)dv_{1x}}{k_y^2\ dx}\right] = \varepsilon(x)v_{1x} \qquad (4.60)$$

for Alfvén disturbances alone, as discussed in Section 6.4.3. The same equation with k_y replaced by k_z also results in the limit when $c_{s0}^2 \gg v_A^2$.

The solutions of Equation (4.59) over some range of x (and subject to the relevant boundary conditions) are quite complicated and, in general, need to be found numerically. If the profile is made up of discrete sections, the solution may be written as a sum of discrete normal modes with different values of ω, including both the surface and body waves. However, if the profile is continuous $v_A(x)$ or $c_T(x)$ equals ω/k at some location, and so the differential equation possesses a singularity where $\varepsilon/(m_0^2 + k_y^2)$ vanishes. This leads to a *continuous spectrum* of frequencies and the general solution consists of an integral over some range of ω in addition to a possible sum of discrete modes.

Consider for simplicity the case of an interface at $x = 0$ separating uniform media, so that

$$B_0(x) = \begin{cases} B_0, & x < 0, \\ B_e, & x > 0, \end{cases}$$

where B_e and B_0 are constants. Supposing further that $k_y = 0$, we see that Equation (4.59) reduces to either $\varepsilon(x) \equiv 0$ or

$$\frac{d^2 v_{1x}}{dx^2} = m_0^2 v_{1x} \quad \text{for } x < 0 \tag{4.61}$$

and

$$\frac{d^2 v_{1x}}{dx^2} = m_e^2 v_{1x} \quad \text{for } x > 0,$$

where m_e is defined in a similar way to m_0 except that the Alfvén and sound speeds appropriate to $x > 0$ are taken. There are two basic classes of solution to Equation (4.61) in the region $x < 0$:

(i) *Unbound states* represent waves that are propagating from minus infinity and are being reflected and transmitted at the interface. They occur for frequencies such that $m_0^2 < 0$ and have $v_{1x} \sim \exp i(-m_0^2)^{1/2} x$, for $x < 0$. In the absence of the interface, they become the normal (*body*) *waves* of a uniform medium.

(ii) *Bound states* decay away spatially to zero at minus infinity like $e^{m_0 x}$ and so they occur for $m_0^2 > 0$. These *surface waves* owe their existence to the interface and cannot exist when it is absent ($B_0 = B_e, \rho_0 = \rho_e$).

When $m_0^2 > 0$ and $m_e^2 > 0$ there are discrete *magnetoacoustic surface-mode* solutions such that

$$v_{1x} \propto \begin{cases} e^{m_0 x}, & x < 0, \\ e^{-m_e x}, & x > 0. \end{cases}$$

The boundary conditions at the interface (continuity of v_{1x} and total pressure) yield their dispersion relation as

$$\frac{\omega^2}{k^2} = v_A^2 - \frac{\rho_e m_0}{\rho_e m_0 + \rho_0 m_e}(v_A^2 - v_{Ae}^2), \tag{4.62}$$

where $v_{Ae}^2 = B_e^2/(\mu \rho_e)$. It can be seen that their phase speed lies between the Alfvén speeds (v_A and v_{Ae}) in the two media, but the fact that m_0 and m_e are functions of ω means that there can be several such solutions.

In the particular case when the second medium has a negligible field strength

($B_e = 0$), Equation (4.62) reduces to

$$\frac{\omega^2}{k^2} = \frac{\rho_0 m_e}{\rho_0 m_e + \rho_e m_0} v_A^2, \tag{4.63}$$

where m_e^2 becomes $m_e^2 = k_z^2 - \omega^2/c_{se}^2$. There is then always a slow surface wave solution to Equation (4.63) with ω/k_z smaller than both c_T and c_{se}. However, when $c_{se} > c_{s0}$ and $v_A > c_{s0}$, a fast surface wave also exists with ω/k_z larger than c_{s0} but less than both c_{se} and v_A.

There are several ways in which the above analysis can be extended. Surface waves on a slab, bounded by two interfaces separating three uniform regions, have been treated by Roberts (1981b), as summarised in Sections 8.7.4; they possess similar features to the above but are arithmetically more complicated. Also, in a cylindrical geometry the equation for waves in a flux tube is (4.64) below; it is close to Equation (4.59) in form and possesses very similar properties. Finally, if the discontinuous interface is replaced by an interface of finite width with a continuous magnetic profile, a surface wave (with a single frequency) no longer exists! This is because the presence of the singularity of Equation (4.59) at $\varepsilon(x) = 0$ (i.e., $v_A(x)^2 = \omega^2/k_z^2$) implies that there is no continuous nontrivial solution which decays away exponentially as x approaches $\pm \infty$. However, although the normal-mode technique fails, the initial-value problem may be solved by constructing the Green's functions in the usual way (Sedlacek, 1971; Rae and Roberts, 1981; Lee *et al*, 1984). The result is that a surface disturbance (with a continuum of frequencies) decays away in time and feeds energy into the interface, where longitudinal motions having a continuum of frequencies grow with an oscillating amplitude. The resulting build-up of large gradients eventually produces substantial viscous or ohmic dissipation, which may provide heat for the corona (see also Section 6.4.3).

4.10.2. A TWISTED MAGNETIC FLUX TUBE

Consider a cylindrically symmetric flux tube, whose pressure varies with radial distance (R) from some axis and whose magnetic field lines lie on cylindrical surfaces. In other words, the equilibrium state is specified by

$$\mathbf{B}_0 = (0, B_{0\phi}(R), B_{0z}(R)), \qquad p_0 = p_0(R), \qquad \rho_0 = \rho_0(R).$$

The gravitational and Coriolis forces are neglected, and the equilibrium variables satisfy the magnetostatic condition

$$0 = \frac{dp_0}{dR} + \frac{d}{dR}\left(\frac{B_0^2}{2\mu}\right) + \frac{B_{0\phi}^2}{R}.$$

The basic Equations (4.1)–(4.5) are linearised for small disturbances from equilibrium, and each of the perturbation variables is written in the form

$$f_1(R, \phi, z, t) = f_1(R) e^{i(\omega t - m\phi - kz)},$$

since the azimuthal and axial variations can be Fourier analysed whereas the radial variations cannot. Then, according to Choe *et al.* (1977), the equations may be reduced to one equation for the radial velocity perturbation $v_{1R}(R)$, namely,

$$\frac{d}{dR}\left(\frac{A_m C_m}{RD}\frac{d}{dR}(Rv_{1R})\right) + Fv_{1R} = 0, \tag{4.64}$$

where

$$A_m = -\rho_0 \omega^2 + (\mathbf{k}\cdot\mathbf{B}_0)^2/\mu,$$
$$C_m = -\rho_0 \omega^2(\gamma p_0 + B_0^2/\mu) + \gamma p_0(\mathbf{k}\cdot\mathbf{B}_0)^2/\mu,$$
$$D = \rho_0^2 \omega^4 + (m^2/R^2 + k^2)C_m,$$
$$\mathbf{k}\cdot\mathbf{B}_0 = kB_{0z} + (m/R)B_{0\phi},$$

$$RF = A_m - \frac{2B_{0\phi}}{\mu}\frac{d}{dR}\left(\frac{B_{0\phi}}{R}\right) - \left(\frac{2B_{0\phi}}{R\mu}\right)^2\frac{(\mathbf{k}\cdot\mathbf{B}_0)^2}{A_m} -$$
$$-\left(\frac{2B_{0\phi}}{R\mu}\right)^2\frac{\rho^2\omega^4 B_{0\phi}^2}{A_m C_m} + \frac{4C_m}{A_m D}\frac{B_{0\phi}^2}{R^2\mu^2}\left(\frac{m}{R}(\mathbf{k}\cdot\mathbf{B}_0) + \frac{\rho^2\omega^2}{C_m}B_{0\phi}\right)^2 -$$
$$- R\frac{d}{dR}\left(\frac{2C_m}{RD}\frac{B_{0\phi}}{R\mu}\left(\frac{m}{R}(\mathbf{k}\cdot\mathbf{B}_0) + \frac{\rho^2\omega^4}{C_m}B_{0\phi}\right)\right). \tag{4.65}$$

(In the particular case when there is no twist in the field, $B_{0\phi}$ vanishes and we have $RF = A_m$; and, if further the field is uniform, Equation (4.63) reduces to $A_m D = 0$, where

$$D = \rho_0^2\omega^4 - \rho_0\omega^2 k^2(\gamma p_0 + B_0^2/\mu) + k^2\gamma p_0(\mathbf{k}\cdot\mathbf{B}_0)^2/\mu.$$

Here the solution $A_m = 0$ gives the Alfvén-wave dispersion relation, while $D = 0$ corresponds to magnetoacoustic waves. As k approaches infinity with ω bounded, so D tends to $k^2 C_m$ and $C_m = 0$ gives the cusp speed for the slow mode.)

The locations where A_m or C_m vanish are singularities of the differential Equation (4.64), and they lead to continuous spectra. When ω is such that A_m (or C_m) vanishes identically across the cylindrical tube, ω is said to be in the Alfvén (or cusp) continuum. In addition, there are discrete solutions to Equation (4.64) that are magnetoacoustic in nature. One interesting feature is that energy may be resonantly absorbed at singular surfaces where A_m vanishes, as described in Section 6.4.3.

4.10.3. A STRATIFIED ATMOSPHERE

Consider an *isothermal* plasma, stratified by gravity $-g\hat{\mathbf{z}}$ and situated in a *uniform, vertical* magnetic field $(B_0\hat{\mathbf{z}})$. According to Equation (3.6) the hydrostatic equilibrium has a pressure and density that fall off exponentially with height like $\exp(-z/\Lambda)$, where Λ is the scale-height. Hollweg (1979) linearises the basic Equations (4.1) to (4.5) about this non-uniform equilibrium (with $\mathbf{\Omega} = 0$), and assumes perturbations of the form

$$f_1(x, z, t) = f_1(z) e^{i(\omega t + k_x x)}.$$

Alfvén waves are excluded by taking a perturbed velocity with no y-component and the magnetic field is assumed to be so strong that $\beta_0 \equiv 2\mu p_0/B_0^2 \ll 1$.

The horizontal component of the equation of motion may be combined with the induction equation to give

$$\frac{d^2 v_{1x}}{dz^2} + \left(\frac{\omega^2}{v_A^2} - k_x^2\right) v_{1x} = \frac{ik_x g}{v_A^2} v_{1z} \qquad (4.66)$$

for the horizontal velocity perturbation (v_{1x}), where the Alfvén speed,

$$v_A(z) = \frac{B_0}{(\mu \rho_0(z))^{1/2}},$$

increases with height. (For wavelengths much smaller than a scale-height, the local approximation to Equation (4.66) gives the dispersion relation

$$\omega^2 = (k_x^2 + k_z^2) v_A^2$$

for *compressional Alfvén waves*.) The vertical component of the equation of motion together with the continuity and energy equations imply

$$\frac{d^2 v_{1z}}{dz^2} - \frac{1}{\Lambda} \frac{dv_{1z}}{dz} + \frac{\omega^2}{c_s^2} v_{1z} = -ik_x \left(\frac{dv_{1x}}{dz} - \frac{\gamma-1}{\gamma \Lambda} v_{1x}\right) \qquad (4.67)$$

for the vertical velocity component v_{1z}. (The homogeneous form of this equation gives the dispersion relation for vertically propagating *acoustic-gravity waves*.) When v_{1x} is eliminated between Equations (4.66) and (4.67) we then find a fourth-order equation for v_{1z} which possesses a singularity at the height where

$$A(z) = \left[\frac{\omega^2}{v_A^2} - k_x^2 + \left(\frac{\gamma-1}{\gamma \Lambda}\right)^2\right]$$

vanishes, and so the possibility arises of heating or some resonance there. However, its properties have not at this time been fully studied.

CHAPTER 5

SHOCK WAVES

5.1. Introduction

The Sun is in a far-from-static state, and often the motions that are universally present produce shock waves. For example, *magnetic reconnections* are likely to be taking place continually when distinct magnetic flux systems press up against one another, and these reconnections generate magnetic shock waves (Section 10.1). Also, the eruption of a huge quiescent *prominence* is likely to generate a wave that propagates ahead, whereas a violent *solar flare* can produce a shock whose presence is revealed by a Type II radio burst. On a smaller scale, shock waves are probably generated by *surges*, *spicules* and the continual granular motion of the solar surface (Sections 6.3 and 6.4.1). Some of the applications are discussed elsewhere in the book, while the present chapter develops the basic theory for both hydrodynamic and magnetic shocks.

5.1.1. Formation of a Hydrodynamic Shock

Consider wave propagation in a non-conducting gas. When the amplitude is so small that linear theory applies, we saw in Section 4.2 that a disturbance propagates as a *sound wave*. If the gas has a uniform pressure (p_1) and density (ρ_1), the speed of propagation is the sound speed

$$c_{s1} = \left(\frac{\gamma p_1}{\rho_1}\right)^{1/2},$$

and the wave profile maintains a fixed shape, since each part of the wave moves with the same speed. However, when the wave possesses a finite amplitude, so that nonlinear terms in the equations become important, the crest of the sound wave moves faster than its leading or trailing edge. This causes a progressive steepening of the front portion of the wave as the crest catches up and, ultimately, the gradients of pressure, density, temperature and velocity become so large that dissipative processes, such as viscosity or thermal conduction, are no longer negligible (Figure 5.1). Then a steady wave-shape is attained, called a *shock wave*, with a balance between the steepening effect of the nonlinear convective terms and the broadening effect of dissipation. The shock wave moves at a speed in excess of the sound speed (c_{s1}), so that information cannot be propagated ahead to signal its imminent arrival, since such information would travel at only c_{s1} relative to the undisturbed medium ahead of the shock. When the shock speed greatly exceeds c_{s1} we have a *strong shock*, but when it is only slightly larger than c_{s1} the shock is said to be *weak*. The dissipation inside the shock front leads to a gradual conversion of the energy being carried by the wave into heat. Thus, the effects of the passage of a shock wave are to convert ordered

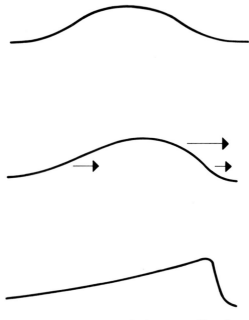

Fig. 5.1. The steepening of a finite-amplitude wave profile to form a shock wave.

energy into random (thermal) energy through particle collisions and also to compress and heat the gas.

In water, for example, surface gravity waves approaching a sloping beach exhibit the increase in wave-amplitude and steepening of the wave-front that we have just described. These processes are expected to occur also in the solar atmosphere, where acoustic waves increase in amplitude as they propagate upwards from the photosphere. They eventually form into shock waves, which probably dissipate their wave energy in the low chromosphere (see Section 6.3).

One way of producing sound waves is to move the walls of a gas container. For example, suppose a long tube contains gas initially at rest and that a piston at one end of the tube is accelerated into uniform motion. If the piston is being withdrawn from the tube, an *expansion wave* travels from the initial location of the piston into the gas. A fall in pressure occurs as the expansion wave passes by. However, nonlinear effects make the wave flatten out as it propagates (Figure 5.2) and so no shock wave develops. By contrast, if the piston is pushed into the tube, a *compression wave* is generated, across which the pressure is increased, and this eventually steepens into a *shock wave*.

The shock front itself is in reality a very thin transition region. Its width is typically only a few mean-free paths, with particle collisions establishing the new uniform state behind the shock. Normally, one models a shock front mathematically by a plane discontinuity, and the two states that it separates are taken to be uniform (and hence dissipationless). They are denoted here by subscripts 1 for the undisturbed gas (ahead of the shock) and 2 for the shocked gas (behind the shock). In a frame of reference at rest, the shock speed is U, while the speed of the shocked gas is $U_2 (< U)$,

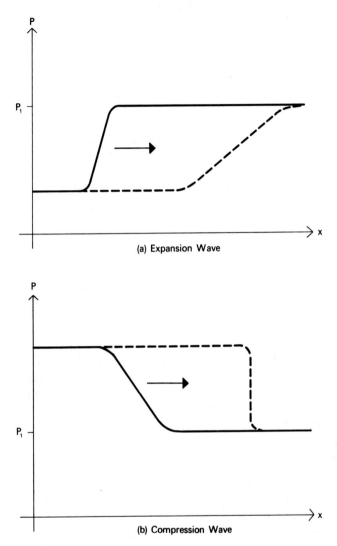

Fig. 5.2. The pressure profiles at two times caused by a piston which is accelerated from $x = 0$ in the directions (a) $x < 0$ and (b) $x > 0$. The profile at the later time is shown dashed, and p_1 is the initial uniform gas pressure.

say (Figure 5.3). However, it is more convenient to use a frame of reference moving with the shock wave, so that the undisturbed gas enters the front of the shock with speed

$$v_1 = U, \qquad (5.1)$$

while the shocked gas leaves the back of the shock with speed

$$v_2 = U - U_2.$$

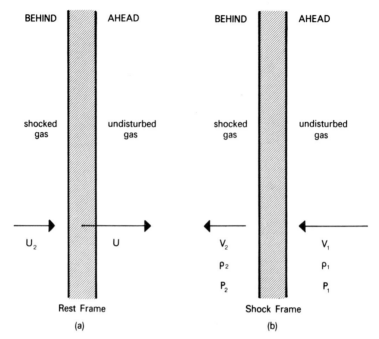

Fig. 5.3. The notation for a plane hydrodynamic shock wave moving to the right with speed u into a gas at rest. Properties ahead of the shock are denoted by 1 and those behind by 2.

Since U_2 is positive, we see from (5.1) and (5.2) that

$$v_2 \leqslant v_1, \tag{5.3}$$

equality occurring when there is no shock.

In Section 5.2 a set of *conservation relations* (or *jump conditions*) is derived. They relate the properties on both sides of the shock front, regardless of the detailed structure within the shock. To obtain physically relevant solutions, these relations are supplemented by an *entropy condition*, namely that the entropy increases following the flow.

A detailed determination of the thickness of the shock and its internal structure is in practice very complicated. For example, in addition to the effects of viscosity and thermal conduction, it often transpires that the electrons heat up first and then share their energy with the ions, a process that suggests at least a *two-fluid model*. Also, the effect of *ionisation* in the shock may need to be included if the unshocked gas is not completely ionised, as for instance near the *temperature-minimum region* of the solar atmosphere. However, if the dominant dissipation mechanism is known, an order-of-magnitude estimate of the *shock width* (δx) may be obtained. For instance, in the case of viscous dissipation, the amount of energy (δE) dissipated during a small time (δt) is given by

$$\frac{\delta E}{\delta t} \approx \rho v \left(\frac{\delta v}{\delta x}\right)^2, \tag{5.4}$$

where v is the kinematic viscosity. After taking

$$\delta t \approx \frac{\delta x}{v_1} \tag{5.5}$$

as the time for the shock front to move a distance δx and also putting $\delta v \approx v_1 - v_2$, Equation (5.4) gives

$$\delta x \approx \frac{\rho v (v_1 - v_2)^2}{v_1 \delta E}, \tag{5.6}$$

where δE can be found by comparing the energies on both sides of the shock front. In particular, it transpires that for a strong shock (Section 5.2) $\delta E \approx \frac{1}{2}\rho_1 v_1^2$, so that

$$\delta x \approx \frac{v}{v_1}. \tag{5.7}$$

In other words, the *Reynolds number* $(v_1 \delta x/v)$ is of order unity (see Section 2.5).

5.1.2. Effects of a Magnetic Field

In a conducting gas, a magnetic field can interact strongly with the flow. The analysis of shock waves therefore becomes more complex, but the basic principles remain the same as for a non-conducting gas where only sound waves are present. We have seen (Section 4.3) that the magnetic field introduces two extra waves, and so, in the absence of rotation and gravity, three wave modes become possible. They are classified according to their phase speeds as *slow, intermediate* and *fast*. When the wave amplitude is large, the intermediate mode (i.e., an Alfvén wave) can propagate without steepening, whereas the slow and fast magnetoacoustic modes steepen to form *slow* and *fast magnetoacoustic shock waves*, respectively. Furthermore, in the motion of a piston, the three degrees of freedom (two transverse and one longitudinal) give rise to the three distinct modes of wave propagation, two of which can steepen as in Figure 5.2(b). A discussion of magnetohydrodynamic shock waves can be found in Kantrowitz and Petschek (1966), as well as in many of the standard text books on magnetohydrodynamics such as Jeffrey (1966) and Ferraro and Plumpton (1961).

A set of jump conditions may again be derived (Section 5.4), but they are considerably more complicated than in the purely hydrodynamic case. The extra complexity arises both from the presence of an extra variable, namely the magnetic field strength, and also from the fact that the magnetic field and plasma velocity may be inclined away from the shock normal. Furthermore, the entropy condition that supplemented the jump conditions for the hydrodynamic shock is now insufficient. The system is no longer isolated, since the magnetic field may do work from the outside on the plasma, with the result that the entropy may, in some cases, decrease. A (stronger) so-called *evolutionary condition* must be employed instead (Jeffrey and Taniuti, 1964), which ensures that the perturbation caused by small disturbances incident on the shock front is both small and unique. For simplicity, just the results of its application will be quoted here. In practice, it implies that the tangential magnetic field component cannot reverse its sign across a shock.

For wave propagation normal to the magnetic field, the slow magnetoacoustic mode ceases to exist, while the fast mode propagates with speed $(c_{s1}^2 + v_{A1}^2)^{1/2}$ in a medium with sound speed c_{s1} and Alfvén speed v_{A1}. Thus, only one type of shock wave can propagate directly across the magnetic field, namely the fast shock, and its speed exceeds $(c_{s1}^2 + v_{A1}^2)^{1/2}$. It is sometimes called a *perpendicular shock*. In any other direction, both the slow and fast shocks can propagate. These *oblique shocks* travel with speeds in excess of the slow and fast wave-speeds, respectively. The slow shock has the effect of decreasing the magnetic field strength as it passes and making the magnetic field rotate towards the shock normal, whereas the fast shock has the opposite effect (Figure 5.6). In the particular case when the magnetic field *behind* the shock is aligned with the shock normal, so that the tangential field component has been 'switched off', the slow shock is called a *switch-off shock*. In the case when the magnetic field *ahead* of the shock is directed along the normal (and the incident speed is Alfvénic), so that a tangential component is 'switched on' by the passage of the shock, the fast shock is termed a *switch-on shock* (Figure 5.7).

Another effect in a conducting (or ionised) gas is the presence of an extra dissipative mechanism, namely *ohmic heating*, due to a finite electrical conductivity. This complicates the shock structure but has no influence on the jump relations, provided ohmic dissipation and the slipping of field lines can be neglected on both sides of the shock front. If ohmic heating dominates the energetics, the shock thickness (δx) may be estimated as follows. The rate of energy dissipation is

$$\frac{\delta E}{\delta t} \approx \frac{j^2}{\sigma}, \tag{5.8}$$

where, according to Ampère's law, the shock current is

$$j \approx \frac{B_{1y} - B_{2y}}{\mu \, \delta x},$$

in terms of the transverse magnetic field components (B_{1y} and B_{2y}) ahead of and behind the shock (Figure 5.5). Thus, with δt given by Equation (5.5), Equation (5.8) implies

$$\delta x \approx \frac{(B_{1y} - B_{2y})^2}{\mu^2 \sigma v_1 \, \delta E},$$

where δE can be determined from the jump conditions. For example, a strong shock propagating perpendicular to the magnetic field (Section 5.3) has $\delta E \approx \tfrac{1}{2} \rho_1 v_1^2$ and $B_{2y} \approx 4 B_{1y} = 4 B_1$, so that its width is

$$\delta x \approx \frac{18 B_1^2}{(\mu \sigma v_1)(\mu \rho_1 v_1^2)}. \tag{5.9}$$

In terms of the *magnetic Reynolds number* $R_{m1}(= \mu \sigma v_1 \, \delta x)$ based on the shock width and the *Alfvén Mach number* $M_{A1}(= v_1 (\mu \rho_1)^{1/2}/B_1)$, this may be written $R_{m1} \approx 18/M_{A1}^2$.

Finally, it must be added that an ionised gas may support shock waves with a thickness much smaller than a collision mean-free path. They are known as *collisionless shocks*, and a magnetohydrodynamic description is inadequate to describe their structure. The conversion of ordered energy into random motion occurs not by particle collisions but by either plasma oscillations (that are subsequently damped) or plasma microinstabilities that generate a turbulent state. In the former case, the shocked plasma is in an oscillatory state rather than a uniform one. The thickness of a collisionless shock is typically of the order of one of the plasma length-scales, such as an ion gyro-radius or the geometric mean of the ion and electron gyro-radii.

5.2. Hydrodynamic Shocks

Consider a plane shock wave propagating steadily with constant speed into a non-conducting, stationary gas, with density ρ_1 and pressure p_1. Suppose that, in a frame of reference moving with the shock, the speed of the shocked gas is v_2 and its density and pressure are ρ_2 and p_2 (Figure 5.3). Then ρ_2, v_2, p_2 are determined in terms of ρ_1, v_1, p_1 by the equations of *conservation of mass, momentum and energy*, namely

$$\rho_2 v_2 = \rho_1 v_1, \tag{5.10}$$

$$p_2 + \rho_2 v_2^2 = p_1 + \rho_1 v_1^2, \tag{5.11}$$

$$p_2 v_2 + (\rho_2 e_2 + \tfrac{1}{2}\rho_2 v_2^2)v_2 = p_1 v_1 + (\rho_1 e_1 + \tfrac{1}{2}\rho_1 v_1^2)v_1, \tag{5.12}$$

where (for a perfect gas) the internal energy per unit mass (2.29b) is $e = p/[(\gamma - 1)\rho]$, so that Equation (5.12) reduces to

$$\frac{\gamma p_2}{(\gamma - 1)\rho_2} + \tfrac{1}{2}v_2^2 = \frac{\gamma p_1}{(\gamma - 1)\rho_1} + \tfrac{1}{2}v_1^2. \tag{5.13}$$

These equations are referred to as the *jump* (or *Rankine-Hugoniot*) relations.

Equation (5.10) arises because ρv is the mass that crosses a unit area in unit time. The significance of the terms in Equation (5.11) is that $(\rho v)v$ gives the rate at which momentum is transported across a unit surface area, while p is the force which acts on that area. In Equation (5.12), $(\rho e + \tfrac{1}{2}\rho v^2)v$ represents the rate of transport of internal and kinetic energy, whereas pv is the rate at which the gas pressure does work. The jump relations may be derived from the fundamental Equations (2.19), (2.21) and (2.40b) of mass, momentum and energy (with losses $\mathscr{L} \equiv 0$ and $\mathbf{B} = \mathbf{F} = \mathbf{0}$). The procedure is to assume that each variable is purely a function of distance through the shock and then to integrate from one side of the shock front to the other.

The nontrivial solution of Equations (5.10), (5.11) and (5.13) can be written

$$\frac{\rho_2}{\rho_1} = \frac{(\gamma + 1)M_1^2}{2 + (\gamma - 1)M_1^2}, \tag{5.14}$$

$$\frac{v_2}{v_1} = \frac{2 + (\gamma - 1)M_1^2}{(\gamma + 1)M_1^2}, \tag{5.15}$$

$$\frac{p_2}{p_1} = \frac{2\gamma M_1^2 - (\gamma - 1)}{\gamma + 1}, \tag{5.16}$$

in terms of the shock *Mach number* $M_1 \equiv v_1/c_{s1}$, where $c_{s1} \equiv (\gamma p_1/\rho_1)^{1/2}$. Furthermore, the second law of thermodynamics states that the *entropy* (s) of an isolated system must increase, where, for a perfect gas (2.31), $s \equiv c_v \log(p/\rho^\gamma) +$ constant. Thus, Equations (5.14) to (5.16) need to be supplemented for our shock wave by

$$s_2 \geq s_1; \qquad (5.17)$$

equality arises only when conditions are the same on both sides of the shock front, so that it ceases to exist.

Several interesting properties arise from Equations (5.14) to (5.17). For instance, the *shock speed* (v_1) *must exceed the sound speed* (c_{s1}) *ahead of the shock*, so that

$$M_1 \geq 1. \qquad (5.18)$$

Furthermore, we find that

$$v_2 \leq c_{s2}, \qquad (5.19)$$

and so, in the shock frame of reference, the flow is supersonic in front of the shock but subsonic behind it. This means that information cannot be transmitted ahead of the shock front but can catch it up from behind. Another property is that *the shock must be compressive* with

$$p_2 \geq p_1, \qquad (5.20)$$

$$\rho_2 \geq \rho_1. \qquad (5.21)$$

Equations (5.21) and (5.10) imply in turn that $v_2 \leq v_1$, while Equations (5.21) and (5.17) lead to $T_2 \geq T_1$, so that the *shock wave slows the gas down but heats it up*, converting flow energy into thermal energy in the process.

The above results may be established as follows. From the expression for s, the entropy jump across the shock is

$$s_2 - s_1 = c_v \log \frac{p_2}{p_1} - \gamma c_v \log \frac{\rho_2}{\rho_1},$$

where s_1, p_1, ρ_1, may be regarded as prescribed constants, while p_2 and ρ_2 vary with the Mach number (M_1) according to Equations (5.16) and (5.14). Differentiating with respect to M_1 gives

$$\frac{ds_2}{dM_1} = \frac{c_v}{p_2} \frac{dp_2}{dM_1} - \frac{\gamma c_v}{\rho_2} \frac{d\rho_2}{dM_1},$$

or, after using Equations (5.16) and (5.14),

$$\frac{ds_2}{dM_1} = \frac{4\gamma(\gamma-1)(M_1^2-1)^2 c_v}{M_1[2\gamma M_1^2 - (\gamma-1)][2+(\gamma-1)M_1^2]}. \qquad (5.22)$$

But γ exceeds unity in value and the pressure p_2 must be positive, so that Equation (5.16) gives $2\gamma M_1^2 - (\gamma - 1) > 0$ and Equation (5.22) implies

$$\frac{ds_2}{dM_1} \geq 0. \qquad (5.23)$$

Now, when $M_1 = 1$, we have $p_2 = p_1, \rho_2 = \rho_1$ and $s_2 = s_1$, so the Equations (5.17) and (5.23) lead to the conclusion that $M_1 \geq 1$, as required. The results (5.20) and (5.21) then follow immediately from Equations (5.16) and (5.14), respectively, while Equation (5.19) is a consequence of Equations (5.14) to (5.16), which give

$$\frac{v_2^2}{c_{s2}^2} = \frac{2 + (\gamma - 1)M_1^2}{2\gamma M_1^2 - (\gamma - 1)} = 1 - \frac{(\gamma + 1)(M_1^2 - 1)}{2\gamma M_1^2 - (\gamma - 1)} \leq 1.$$

As the Mach number (M_1) increases from unity to infinity, so the pressure ratio p_2/p_1 (from Equation (5.16)) increases without bound like $2\gamma M_1^2/(\gamma + 1)$, but the density jump (5.14) varies only in the range

$$1 \leq \frac{\rho_2}{\rho_1} < \frac{\gamma + 1}{\gamma - 1}.$$

(For example, a monatomic gas with $\gamma = 5/3$ has a maximum density jump of only 4.)

5.3. Perpendicular Shocks

The simplest type of magnetic shock wave is the perpendicular shock. In this case, the velocities of both the shock and plasma are perpendicular to the magnetic field, which itself is unidirectional and parallel to the shock front. In a frame of reference moving with the shock front (Figure 5.4), the properties of the shocked plasma (v_2, ρ_2, p_2, B_2) are related to those of the unshocked plasma (v_1, ρ_1, p_1, B_1) by the equations for conservation of mass, momentum, energy and magnetic flux, namely,

$$\rho_2 v_2 = \rho_1 v_1, \tag{5.24}$$

$$p_2 + B_2^2/(2\mu) + \rho_2 v_2^2 = p_1 + B_1^2/(2\mu) + \rho_1 v_1^2, \tag{5.25}$$

$$(p_2 + B_2^2/(2\mu))v_2 + (\rho_2 e_2 + \tfrac{1}{2}\rho_2 v_2^2 + B_2^2/(2\mu))v_2$$
$$= (p_1 + B_1^2/(2\mu))v_1 + (\rho_1 e_1 + \tfrac{1}{2}\rho_1 v_1^2 + B_1^2/(2\mu))v_1, \tag{5.26}$$

$$B_2 v_2 = B_1 v_1, \tag{5.27}$$

where the internal energy (e) is given by Equation (2.29b). The first three conservation equations are natural extensions of Equations (5.10) to (5.12) for the hydrodynamic shock, with the gas pressure replaced by the total pressure (plasma plus magnetic) and extra terms $B^2/(2\mu)v$ in Equation (5.26) representing the transport of magnetic energy.

In Equation (5.27) the quantity Bv gives the rate at which magnetic flux is transported across a unit surface area, and this is assumed to be conserved. (An alternative derivation of Equation (5.27) may be obtained as follows. For our steady flow we have curl $\mathbf{E} = 0$, which implies that the tangential electric field component is continuous across the shock, $E_2 = E_1$. But, if the magnetic field is frozen to the plasma on both sides of the shock, the total electric field ($\mathbf{E} + \mathbf{v} \times \mathbf{B}$) vanishes, so that

$$E_2 = -v_2 B_2, \quad E_1 = -v_1 B_1, \tag{5.28}$$

and Equation (5.27) follows.) It is also of interest to note that on each side of the energy equation (5.26) the two terms involving the magnetic field may be combined to give $(B^2/\mu)v$, which by virtue of Equations (5.28) is just the magnitude of the Poynting vector $\mathbf{E} \times \mathbf{H}$.

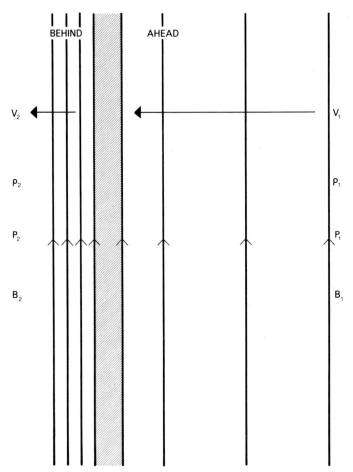

Fig. 5.4. The notation for a plane perpendicular shock wave in the shock frame of reference. The magnetic field is parallel to the wave front and perpendicular to the flow velocity.

After some manipulation, the solution to the set (5.24) to (5.26) may be written in terms of the density ratio

$$\rho_2/\rho_1 \equiv X, \tag{5.29}$$

the shock Mach number,

$$M_1 \equiv v_1/c_{s1}, \tag{5.30}$$

and the plasma beta

$$\beta_1 \equiv \frac{2\mu p_1}{B_1^2} \equiv \frac{2c_{s1}^2}{\gamma v_{A1}^2}, \tag{5.31}$$

where c_{s1} is the sound speed and $v_{A1} = B_1/(\mu\rho_1)^{1/2}$ is the Alfvén speed for the unshocked plasma. The result is

$$v_2/v_1 = X^{-1}, \tag{5.32}$$

$$B_2/B_1 = X, \tag{5.33}$$

$$p_2/p_1 = \gamma M_1^2(1 - X^{-1}) + \beta_1^{-1}(1 - X^2), \tag{5.34}$$

where X is the positive solution of

$$f(X) \equiv 2(2-\gamma)X^2 + (2\beta_1 + (\gamma-1)\beta_1 M_1^2 + 2\gamma)X - \gamma(\gamma+1)\beta_1 M_1^2 = 0. \tag{5.35}$$

The fact that γ lies between 1 and 2 implies that this equation has just one positive root. Furthermore, we note that the solution reduces to the hydrodynamic value (5.14) in the limit of large β_1. The effect of the magnetic field is to reduce X below its hydrodynamic value, since the flow energy can be converted into magnetic energy as well as heat.

Comparable results to those for the hydrodynamic shock can be established. For example, one can show that *the shock is compressive with $X \geq 1$*, which, since $f(X)$ is a quadratic function with a single minimum, implies that $f(1) \leq 0$, or $M_1^2 \geq 1 + (2/(\gamma\beta_1))$.

In terms of the sound and Alfvén speeds this may be written $v_1^2 \geq c_{s1}^2 + v_{A1}^2$, so that the *shock speed (v_1) must exceed the fast magnetoacoustic speed $(c_{s1}^2 + v_{A1}^2)^{1/2}$ ahead of the shock*. The latter speed therefore plays the same role as the sound speed does for the hydrodynamic shock. Furthermore, as M_1 increases without bound so the compression ratio (X) increases to the limiting value $(\gamma+1)/(\gamma-1)$ and the magnetic compression is therefore restricted to the range

$$1 < \frac{B_2}{B_1} < \frac{\gamma+1}{\gamma-1},$$

or for $\gamma = \tfrac{5}{3}$

$$1 < \frac{B_2}{B_1} < 4.$$

5.4. Oblique Shocks

5.4.1. Jump relations

Apart from the special case of the perpendicular shock, the magnetic field contains components both parallel and normal to the shock front. As before, axes are set up in a frame of reference moving with the shock front, and variables ahead of and behind the front are denoted by subscripts 1 and 2, respectively, as indicated in Figure 5.5. Further, the velocity and magnetic field vectors are assumed to lie in the xy plane. Then the equations for conservation of mass, x-momentum, y-momentum, energy and magnetic flux may be written

$$\rho_2 v_{2x} = \rho_1 v_{1x}, \tag{5.36}$$

$$p_2 + B_2^2/(2\mu) - B_{2x}^2/\mu + \rho_2 v_{2x}^2 = p_1 + B_1^2/(2\mu) - B_{1x}^2/\mu + \rho_1 v_{1x}^2, \tag{5.37}$$

$$\rho_2 v_{2x} v_{2y} - B_{2x} B_{2y}/\mu = \rho_1 v_{1x} v_{1y} - B_{1x} B_{1y}/\mu, \tag{5.38}$$

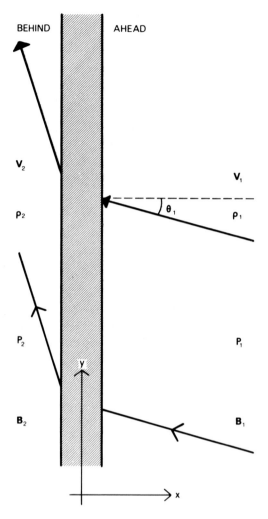

Fig. 5.5. The notation for an oblique shock wave in a frame of reference with two velocity components: the component along the shock normal is the same as the shock front, whereas the speed along the shock front is chosen to make the plasma velocity (**v**) parallel to the magnetic induction (**B**).

$$(p_2 + B_2^2/(2\mu))v_{2x} - B_{2x}(\mathbf{B}_2 \cdot \mathbf{v}_2)/\mu + (\rho_2 e_2 + \tfrac{1}{2}\rho_2 v_2^2 + B_2^2/(2\mu))v_{2x}$$
$$= (p_1 + B_1^2/(2\mu))v_{1x} - B_{1x}(\mathbf{B}_1 \cdot \mathbf{v}_1)/\mu + (\rho_1 e_1 + \tfrac{1}{2}\rho_1 v_1^2 + B_1^2/(2\mu))v_{1x}, \quad (5.39)$$

$$B_{2x} = B_{1x}, \quad (5.40)$$

$$v_{2x}B_{2y} - v_{2y}B_{2x} = v_{1x}B_{1y} - v_{1y}B_{1x}, \quad (5.41)$$

where the internal energy (e) is given by Equation (2.29b). There are some extra terms by comparison with the jump relations (5.24) to (5.27) for the perpendicular shock, in which the components v_y and B_x were absent. In Equation (5.37), B_x^2/μ represents the x-component of the tension force $(B_x \mathbf{B}/\mu)$ acting across a plane $x =$ constant, whereas in Equation (5.38) $B_x B_y/\mu$ represents the y-component of that force and $\rho v_x v_y$ is the rate of transport of y-momentum (ρv_y) across a unit surface area of the shock front. The new term $B_x(\mathbf{B}\cdot\mathbf{v})/\mu$ in Equation (5.39) arises from the rate of working of the magnetic tension, while $(p + B^2/(2\mu))v_x'$ gives the rate of working of the total pressure and $(\rho e + \frac{1}{2}\rho v^2 + B^2/(2\mu))v_x$ represents the rate of transfer of energy (internal, kinetic and magnetic) across the shock front. The conservation of the normal component of magnetic flux B_x (Equation (5.40)) is a consequence of div $\mathbf{B} = 0$. Finally, the continuity of $|\mathbf{v} \times \mathbf{B}|$ (Equation (5.41)) follows from the results that the tangential electric field component is continuous and that the total electric field $(\mathbf{E} + \mathbf{v} \times \mathbf{B})$ vanishes outside the shock front where dissipation is negligible.

An analysis of the jump relations can be considerably simplified by choosing axes moving along the y-axis parallel to the shock front at such a speed that

$$v_{1y} = v_{1x}\frac{B_{1y}}{B_{1x}}. \tag{5.42}$$

(This choice is not possible in the case of the perpendicular shock (Section 5.3), for which B_{1x} vanishes.) The result is that in this frame of reference both sides of Equation (5.41) vanish, and the plasma velocity becomes parallel to the magnetic field on both sides of the shock front. This in turn implies the vanishing of both the electric field (E_z) and the Poynting vector $(\mathbf{E} \times \mathbf{H})$, which represents the total flux of magnetic energy across a surface. But on either side of Equation (5.39) all the terms involving the magnetic field may be combined to give $\mathbf{E} \times \mathbf{H}$, so that their sum vanishes and Equation (5.39) reduces to the purely hydrodynamic form

$$\frac{\gamma p_2}{(\gamma-1)\rho_2} + \frac{v_2^2}{2} = \frac{\gamma p_1}{(\gamma-1)\rho_1} + \frac{v_1^2}{2}.$$

Once the solutions to the jump relations have been derived, a whole family of extra solutions may be obtained by adding the same constant value to both v_{2y} and v_{1y}. Such solutions are discussed by, for instance, Lynn (1966) and Cargill and Priest (1982a).

In terms of the compression ratio $(X \equiv \rho_2/\rho_1)$, the sound speed $(c_{s1} \equiv (\gamma p_1/\rho_1)^{1/2})$ and the Alfvén speed $(v_{A1} \equiv B_1/(\mu\rho_1)^{1/2})$, Equations (5.36) to (5.42) may be combined to give

$$v_{2x}/v_{1x} = X^{-1}, \tag{5.43}$$

$$\frac{v_{2y}}{v_{1y}} = \frac{v_1^2 - v_{A1}^2}{v_1^2 - Xv_{A1}^2}, \tag{5.44}$$

$$B_{2x}/B_{1x} = 1, \tag{5.45}$$

$$\frac{B_{2y}}{B_{1y}} = \frac{(v_1^2 - v_{A1}^2)X}{v_1^2 - Xv_{A1}^2}, \tag{5.46}$$

$$\frac{p_2}{p_1} = X + \frac{(\gamma-1)Xv_1^2}{2c_{s1}^2}\left(1 - \frac{v_2^2}{v_1^2}\right). \tag{5.47}$$

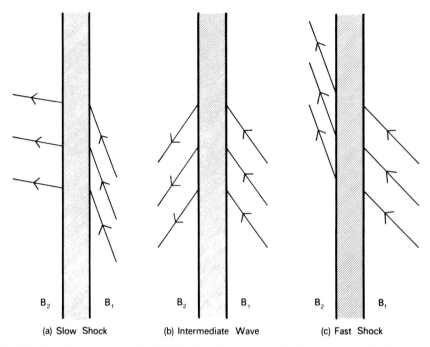

Fig. 5.6. The changes in magnetic field direction that are caused by the three types of oblique wave.

Here X is a solution of

$$(v_1^2 - Xv_{A1}^2)^2 \{Xc_{s1}^2 + \tfrac{1}{2}v_1^2 \cos^2\theta(X(\gamma-1)-(\gamma+1))\} + \tfrac{1}{2}v_{A1}^2 v_1^2 \sin^2\theta\, X \times$$
$$\times \{(\gamma + X(2-\gamma))v_1^2 - Xv_{A1}^2((\gamma+1) - X(\gamma-1))\} = 0, \qquad (5.48)$$

and θ is the inclination of the upstream magnetic field to the shock normal (Figure 5.5) such that $v_{1x} = v_1 \cos\theta$.

Corresponding to the three solutions of Equation (5.48), there are three waves, namely a *slow shock*, an *intermediate (or Alfvén) wave* and a *fast shock* (Figure 5.6). Indeed, in the limit as $X \to 1$, they reduce to the three infinitesimal waves, since Equation (5.48) becomes $v_1^2 = v_{A1}^2$, for an Alfvén wave, together with $v_{1x}^4 - (c_{s1}^2 + v_{A1}^2)v_{1x}^2 + c_{s1}^2 v_{A1}^2 \cos^2\theta = 0$, for the propagation speeds of slow and fast magnetoacoustic waves (see Equation (4.51)).

5.4.2. Slow and Fast Shocks

A full derivation of the properties of oblique shock waves can be found in, for instance, Bazer and Ericson (1959) or Jeffrey and Taniuti (1964). Consider first the slow and fast shocks. They are *compressive*, with

$$X > 1, \qquad (5.49)$$

which from Equation (5.47) implies that $p_2 > p_1$. They also conserve the sign of the tangential magnetic field component, so that B_{2y}/B_{1y} is positive. This makes both the numerator and denominator on the right-hand side of Equation (5.46) either negative or positive. In the first case, we have

$$v_1^2 \leqslant v_{A1}^2 (< X v_{A1}^2), \tag{5.50}$$

and so Equation (5.46) implies that $B_2 < B_1$. This is the basic property of a *slow shock*, for which the magnetic field is refracted *towards* the shock normal and its strength *decreases* as the shock front passes by (Figure 5.6(a)). In the second case,

$$v_1^2 \geqslant X v_{A1}^2 (> v_{A1}^2), \tag{5.51}$$

and we find from Equation (5.46) that $B_2 > B_1$ for the *fast shock*. Here the shock front has the effect of refracting the magnetic field away from the normal and increasing its strength. It can be shown from the evolutionary condition that the *shock speed* (v_{1x}) *relative to the unshocked plasma (ahead) must exceed the characteristic wave speed* for that plasma, namely the slow magnetoacoustic wave speed in the case of a slow shock and the fast magnetoacoustic speed in the case of a fast shock. Also, behind the shock front the flow speed (v_{2x}) must be *less* than the relevant characteristic wave speed. Furthermore, we see from Equations (5.43) and (5.49) that the effect of the shock is to *slow down the flow in the x-direction* $(v_{2x} < v_{1x})$. The flow in the y-direction is slowed down for a slow shock $(v_{2y} < v_{1y})$ but speeded up for a fast shock $(v_{2y} > v_{1y})$.

In the limit as the normal magnetic field component (B_x) approaches zero, so that the field becomes purely tangential, the fast shock becomes a perpendicular shock. On the other hand, the slow shock reduces to a *tangential discontinuity*, for which both the flow velocity and magnetic field are tangential to the plane of discontinuity, since $v_{2x} = v_{1x} = B_{2x} = B_{1x} = 0$. It is therefore simply a boundary between two distinct plasmas, at which the jumps in the tangential components of velocity (v_y) and magnetic field (B_y) are arbitrary, subject only to the condition

$$p_2 + \frac{B_2^2}{2\mu} = p_1 + \frac{B_1^2}{2\mu},$$

that the total pressure be continuous.

Consider lastly the case

$$B_{2x} = B_{1x} \neq 0, \quad v_{2x} = v_{1x} = 0,$$

so that magnetic field lines cross the boundary but there is no flow across it. One trivial solution has the velocity, magnetic field and pressure continuous. But also the density and temperature may be discontinuous in such a way that $p_2 = p_1$. This boundary between plasmas at different densities and temperatures is known as a *contact* (or *entropy*) *discontinuity*.

5.4.3. Switch-off and Switch-on Shocks

Two special cases of slow and fast shocks are of particular interest, namely the so-called *switch-off* and *switch-on shocks* (Figure 5.7). They occur in the limit when the equality holds in Equation (5.50) or (5.51), respectively. When

$$v_1 = v_{A1} \tag{5.52}$$

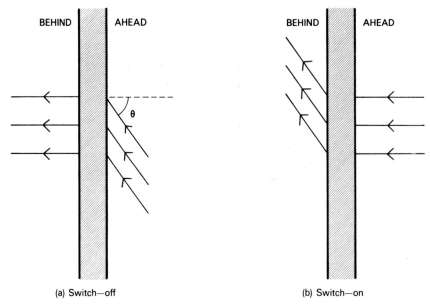

Fig. 5.7. The magnetic field changes for switch-off and switch-on shock waves.

and $X \neq 1$, Equation (5.46) implies that the tangential magnetic field component behind the shock (B_{2y}) must vanish, and so, if B_{1y} is non-zero, we have the *switch-off shock*. Since $\mathbf{v}_1$ and $\mathbf{B}_1$ are parallel by Equation (5.42), the condition (5.52) is equivalent to $v_{1x} = B_{1x}/(\mu\rho_1)^{1/2}$. In other words, a switch-off shock propagates at a speed (v_{1x}) equal to the Alfvén speed based on the normal magnetic field component. Furthermore, Equation (5.48) for the compression ratio $X(\neq 1)$ reduces to

$$(2c_{s1}^2/v_{A1}^2 + \gamma - 1)X^2 - (2c_{s1}^2/v_{A1}^2 + \gamma(1 + \cos^2\theta))X + (\gamma + 1)\cos^2\theta = 0,$$

with exactly one solution greater than unity. When $c_{s1}^2/v_{A1}^2 > \frac{1}{2}$, the solution for X increases from 1 to $1 + (2c_{s1}^2/v_{A1}^2 + \gamma - 1)^{-1}$ as the angle of incidence θ increases from 0 to $\frac{1}{2}\pi$. When $c_{s1}^2/v_{A1}^2 < \frac{1}{2}$, X decreases from $(\gamma + 1)/(2c_{s1}^2/v_{A1}^2 + \gamma - 1)$ to $1 + (2c_{s1}^2/v_{A1}^2 + \gamma - 1)^{-1}$ as θ increases from 0 to $\frac{1}{2}\pi$.

For a shock propagating along the magnetic field (so that B_{1y} and θ vanish), Equation (5.48) reduces to

$$(v_1^2 - Xv_{A1}^2)^2 \{Xc_{s1}^2 + \tfrac{1}{2}v_1^2(X(\gamma - 1) - (\gamma + 1))\} = 0.$$

The slow shock solution for the compression ratio (X) is here given by the vanishing of the expression in curly brackets, and so it represents a purely hydrodynamic shock (Section 5.2). The fast shock solution, on the other hand, is

$$X = \frac{v_1^2}{v_{A1}^2}, \tag{5.53}$$

corresponding to a *switch-on shock*. Since $X > 1$, this occurs only when the shock speed exceeds the Alfvén speed $(v_1 > v_{A1})$ by Equation (5.53). Furthermore, $B_{2x} = B_1$,

and an elimination of p_2 between Equations (5.37) and (5.39) yields
$$B_{2y}^2/B_{2x}^2 = (X-1)\{(\gamma+1)-(\gamma-1)X - 2\mu\gamma p_1/B_1^2\}.$$
Since the right-hand side must be positive, the density ratio has to lie in the range
$$1 < X \leqslant \frac{\gamma + 1 - 2c_{s1}^2/v_{A1}^2}{\gamma - 1}. \tag{5.54}$$

The upper limit is at its largest, namely $(\gamma+1)/(\gamma-1)$, when $v_{A1} \gg c_{s1}$. Moreover, Equation (5.54) implies $v_{A1} > c_{s1}$, which means that a switch-on shock can exist *only when the Alfvén speed exceeds the sound speed* in the unshocked plasma. As X increases from unity, so the deflection of the field lines, as measured by B_{2y}^2/B_{2x}^2, increases from zero to a maximum value of $4(1 - c_{s1}^2/v_{A1}^2)^2/(\gamma-1)^2$ at $X = (\gamma - c_{s1}^2/v_{A1}^2)/(\gamma - 1)$, and thereafter it decreases to zero at
$$X = \frac{\gamma + 1 - 2c_{s1}^2/v_{A1}^2}{\gamma - 1}.$$

5.4.4. THE INTERMEDIATE WAVE

When the wave-front propagates at the Alfvén speed in the unshocked plasma, so that $v_1 = v_{A1}$, one solution to Equation (5.48) is simply $X = 1$. In this case, Equations (5.44) and (5.46) become meaningless, but the more fundamental Equations (5.41) and (5.42) imply
$$\frac{v_{2y}}{v_{1y}} = \frac{B_{2y}}{B_{1y}},$$
while Equations (5.37) and (5.39) give
$$p_2 = p_1, \qquad B_{2y}^2 = B_{1y}^2.$$
Thus, in addition to the trivial solution $\mathbf{B}_2 = \mathbf{B}_1$, we have
$$B_{2y} = -B_{1y}, \qquad B_{2x} = B_{1x},$$
$$v_{2y} = -v_{1y}, \qquad v_{2x} = v_{1x},$$
for an *intermediate* (or *transverse*) *wave* (or *rotational discontinuity*). The tangential magnetic field component is reversed by the wave, and within the wave front the magnetic field simply rotates out of the plane of Figure 5.6(b) while maintaining a constant magnitude. This is therefore just a finite-amplitude Alfvén wave (Section 4.3.1), and, since no change in pressure or density is produced, it is not a shock at all. Also, by contrast with a shock wave, it broadens in time due to ohmic dissipation, and so it may be regarded as possessing a constant width only over times small by comparison with the ohmic decay-time.

CHAPTER 6

HEATING OF THE UPPER ATMOSPHERE

6.1. Introduction

It was Biermann (1946) and Schwarzschild (1948) who first suggested the heating of the upper atmosphere (the chromosphere and corona) by *sound waves* that are generated from turbulence in the convection zone and then steepen to form *shock waves* as they propagate upwards. Until relatively recently this was universally accepted but now it is thought to be important only for the low chromosphere. This chapter first gives a summary of some energy-balance models that have been proposed for the upper atmosphere (assuming a simple form for the heating), and then it proceeds to discuss the processes that may produce the heating. Qualitatively, it is clear that a source of heat is needed to balance not only the energy radiated away in the chromosphere but also the energy removed by conduction from the temperature maximum. Quantitatively, however, it is still uncertain how the heating varies with altitude, and the detailed nature of the heating mechanism is highly controversial. It is also probable that the heating mechanism in the outer corona is collisionless and so beyond the scope of this book.

According to Withbroe and Noyes (1977), in the chromosphere the heating needed to balance radiation is about 4×10^3 W m^{-2} for quiet regions or coronal holes and 2×10^4 W m^{-2} for active regions. In the corona, the required energy input drops to only 3×10^2 W m^{-2} for quiet regions and 5×10^3 W m^{-2} for active regions. Down at the photosphere, an enormous input of wave flux of between 10^4 and 10^6 W m^{-2} is believed to exist (1 W m$^{-2} \equiv 10^3$ erg cm^{-2} s^{-1}), but it is not clear how much of this reaches higher levels.

Indirect observational support for the classical picture of atmospheric heating by acoustic waves had come from spectral line profiles; these are broadened by the presence of *nonthermal velocities* that increase with height from a few km s^{-1} in the low chromosphere to 25 to 30 km s^{-1} in the transition region and 10 to 30 km s^{-1} in the corona. The nonthermal broadening was thought possibly due to waves that were propagating upwards rather than remaining stationary. Indeed, oscillatory motions with a period of 300 s or shorter had been observed through the photosphere and chromosphere up to the low transition region (but not in the corona). Furthermore, the observations by Deubner (1976) suggested the presence near the temperature minimum of an energy flux of at most 10^5 to 10^6 W m^{-2}, carried by waves with periods 10 to 300 s. Even though this wave flux is radiatively damped in the photosphere, only a small fraction needed to propagate upwards to supply enough energy to heat the chromosphere and corona.

More recently, direct evidence from OSO 8 that only the *low chromosphere may be heated by acoustic waves* has been presented by Athay and White (1977). Mein *et al.*

(1980) and Mein and Schmieder (1981) have summarised other observations. At the temperature minimum a 5-min wave-train is observed, with possibly enough energy to heat the low chromosphere; but, by the time the upper chromosphere is reached, its flux has been reduced and its coherence destroyed (Section 6.3.2). *Heating of the upper chromosphere and corona is therefore probably magnetic in nature* (Section 6.4). Magnetic waves present difficulties because they are both scattered off inhomogeneities and refracted as the Alfvén speed increases with altitude; but, provided enough wave-flux can reach the corona, *short-period ($\approx$ 10 s) Alfvén waves appear to be a viable heating mechanism for most coronal loops* (Section 6.4.2), where the magnetic field is less than about 20 G. However, they are difficult to dissipate in stronger magnetic field regions, such as *X-ray bright points* and *active-region cores*. An alternative heating mechanism that does work in a strong field is *magnetic dissipation in current sheets or filaments* (Section 6.4.4). Here the energy is transferred to the corona by motions of such a low frequency that the magnetic field evolves through a series of *quasi-static configurations*. In the corona the energy then dissipates ohmically in the classical way that has long been proposed for solar flares. A continuous occurrence of such 'mini-flares' is proposed as the means by which the corona is heated generally; but a detailed analysis of the process remains to be carried out.

6.2. Models for Atmospheric Structure

6.2.1. Basic model

The temperature increases dramatically from the chromosphere up through the narrow transition region to the corona, in a manner that is shown schematically in Figure 6.1. The gradient of the temperature increases from small values in the chromosphere to extremely large values in the transition region and then decreases to zero at the temperature maximum. The inflexion point in the temperature profile corresponds to the place of maximum temperature gradient and is located typically near the base of the transition region ($\approx 2 \times 10^4$ K), whereas the inflexion point in the profile of $T^{7/2}$ (at T_i) gives the maximum heat flux and occurs typically at about 10^6 K.

At each location a thermal equilibrium,

$$C = H - R, \tag{6.1}$$

has often been assumed to hold between some kind of heating (H), radiative losses ($-R$) and a downwards conductive flux ($\mathbf{F}_c = -\kappa_0 T^{5/2} \nabla T$), whose divergence is the conductive loss (C). Below T_i, C is negative, so that conduction deposits heat and the radiation (R) exceeds heating (H), whereas above T_i C is positive and heating dominates radiation.

When Equation (6.1) holds, the temperature structure is determined by the relative sizes of R and H and by the response of C in maintaining the thermal balance. In the *chromosphere*, R and H are both relatively large, while their difference C is small, so that the spatial change in temperature gradient is slow. As the temperature rises, so the radiation (R) increases to a maximum between 10^4 and 10^5 K (see Table 2.2); here R greatly exceeds the heating (H) and is balanced by C. This in turn forces the temperature gradient to increase to large values in the *lower transition region* (which

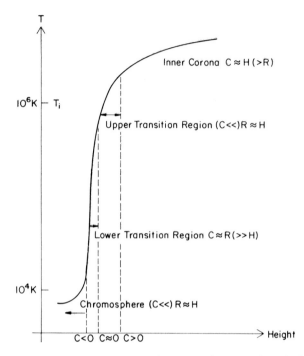

Fig. 6.1. A sketch of the temperature structure in the upper solar atmosphere, indicating the relative roles of conduction (C), radiation (R) and heating (H).

is therefore being heated by conduction from above and cooled by radiation). Above about 10^5 K the radiation falls dramatically and eventually it reaches equality with heating. Then, through the *upper transition region*, the conductive flux $\mathbf{F}_c$ stays relatively constant and at a high value. In the *lower corona* R and H have fallen to much smaller values and heating is balanced mainly by conductive losses, the relatively small value of C implying slow changes of temperature gradient with height. Above the temperature maximum of an open magnetic field region, the energy transport by the solar wind becomes increasingly important and eventually it dominates conduction of heat outwards.

Thus it can be seen that the cause of the extremely steep rise of temperature in the lower transition region is the fact that the *radiation around 10^4 to 10^5 K is so large that it cannot be supplied by mechanical heating* but must be provided by conduction from above. Furthermore, at greater altitudes the energy that is deposited as *heat cannot be radiated away* and so it must be conducted both inwards and outwards from the temperature maximum.

Typical values for the coronal temperature and transition-region pressure are given in Table 6.1 (from Withbroe and Noyes, 1977) along with estimates for the conductive and radiative losses (C and R) per unit area at different levels. The heating (H) that is required may be obtained by summing the losses (C and R). The energy necessary to heat the corona is typically only a few percent of that needed down in the chromosphere, so a comprehensive model of atmospheric heating would have to

TABLE 6.1.
Energy losses from the upper atmosphere ($1 \text{ W m}^{-2} \equiv 10^3 \text{ erg cm}^{-2} \text{ s}^{-1}$)

	Conduction (W m^{-2})	Radiation (W m^{-2})	Temperature (K)	Pressure (N m^{-2})
Quiet region:				
Lower and middle chromosphere		4×10^3		
Upper chromosphere		3×10^2		2×10^{-2}
Corona.	2×10^2	10^2	$1.1 - 1.6 \times 10^6$	
Coronal hole:				
Lower and middle chromosphere		4×10^3		
Upper chromosphere		3×10^2		7×10^{-3}
Corona	6×10	10	10^6	
Active region:				
Lower and middle chromosphere		2×10^4		
Upper chromosphere		2×10^3		2×10^{-1}
Corona	$10^2 - 10^4$	5×10^3	2.5×10^6	

treat the generation, propagation and dissipation of energy through the whole region to a high degree of accuracy. Table 6.1 also shows the extent to which a coronal hole region is both cooler and less dense than a normal quiet region of the Sun, while an active region is both hotter and denser.

A simple numerical model for the upper atmosphere may be computed by assuming forms for the terms in (6.1). For example, Wragg and Priest (1981b) have adapted an earlier model of McWhirter *et al.* (1975). They solve

$$\frac{d}{dz}\left(\kappa_0 T^{5/2} \frac{dT}{dz}\right) = \chi n_e^2 T^\alpha - H \tag{6.2}$$

for the temperature profile ($T(z)$) as a function of height, and use the standard values for the conduction coefficient (κ_0) (Equation 2.34) and the optically thin radiative-loss coefficients χ and α (Table 2.2). The heating (H) is assumed uniform per unit volume for simplicity, and Equation (6.2) is coupled with the equation of hydrostatic equilibrium for a fully-ionised hydrogen plasma, namely

$$\frac{dp}{dz} = -m_p n_e g, \tag{6.3}$$

where $p = 2n_e k_B T$. Three boundary conditions are needed to solve Equations (6.2) and (6.3), such as the prescription at the base of temperature, temperature gradient and electron density. The resulting profiles depend on one parameter alone, namely H, some examples being shown in Figure 6.2(a). It is particularly interesting to discover how the temperature maximum and the height at which it occurs vary with

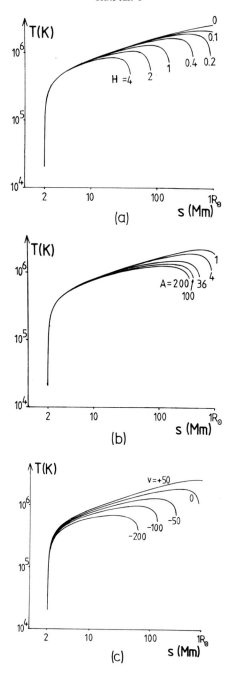

Fig. 6.2. The temperature (T) of a model coronal atmosphere as a function of height s (1 Mm $\equiv 10^6$ m), showing the effect of varying: (a) the heating strength H, (b) the flux tube divergence for $H = 0.2$, (c) a flow for $H = 0.2$. The heating H is measured in units of the radiation at a temperature of 10^6 K and a density of 5×10^{14} m^{-3}. A is the ratio of the loop area at the summit to that at the base. v is the flow speed in m s^{-1} (+ represents an upflow and − a downflow) (from Wragg and Priest, 1981b).

H. If these models were linked to a coronal hole model, the rapid fall in temperature after the temperature maximum would be replaced by a more gradual decline. In Figures 6.2(b) and 6.2(c) the effects of varying the cross-sectional area and flow speed are also displayed.

6.2.2. MAGNETIC FIELD EFFECTS

It is clear from eclipse or X-ray pictures of the Sun, such as Figures 1.3(d), 1.4 and 1.15, that coronal structure is dominated by the magnetic field. Regions of open field show up as dark *coronal holes*, whereas closed-field regions are seen as bright *coronal loops*. The influence of the magnetic field on coronal plasma is threefold.

(i) *It exerts a force.* The **j** × **B** force is able to act inward and so contain plasma with an enhanced pressure in features such as *X-ray bright points*, coronal loops and active regions.

(ii) *It stores energy.* The energy ($B^2/2\mu$) per unit volume) that is stored in the magnetic field may provide an extra source of heating, either by allowing *additional wave modes* (Section 4.3) that eventually dissipate or by being released directly by ohmic dissipation (j^2/σ) in regions where electric currents are strong.

(iii) *It channels heat.* The coefficient of thermal conduction ($\kappa_\parallel$) along the field is much larger than the coefficient ($\kappa_\perp$) across the field, so the magnetic field acts as a 'blanket' and thermally insulates the plasma very effectively. Heat is constrained to flow largely along the field, which means that in the transition region and corona, where conduction is an important means of energy transport, the temperature and density are strongly affected by the structure of the magnetic field. This is one reason why the *coronal structures in eclipse and X-ray pictures probably outline the magnetic field*. A discussion of such coronal loops can be found in Section 6.5.

A model has been developed by Gabriel (1976) (from an earlier attempt of Kopp and Kuperus, 1968) for the atmosphere above a (quiet-region) *supergranule cell*, typically 30 000 km in diameter. In the photosphere, magnetic flux is concentrated at supergranule boundaries by the convective flow and instability (Section 8.7), but higher up in the atmosphere the flux expands until it has become relatively uniform at the corona (Figure 6.3). Thus, images formed in transition-region lines follow the supergranulation pattern, with intensities over supergranule boundaries about a factor of ten higher than those over cell centres, while at the corona the pattern has disappeared (Figure 1.14). Gabriel assumes for simplicity that **∇** × **B** = **0**, so that the magnetic field is a potential one. He next assumes the plasma to be in thermal equilibrium under a balance

$$\frac{\mathrm{d}}{\mathrm{d}s}\left(\kappa_\parallel A(s)\frac{\mathrm{d}T}{\mathrm{d}s}\right) = \chi n_e^2 T^\alpha A(s)$$

between only conduction and radiation, where *A* is a flux-tube cross-sectional area and local heating has been ignored; heat is supposedly deposited at greater heights. This equation is solved along each field line together with the Equation (6.3) of hydrostatic equilibrium, subject to the boundary conditions that the density, temperature and conductive flux be the same on each field line in the corona at an altitude of 30 000 km. A coronal conductive flux of 360 W m^{-2} is found to give closest agree-

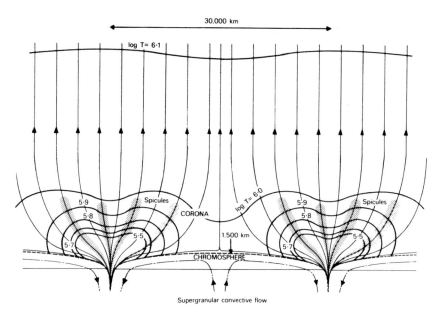

Fig. 6.3. Magnetic field lines and temperature contours for the atmosphere above a supergranule cell in a quiet region (after Gabriel, 1976).

ment with observation for the network width, and the resulting isotherms are sketched in Figure 6.3. By comparison with a plane-parallel model, the effect of the above flux-tube divergence (through $A(s)$) is to increase the temperature gradient at transition-region temperatures and so lower the height at which coronal temperatures are attained. Furthermore, the observed intensities of optically thin lines may be used to derive the differential emission measure ($n_e^2 \, T \, \mathrm{d}h/\mathrm{d}T$) as a function of temperature. At temperatures between $10^{5.2}$ and $10^{6.2}$ K, agreement with this emission measure is much better for Gabriel's model than for McWhirter et al. (1975)'s previous spherically symmetric models. Below $10^{5.2}$ K there is a need to include heating of amount 2×10^3 W m^{-2}.

Gabriel's model has recently been extended by Athay (1981b), who includes gravitational energy and enthalpy flux but no mechanical heating. With a downflow he obtains good agreement with observations for $3 \times 10^5 \text{ K} \leqslant T \leqslant 10^6 \text{ K}$, and so concludes that there is no need for mechanical heating. In future, there is a need to calculate a wider range of models and to couple the energy balance with a magnetostatic force-balance, since the plasma beta is probably of order unity.

6.2.3. Additional effects

Several effects may seriously modify the energy balance in the upper solar atmosphere but they are normally omitted from the models. For example, the waves that may be propagating up from below and heating the atmosphere exert a turbulent pressure, $\langle \rho v^2 \rangle$, which is just the time-average over a wave period of the momentum flux

(ρv^2). This *wave pressure* may well exceed the plasma pressure, and so it needs to be included in calculating the overall hydrostatic equilibrium. It may even overcome gravity and cause the plasma to flow outwards.

Most models of the thermal energy balance assume a static plasma and give a transition-region thickness of only 10 km or so, but in practice the region is probably in a *dynamic state*, with an effective thickness of a few thousand km. For example, material is continually ejected upwards as *spicules* from the chromosphere at supergranule boundaries, and persistent *downflows* of hotter plasma at 5 to 100 km s^{-1} are observed over supergranule boundaries, sunspots and plages. (It may even be the case that spicules are not just superimposed on a static atmosphere, but rather that essentially all of the transition-region emission is from the hot surroundings of spicules.) As cool spicular material moves upwards, it is heated and so extracts energy from its surroundings; when it falls back down, it carries thermal energy with it. Typically, one finds a downward mass flux at 10^5 K over supergranule boundaries of 5×10^{19} m^{-2} s^{-1} for a density of 5×10^{15} m^{-3} and a speed of 10 km s^{-1}. Pneuman and Kopp (1978) estimate that the heat carried down by such motions can exceed the heat transport by conduction, and so they construct a transition-region model based on a balance between heat downflow and radiation. The effect of a downflux (q) on the model of Wragg and Priest (1981b) can be seen in Figure 6.2(c), for which the term $d/dz(5kTq)$ has been added to the right-hand side of Equation (6.2).

6.3. Acoustic Wave Heating

Acoustic waves are believed to be *generated* near the photosphere (Section 4.9.3) and to *steepen* into shock waves at an altitude of a few hundred kilometres. They continue to *propagate* upwards and (according to some models) they may *dissipate* enough energy to balance radiation from the chromosphere (e.g., Schatzman, 1949; Kuperus, 1969; Ulmschneider, 1971, 1974, 1979; Kuperus and Chiuderi, 1976). The models predict that waves with a period of only a few tens of seconds heat the lower chromosphere. At one time, it was thought that 300 s waves could heat the upper chromosphere, but now their flux is thought to be too low (Ulmschneider, 1976). Also, by the time the corona is reached, the uncertainties are very large, since weak shock theory is no longer valid and there is considerable reflection and refraction in the transition region.

6.3.1. Steepening

Section 5.1 describes the way in which acoustic waves steepen to form shocks, and Section 5.2 includes a derivation of the jump conditions across a shock front. In a *uniform medium*, sound waves steepen because every part of the wave profile moves with a different speed. The crests possess a higher temperature than the troughs, and so they propagate faster. If c_s is the ambient sound speed and v_1 the velocity amplitude (Section 4.2), the crest of the wave moves with speed $c_s + v_1$, while the trough moves at only $c_s - v_1$. The crest therefore catches up the trough when the relative speed of $2v_1$. Since the trough is initially half a wavelength ($\frac{1}{2}\lambda$) ahead of the crest, the time it takes the trough to be overtaken is just $\lambda/(4v_1)$, and so the distance that a sound wave can travel at speed c_s before shocking is

$$d = \frac{\lambda c_s}{4v_1}, \tag{6.4}$$

or, in terms of the wave-period ($\tau = \lambda/c_s$),

$$d = \frac{\tau c_s^2}{4v_1}. \tag{6.5}$$

From this result, it can be seen that short-period waves evolve into shock waves over much smaller distances than long-period waves, which is why the former were thought to heat the low chromosphere and the latter the higher atmosphere.

For a *vertically stratified atmosphere*, rather than a uniform medium, the distance for shock formation is greatly reduced because the wave-amplitude increases rapidly with altitude. The density for an isothermal atmosphere decreases with height like $\rho(z) \sim e^{-z/\Lambda}$, where Λ is the scale-height. With no dissipation, the total *wave energy* is proportional to $\frac{1}{2}\rho v_1^2$; this remains constant as the wave propagates up, and so the wave amplitude increases as

$$v_1(z) \sim e^{z/(2\Lambda)}. \tag{6.6}$$

Thus, for example, taking a scale-height of 100 km, a wave starting in the low chromosphere would find its initial amplitude of only 0.2 km s^{-1}, say, grow to the sound speed (about 7.5 km s^{-1}) at an altitude of 1000 km. The distance required for a shock to form in the stratified medium becomes

$$d = 2\Lambda \log_e \left(1 + \frac{\tau c_s^2}{2(\gamma + 1)\Lambda v_1}\right)$$

in place of Equation (6.5), according to Stein and Leibacher (1974). For instance, setting $\gamma = 5/3$, $c_s = 6$ km s^{-1}, $\Lambda = 130$ km, a wave with an initial amplitude (v_1) of 0.6 km s^{-1} at the base of the chromosphere would develop into a shock at an altitude of 500 km if its period were 10 s or 800 km for a period of 30 s. This makes it most unlikely that the heating of the chromosphere is due to damping (by thermal conduction) of small-amplitude linear waves, since they develop into shocks too quickly.

The height of shock formation is in practice governed by the extent to which the wave is damped (as well as the height of generation, the initial amplitude and the wave-period). A wave can dissipate its energy through viscous, thermal or radiative losses, the last being the most important up to a height of about 1000 km (Osterbrock, 1961). Ulmschneider (1971) has estimated the radiative damping time to increase from 30 s at $h = 0$ in the photosphere to 750 s at $h = 400$ km and thence to decrease to 300 s at $h = 1000$ km, for the Harvard–Smithsonian Reference Atmosphere. His calculations demonstrate that sound waves with a period less than 100 s can propagate into the chromosphere and steepen into shocks below a height of 1000 km. Later, Ulmschneider and Kalkofen (1977) found that the effect of radiative damping is to make the height of shock formation independent of period.

6.3.2. Propagation and dissipation

Three approximate theories have been developed to describe the propagation of a shock wave through an inhomogeneous atmosphere. *Geometrical acoustics* (Jeffrey and Taniuti, 1964) makes the assumption that the shock energy remains constant, although a correction to include the dissipation can be added easily when the atmosphere has weak gradients. The *Chisnell–Witham method* (Witham, 1974) replaces an inhomogeneous atmosphere by a series of homogeneous layers. It neglects the reflected waves that are generated at each interface as the shock passes through, and also it does not include the effect of dissipation on the shock strength. However, a more useful theory for obtaining the dissipation from *periodically generated shocks* is that of Brinkley and Kirkwood, as developed by Schatzman (1949), Osterbrock (1961) and Ulmschneider (1971). It has been described by Bray and Loughhead (1974) and is outlined below.

Consider a train of shock waves that have developed from waves of frequency v, so that the time interval between successive shocks is v^{-1}. The flux of energy transmitted per second by the shocks at an altitude z_1 is the work done by the pressure, namely $F(z) = v \int (p - p_1) v \, dt / \int dt$, where the integrals are performed over one period; the front of the shock and its rear are denoted by subscripts 1 and 2, respectively. The energy flux may be rewritten

$$F(z) = v(p_2 - p_1)v_2 t_0 / \tau \tag{6.7}$$

in terms of a characteristic time (t_0) that is assumed (for self-similarity) to be independent of height. t_0 depends on the shock profile and is taken as $\tau/12$ by Ulmschneider (1971), where τ is the duration of the shock pulse. The jump conditions (5.14) to (5.16) may be used to write the energy flux (6.7) in terms of the fractional compression $\bar{\eta} = (\rho_2 - \rho_1)/\rho_1$. For *weak shocks* $(\rho_2 \approx \rho_1)$, it becomes

$$F(z) = v \rho_1(z) c_{s1}(z)^3 \bar{\eta}(z)^2 / 12 . \tag{6.8}$$

Now, following the passage of a shock, some energy is used in returning the plasma to its initial state. Schatzman suggested that it first expands adiabatically to its original pressure and then cools by radiation to its original density, ready for the passage of the next shock. For weak shocks, the resulting rate of energy dissipation is

$$\frac{-v\gamma(\gamma+1)p_1\bar{\eta}^3}{12} \equiv \frac{dF}{dz}. \tag{6.9}$$

Since this is proportional to the wave frequency, short-period waves will dissipate faster than long-period ones. The corresponding rate of decrease of the peak pressure is

$$\frac{dp_2}{dz} = \frac{dp_1}{dz} - \frac{v\gamma^2(\gamma+1)p_1^2\bar{\eta}^4}{12F},$$

which, when solved simultaneously with Equation (6.9), leads to a decrease of both p_2 and F with altitude like $z^{-1/2}$. An alternative procedure is to differentiate Equation (6.8) with respect to z and equate it to Equation (6.9). The resulting differential equation for the shock strength $(\bar{\eta}(z))$ has been solved numerically by Ulmschneider (1971) for a model atmosphere, including the additional effect of shock refraction away from the vertical. He deduces the dissipation as a function of height for a range of periods, and he finds agreement between this energy dissipation and the radiative loss in the *lower chromosphere* only if the shock period lies between 10 and 30 s. The necessary shock strengths ($\bar{\eta}$) lie between 0.1 and 0.5, so that the weak shock approximation is justified. In order of magnitude, the distance it takes a shock to dissipate significantly, the so-called *shock damping-length*, is

$$d \approx \frac{F}{dF/dz} \approx \frac{c_s t_0}{\bar{\eta}} \tag{6.10}$$

from Equations (6.7) and (6.9).

It may be noted that, in order to use the general energy equation (2.30) in modelling, one needs to insert a form for the heating. When heating is from weak acoustic shocks, it is given by Equation (6.9) in terms of the shock strength ($\bar{\eta}(z)$). If, furthermore, the shock *Mach number* ($M_1(z)$) is known as a function of altitude, the heating may be deduced from

$$H = \tfrac{2}{3} v \gamma (\gamma + 1) p_1 (M_1 - 1)^3, \tag{6.11}$$

and so, for constant M_1, it is proportional to the plasma pressure. This differs from the heating forms that are sometimes used in static models, namely heating uniform or proportional to density. There is always a definite relation between $\bar{\eta}$ and M: in general, it can be obtained by integrating Equations (6.8) and (6.9), but for weak shocks $\bar{\eta} = (4(M_1 - 1)/(\gamma + 1))$.

Short-period acoustic waves (10 to 50 s) are, according to the above theory, expected to develop into weak shocks within a few hundred kilometres of the chromospheric base and to heat the low chromosphere. This heating mechanism has recently been put on a firmer foundation by some numerical calculations of Ulmschneider and Kalkofen (1977) with an improved hydrodynamic code including radiative damping. The height of shock formation agrees with the position of the temperature minimum (namely 500 km for the Harvard–Smithsonian Reference Atmosphere) provided the wave period lies between 25 and 45 s and the initial acoustic flux is 3 to 6×10^4 W m^{-2}. Furthermore, the acoustic flux at the height of shock formation agrees with the chromospheric radiation provided the period is less than 35 s and the initial flux is 2 to 6×10^4 W m^{-2}.

Longer-period waves were once expected to increase in amplitude as radiative damping becomes less effective above 200 km and to develop into *strong shocks* with $\bar{\eta} > 1$ in the upper chromosphere. (Indeed, the shocks may well be strong lower in the atmosphere, so that weak-shock theory is invalid and gives no more than a qualitative picture.) Self-consistent models for heating the upper chromosphere and corona by strong shocks have been computed by, for instance, Kuperus (1965) and Ulmschneider (1971) assuming hydrostatic equilibrium and a thermal balance between conduction, radiation and heating. However, observations from the OSO 8 satellite (Athay and White, 1977) give strong evidence that 300-s waves have insufficient energy to heat the upper chromosphere, because their amplitudes have been reduced either by scattering from chromospheric inhomogeneities such as spicules and fibrils or by refraction in the region of rapidly increasing sound speed (see Section 6.4.1). In the low transition region at a temperature of 10^5 K the observed fluctuations are largely aperiodic and the shocks are very weak with an energy flux of only 10 W m^{-2}. Thus, although *short-period acoustic shock waves may well heat the low chromosphere*, some form of *magnetic heating is probably needed for the upper chromosphere and corona* (i.e., resonant absorption or tearing instability).

6.4. Magnetic Heating

The importance of the magnetic field in heating the solar atmosphere is being increasingly recognised, and one expects this subject to receive a great deal of attention in the future. At first, it was believed that the magnetic field had the secondary role of just enhancing the heating in active regions and in the network. But the discoveries of *intense kilogauss fields* (Sections 1.3.2B and 8.7) at *supergranulation boundaries* in the photosphere, and of coronal loops and bright points in soft X-ray photographs, have emphasised the dominance of the magnetic field in these regions too. Furthermore, the probability that insufficient acoustic flux reaches the upper chromosphere and corona has led to the suggestion that, even for the 'quiet' corona, the dominant heating mechanism is due to the magnetic flux that spreads upwards from supergranulation boundaries. The heating of both quiet and active regions would then be caused by the same mechanism, with active regions receiving more heat simply because the concentration of intense flux elements is higher. Two of the possibilities for magnetic heating are *magnetic waves* and direct *magnetic dissipation*. The actual cause of heating is the same in both cases, namely ohmic or viscous dissipation in small-scale regions, but the means of producing the regions is different, so that in one case the current sheets are propagating (i.e., magnetic shocks) and in the other case they are non-propagating.

Magnetic field disturbances are generated by the motion of the footpoints of field lines in the photosphere. For instance, the *intense fields* at supergranulation boundaries are continually being buffeted by granulation with a period of roughly 5 min and a scale size of about 1000 km. Also, the footpoints are being shuffled around on a supergranular time-scale of many hours. A general photospheric disturbance produces waves of several types (such as fast and slow magnetoacoustic waves and Alfvén waves) that may propagate upwards. The *magnetoacoustic waves* steepen into shocks and dissipate in a similar manner to pure acoustic waves, with ohmic heating providing extra dissipation. However, *Alfvén waves* dissipate much less readily. They require a nonlinear interaction to produce magnetoacoustic waves that subsequently relinquish their energy. Areas in need of further study are the way in which the generation of wave flux is affected by the magnetic field, the nonlinear coupling of the different wave modes and the propagation of magnetic waves in an inhomogeneous medium.

The classical linear treatment for the propagation of a photospheric disturbance as a wave through a uniform or slowly varying medium becomes inadequate in four situations:

(i) when the initial disturbance is so large that nonlinear effects are important;
(ii) when the source of the disturbance is closer than a few wavelengths, so that a wave-train has not developed;
(iii) when the disturbance is so slow that its wavelength exceeds a scale-height, i.e., $\lambda > \Lambda$, in which case the ambient medium cannot be considered slowly-varying;
(iv) when the footpoints of magnetic field lines move more slowly than the Alfvén travel-time (τ_A); in other words

$$\tau > \tau_A = \frac{L}{v_A}, \tag{6.12}$$

or, equivalently, the wavelength exceeds the dimension (L) of the configuration:

$$\lambda > L. \tag{6.13}$$

In case (iv) the magnetic configuration evolves passively through a series of equilibria, set up by the relatively fast propagation of magnetic waves. However, the new equilibria can contain *current sheets*, which then dissipate magnetic energy ohmically and allow a *reconnection of the magnetic field lines* in the process. This has long been proposed as a mechanism for releasing magnetic energy as heat and kinetic energy in solar flares (see Chapter 10), but it may be taking place more often on a smaller scale throughout the corona. Such mechanisms for heating plasma in a magnetic field by waves or current-sheet dissipation are sketched in the sections that follow, and they are also described in the reviews by Heyvaerts and Schatzman (1980), Chiuderi (1981), Kuperus *et al.* (1981) and Hollweg (1981a).

6.4.1. Propagation and dissipation of magnetic waves

In a classic paper, Osterbrock (1961) analysed the effect of the magnetic field on the generation, propagation and dissipation of the waves that may heat the chromosphere and corona. The analysis is limited to low-amplitude waves, with a wavelength much smaller than a scale-height. For quiet regions, where the mean magnetic field may be 2 G, upward-moving sound waves become increasingly magnetohydrodynamic in character as the fall-off in density makes the Alfvén speed increase; indeed, above a height of about 2000 km the Alfvén speed (v_A) exceeds the sound speed (c_s). For *plage regions*, where the mean field strength is typically 50 G, the magnetic field dominates even more and may cause the enhanced heating that is observed. Furthermore, the recent discovery of *kilogauss fields* at supergranulation boundaries (even in 'quiet' regions) makes the magnetic field less homogeneous than Osterbrock supposed and increases its effect on wave propagation.

It will be remembered that *fast magnetoacoustic waves* can propagate in any direction, with a phase speed that varies from the maximum of c_s and v_A along the magnetic field to $(c_s^2 + v_A^2)^{1/2}$ across the field. Thus, in a region where $v_A \gg c_s$, propagation is at v_A equally in all directions. The *slow mode*, by comparison, can transmit energy only in directions that are close to that of the magnetic field, the speed of propagation being the smaller of c_s and v_A. The third type of wave, namely the Alfvén mode, possesses a group velocity of v_A along the field and produces no change in either density or pressure.

Osterbrock points out that, in regions where the magnetic field strength is below the equipartition value of a few hundred Gauss, most of the wave energy that is generated by isotropic turbulence in the convective zone takes the form of fast-mode waves. Much smaller fluxes of slow-mode or Alfvén waves are generated, and so he proposes the *dissipation of fast shocks* as the dominant heating mechanism in the chromosphere.

The propagation and dissipation of *fast waves* differs quantitatively from that of ordinary sound waves in several respects, according to Osterbrock. Consider, for example, the refraction of waves as they propagate upwards in the stratified solar atmosphere. Since the acoustic speed (c_s) increases with height (from, say, 10 km s^{-1}

at the photosphere to 200 km s^{-1} in the corona), sound waves find their directions of propagation rotate away from the vertical as they progress upwards, in much the same way that light is refracted away from the normal as it crosses from water to air. However, the effect on the Alfvén speed (v_A) of the fall-off in plasma density away from the solar surface is to make the Alfvén speed increase even more rapidly with altitude (from typically 10 km s^{-1} at the photosphere to 10^3 km s^{-1} in the corona). Thus the refraction of fast-mode waves as they propagate upwards is even more pronounced than that of sound waves, so that even less energy is likely to reach coronal layers. Osterbrock calculated the *ray paths* in a vertically stratified plasma as follows. As the phase speed ($v_p(h)$) increases with height h (with the wave frequency remaining constant), so the wavelength increases. But, by Snell's Law, the horizontal component of **k** must remain constant, and this makes $\theta(h)$, which is the inclination to the vertical of the propagation direction, increase in such a way that $\sin \theta(h)/v_p(h) = $ constant. The resulting ray path ($h(x)$) taken by the wave in this approximation follows from integrating $dh/dx = \tan \theta$.

The damping, prior to shocking, of upward-propagating *magnetic waves* may also differ from that of acoustic waves. It is caused by *ohmic* as well as *viscous dissipation* and, near the temperature-minimum region, by *ambipolar diffusion* too. For a wavelength λ, the time-scale for *ohmic dissipation* is simply $\tau_d = \lambda^2/\eta$, where η is the magnetic diffusivity, and so the distance that a wave can travel at the Alfvén speed, say, before it dissipates (the so-called *damping length*) is

$$L_d = v_A \tau_d = \frac{v_A \lambda^2}{\eta}.$$

In terms of the wave frequency (ω), this may be written $L_d = v_A^3/(\eta \omega^2)$, from which it follows that dissipation is smallest for waves at the lowest frequency and in regions of the highest magnetic field strength (and so largest Alfvén speed). Osterbrock found that in the *quiet Sun* a *weak field* of 2 G would imply strong damping of Alfvén waves or slow-mode waves (with a frequency $\omega = 1.2 \times 10^{-2}$ Hz) at the low chromosphere. However, the fast-mode waves suffer negligible attenuation, since they are basically acoustic in the photosphere and therefore subject to only viscous dissipation, which is less effective than ohmic dissipation. In *plage regions*, where the field strength exceeds 50 G, Alfvén waves are able to propagate up through the chromosphere with negligible damping and so contribute to the heating above, provided an efficient dissipation mechanism exists there.

There are two effects which make *fast magnetoacoustic waves steepen into shock waves more slowly* than sound waves. When $v_A \gg c_s$, the distance that a fast wave needs to travel before shocking in a uniform medium is $d = \tau v_A^2/(4v_1)$, by analogy with Equation (6.5), and so it is larger than for sound waves. Furthermore, in a stratified isothermal atmosphere the density falls off as $\rho(z) \sim e^{-z/\Lambda}$, and the wave energy flux is proportional to $(\frac{1}{2}\rho v_1^2)v_A$, where v_A is the group velocity. Since this flux remains constant as the wave propagates up and the wave speed (v_A) is proportional to $\rho^{-1/2}$, the wave velocity amplitude increases as $v_1(z) \sim \rho^{-1/4} \sim e^{z/(4\Lambda)}$, which represents a much slower rate than that given by Equation (6.6) for sound waves.

Osterbrock extended the Brinkley–Kirkwood analysis for weak acoustic-shock dissipation to fast-mode shocks in a straight-forward manner. One effect of the magnetic field is to introduce extra terms into the expression (6.7) for the energy flux, due to the additional work done by the magnetic tension and pressure. The resulting *damping length* for fast shocks is just $d \approx v_f t_0 / \bar{\eta}$, in which the sound speed (c_s) in Equation (6.10) has been replaced by the fast-mode speed (v_f). For a magnetic field of 2 G, Osterbrock found the dissipation peaks at heights of 1000 to 2000 km and is sufficient to account for heating of the low chromosphere. By the time the corona is reached, however, the calculated direct fast-mode flux is far too small to provide the necessary dissipation, and so he suggested that heating of the upper chromosphere may be by slow-mode waves that are generated from the interaction of the fast-mode shocks.

The heating of a coronal loop by fast magnetoacoustic waves of high frequency ($\approx 3 \text{ s}^{-1}$) has been suggested by Habbal *et al.* (1979). Such waves propagate across the magnetic field and are refracted into regions of low Alfvén speed, where they suffer significant collisionless damping if β is larger than about 0.1. The required wave flux for such (as yet unobserved) waves is about 100 W m^{-2} at the coronal base. Provided that the wavelength is much smaller than the length-scale for Alfvén speed variations, the ray paths ($\mathbf{r}(t)$) satisfy

$$\frac{d\mathbf{r}}{dt} = \frac{\partial \omega}{\partial \mathbf{k}}, \quad \frac{d\mathbf{k}}{dt} = -\frac{\partial \omega}{\partial \mathbf{r}},$$

where the dispersion relation is $\omega = k v_A$ when $\beta \ll 1$. The authors extend Osterbrock's analysis for a plane-parallel atmosphere by considering propagation in a dipole magnetic field that contains isothermal plasma in hydrostatic equilibrium. Sample results are presented for a large loop that reaches a height of half a solar radius and contains plasma at about twice the ambient density. If β is large enough in the loop, the required dissipation occurs in a loop-shaped region coming up from the magnetic loop foot-points and passing below its summit. However, the effect of gravity is to make these waves evanescent, and so the coronal amplitudes are significant only for large horizontal wavelengths; furthermore, for such long waves a non-local analysis is required. Zweibel (1980) has evaluated the damping of fast waves in more detail and considered the energy balance between radiation and heating alone; she finds the equilibrium is unstable thermally (Section 7.5.7) to the formation of cool filaments parallel to the field. Furthermore, Hollweg (1981b) has pointed out two difficulties with fast modes: they are efficiently reflected at the transition region and so may never reach the corona; also, for values of ω and horizontal wavenumber corresponding to observed photospheric motions, fast waves should be evanescent.

6.4.2. Nonlinear Coupling of Alfvén Waves

The corona may possibly be heated by *Alfvén waves*. Osterbrock considered such low magnetic field strengths that the Alfvén waves would be strongly damped in the photosphere and so would need to be generated higher up by interactions of magnetoacoustic waves. However, it is now thought likely that a large flux of Alfvén waves is generated at *supergranulation boundaries*, where the presence of *intense kilogauss fields* allows them to penetrate the photosphere. They then propagate way up into the corona with hardly any attenuation at all. In fact, the damping of Alfvén waves for fields larger than 10 to 100 G is so small that the problem is to explain how they give up their energy before propagating away, either into the solar wind along open field lines or back down to the photosphere along closed fields.

An Alfvén wave is likely to dissipate in practice because of its *nonlinear interaction* with either the non-uniform ambient field or another Alfvén wave (Wentzel, 1974). The method used to analyse the *effect of magnetic field inhomogeneity* depends on the wave-period $\tau = \lambda/v_A$. For a magnetic field of 10 G and a density of 10^{16} m^{-3}, characteristic of the active-region corona, the Alfvén speed (v_A) is about 300 km s^{-1} according to Equation (2.48b). Wavelengths (λ) of 100 000 km, comparable with the coronal scale-height, then correspond to wave-periods of 5 min. For periods much less than 5 min, the wavelength is much smaller than a scale-height and so the Alfvén waves propagate in a 'slowly varying' medium. As they propagate around a bend in the magnetic field much of their energy is converted to fast-mode waves (provided the waves are not well collimated along the field). They in turn decay rapidly and would *brighten the bends in a magnetic configuration* such as a coronal loop or the region above a supergranule cell. For wave periods of order or greater than 5 min, there are large variations in the ambient medium over a wavelength and the 'slowly-varying' approximation fails. Significant dissipation is still expected, but the details have not yet been worked out.

The *nonlinear interaction of magnetohydrodynamic waves* has been treated in detail by Kaburaki and Uchida (1971), Chiu and Wentzel (1972), Uchida and Kaburaki (1974). When the magnetic field is so weak that $v_A < c_s$, it is found that two Alfvén waves travelling in opposite directions along a magnetic field line can couple nonlinearly to give an acoustic wave, which in turn dissipates relatively quickly. Suppose that the frequency (ω) and wavenumber (k) of the two Alfvén waves and the acoustic wave are denoted by subscripts 0, 1, 2, respectively, so that

$$\omega_0 = v_A k_0, \qquad \omega_1 = v_A k_1, \qquad \omega_2 = c_s k_2, \tag{6.14}$$

where all the frequencies and wavenumbers are assumed positive. If a coupling of the two incident waves is to occur, the resulting acoustic wave must possess a frequency

$$\omega_2 = \omega_0 + \omega_1 \tag{6.15}$$

and wavenumber

$$k_2 = k_0 - k_1, \tag{6.16}$$

the minus sign resulting from the fact that the two Alfvén waves are propagating in opposite directions. After substituting for the wavenumbers from Equation (6.14) and eliminating ω_2 between Equations (6.15) and (6.16), we find that the two Alfvén waves can interact in this way only if their frequencies are in the ratio $\omega_1/\omega_0 = (c_s - v_A)/(c_s + v_A)$. Furthermore, the resulting acoustic frequency is $\omega_2 = 2\omega_0 c_s/(c_s + v_A)$.

In regions of *strong magnetic field* such that $v_A > c_s$, one Alfvén wave (ω_0, k_0) can decay into another Alfvén wave (ω_1, k_1) travelling in the opposite direction together with a sound wave (ω_2, k_2) travelling in the same direction. The interaction takes place provided the selection rules $\omega_1 + \omega_2 = \omega_0$, $-k_1 + k_2 = k_0$, are obeyed, giving Alfvén and acoustic frequencies of

$$\omega_1 = \omega_0 \frac{v_A - c_s}{v_A + c_s}, \qquad \omega_2 = \frac{2\omega_0 c_s}{v_A + c_s}.$$

The resulting Alfvén wave has a frequency smaller than the original one and it can in turn decay to another lower-frequency Alfvén wave plus an acoustic wave. The cascade continues until all the Alfvénic energy has been converted to acoustic waves that dissipate rapidly.

Wentzel (1974, 1977) has calculated the rates at which the waves interact and has estimated the heating. He finds a significant production of acoustic waves from Alfvén waves in regions where v_A/c_s lies between 1/30 and 30. Dissipation occurs over distances comparable with coronal loop lengths, and it is strongest for wave periods of about a minute and at locations where the magnetic field is greatest. For an Alfvén wave of velocity amplitude v_1, the wave energy flux is given by

$$F = \tfrac{1}{2}\rho v_1^2 v_A, \tag{6.17}$$

where ρ is the ambient plasma density. Furthermore, the dissipation length for waves of wavenumber k generating a density perturbation ρ_1 can be written in order of magnitude as $d = \rho/(k\rho_1)$. For sound waves $\rho_1/\rho \approx v_1/c_s$, which gives Equation (6.4), but Alfvén waves produce much smaller density changes, namely $\rho_1/\rho \approx 2\pi(v_1/v_A)^2$. (The factor 2π is appropriate to the case when oppositely travelling Alfvén waves interact to give sound waves.) The dissipation length then becomes

$$d = \frac{\tau v_A}{(2\pi)^2} \left(\frac{v_A}{v_1}\right)^2, \tag{6.18}$$

in terms of the wave-period (τ).

Consider now whether Alfvén waves of period 10 s, say, may produce the required heating for *coronal loops*. For *interconnecting loops* or *quiet-region loops* let us adopt values for the necessary wave flux of 300 W m^{-2} and for the magnetic field of 12 G. Then Equations (6.17) and (6.18) imply a wave amplitude of 20 km s^{-1} and a dissipation length of 200 000 km, which is comparable with the half-length of such loops. For weak *active-region loops* a wave flux of 5000 W m^{-2} and a magnetic field of 20 G give a rather large wave amplitude of 60 km s^{-1} and a dissipation length of 110 000 km, again comparable with the half-length. We thus conclude that *short-period* (≈ 10 s) *Alfvén waves provide a viable means of heating coronal loops outside or on the edge of active regions*. Deep within active regions, the magnetic field (B) is stronger, and so the dissipation length (proportional to B^4 for constant F) is too long. In this case longer-period waves or magnetic field dissipation may be the answer.

When equal fluxes of Alfvén waves propagate up the two legs of a loop, the heating due to their interaction is concentrated near the summit. Wentzel (1976, 1978) has extended the above discussion to include asymmetric heating due to unequal fluxes and also wave reflection at the transition region, which tends to equalise the fluxes along a coronal loop and to disorder the waves below the transition region.

Hollweg (1979, 1981a, b) has considered heating by Alfvén waves in some detail. Besides being tractable mathematically, the advantage of these waves is that they propagate so easily without becoming evanescent or being internally reflected. Also, the energy propagates along the magnetic field, so heating the strongly magnetic regions, and they are observed to dominate solar wind fluctuations at 1 AU. Hollweg shows that the propagation of axisymmetric twists near the axis of a vertical flux tube obeys

$$\left(v_A^2 \frac{\partial^2}{\partial s^2} + \omega^2\right)(B_0^{1/2}\,\delta v) = 0, \qquad i\omega\,\delta B = B_0^{1/2} \frac{\partial}{\partial s}(B_0^{1/2}\,\delta v),$$

where ω is the frequency and s is the distance along the ambient field (B_0). He solves these equations for the perturbed velocity (δv) and magnetic field (δB) in a model atmosphere, in response to imposed photospheric motions. Two cases of wave propagation are considered, namely in the *open field* of a coronal hole and in the *closed field* of an *active-region* loop.

In the open region, *long-period* waves (with a period (τ) in excess of 10 min) possess an energy flux of typically $10\ \mathrm{W\ m^{-2}}$, and they may drive the solar wind (Sections 12.3 and 12.4.2). *Short-period* waves ($10\mathrm{s} < \tau < 5$ min) have an energy flux of 10^3 to $10^4\ \mathrm{W\ m^{-2}}$. They may drive spicules, and their energy is enough to account for chromospheric and coronal heating. The problem used to be how to damp them, but now there appear to be several viable dissipation mechanisms, although their details still need to be worked out. *Joule damping* is important in the middle chromosphere at high frequencies (Section 6.4.1). *Nonlinear damping* may occur by local velocity shears inducing *Kelvin-Helmholtz instability* (Section 7.5.4) or by local magnetic shears driving *tearing modes* (Section 7.5.5). *Nonlinear coupling* to fast and slow modes (which damp efficiently) takes place especially in the chromosphere, because the wave pressure is large (unlike the photosphere) and the waves are nonlinear (unlike the corona). Finally, linear *geometric wave coupling* turns Alfvén waves into fast waves (in the low-β corona), as they refract or propagate around curved field lines. Solutions to the *nonlinear* equations for propagation up into a realistic atmosphere by Hollweg et al. (1982) show that Alfvén waves can steepen into fast shocks in the chromosphere, provided their periods are smaller than a few minutes and the photospheric velocity amplitudes are of order $1\ \mathrm{km\ s^{-1}}$ (or greater). They suggest that such waves can drive upward flows (spicules) and can heat the upper chromosphere and corona.

In a closed loop, *resonant frequencies* appear at multiples of $v_A/(2L)$, where L is the coronal length of the loop. For example, a short loop with $L = 20\,000$ km, $B = 100$ G and $n = 10^{16}\ \mathrm{m^{-3}}$ has resonant periods of 20 s, 10 s, 7 s .... The resonances occur because of reflections off the transition regions at the ends of the loop. They act like windows, which allow a large energy flux (typically $1.5 \times 10^4\ \mathrm{W\ m^{-2}}$) to pass unimpeded up into the corona, rather than being reflected off the steep Alfvén-speed gradient. Leroy (1980) has used methods from optics to analyse the reflection of Alfvén waves propagating up a vertical magnetic field ($B_0\,\hat{\mathbf{z}}$) containing an isothermal plasma, for which the perturbed field (B_{1x}) satisfies $\partial^2 B_{1x}/\partial t^2 = (B_0^2/\mu)\partial/\partial z\,(\rho_0(z)^{-1}\,\partial B_{1x}/\partial z)$. He finds that in a 1 G field, waves with periods of less than an hour can reach the corona unreflected, but in a 3000 G field only those with periods lower than 1 s can propagate to the corona. This suggests that only short-period waves are able to make use of intense flux tubes to reach the corona.

6.4.3. RESONANT ABSORPTION OF ALFVÉN WAVES

As pointed out in Section 4.10.1, when the ambient medium is *nonuniform* a continuous spectrum of Alfvén waves may exist. The *resonant absorption* of such waves at singular surfaces in the plasma has been suggested as a means of heating in laboratory plasma devices and has been analysed by Grossman and Tataronis (1973) and Hasegawa and Chen (1974). It may also provide a mechanism for absorbing energy in the solar corona.

Consider a *force-free flux tube* with magnetic field components $\mathbf{B}_0 = (0, B_{0\phi}(R), B_{0z}(R))$, and suppose there is a wave-like disturbance of the form $f_1(R, \phi, z, t) = f_1(R) \exp(i(\omega t - m\phi - kz))$. Then, by putting $p_0 \equiv 0$ in Equation (4.64), we can see that the radial velocity perturbation $(v_{1R}(R))$ satisfies

$$\frac{d}{dR}\left(\frac{(\rho_0\omega^2 - (\mathbf{k}\cdot\mathbf{B}_0)^2/\mu)B^2/\mu}{(\rho_0\omega^2 - (m^2/R^2 + k^2)B^2/\mu)R}\frac{d}{dR}(Rv_{1R})\right) + Fv_{1R} = 0,$$

where $F(R)$ is given by Equation (4.65) and $\mathbf{k}\cdot\mathbf{B}_0 = kB_{0z} + mR^{-1}B_{0\phi}$.

When $m^2/R^2 \gg \rho_0\omega^2 - k^2B^2/\mu$ this simplifies to the equation

$$\frac{d}{dR}\left((\rho_0\omega^2 - (\mathbf{k}\cdot\mathbf{B}_0)^2/\mu)R\frac{d}{dR}(Rv_{1R})\right) + m^2 Fv_{1R} = 0$$

for Alfvén waves alone.

Consider also a *unidirectional field* and a plasma pressure which vary with x

$$\mathbf{B}_0 = B_0(x)\hat{\mathbf{z}}, \qquad p_0 = p_0(x),$$

and suppose the disturbances behave like

$$f_1(x, y, z, t) = f_1(x) \exp(i(\omega t - k_y y - k_z z)).$$

Then, according to Equation (4.60), the perturbation equation for the transverse velocity component (v_{1x}) in the limit when k_y is large enough becomes,

$$\frac{d}{dx}\left(\varepsilon(x)\frac{dv_{1x}}{dx}\right) - k_y^2 \varepsilon(x) v_{1x} = 0,$$

for Alfvén waves alone, where $\varepsilon(x) = \rho_0(x)\omega^2 - k_z^2 B_0(x)^2/\mu$.

Now, suppose that the footpoints of either the force-free flux tube or the unidirectional field are forced to vibrate at a given frequency (ω). In general, the local Alfvén frequency $\omega_A = \mathbf{k}\cdot\mathbf{B}_0/(\mu\rho_0)^{1/2}$ is not uniform but varies with R (or x). If there exists a radius R^* (or distance x^*) at which $\omega_A = \omega$, the coefficient of the second-order term in the above differential equations vanishes; the equations will therefore be singular at that point and so a *singular surface* (or *resonant absorption sheath*) will form there. The radial velocity component v_{1R} (or v_{1x}) possesses a logarithmic singularity, while $v_{1\phi}$ (or v_{1y}) has a hyperbolic singularity and the plasma energy becomes infinite. However, analytic continuation of v_{1R} (or v_{1x}) through the singularity shows that *energy is continuously accumulating at the singular surface, and so it may heat the plasma there*. As the energy builds up, so the width of the surface decreases until a steady state is reached where the energy flux is dissipated ohmically (or viscously).

The thickness of the singular layer is typically 10 ion-Larmor radii, but heating may possibly occur over a much larger region if the Alfvén wave is first converted linearly to a *kinetic Alfvén wave*. It has a dispersion relation

$$\omega^2 = k_{\parallel}^2 v_A^2 (1 + k_{\perp}^2 \rho_i^2),$$

where $k_{\parallel}$ and $k_{\perp}$ are wavenumbers parallel and perpendicular to the magnetic field and ρ_i is the ion Larmor radius. Dissipation of the wave is by kinetic effects such as Landau damping. The above ideas have been examined by Ionson (1978) as a mechanism for heating *coronal loops*. He suggests that 5-min chromospheric oscillations shake the footpoints of a loop and cause *Alfvénic surface waves* (Section 4.10.1) to propagate upwards along the surface of the loop. They in turn couple to the kinetic Alfvén waves, which dissipate in an extremely thin sheath only 1 km thick.

Resonant absorption of more general magnetoacoustic waves is described by Equation (4.59), which possesses resonance points where $\omega_A(x) = \omega$ or $\omega_T(x) = \omega$ and cut-off points where $\omega_+(x) = \omega$ or $\omega_-(x) = \omega$. At the resonance points, v_{1x} possesses a logarithmic singularity and energy is resonantly absorbed as short wavelength oscillations build up until they are limited by viscous or ohmic dissipation. At the cut-off points, the wave becomes evanescent and most of the energy is reflected while some can tunnel through. Thus, as a magnetoacoustic wave propagates in the solar atmosphere with a certain frequency ω, so the values of $\omega_T, \omega_-, \omega_A, \omega_+$ will vary and its nature will change. When $\omega_+ < \omega$ or $\omega_T < \omega < \omega_-$ it propagates, but when $\omega_- < \omega < \omega_+$ or $\omega < \omega_T$ it is evanescent. The wave is evanescent both sides of the Alfvén resonance, but it changes from being propagating to evanescent at the cusp resonance, which is therefore likely to have more energy being fed into it.

6.4.4. Magnetic Field Dissipation

When photospheric motions are *sufficiently slow* and the wavelength *sufficiently long* that the conditions (6.12) and (6.13) hold, a wave description ceases to be helpful. Instead the coronal magnetic configuration *evolves passively through a series of equilibria*, which store energy in excess of potential. This energy has come originally from the photospheric motion. The electric currents associated with such large-scale equilibria produce ohmic heating, which is, however, completely negligible, since the coronal conductivity (σ) is so large. This is true even when allowance is made for the fact that the corona is probably in a permanent state of weak turbulence, with an anomalous electrical conductivity (σ^*) that is a factor of a hundred or so lower than the classical value (Benz, private communication). The only way that magnetic field (i.e., ohmic) dissipation can produce the necessary coronal heating is for the magnetic field changes and accompanying electric currents to be concentrated in

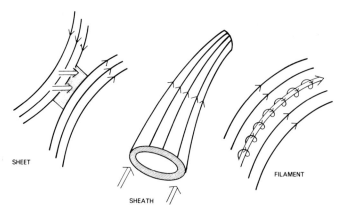

Fig. 6.4. Three geometries for a current concentration in which enhanced magnetic field dissipation may occur. Light arrows label magnetic field lines, whereas large arrows indicate electric current directions.

extremely intense *current sheets, current sheaths* (around flux tubes) or *current filaments* (Figure 6.4). If the current density is so strong that the width of such a current concentration is less than (typically) a few metres, the dissipation may be considerably enhanced by the presence of plasma turbulence (Section 2.1.5).

Provided current sheets, sheaths or filaments can be formed, they produce a rapid conversion of magnetic energy into heat (by ohmic dissipation), bulk kinetic energy and fast-particle energy, in a manner that has been studied extensively in connection with the more violent heating of a solar flare (Section 10.1). This suggests that, especially in the strong magnetic field of an active region, the corona is in a state of *ceaseless activity* and is being heated by many tiny micro-flarings that are continually generated by the photospheric motion below. The coronal loops that stand out in soft X-ray pictures are those in which most heat is being released and then conducted efficiently along the magnetic field.

The features of heating by *magnetic* (or *current*) *dissipation* that need to be understood concern the *way in which current sheets, sheaths or filaments are formed, are maintained* (if necessary) *and decay*. The order-of-magnitude estimates of Tucker (1973) and others are described below, to determine how thin the resistive regions need to be to provide the necessary heating. A discussion of the formation of current sheets or filaments is also given, but no convincing explanation for sheaths has yet been put forward. Of the three proposed alternatives, *current sheets* have received by far the most attention; they may be formed either by pushing topologically distinct regions against one another or by magnetic non-equilibrium. In the former case, they are maintained for as long as the external footpoint motion continues. Current filaments may be created as a result of *tearing-mode instability* (Section 7.5.5) or *thermal instability* (Section 6.4.4C).

6.4.4A. Order of Magnitude

Tucker (1973) and Levine (1974) were among the first to suggest coronal heating by the dissipation of non-potential magnetic fields. They considered neutral current sheets dispersed throughout active regions, and they established qualitatively that *current dissipation* could provide enough heat for the corona. Tucker supposes that magnetic energy is being stored at a rate

$$\frac{dW_m}{dt} \approx \frac{vB^2}{2\mu} L \qquad (6.19)$$

by photospheric motions (v) that twist a magnetic field of strength B over an area L^2. The energy is at the same time being dissipated ohmically, at a rate

$$D \approx \frac{j^2}{\sigma} L^3 \qquad (6.20)$$

for currents (j) distributed uniformly through the volume (L^3) of the active region. If the magnetic field is being twisted up faster than it is relaxing ohmically, the excess energy will be stored until it is released as, for instance, a *solar flare*. But, if the two rates (6.19) and (6.20) are equal, the active region will maintain a steady state. The effective twisting speed (v) that is needed to provide a heat input of, say, 3000 W m^{-2} to the corona can be found from Equation (6.19) as $v \approx 100$ m s^{-1} for a photospheric field strength of 100 G. Furthermore, uniform dissipation throughout the active region with a classical Coulomb electrical conductivity (σ) requires a current density that can be estimated by equating DL^{-3} from Equation (6.20) to 3000 W m^{-2}. With $L \approx 10\,000$ km, Tucker finds $j \approx 30$ A m^{-2}. Since this corresponds to the rather large magnetic field gradient of 0.4 G m^{-1}, he suggests that the dissipation is concentrated at thin current sheets rather than distributed uniformly. The ohmic dissipation inside sheets may be greatly enhanced above normal because of the much larger electric currents and the possibility of plasma turbulence, but the sheets occupy only a small fraction of the active-region volume. For each sheet of thickness l^* and area L^{*2} with an electric current $j^* \approx B/(\mu l^*)$ and a turbulent electrical conductivity (σ^*), the rate of heat generation is

$$D^* \approx \frac{j^{*2}}{\sigma^*} L^{*2} l^* \quad \text{or} \quad D^* \approx \frac{B^2}{\mu^2 \sigma^*} \frac{L^{*2}}{l^*}.$$

Tucker adopts a turbulent conductivity that is about a million times smaller than the classical value and assumes a sheet width of 10 m, consistent with the critical current for turbulence onset. He finds that only a few current sheets of length $L^* = 1000$ km are necessary to generate the heat that is required for an active region. Levine (1974) suggested that the tangled nature of coronal magnetic fields produces many small current sheets that are collapsing. During the collapse, particles are accelerated and then thermalised by Coulomb collisions in the surrounding region.

Rosner *et al.* (1978) support Tucker's ideas for direct coronal heating by magnetic dissipation. They point out that the observed intensity of active regions in X-rays appears to be directly related to the level of photospheric magnetic activity. Early in the life of an active region, the magnetic field is complex and the corona bright, whereas one rotation later the region often possesses a more dispersed field with a coronal plasma whose pressure is an order of magnitude lower. Instead of heating in current sheets, Rosner *et al.* suggest that the heating is concentrated in current *sheaths*, which are thin, annular regions near the edges of coronal flux tubes or loops. The loops are stressed by twisting motions at the footpoints, and, in a steady state, the work done by photospheric motions balances the coronal dissipation. Once the heat has been released, it is conducted *along* the magnetic field very efficiently (and so produces loop-like structures), but the distance that it is transported *across* the field depends on the nature of the small-scale instabilities that may be present. Golub *et al.* (1980) discussed the problem further and suggested that heating due to twisting motions (v_ϕ) in a loop is $E_H \sim B_\phi B_z v_\phi / L$ per unit volume. Equating this to Equation (6.40) gives a scaling law $p \sim B^{12/7}$, which relates the loop pressure and field strength and agrees reasonably with observations.

The main uncertainty with the model concerns the way in which the necessary strong currents are created and maintained. In order to provide sufficient heating, the current sheath in the loop needs to be typically 50 m thick and the sheath plasma must be turbulent, with an electrical conductivity smaller than normal by a factor of 10^4. However, (apart from flaring situations (Chapter 10)) it seems unlikely that large enough current densities can be produced in a loop to create and maintain a strongly turbulent plasma. A more likely alternative is magnetic dissipation in the loop by means of *resistive instabilities* (Section 7.5.5), such as the collisionless (or collisional) tearing mode, which occurs for much smaller current densities.

The possibility of magnetic dissipation was later considered in more detail. Sakurai and Levine (1981) analysed the generation of force-free fields and the storage of magnetic energy in the corona due to small photospheric motions at the footpoints of both a uniform and a bipolar configuration. Sturrock and Uchida (1981) estimated the rate of increase of magnetic energy due to a *stochastic* motion of the photospheric footpoints of a coronal loop. They obtained scaling laws for a loop's temperature in terms of its length and magnetic field strength. Also, Ionson (1981, preprint) set up an interesting LRC circuit analogue for coronal loops.

6.4.4B. Current Sheets

Current sheets may be formed in several ways. One is by the *interaction of topologically separate parts* of the magnetic configuration of, say, an active region. High-resolution observations of the *photospheric* magnetic field (e.g., Figure 1.10) exhibit a highly complex magnetic pattern with frequent changes of polarity. The *coronal* field is also complex, with many distinct magnetic flux tubes shown up by X-ray and EUV pictures (e.g., Figure 1.15). As the photospheric footpoints of coronal loops move, so the neighbouring coronal flux-tubes will respond and interact with one another, either moving further apart or coming closer together. At the interface between the two tubes, a *current sheet* is formed, the magnetic field reconnects, and magnetic energy is released in the process. Such magnetic dissipation takes place not only when neighbouring magnetic field lines are oppositely directed, as in Figure 6.5, but also when the field lines are inclined at a non-zero angle (Priest and Sonnerup, 1975).

The formation of current sheets when *new magnetic flux is emerging* from below the photosphere has been studied in connection with solar flares by several authors (see Section 10.2.1), but the same calculations are applicable when *magnetic flux is evolving* rather than emerging. In particular, it must be stressed that the current sheet is a response to the applied photospheric motions. If the neighbouring footpoints move relative to one another at a certain speed, then the *corona will just respond*

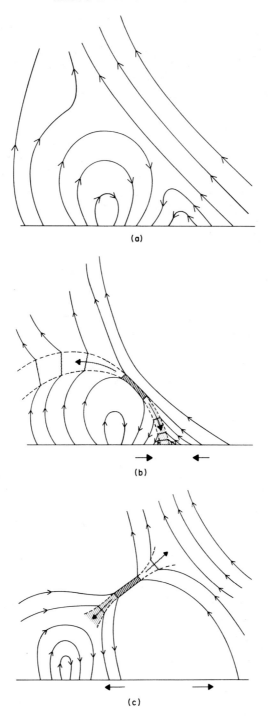

Fig. 6.5. Magnetic dissipation due to the relative motion of: (a) two neighbouring flux tubes when they (b) approach one another or (c) move further apart.

by creating a current sheet and allowing *magnetic reconnection* at that speed (provided the speed is less than a maximum value (Section 10.1)). Furthermore, the reconnection and associated dissipation is maintained as long as the relative footpoint-motion continues, with the dimensions of the current sheet depending on the magnetic field strength and photospheric speed. Conditions inside the sheet will only be turbulent if the resulting sheet width is small enough. Indeed, it is when the width becomes less than a critical value that a subflare or flare is triggered, according to the *emerging flux mechanism* (Section 10.2.1). It should also be noted that slow magnetoacoustic shock waves radiate from the ends of the current sheet and that fine jets of plasma are emitted between pairs of shocks. As plasma comes in slowly from the sides, the *bulk of the heat is released at these shock waves* rather than in the central current sheet itself.

Current sheets may also develop when magnetostatic equilibrium becomes *unstable* or even ceases to exist, a situation known as *non-equilibrium*. In a simple bipolar magnetic field when the photospheric footpoints move slowly, the low-β (Section 2.5) corona responds by establishing a series of *force-free* configurations (Section 3.5). In general, however, the coronal magnetic field is much more complex than this, and it contains topologically distinct flux systems. Parker (1972) and Syrovatsky (1978) have demonstrated that, as the footpoints of such a magnetic field move, the *corona cannot adjust to a new force-free equilibrium* and current sheets are formed instead. These current sheets are themselves not in equilibrium, since they allow a rapid reconnection at some fraction of the Alfvén speed (Section 10.1), and the magnetic configuration reduces to the state of lowest potential energy. Parker referred to such a process as *topological dissipation*. Continual footpoint motion means that the coronal field is all the time responding by reconnecting and so converting magnetic energy into heat.

Parker (1972) establishes that, *if the pattern of small-scale variations is not uniform* along a large-scale field, then the field *cannot be in magnetostatic equilibrium*. In other words, equilibrium exists only if the field variations consist of a simple twist extending from one footpoint to another. More complex topologies (such as braided flux tubes with several field lines wrapped around each other) are not in equilibrium. To obtain his result, Parker considers a uniform configuration, having a plasma pressure p_0 and magnetic field $B_0\hat{z}$, with the footpoints anchored at the planes $z = \pm L$ (Figure 6.6). Suppose that a displacement of the footpoints by at most $\lambda (\ll L)$ leads to small deviations ($\mathbf{B}_1$ and p_1) from the uniform field and pressure, respectively, so that $\mathbf{B} = B_0\hat{z} + \mathbf{B}_1(x, y, z)$, $p = p_0 + p_1(x, y, z)$, where $B_1/B_0 \approx p_1/p_0 \ll 1$. Then the equation $-\nabla p + (\nabla \times \mathbf{B}) \times \mathbf{B}/\mu = 0$ for magnetostatic equilibrium (Section 3.1) becomes, to first order,

$$-\nabla\left(p_1 + \frac{B_0 B_{1z}}{\mu}\right) + \frac{B_0}{\mu}\frac{\partial \mathbf{B}_1}{\partial z} = 0. \tag{6.21}$$

But the basic equation $\nabla \cdot \mathbf{B} = 0$ gives $\nabla \cdot \mathbf{B}_1 = 0$, and so the divergence of Equation (6.21) implies

$$\nabla^2\left(p_1 + \frac{B_0 B_{1z}}{\mu}\right) = 0. \tag{6.22}$$

Now, variations of $(p_1 + B_0 B_{1z}/\mu)$ extend at most a distance of order $\lambda (\ll L)$ into

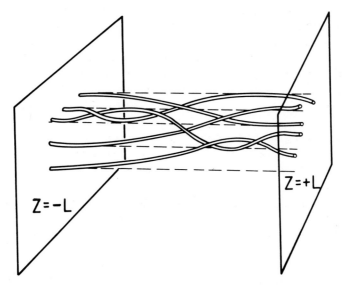

Fig. 6.6. Schematic drawing of the topology of magnetic tubes of force following a displacement of the ends of the tubes where they intersect $z = +L$ (from Parker, 1972).

the region $-L < z < L$ from the end-planes ($z = \pm L$), so the solution to Laplace's equation (6.22) is

$$p_1 + \frac{B_0 B_{1z}}{\mu} = \text{constant}, \tag{6.23}$$

save in boundary layers of width λ near $z = \pm L$. However, the z-component of Equation (6.21) is $\partial p_1/\partial z = 0$, and hence the z-derivative of Equation (6.23) implies $\partial B_{1z}/\partial z = 0$. Thus, if a field is in magnetostatic equilibrium, its pattern does not vary along the general direction of the field, which establishes Parker's result. Sakurai and Levine (1981) have since established that the perturbed field *is* determined uniquely by a small motion of the footpoints*. Nevertheless, the detailed consequences of Parker's result remain to be seen when finite-amplitude displacements at the boundary are taken into account. Later, Parker (1981) suggested that large displacements of the feet of a flux tube could cause the tube to become dislocated from its initially neighbouring field. In its new quasi-equilibrium position the tube would become flattened and eventually dissipated.

Green (1965) and Syrovatsky (1971) demonstrated that the slow, continuous deformation of a two-dimensional potential magnetic field containing *neutral points* leads to the production of neutral current sheets in the *perfectly conducting* limit. Consider the initial magnetic field $B_x = y$, $B_y = x$, which may be written in terms of the complex variable $z = x + iy$ as

$$B_y + iB_x = z, \tag{6.24}$$

and contains a neutral point at the origin (Figure 6.7). Next, suppose that this configuration is deformed by the imposition of a uniform electric field (perpendicular

---

*All the variations occur in the boundary layers near $\pm L$, and so Figure 6.6 is misleading.

to the z-plane) that drives a motion normal to the magnetic field. If the plasma is regarded as perfectly conducting, the field lines are carried with the motion and Ohm's Law may be written

$$E + vB = 0. \tag{6.25}$$

It is clear that the instantaneous plasma speed is determined by Equation (6.25) everywhere except at the neutral point where B vanishes. Furthermore, suppose that the flow speed and plasma pressure are so small that the configuration passes through a series of equilibria with vanishing Lorentz force, so that in this two-dimensional situation

$$0 = jB. \tag{6.26}$$

The peculiar role of the neutral point is evident in both Equations (6.25) and (6.26), which imply that a continuous deformation of the original field (6.24) through a series of potential configurations with $j = 0$ is possible everywhere *save in the vicinity of the neutral point*. At the neutral point itself, Equation (6.26) allows non-zero currents. Indeed, a solution to the problem is that a current sheet develops, represented by a cut from $z = -iL$ to $z = +iL$, say, in the complex plane. The resulting magnetic field components are given by $B_y + iB_x = (z^2 + L^2)^{1/2}$, and the field is sketched in Figure 6.7. This complex variable technique for obtaining the position of the current sheet that forms as oppositely directed fields approach one another has been extended to the case of approaching bipolar fields that are equal (Priest and Raadu, 1975) or unequal (Tur and Priest, 1976). In the latter case the sheet is curved. All these calculations use the assumption that the plasma remains frozen to the field everywhere, but in practice this approximation fails inside the current sheet. During the approach of the two flux systems, the current sheet bifurcates into two slow magnetoacoustic shocks and reconnection occurs in the manner described in Section 10.1. When the footpoint motion ceases, the magnetic configuration rapidly reduces (over an Alfvén travel-time) to its lowest energy state, namely a potential field.

Syrovatsky (1978) generalised his previous result for potential fields to show that a

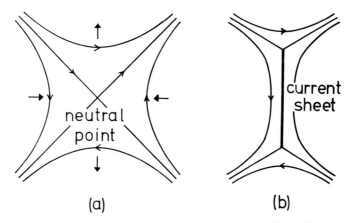

Fig. 6.7. (a) A potential magnetic field near an X-type neutral point. (b) The field produced by the slow motion indicated in (a) by solid-headed arrows. The plasma is assumed perfectly conducting.

continuous deformation of a *force-free* field in general leads to the production of *current sheets*. It is only for rather simple fields and simple footpoint motions that no current sheets are generated (see Section 3.5). Suppose a magnetic configuration evolves through a series of force-free states such that

$$(\nabla \times \mathbf{B}) \times \mathbf{B} = 0, \tag{6.27}$$

with $\nabla \cdot \mathbf{B} = 0$. Then, in general, this evolution cannot maintain the condition of frozen-in flux, since, given a time-sequence of magnetic fields satisfying Equation (6.27), the solution of the induction equation

$$\frac{\partial \mathbf{B}}{\partial t} = \nabla \times (\mathbf{v} \times \mathbf{B}) \tag{6.28}$$

for $\mathbf{v} \times \mathbf{B}$ requires it to have (in general) a component parallel to $\mathbf{B}$. This is unacceptable since $\mathbf{v} \times \mathbf{B}$ must be normal to $\mathbf{B}$ by definition. The difficulty in solving Equation (6.28) for a given magnetic field shows up particularly clearly when a closed magnetic field line (C) exists. Simply integrate over a surface (S) bounded by such a C to give

$$\frac{\partial}{\partial t} \int_S \int \mathbf{B} \cdot d\mathbf{S} = \int_C (\mathbf{v} \times \mathbf{B}) \cdot d\mathbf{s}. \tag{6.29}$$

Then, for a given sequence of solutions to Equation (6.27), there is in general no reason to suppose that the flux through S should remain constant. This implies that the left-hand side of Equation (6.29) is non-zero, whereas the right-hand side must vanish (since $\mathbf{v} \times \mathbf{B}$ is normal to the field-line curve (C)).

More details of the formation and properties of curent sheets can be found in the reviews by Priest (1976, 1981b).

6.4.4C. *Current Filaments*

A magnetic configuration that is non-potential, such as a sheared force-free structure, may become unstable in several ways with the electric current concentrating into filaments. One mechanism is the tearing-mode instability, which is described in Section 7.5.5 (see also Galeev *et al.* (1981), who suggest nonlinear kinetic tearing as an effective heating mechanism for coronal loops), and another is thermal instability. Heyvaerts (1974) has described two such types of instability, namely the *Joule mode* and the *antidiffusion mode*. They both cause a uniform electric current to concentrate into many small *current threads* (or filaments) parallel to the magnetic field. The main conditions for the validity of his analysis are that the temperature be about 10^5 K, so that the radiative loss function $(Q(T))$ be approximately constant (Section 2.3.3), and that the perturbation wave-number (k) be both small enough that thermal conduction is swamped by Joule heating and also large enough that a local stability analysis is applicable.

For a disturbance propagating at an angle (θ) to the ambient magnetic field with $\gamma = \frac{5}{3}$, and an electrical conductivity (σ) proportional to $T^{3/2}$, the *Joule mode* has a growth-rate

$$\omega = (\sin^2 \theta - \cos^2 \theta) \frac{j_0^2}{\sigma p_0}, \tag{6.30}$$

where j_0 is the equilibrium current density and p_0 the equilibrium pressure. It occurs most effectively for $\theta = \frac{1}{2}\pi$, and so it forms fine structures aligned along the magnetic field and consisting of hotter current concentrations separated by cooler regions. The mode is, however, restricted to wavenumbers $k \gg (\omega\mu\sigma)^{1/2}$, such that the plasma is not frozen to the field, and another limitation in applicability is that the current density needs to be rather large for the growth-rate to be reasonably high. For example, suppose we adopt $T_0 = 10^5$ K, $N_0 = 10^{15}$ m^{-3}, together with a turbulent conductivity that is a factor of 100 smaller than the Coulomb value (2.17) and a current density $j_0 = 1$ A m^{-2} (corresponding to a magnetic field change of 100 G over 1000 km). Then (6.30) gives a growth-rate of only 4×10^{-4} s^{-1}.

The *antidiffusion mode* causes the magnetic flux to concentrate rather than diffuse, and it occurs only for θ close to $\frac{1}{2}\pi$. Its growth-rate is

$$\omega = \frac{k^2}{\mu\sigma(\theta - \frac{1}{2}\pi)^2}, \qquad (6.31)$$

so that $\omega \gg k^2/(\mu\sigma)$ and the plasma is almost frozen to the field. A local increase in temperature enhances the electrical conductivity and current density, which in turn produces more heating and drives the instability.

6.5. Coronal Loops

The solar atmosphere, which has a vertical stratification produced by the force of gravity, is by no means uniform in the horizontal direction and possesses a complex

Fig. 6.8. An active-region loop system in the EUV line of O VI (3×10^5 K) (courtesy R. Levine, Centre for Astrophysics, Harvard).

TABLE 6.2.

Typical length $2L(\times 1000$ km), temperature $T(K)$ and density $n(\text{m}^{-3})$ for the different kinds of coronal loop

	Interconnecting	Quiet-region	Active-region	Post-flare	Simple-flare
$2L$	20–700	20–700	10–100	10–100	5–50
T	$2-3 \times 10^6$	1.8×10^6	$10^4 - 2.5 \times 10^6$	$10^4 - 4 \times 10^6$	$\lesssim 4 \times 10^7$
n	7×10^{14}	$0.2 - 1.0 \times 10^{15}$	$0.5 - 5.0 \times 10^{15}$	10^{17}	$\leqslant 10^{18}$

structure dominated by the magnetic field. X-ray and EUV observations such as those from Skylab have indicated that the corona (outside *coronal holes*) consists largely of loop structures that presumably outline the magnetic field. Five morphological types of loop are found, namely, *interconnecting loops, quiet-region loops, active-region loops* (an example being shown in Figure 6.8), '*post*'-*flare loops* and *simple (compact) flare loops*. General properties of their structure have been discussed, for instance, by Priest (1978) and Chiuderi *et al.* (1981), and their observational characteristics are summarised in Sections 1.3.4B and 1.4.1C and Table 6.2. We describe below a model for the temperature-density structure of coronal loops, together with some general comments about the types of flow that may be expected in them.

6.5.1. Static energy-balance models

For a loop in *hydrostatic equilibrium* and in *thermal equilibrium* between conduction, radiation and heating, the temperature (T) and electron density (n_e) satisfy (see Equations (2.32) and (3.5))

$$\frac{1}{A}\frac{d}{ds}\left(\kappa_0 T^{5/2} \frac{dT}{ds} A\right) = \chi n_e^2 T^\alpha - H \tag{6.32}$$

and (for fully-ionised hydrogen)

$$\frac{1}{\cos\theta}\frac{dp}{ds} = -m_p n_e g, \tag{6.33}$$

where the pressure

$$p = 2n_e k_B T; \tag{6.34}$$

$A(s)$ is the cross-sectional area of the loop at a distance s along it from the base, and $\theta(s)$ is the inclination of the loop to the vertical (Figure 6.9). For given forms of $A(s)$ and $\theta(s)$, the Equations (6.32) and (6.33) are to be solved subject to: (i)

$$p = p_0, \qquad T = T_0, \tag{6.35}$$

at the base $(s = 0)$ of the loop, and (ii) the symmetry condition

$$\frac{dT}{ds} = 0 \tag{6.36}$$

at the summit $(s = L)$.

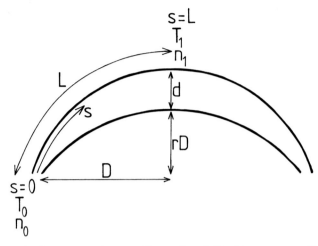

Fig. 6.9. The notation for a symmetric coronal loop of length $2L$ with temperature T_0 and density n_0 at the footpoint ($s = 0$), and T_1 and n_1 at the summit ($s = L$). r is the ratio of loop height to half the base length (D), and d is the ratio of the diameter of the loop cross-section at the top to that at the footpoint.

With T_0 fixed, the temperature profiles and, in particular, the *summit temperatures* (T_1) are determined by three parameters, namely the loop length ($2L$), the base pressure (p_0) and the heating rate (H), so that $T_1 = T_1(L, p_0, H)$. The value of T_0 is fairly arbitrary, but, when a value near the temperature minimum or a chromospheric plateau is taken, we have a so-called *thermally isolated loop*, for which the above three boundary conditions must be supplemented by

$$\frac{dT}{ds} = 0 \quad \text{at } s = 0. \tag{6.37}$$

In this case, the summit temperature is a function of only two parameters: $T_1 = T_1(L, H)$, since the base pressure cannot be freely prescribed as in Equation (6.35) but must be adjusted to a value $p_0 = p_0(L, H)$ determined by the remaining boundary conditions. Of course, in practice there may be some complicated feedback if H itself depends on p_0.

6.5.1A. Uniform Pressure Loops

For very *low-lying loops*, whose summits are much below a coronal scale-height of roughly 80 000 km (Equation 3.10), the loop pressure is uniform, and so one needs to solve just Equation (6.32), with n_e given by Equation (6.34) and p constant.

Consider first the *thermally isolated loops* with uniform cross-sectional areas (Rosner *et al.*, 1978). Their summit temperatures may be estimated in order of magnitude as follows, by using the fact that, whereas the relative sizes of C, R, H vary locally, their global (or integral) values are similar. Since the heating and radiation in Equation (6.32) are globally of the same order, the heating is roughly

$$H \approx \frac{p^2 \chi}{4k_B^2} T_1^{-5/2}, \tag{6.38}$$

in terms of the summit temperature. (Here α is approximated by $-\frac{1}{2}$.) Furthermore, the thermal conduction term has a similar size to the radiation term, and so Equation (6.32) gives, in order of magnitude,

$$\frac{\kappa_0 T_1^{7/2}}{L^2} \approx \frac{p^2 \chi}{4k_B^2} T_1^{-5/2},$$

or, after rearranging,

$$T_1 \sim (pL)^{1/3}. \tag{6.39}$$

The constant of proportionality, namely $(4k_B^2 \kappa_0/\chi)^{-1/6}$, is about 10 000 in m.k.s. units (when $\kappa_0 = 10^{-11}$, $\chi = 10^{-32}$, $k_B = 1.4 \times 10^{-23}$). Substitution of T_1 from Equation (6.39) in (6.38) then gives

$$H \sim p^{7/6} L^{-5/6} \tag{6.40}$$

for the mechanical heating. If one regards the loop heating as being prescribed, so that its pressure just responds to preserve equilibrium, Equations (6.39) and (6.40) may be rearranged to give

$$T_1 \sim H^{2/7} L^{4/7}, \qquad p \sim H^{6/7} L^{5/7}. \tag{6.41}$$

This implies that both *the temperature and pressure are increased by either stretching a thermally isolated loop or enhancing its heating.*

When $T_1 \gg T_0$, the above order-of-magnitude expressions (6.39), (6.40) for heating and summit temperature may be derived rigorously from Equation (6.32), subject to Equations (6.35) to (6.37), as follows. Multiply Equation (6.32) by $T^{5/2} dT/ds$ and integrate from the loop summit, so that

$$\frac{1}{2}\kappa_0 \left(T^{5/2} \frac{dT}{ds}\right)^2 = \int_{T_1}^{T} T^{5/2} \left(\frac{p^2 \chi}{4k_B^2} T^{-5/2} - H\right) dT,$$

or

$$\frac{1}{2}\kappa_0 T^5 \left(\frac{dT}{ds}\right)^2 = \frac{p^2 \chi}{4k_B^2}(T - T_1) - \frac{2H}{7}(T^{7/2} - T_1^{7/2}). \tag{6.42}$$

Now, since $T_1 \gg T_0$ and the conductive flux ($\kappa_\parallel dT/ds$) vanishes at the base, Equation (6.42) gives

$$H = 3.5 \frac{p^2 \chi}{4k_B^2} T_1^{-5/2},$$

which is just Equation (6.38), apart from the factor 3.5. After substituting for H and taking the square-root, Equation (6.42) then reduces to

$$T^2 \frac{dT}{ds} = \left(\frac{p^2 \chi}{2k_B^2 \kappa_0}\right)^{1/2} \left(1 - \frac{T^{5/2}}{T_1^{5/2}}\right)^{1/2},$$

which in turn integrates to give the loop length as

$$Lp = \left(\frac{2k_B^2 \kappa_0}{\chi}\right)^{1/2} I T_1^3,$$

where $I = \int_0^1 t^2(1-t^{5/2})^{-1/2} dt$ is approximately 0.72. Finally, taking the cube-root gives the same scaling law as Equation (6.39), but with a constant of proportionality larger than before by a factor of only $2^{1/6} I^{-1/3}$. Full numerical solutions of the equations of energy balance and hydrostatic equilibrium by Wragg and Priest (1982) show that the scaling law is accurate for short loops. However, for sufficiently long (or rare) loops, the temperature may be lower by a factor of two or so than the scaling law predicts, and too large an increase in loop length may actually cause the temperature to fall.

Now, in general, the conductive flux may not vanish at the temperature (T_0) that is chosen as the base of the loop. This means that the loop is not thermally isolated, and it leads not only to *more general expressions* for the summit temperature than Equation (6.39), with the heating rate as an extra parameter, but also to the possibility of *thermal non-equilibrium* and *instability*. These more general loop models are especially useful if the loop base is taken at, say, 5×10^5 K, in view of the difficulties and uncertainties in modelling the transition region. Starting at a value of, say, 2×10^4 K for T_0 (rather than, say, 10^6 K) may give significant errors if Equation (6.32) describes the transition-region physics inadequately. It would be invaluable to extend the models down to the temperature-minimum region where the conductive flux vanishes, but, at such low temperatures, radiative transfer effects would have to be included and these introduce a coupling to the region below.

Numerical solutions for uniform-pressure loops have been found by many authors, including Hood and Priest (1979a), who treat the case when T_0 is 10^6 K and the heating is proportional to density: $H = hn_e$. These solutions depend on the three parameters

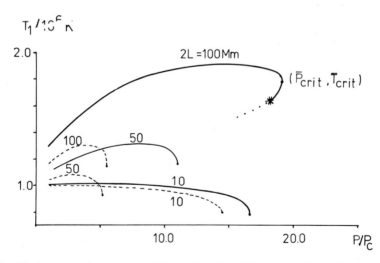

Fig. 6.10. The loop summit temperature (T_1) as a function of the pressure (p) and half-length (L) for a low-lying static coronal loop. p_c is the pressure for a standard plasma of density 5×10^{14} m^{-3} and temperature 10^6 K. The solid (or dashed) curves are for mechanical heating ten (or five) times larger than the radiation from the standard plasma. The curves end at critical conditions (p_{crit}, T_{crit}), indicated here for hot loops of length 100 000 km. The lower unstable solutions are also included for this loop; the star indicates a thermally isolated loop and the dots show oscillatory solutions (from Hood and Priest, 1979a).

L, p, h through just the two combinations Lp and h/p. (The base conductive flux can be regarded as an alternative parameter to p, and in practice there may be some limitation on the range of values of this flux which the atmosphere below T_0 can cope with statically.) Some typical numerical results are shown in Figure 6.10. They suggest explanations for many observed loops properties. For example, the fact that *shorter loops often appear brighter* may be because they possess consistently higher heating rates, or it may be caused by the higher pressures (p_{crit}) that are allowed for shorter loops. Furthermore, the relatively *small variation in observed X-ray temperature compared with pressure* is clearly present in the results. In order to produce a temperature range of 2.2 to 2.8×10^6 K typical of active-region loops, one requires a heating rate that is between 10 and 15 times bigger than the standard radiation. Quiet-region loops need a heating rate about half as big as this. For the above numerical modelling, the energy balance Equation (6.32) was solved along a single magnetic field line, but it has also been solved more generally in the two cases of a cylindrically symmetric structure (Hood and Priest, 1979a) and a force-free arcade of loops (Priest and Smith 1979), as described in Sections 11.1.1 and 11.1.2.

Chiuderi *et al.* (1981) suggest that the observational errors in L, p_0 and T_1 are so great that no meaningful information can yet be inferred about the heating processes. Since the temperature profile is flat over most of a loop's length, a good approximate scaling law can be obtained by neglecting the radiative loss, in agreement with Roberts and Frankenthal (1980).

6.5.1B. Cool Cores

A striking feature of the results in Figure 6.10 is that, as the loop pressure slowly increases (with its heating and length held fixed), so the summit temperature (T_1) rises to a maximum and then decreases to a critical value (T_{crit}) at which dT_1/dp becomes infinite and a *catastrophe* occurs, as indicated schematically in Figure 6.11. If the *pressure exceeds the value* p_{crit}, the loop is therefore in a state of *thermal nonequilibrium*. There is no neighbouring equilibrium, and so the plasma cools along the dotted line seeking a new equilibrium below 10^5 K. As it cools, much of the plasma is likely to drain out of the loop, since it cannot all maintain hydrostatic equilibrium at lower temperatures. The existence of a lack of equilibrium and the consequent cooling is also present when a *loop is (slowly) stretched* far enough at constant p and H or when its *heating is slowly decreased* at constant p and L. It provides an explanation for the existence of extremely cool cores (Section 1.4.1C) that Foukal (1975, 1976) and Jordan (1975) have observed in some coronal loops. (In addition it may be the cause of the sudden loop evacuation that Levine and Withbroe (1977) described.) In extreme cases, Hood and Priest (1979a) propose this as the mechanism for creating active-region filaments or prominences (Section 11.1.1). The idea that active-region filaments are just stretched magnetic flux-ropes containing cool plasma is consistent with the frequently reported observation of motions along filaments, presumably guided by the magnetic field. The cool cores contain too much plasma to be in hydrostatic equilibrium at such large heights, and so they must be dynamic. The cool plasma may have been injected up as spicules, or its temperature may have decreased by thermal instability.

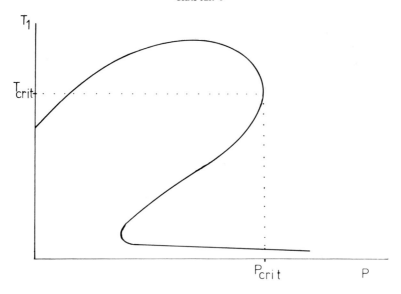

Fig. 6.11. The summit temperature (T_1) for a static coronal loop shown schematically as a function of its pressure (p). When p_{crit} is reached, the plasma cools along the dotted line to a new equilibrium well below T_{crit}.

Another feature of the numerical solutions (Figure 6.10) is that thermally isolated loops, in particular, are *thermally unstable* to perturbations that preserve the base temperature (Antiochos, 1979; Hood and Priest, 1980a Chiuderi *et al.*, 1981). This may provide an explanation for the *ceaseless activity* of the solar atmosphere and, in particular, for *spicules*. However, the loops are close to marginally stable conditions with an extremely small growth-rate (Craig *et al.*, 1982), although an adequate coupling to the chromosphere has not yet been incorporated.

6.5.1C. Hydrostatic Equilibrium

For loops that are about a coronal scale-height or greater in vertical extent, the pressure decreases substantially from the loop base to its summit, and so the full equations (6.32) to (6.34) need to be solved. Wragg and Priest (1981a) have done so for a loop that is an arc of a circle. The ratio of height to foot-point separation is denoted by $\frac{1}{2}r$, and the increase in cross-sectional diameter from base to summit is d. Figure 6.12 gives the results for the summit temperature of thermally isolated loops and shows the effect on the summit temperature of increasing the loop height or divergence. See also Vesecky *et al.* (1979) and Serio *et al.* (1981). The latter have modelled thermally isolated loops in hydrostatic equilibrium with a heating that declines away from the base like $\exp(-s/s_H)$; they derive scaling laws and find that loops longer than $2s_H$ develop a temperature minimum at the summit, which may be relevant to prominence formation (Section 11.1).

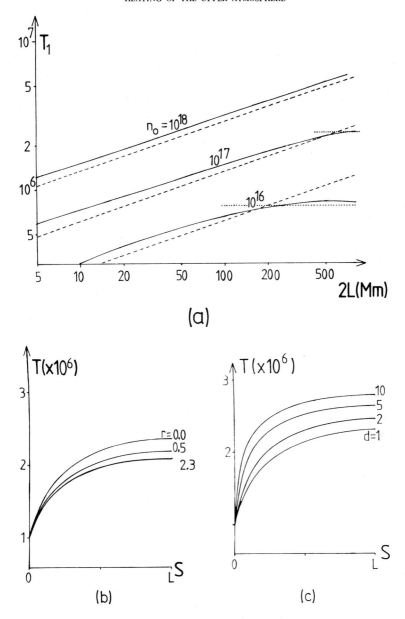

Fig. 6.12. Coronal loops in hydrostatic equilibrium. (a) The summit temperature (T_1) as a function of loop length ($2L$) in Mm ($= 10^6$ m) and base density (n_0) in m^{-3} for thermally isolated loops. The scaling laws $T_1 \sim (p_0 L)^{1/3}$ for short loops and $T_1 \sim p_0^{1/2}$ for long loops are shown dashed and dotted, respectively. (b) The effect of loop geometry on the temperature profile of an interconnecting loop of length 225 Mm and heating $\bar{h} = 7$. The ratio of loop height to footpoint separation is denoted by $\frac{1}{2}r$: $r = 0$ gives the uniform pressure case, while $r = 2.3$ gives the loop of maximum height for a given L. (c) The effect of loop divergence on the temperature profile $T(s)$ for a semicircular active-region loop of length 80 Mm and heating $\bar{h} = 20$. The ratio of summit diameter to base diameter is denoted by d. (From Wragg and Priest, 1981a.)

6.5.2. Flows in coronal loops

For the most part the effect of flows in models of the solar atmosphere has been neglected, and yet both steady and unsteady flows are universally present (Figure 1.23). The present rudimentary state of the theory of such flows is summarised by Priest (1981c) and observations of the many different types have been described in Section 1.4.1D. Ground-based observations reveal Evershed outflow (6 to 7 km s^{-1}), Evershed inflow (20 km s^{-1}), network downflow (0.1 to 2 km s^{-1}), surges (20 to 30 km s^{-1}), spicules (20 to 30 km s^{-1}) and coronal rain (50 to 100 km s^{-1}). Space observations show both transient, small-scale, fast flows (0 to 150 km s^{-1}), lasting for minutes or less, and persistent, large-scale slow flows (2 to 10 km s^{-1}), lasting for an hour or more.

From a theoretical viewpoint there are several ways of generating a flow in a coronal loop, which are briefly outlined in the remainder of this section (Figure 6.13). *Siphon flow* would be driven by a pressure difference between the two footpoints. It has been invoked by Meyer and Schmidt (1968) to explain Evershed motions along low-lying photospheric and chromospheric loops, but it may also occur along coronal loops. If one starts with a static loop and switches on a pressure difference, an *accelerated flow* will be driven from the high-pressure footpoint. But, if one starts with a loop containing a flow and then a small pressure difference is imposed in opposition to the flow, it is possible for a *decelerated* flow to be set up towards the higher pressure. There are several ways in which different footpoint pressures may be maintained. For example, the constancy of total base pressure (plasma plus magnetic) would imply that regions of high magnetic field strength possess a low plasma pressure. Also, a converging photospheric flow could compress both magnetic field and plasma, and so enhance the pressure locally. Again, a supergranular flow could drive a downflow by viscous coupling in the intense tubes (Section 8.7) that make up the boundary of a supergranule cell. Finally, the pressure at a loop footpoint may be increased by enhancing the heating there.

Coronal siphon flow has been analysed by Cargill and Priest (1980) and Noci (1981). A simple case is that of steady flow along a loop of uniform cross-section, satisfying conservation of mass, momentum and energy in the form

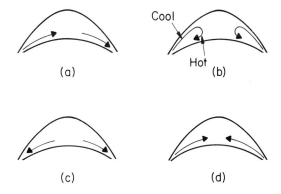

Fig. 6.13. The main types of flow in coronal loops are: (a) siphon flow, (b) spicule flow, (c) loop draining, (d) loop filling.

$$\frac{d}{ds}(\rho v) = 0; \quad \rho v \frac{dv}{ds} = -\frac{dp}{ds} - \rho g \cos\theta, \quad \frac{d}{ds}\left(\frac{p}{\rho^\gamma}\right) = 0,$$

where s is the distance measured along a loop of length $2L$ and $\theta(s)$ is the local inclination of a section of the loop to the vertical, so that for a semicircular loop $\theta = \pi s/(2L)$. The adiabatic law has been assumed for simplicity. Eliminating p and ρ between the three equations and writing $c_s(s)$ as the sound speed yields

$$\left(v - \frac{c_s^2}{v}\right)\frac{dv}{ds} = -g \cos\frac{\pi s}{2L},$$

which is similar in form to the solar wind equations (Section 12.2) and possesses a critical point ($v = c_s$) at the loop summit ($s = L$). More general solutions for a varying cross-sectional area and a full energy equation have also been produced (Cargill and Priest 1982b). The main feature is that for small pressure differences the flow is *subsonic*, but for larger pressure differences the flow becomes *supersonic* near the loop summit and is then slowed down by a shock wave in the downflowing leg (Figure 6.14).

The cause of *spicular motions* (Figure 6.13(b)), in which cool plasma is propelled up a loop leg and then falls back down, has not yet been adequately explained. One possibility is that they are driven by granular buffeting (Section 8.7.3), with a resonance between the forcing granular motion at the edge of an intense magnetic flux tube and the vertical plasma motion within the tube (Roberts, 1979). Alternatively, a similar resonance may occur when wave motions within a supergranule cell impinge on its boundaries, or spicules may be a result of a lack of thermal stability in

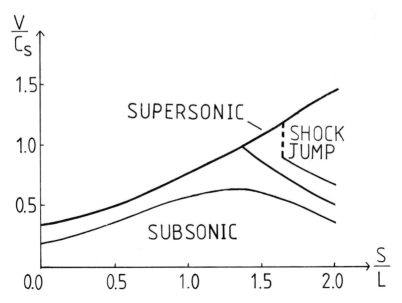

Fig. 6.14. The siphon flow speed (v) at a distance s along a converging coronal loop of length $2L = 100\,000$ km (from Cargill and Priest, 1980).

the loop plasma (Section 6.5.1B, Hood and Priest, 1980a).

Surges may be caused by the reconnection between *newly emerging* or *evolving* satellite flux and the ambient sunspot field (Heyvaerts et al., 1977). Instead, they may be the result of *non-equilibrium*, when the pressure within a closed magnetic structure exceed a critical value.

Evershed flow may be evidence of siphon flow, or it may be a result of the interaction between the convection rolls and the magnetic field within the penumbra of a sunspot.

Downflow in both legs of a loop (Figure 6.13(c)) may occur if there is a *condensation* or *prominence* at the summit of the loop. It may, alternatively, be simply plasma that is returning to the lower atmosphere from spicules and surges and it may be heated adiabatically in the process (Poletto (1980) *Ap.J.* **240**, L 69). Downflow may also take place during the formation of *cool cores* as follows. Suppose a hot loop is stretched (or its pressure increased or heating reduced) until critical conditions for thermal non-equilibrium are reached. Plasma in the core of the loop near the summit will then cool down. Since it is no longer in hydrostatic equilibrium, most of the cool plasma will drain out of the core until a new hydrostatic equilibrium is reached. Once a cool core is produced, small-scale magnetohydrodynamic instabilities may drive a circulation of plasma from the ambient corona across the interface and into the core where it falls down. The generation mechanism for this downflow may instead operate in the photosphere rather than the corona. For example, supergranular flow may drag plasma downwards in the network, or a downflow may be associated with the transient formation of an intense tube by the *intense magnetic field instability* (Section 8.7.2).

A final category of flow is upflow in both legs (Figure 6.13(d)), which may be part of an overall circulation within a coronal arcade containing a *quiescent prominence* (Section 11.1.2). It is also driven during the rise phase of a *simple loop (or compact) flare* (Section 10.2), when the presence of an extra source of heating in the loop means there is insufficient plasma for hydrostatic equilibrium, and so extra material is sucked up from below. Another way of driving an upflow, for example above a sunspot, is by Alfvén or magnetosonic waves that are propagating upwards and dumping their momentum as well as their energy in the plasma.

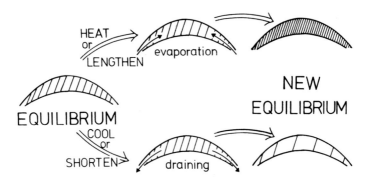

Fig. 6.15. The evolution of a coronal loop from one equilibrium to another by means of an evaporation or draining, depending on whether the heating rate (or loop length) increases or decreases in value.

Draining or filling (i.e., *evaporation*) may also take place if a loop passes through a series of equilibria with a different heating (H) or loop length (L). Suppose H or L increases in value; then Equation (6.41) implies that the new equilibrium possesses a higher density, and so extra material must be brought up (or *evaporated*) from below along the loop (Figure 6.15). Similarly, if the heating or length are reduced in value, there is too much plasma in the loop for equilibrium; some of it must drain down until the pressure gradient balances gravity and all the energy terms balance.

There are many other ways in which flows may arise. As the large-scale magnetic field evolves through a series of largely force-free states in response to the motion of the photospheric footpoints, it may occasionally find that the threshold for the onset of a *magnetohydrodynamic instability* is passed (Section 7.5); instead, there may be no magnetic equilibrium at all and a state of non-equilibrium arises (Section 6.4.4).

CHAPTER 7

INSTABILITY

7.1. Introduction

Questions of stability and instability are important for many solar phenomena. Sometimes one needs to explain how a structure can remain *stable* for a long period of time, when preliminary theoretical considerations may suggest it should be unstable. For instance, the Sun seems to maintain stability of a solar prominence or a coronal loop with great ease, in contrast to the immense difficulty of containing plasmas in the laboratory. At other times, one wants to understand why magnetic structures on the Sun suddenly become *unstable* and produce events of great beauty such as *erupting prominences* or *solar flares*.

The methods employed to investigate the linear stability of a hydromagnetic system are natural generalisations of those for studying a particle in one-dimensional motion. Suppose such a particle has a mass m and moves along the x-axis under the action of a conservative force

$$F(x)\hat{\mathbf{x}} = -\frac{dW}{dx}\hat{\mathbf{x}},$$

where $W(x)$ is the potential energy. Its equilibrium position is taken as $x = 0$ and its equation of motion is simply

$$m\ddot{x} = F(x) \equiv -\frac{dW}{dx}.$$

For small displacements, this reduces to the linear form

$$m\ddot{x} = F_1(x) \equiv -x\left(\frac{d^2W}{dx^2}\right)_0,$$

where $F_1(x)$ is the first-order approximation to $F(x)$.

One approach is to seek *normal-mode solutions* of the form $x = x_0 e^{i\omega t}$, so that the equation of motion gives

$$\omega^2 = \frac{1}{m}\left(\frac{d^2W}{dx^2}\right)_0.$$

If $W(x)$ has a minimum at the origin (Figure 7.1(a)), we have $(d^2W/dx^2)_0 > 0$, which implies that $\omega^2 > 0$, and so the particle oscillates about $x = 0$. The force acting on the particle tends to restore its equilibrium, which is therefore said to be *stable*. If, on the other hand, $W(x)$ possesses a maximum, $(d^2W/dx^2)_0 < 0$, so that $\omega^2 < 0$ and the displacement ($|x|$) increases in time from the equilibrium position, which is now *unstable*. When $(d^2W/dx^2)_0 = 0$, the particle is said to be *neutrally stable*.

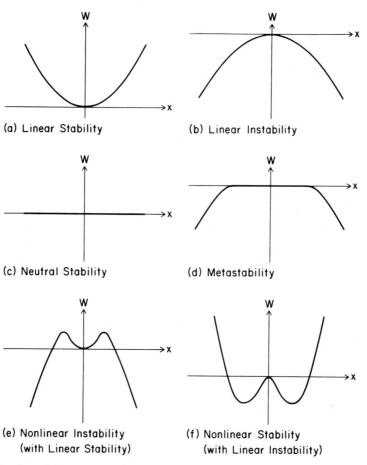

Fig. 7.1. Potential energy curves for a one-dimensional system that is in equilibrium at $x = 0$.

An alternative approach for tackling particle stability is to consider the change (δW) in *potential energy* due to a displacement (x) from equilibrium. To first order in x, $\delta W = x(\mathrm{d}W/\mathrm{d}x)_0$, which vanishes by assumption. To second order,

$$\delta W \equiv W(x) - W(0) = \frac{x^2}{2}\left(\frac{\mathrm{d}^2 W}{\mathrm{d}x^2}\right)_0,$$

which may be derived alternatively by noting that the change in potential energy is just minus the work done by the linear force ($F_1(x)$), namely,

$$\delta W = -\int_0^x F_1(x)\,\mathrm{d}x = -\tfrac{1}{2}xF_1(x). \qquad (7.1)$$

The particle is in *stable* equilibrium if $\delta W > 0$ for *all* small displacements from $x = 0$, both with $x > 0$ and $x < 0$. It is *unstable* if $\delta W < 0$ for *at least one* small displacement, either with $x > 0$ or $x < 0$. The frequency (ω) can be written in terms of δW by eliminating $(d^2W/dx^2)_0$ between the above expressions for ω^2 and δW with the result that

$$\tfrac{1}{2}m\omega^2 x^2 = \delta W. \tag{7.2}$$

The stability of a magnetohydrodynamic system is studied in a similar way. One first linearises the equations, and then either looks for normal modes or considers the variation of the energy. Each method has its advantage. With the first, one can find a *dispersion relation* linking the frequency (ω) to the wavenumber (k) of the disturbance, whereas the variational method may be applied to more complex equilibrium states. (Occasionally, the normal mode method may fail, in which case a treatment of the initial-value problem may succeed instead, e.g., Section 4.10.1 and Roberts (1967), p. 154.)

Most of this chapter refers to *linear* stability, but, by considering deviations from equilibrium that are not small, it is possible to investigate the *nonlinear* stability of a system. For instance, as indicated in Figure 7.1, a particle able to move in one dimension may be linearly stable but nonlinearly unstable (Figure 7.1 (e)), or linearly unstable but nonlinearly stable (Figure 7.1(f)). In the former case, the instability is promoted by nonlinear effects and is said to be *explosive*. The term *metastability* refers to a system that is neutrally stable ($d^2W/dx^2 = 0$) to small-amplitude (linear) perturbations but is unstable to large (finite-amplitude) ones (Figure 7.1(d)), so that, for instance, $d^3W/dx^3 = 0$ and $d^4W/dx^4 < 0$. As one varies the parameters of the system, the transfer from stability to instability occurs via a state of *marginal* (or *neutral*) *stability* in two possible ways. If ω^2 is real and it decreases through zero, we have the onset of a monotonic growth in the perturbation (*the principle of the exchange of stabilities*); the marginal state is then stationary ($\omega = 0$). If, on the other hand, the frequency (ω) in Equation (7.2) is complex and its imaginary part decreases from positive to negative values, then a state of growing oscillations appears, called *overstability* by Eddington. The marginal state in this case has oscillatory motions at a certain frequency.

The main sources for the material of this chapter are the fundamental text by Chandrasekhar (1961), which gives a comprehensive treatment of the basic instabilities from a hydrodynamical viewpoint, and Jeffrey and Taniuti (1966), which summarises work relevant to the problem of containing plasma in the laboratory. After a derivation of the linearised equations (Section 7.2), the use of the two main techniques, namely the normal mode method (Section 7.3) and the energy method (Section 7.4) is illustrated, and then the properties of some of the main instabilities are summarised in Section 7.5.

7.2. Linearised Equations

The behaviour of an ideal (dissipationless) hydromagnetic system is governed by the basic Equations (2.13), (2.14), (2.15), (2.19) and (2.21) of Section 2.1 in the limit as the magnetic diffusivity and viscosity approach zero, namely,

$$\frac{\partial \mathbf{B}}{\partial t} = \nabla \times (\mathbf{v} \times \mathbf{B}), \tag{7.3}$$

$$\rho \frac{D\mathbf{v}}{Dt} = -\nabla p + \mathbf{j} \times \mathbf{B} + \rho \mathbf{g}, \tag{7.4}$$

$$\frac{\partial \rho}{\partial t} + \nabla \cdot (\rho \mathbf{v}) = 0, \tag{7.5}$$

where $\mathbf{j} = \nabla \times \mathbf{B}/\mu$ and $\nabla \cdot \mathbf{B} = 0$. Also, the energy equation (2.28d) is approximated here, for simplicity, by

$$\frac{D}{Dt}\left(\frac{p}{\rho^\gamma}\right) = 0.$$

The condition for the validity of this equation is that the time-scale of interest be much smaller than the time it takes heat to be conducted or radiated away. More general forms can easily be adopted instead (see, for instance, Heyvaerts (1974), who considers thermal instability in the presence of a magnetic field).

Suppose the equilibrium is static (with vanishing velocity) and has a magnetic field ($\mathbf{B}_0$), plasma pressure (p_0), density (ρ_0) and electric current ($\mathbf{j}_0$), all independent of time; then Equations (7.3) and (7.5) are satisfied trivially and the remaining equations yield

$$0 = -\nabla p_0 + \mathbf{j}_0 \times \mathbf{B}_0 + \rho_0 \mathbf{g}, \tag{7.6}$$

$$\mathbf{j}_0 = \nabla \times \mathbf{B}_0/\mu, \tag{7.7}$$

$$\nabla \cdot \mathbf{B}_0 = 0. \tag{7.8}$$

Next, perturb the equilibrium by setting

$$\rho = \rho_0 + \rho_1, \quad \mathbf{v} = \mathbf{v}_1, \quad p = p_0 + p_1,$$
$$\mathbf{B} = \mathbf{B}_0 + \mathbf{B}_1, \quad \mathbf{j} = \mathbf{j}_0 + \mathbf{j}_1,$$

and ignore squares and products of the perturbation quantities (denoted by subscript 1); the basic equations then reduce to

$$\frac{\partial \mathbf{B}_1}{\partial t} = \nabla \times (\mathbf{v}_1 \times \mathbf{B}_0), \tag{7.9}$$

$$\rho_0 \frac{\partial \mathbf{v}_1}{\partial t} = -\nabla p_1 + \mathbf{j}_1 \times \mathbf{B}_0 + \mathbf{j}_0 \times \mathbf{B}_1 + \rho_1 \mathbf{g}, \tag{7.10}$$

$$\frac{\partial \rho_1}{\partial t} + \nabla \cdot (\rho_0 \mathbf{v}_1) = 0, \tag{7.11}$$

$$\mathbf{j}_1 = \nabla \times \mathbf{B}_1/\mu, \quad \nabla \cdot \mathbf{B}_1 = 0, \tag{7.12}$$

$$\frac{\partial p_1}{\partial t} = \frac{\gamma p_0}{\rho_0}\frac{\partial \rho_1}{\partial t} - p_0 (\mathbf{v}_1 \cdot \nabla) \log_e\left(\frac{p_0}{\rho_0^\gamma}\right). \tag{7.13}$$

The displacement ($\boldsymbol{\xi}(\mathbf{r}_0, t)$) of a plasma element from equilibrium may be written $\boldsymbol{\xi} = \mathbf{r} - \mathbf{r}_0$, in terms of the (*Eulerian*) position vector ($\mathbf{r}$) and the (*Lagrangian*) equilibrium position ($\mathbf{r}_0$) (Figure 7.2). By writing equations (7.9) to (7.13) in terms of $\boldsymbol{\xi}$ alone and changing from Eulerian variables ($\mathbf{r}, t$) to Lagrangian coordinates ($\mathbf{r}_0, t$), it is possible to reduce the whole set of linearised equations to just one equation, as

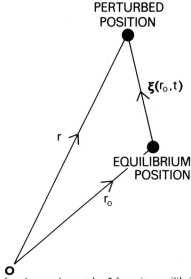

Fig. 7.2. The displacement of a plasma element by ξ from its equilibrium position (r_0) relative to the origin O.

follows. To lowest order, the vector operator ∇ is identical with

$$\nabla_0 \equiv \hat{x}\frac{\partial}{\partial x_0} + \hat{y}\frac{\partial}{\partial y_0} + \hat{z}\frac{\partial}{\partial z_0}.$$

Furthermore, the perturbed velocity vector may be written $v_1 = Dr/Dt$, or, since r_0 is independent of time,

$$v_1 = \frac{\partial \xi}{\partial t}. \tag{7.14}$$

Then, the fact that B_1 and ξ vanish initially may be used to integrate Equation (7.9) with respect to time to give

$$B_1 = \nabla_0 \times (\xi \times B_0). \tag{7.15}$$

Similarly, the integral of Equation (7.11) is

$$\rho_1 = -\nabla \cdot (\rho_0 \xi), \tag{7.16}$$

and, after substituting for j_1, B_1 and ρ_1 from Equations (7.12), (7.15) and (7.16), respectively, Equation (7.10) becomes

$$\rho_0 \frac{\partial^2 \xi}{\partial t^2} = F(\xi(r_0, t)), \tag{7.17}$$

where

$$\begin{aligned} F(\xi) &\equiv -\nabla p_1 + \rho_1 g + j_1 \times B_0 + j_0 \times B_1 \\ &= -\nabla p_1 + \nabla \cdot (\rho_0 \xi) g + (\nabla \times [\nabla \times (\xi \times B_0)]) \times B_0/\mu + \\ &\quad + (\nabla \times B_0) \times [\nabla \times (\xi \times B_0)]/\mu. \end{aligned} \tag{7.18}$$

To first order, the perturbation force (F) is a linear functional of ξ and its spatial derivatives, so that, for a displacement of the form

$$\xi(\mathbf{r}_0, t) = \xi(\mathbf{r}_0) e^{i\omega t}, \tag{7.19}$$

the equation of motion (7.17) reduces to

$$-\omega^2 \rho_0 \xi(\mathbf{r}_0) = \mathbf{F}(\xi(\mathbf{r}_0)). \tag{7.20}$$

Both the normal mode and energy methods may use this equation as a basis.

In general, the time integral of Equation (7.13) determines p_1 as

$$p_1 = \frac{\gamma p_0}{\rho_0} \rho_1 - p_0 (\xi \cdot \nabla) \log_e \frac{p_0}{\rho_0^\gamma},$$

and two special cases are of interest. Firstly, for a *uniform equilibrium*, $\rho_0, p_0, \mathbf{B}_0$ are all constant, and Equation (7.18) reduces to

$$\mathbf{F} = \rho_0 c_s^2 \nabla(\nabla \cdot \xi) - \nabla \cdot (\rho_0 \xi)\mathbf{g} + \rho_0 v_A^2 (\nabla \times [\nabla \times (\xi \times \hat{\mathbf{B}}_0)]) \times \hat{\mathbf{B}}_0, \tag{7.21}$$

where $c_s = (\gamma p_0/\rho_0)^{1/2}$ is the sound speed and $v_A = B_0/(\mu \rho_0)^{1/2}$ is the Alfvén speed. The other special case occurs when density variations are *incompressible* (Section 2.4.2). This is obtained by letting the sound speed (c_s) approach infinity. Then the time integral of Equation (7.13) implies that ρ_1 tends to zero and Equation (7.13) no longer determines p_1.

7.3. Normal Mode Method

Once the boundary conditions and the equilibrium configuration have been prescribed, the perturbed variables ($\rho_1, \mathbf{v}_1, p_1, \mathbf{j}_1$ and $\mathbf{B}_1$) are determined by the set of linear Equations (7.9) to (7.13). Each of these variables may be decomposed into a spectrum of Fourier components, which behave like $e^{i\omega t}$. The resulting 'normal mode' equations may be solved to determine which values of ω are allowed by the boundary conditions. If *all* the normal modes have real frequencies ($\omega^2 > 0$), the system just oscillates about the equilibrium configuration, which is therefore *stable*. If at least one of the frequencies is imaginary ($\omega^2 < 0$), the system is *unstable*, since the corresponding perturbations grow exponentially.

The method of normal modes is applied below to a particular example in which the original set of Equations (7.9) to (7.13) is used; but one could just as well solve the normal mode equation

$$-\omega^2 \rho_0 \xi(\mathbf{r}_0) = \mathbf{F}(\xi(\mathbf{r}_0)) \tag{7.20}$$

for the displacement $\xi(\mathbf{r}_0, t) = \xi(\mathbf{r}_0) e^{i\omega t}$ and deduce the original variables from Equations (7.14) to (7.16), (7.12) and (7.13). Sturm–Liouville theory can be used to show that the eigenvalues (ω^2) of Equation (7.20) are all real. Thus overstability cannot occur in a static, ideal magnetohydrodynamic system; however, the presence of shear flow, rotation or dissipation may produce overstability (e.g., Section 8.1).

7.3.1. Example: Rayleigh–Taylor instability

As an example of the use of the normal mode method, consider a boundary separating two perfectly conducting plasmas, whose properties are denoted by superscripts +

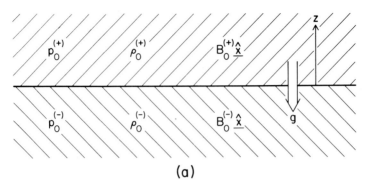

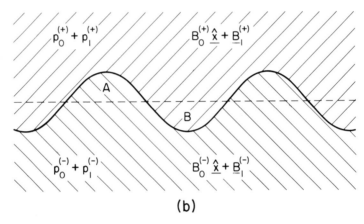

Fig. 7.3. (a) The interface between two plasmas in equilibrium. (b) The perturbed interface.

and $-$, as shown in Figure 7.3(a). The boundary is situated at $z = 0$, and a gravitational force acts normal to it in the direction of the negative z-axis; the undisturbed magnetic field is parallel to the boundary along the x-axis, and discontinuous jumps in the variables are allowed at it.

Consider first the case when no magnetic field is present, and suppose the interface receives the sinusoidal displacement shown in Figure 7.3(b); plasma of density $\rho_0^{(+)}$ now occupies the region B (previously of density $\rho_0^{(-)}$), while plasma of density $\rho_0^{(-)}$ occupies region A (previously of density $\rho_0^{(+)}$). The net result is that plasma of density $\rho_0^{(+)}$ in A has been exchanged with plasma of density $\rho_0^{(-)}$ in B. If $\rho_0^{(+)} < \rho_0^{(-)}$, this represents a gain of gravitational potential energy, since the net displacement is against gravity. But, if $\rho_0^{(+)} > \rho_0^{(-)}$, so that the denser plasma rests on top of the lighter, a loss of gravitational potential energy results and we have instability – the *Rayleigh–Taylor instability*.

The influence of a horizontal magnetic field on this instability depends on the orientation of the field parallel to the interface. For a magnetic field in the plane of Figure 7.3, the perturbation stretches the field lines and so produces a stabilising force.

Magnetic field lines normal to the plane of Figure 7.3, however, are just displaced as a whole without being stretched and therefore have no stabilising effect. In order to derive the above results mathematically and find the quantitative effect of the magnetic field, two special cases will be considered of the equilibrium configuration shown in Figure 7.3. In the first case, plasma above $z = 0$ is supported by a magnetic field below $z = 0$; in the second, the initial magnetic field is uniform, with a dense plasma resting on top of a lighter one. Incompressible variations alone are considered, so that

$$\mathbf{\nabla} \cdot \mathbf{v}_1 = 0, \tag{7.22}$$

which is often the case for the most unstable displacement, since energy is then not wasted in compressing the plasma unnecessarily (see Section 7.4.2 (e)).

7.3.1A. *Plasma Supported by a Magnetic Field*

Consider a plasma of uniform density ($\rho_0^{(+)}$) supported against gravity by a magnetic field ($B_0^{(-)}\hat{\mathbf{x}}$), with the z-axis normal to the plane interface indicated in Figure 7.4(a).

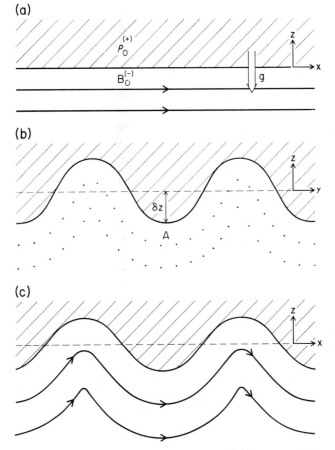

Fig. 7.4. Plasma (shaded) supported by a magnetic field ($B_0^{(-)}\hat{\mathbf{x}}$). (a) Equilibrium configuration. (b) Perturbations rippled in the y-direction. (c) Perturbations rippled in the x-direction.

Assume that the plasma in the region below $z=0$ has a negligible pressure ($p_0^{(-)} \ll B_0^{(-)2}/(2\mu)$) and that the interface itself is in equilibrium under the pressure balance $p_0^{(+)} = B_0^{(-)2}/(2\mu)$. Suppose also that the boundary develops oscillations along the y-direction alone, so that

$$\mathbf{v}_1 = e^{i\omega t} \mathbf{v}_1(z) e^{iky} \tag{7.23}$$

and similarly for all the other variables.

First of all, it is necessary to derive the boundary condition at the interface, which may be regarded as a thin layer where variations in z are much more rapid than those in x or y. The z-component of the time-derivative of Equation (7.10) is

$$\rho_0 \frac{\partial^2 v_{1z}}{\partial t^2} = -\frac{\partial}{\partial t}\left(\frac{\partial p_1}{\partial z}\right) - \frac{\partial j_{1y}}{\partial t} B_{0x} - j_{0y}\frac{\partial B_{1x}}{\partial t} - \frac{\partial \rho_1}{\partial t} g. \tag{7.24}$$

However, from Equations (7.9), (7.11) and (7.12)

$$\frac{\partial \rho_1}{\partial t} = -\frac{d}{dz}(\rho_0 v_{1z}), \quad j_{1y} = \frac{1}{\mu}\frac{\partial B_{1x}}{\partial z} \quad \text{and} \quad \frac{\partial B_{1x}}{\partial t} = 0;$$

and so, for the perturbation (7.23), Equation (7.24) reduces to

$$-\rho_0 \omega^2 v_{1z} = -\omega \frac{dp_1}{dz} + g\frac{d}{dz}(\rho_0 v_{1z}).$$

After being integrated across the interface, this becomes

$$0 = -i\omega[p_1] + g[\rho_0 v_{1z}], \tag{7.25}$$

where square brackets denote the jump in a quantity across the interface, so that, for instance, $[p_1] \equiv p_1^{(+)} - p_1^{(-)}$. Moreover, perturbations of the form (7.23) make Equation (7.22) and the y-component of Equation (7.10) reduce to the pair

$$\frac{dv_{1z}}{dz} + ikv_{1y} = 0, \quad i\rho_0 \omega v_{1y} = -ikp_1,$$

which may be combined to give

$$p_1 = -\frac{i\omega \rho_0}{k^2}\frac{dv_{1z}}{dz}.$$

After using the result from the integral of Equation (7.22) that v_{1z} is continuous and substituting for p_1, Equation (7.25) becomes

$$0 = -\frac{\omega^2}{k^2}\left[\rho_0 \frac{dv_{1z}}{dz}\right] + gv_{1z}[\rho_0],$$

which for our interface reduces finally to the required boundary condition

$$0 = -\frac{\omega^2}{k^2}\rho_0^{(+)}\left(\frac{dv_{1z}}{dz}\right)^{(+)} + gv_{1z}^{(+)}\rho_0^{(+)}. \tag{7.26}$$

The next aim is to derive the solution for v_{1z} on both sides of the interface. Assuming that the initial equilibrium values ($\mathbf{B}_0, \rho_0, p_0$) are uniform and that Equation (7.22) holds, Equations (7.9) and (7.11) give

$$\frac{\partial \mathbf{B}_1}{\partial t} = \mathbf{0}, \qquad \frac{\partial \rho_1}{\partial t} = 0,$$

so that $\mathbf{j}_1$, $\mathbf{B}_1$ and ρ_1 all vanish and Equation (7.10) simplifies to

$$\rho_0 \frac{\partial \mathbf{v}_1}{\partial t} = -\nabla p_1.$$

The curl of this gives $\nabla \times \mathbf{v}_1 = \mathbf{0}$, which, together with Equation (7.22), implies that $\nabla^2 \mathbf{v}_1 = \mathbf{0}$.

For perturbations (7.23), the z-component becomes

$$\frac{d^2 v_{1z}}{dz^2} - k^2 v_{1z} = 0,$$

and the continuous solution that vanishes as z approaches $\pm\infty$ is

$$v_{1z} = \begin{cases} e^{-kz}, & z > 0, \\ e^{kz}, & z < 0. \end{cases} \qquad (7.27)$$

From this solution $v_{1z}^{(+)} = 1$ and $dv_{1z}^{(+)}/dz = -k$, so that the boundary condition (7.26) gives $\omega^2 = -gk$. Ripples of the boundary in the y-direction (Figure 7.4(b)) therefore grow like $\exp\left[(gk)^{1/2} t\right]$, the smallest wavelengths being the fastest.

An alternative explanation for the instability (to the one mentioned above when discussing Figure 7.3) comes from considering the point A in Figure 7.4(b). It is depressed by a distance $|\delta z|$, so that a greater mass of plasma exists above it. The plasma pressure at A is therefore larger than $p_0^{(+)}$ by an amount $\rho g |\delta z|$, and this forces the perturbation to grow.

More general perturbations of the form

$$\mathbf{v}_1 = e^{i\omega t} v_1(z) e^{i(k_x x + k_y y)} \qquad (7.28)$$

lead to the boundary condition

$$0 = -\frac{\omega^2}{k^2} \rho_0^{(+)} \left(\frac{dv_{1z}}{dz}\right)^{(+)} - \frac{k_x^2 B_0^{(-)2}}{k^2 \mu} \left(\frac{dv_{1z}}{dz}\right)^{(-)} + g v_{1z}^{(+)} \rho_0^{(+)}$$

and the corresponding dispersion relation

$$\omega^2 = -gk + \frac{k_x^2 B_0^{(-)2}}{\mu \rho_0^{(+)}}, \qquad (7.29)$$

where $k = (k_x^2 + k_y^2)^{1/2}$. From this it can be seen that ripples along the direction of the field with $k_y = 0$ (Figure 7.4(c)) are unstable at long wavelengths $0 < k < k_{\text{crit}}$, where

$$k_{\text{crit}} = \frac{g\mu\rho_0^{(+)}}{B_0^{(-)2}}; \qquad (7.30)$$

but they are stable at short enough wavelengths ($k > k_{\text{crit}}$), since then the magnetic tension dominates. The wavenumber of the most rapidly growing mode is $\frac{1}{2}k_{\text{crit}}$.

This magnetic analogue of the Rayleigh–Taylor instability is known as the *Kruskal–*

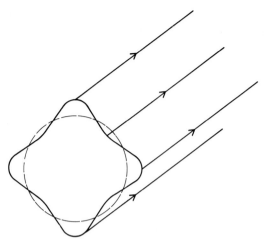

Fig. 7.5. The fluted cross-section of a cylindrical column of plasma that is subject to an interchange instability and whose undisturbed cross-section is circular (shown dashed).

Schwarzschild (or *hydromagnetic Rayleigh–Taylor*) *instability*. Both are examples of a wider class of *interchange instabilities*, which includes the *sausage instability* (Figure 7.10). Another example is the *flute instability* of plasma contained in a flux tube (Section 7.5.1). Other effects that can be included in the above analysis are: dissipation (Furth *et al.*, 1963), nonuniformities in conditions either side of the interface (so that $(dB_0/dz)^{(+)} \neq 0$ in deriving the equivalent of Equation (7.26)), and different inclinations of the magnetic field each side of $z = 0$ (see, for example, Boyd and Sanderson (1969), page 75, Nakagawa and Malville (1969) or Uchida, Y. and Sakurai, T. (1977) *Solar Phys.* **51**, 413). Also, allowance can be made for steady motions in the initial state (Chandrasekhar, 1961) or for compressible variations.

7.3.1B. *Uniform Magnetic Field* ($B_0^{(+)} = B_0^{(-)}$)

Consider now a uniform horizontal magnetic field ($B_0 \hat{x}$) parallel to the interface that separates uniform plasmas with densities $\rho_0^{(+)}$ and $\rho_0^{(-)}$ (see Figure 7.3). The velocity perturbation again has the form given by Equation (7.27), and the boundary condition for perturbations behaving like Equation (7.28) can be shown to be

$$\left(\frac{\omega^2}{k^2}\rho_0^{(+)} + \frac{k_x^2 B_0^2}{k^2 \mu}\right)\left(\frac{dv_{1z}}{dz}\right)^{(+)} + \left(\frac{\omega^2}{k^2}\rho_0^{(-)} - \frac{k_x^2 B_0^2}{k^2 \mu}\right)\left(\frac{dv_{1z}}{dz}\right)^{(-)}$$
$$= -g(v_{1z}^{(+)}\rho_0^{(+)} - v_{1z}^{(-)}\rho_0^{(-)}),$$

where $k = (k_x^2 + k_y^2)^{1/2}$. After substituting $(dv_{1z}/dz)^{(+)} = -(dv_{1z}/dz)^{(-)} = -k$ and $v_{1z}^{(+)} = v_{1z}^{(-)} = 1$, it reduces to

$$\omega^2 = -gk\frac{\rho_0^{(+)} - \rho_0^{(-)}}{\rho_0^{(+)} + \rho_0^{(-)}} + \frac{2B_0^2 k_x^2}{\mu(\rho_0^{(+)} + \rho_0^{(-)})}. \tag{7.31}$$

This dispersion relation implies that, when there is no magnetic field present ($B_0 = 0$), the interface is unstable ($\omega^2 < 0$) provided the heavy fluid rests on top of the light one ($\rho_0^{(+)} > \rho_0^{(-)}$): – the *Rayleigh–Taylor instability*. Furthermore, for disturbances that are uniform along the magnetic field direction ($k_x = 0$), the magnetic field has no effect on stability; whereas undulations of the boundary purely *along* the field ($k_y = 0$, $k_x = k$) make the second term in Equation (7.31) positive and so allow a stabilising effect. In the latter case, when $\rho_0^{(+)} > \rho_0^{(-)}$ the interface is unstable for wavenumbers

$$0 < k < k_c, \qquad (7.32)$$

where

$$k_c = \frac{(\rho_0^{(+)} - \rho_0^{(-)})g\mu}{2B_0^2}, \qquad (7.33)$$

and the wavenumber for the fastest growing mode is $\tfrac{1}{2}k_c$. For large wavelengths ($\lambda > 2\pi/k_c$), the magnetic tension is insufficient to counteract gravity; but, for short wavelengths ($\lambda < 2\pi/k_c$), the magnetic tension is strong enough to make the interface stable. This is similar to the effect of surface tension on the Rayleigh–Taylor instability in a liquid.

7.4. Variational (or Energy) Method

By analogy with the method outlined in Section 7.1 for one-dimensional dynamics, the object here is to construct the (second-order) change (δW) in the potential energy of the system when the plasma element at $\mathbf{r}_0$ is displaced by an amount $\boldsymbol{\xi}(\mathbf{r}_0, t)$ from equilibrium. This follows the classic treatment of Bernstein *et al.* (1958), who extended an earlier energy principle of Lundquist (1951). The system is stable if all possible displacements make $\delta W > 0$, and it is unstable if at least one displacement makes $\delta W < 0$. Stability of our ideal magnetohydrodynamic system guarantees stability in both the 'quasi-magnetohydrodynamic' and 'double-adiabatic' magnetohydrodynamic descriptions (Kruskal and Oberman, 1958; Chew *et al.*, 1956), but not vice versa.

For a plasma in equilibrium, with gravity acting in the negative z-direction, the potential energy is

$$W_0 = \int (B_0^2/(2\mu) + \rho_0 U_0 + \rho_0 g z) \, dV,$$

where the first and last terms represent the magnetic and gravitational energies; U is the internal energy per unit mass, and the integration is performed over the whole volume occupied by the plasma. The first-order perturbations ($\rho_1, p_1, \mathbf{B}_1$) from equilibrium are given in terms of the displacement vector ($\boldsymbol{\xi}$) by Equations (7.16), (7.13) and (7.15), and so the resulting potential energy may be written in terms of $\boldsymbol{\xi}$ by substituting for $\rho_0 + \rho_1, p_0 + p_1, \mathbf{B}_0 + \mathbf{B}_1$ into an integral of the above form. To first order in $\boldsymbol{\xi}$, the change in potential energy vanishes because of the initial equilibrium state. To second order, an expression for δW may be derived most simply in the following way. A displacement of the form

$$\boldsymbol{\xi}(\mathbf{r}_0, t) = \boldsymbol{\xi}(\mathbf{r}_0) e^{i\omega t} \qquad (7.34)$$

turns the linearised equation of motion into
$$-\omega^2 \rho_0 \boldsymbol{\xi}(\mathbf{r}_0) = \mathbf{F}(\boldsymbol{\xi}(\mathbf{r}_0)), \tag{7.35}$$
where $\mathbf{F}$ (given by Equation (7.18)) is the linearised force per unit volume acting on the plasma. The change in potential energy is then just minus the work done by this force during the displacement, namely
$$\delta W = -\tfrac{1}{2} \int \boldsymbol{\xi} \cdot \mathbf{F} \, dV; \tag{7.36}$$
the factor $\tfrac{1}{2}$ arises because the mean force during the displacement from $\mathbf{0}$ to $\boldsymbol{\xi}$ is $\tfrac{1}{2}\mathbf{F}(\boldsymbol{\xi})$. Furthermore, the scalar product of Equation (7.35) with $\boldsymbol{\xi}$ and an integration over the volume give
$$\omega^2 \int \tfrac{1}{2} \rho_0 \xi^2 \, dV = \delta W. \tag{7.37}$$
The left-hand side is minus the plasma kinetic energy, so that, as expected, the kinetic energy gain is positive if the potential energy decreases ($\delta W < 0$) due to a positive amount of work done. Note that Equations (7.36) and (7.37) are analogous to the Equations (7.1) and (7.2) for particle dynamics.

The object of the energy method is to investigate stability subject to a chosen class of perturbations ($\boldsymbol{\xi}$). This is accomplished by minimising δW and determining its sign. For the particular case where gravity is neglected, it can be shown (Cowling, 1976, p. 33) that, for given components of $\boldsymbol{\xi}$ normal to $\mathbf{B}_0$, the component of $\boldsymbol{\xi}$ along $\mathbf{B}_0$ that minimises δW is given by $\nabla \cdot \boldsymbol{\xi} = 0$. In other words, the most unstable perturbation is incompressible, which is to be expected since energy would be needed to compress or expand the original plasma. Thus, when searching for instability, one need usually only consider perturbations which make $\nabla \cdot \boldsymbol{\xi} = 0$ and so forget the term arising from plasma pressure in Equation (7.36).

It proves convenient to rewrite the expression (7.36) for δW, namely
$$\delta W = \tfrac{1}{2} \int \boldsymbol{\xi} \cdot \nabla p_1 - \boldsymbol{\xi} \cdot \rho_1 \mathbf{g} - \boldsymbol{\xi} \cdot (\mathbf{j}_1 \times \mathbf{B}_0 + \mathbf{j}_0 \times \mathbf{B}_1). \tag{7.38}$$
After substituting for $\mathbf{j}_1$ and $\mathbf{B}_1$ from Equations (7.12) and (7.15), the last two terms on the right-hand side of Equation (7.38) may be rearranged, using vector identities as follows:
$$\int \boldsymbol{\xi} \cdot (\mathbf{j}_1 \times \mathbf{B}_0 + \mathbf{j}_0 \times \mathbf{B}_1) \, dV = \int \boldsymbol{\xi} \cdot ((\nabla \times \mathbf{B}_1) \times \mathbf{B}_0/\mu + \mathbf{j}_0 \times \mathbf{B}_1) \, dV$$
$$= -\mu^{-1} \int \boldsymbol{\xi} \times \mathbf{B}_0 \cdot (\nabla \times \mathbf{B}_1) \, dV$$
$$+ \int \boldsymbol{\xi} \cdot \mathbf{j}_0 \times \mathbf{B}_1 \, dV$$
$$= -\mu^{-1} \int \nabla \cdot [(\boldsymbol{\xi} \times \mathbf{B}_0) \times \mathbf{B}_1]$$
$$+ [\nabla \times (\boldsymbol{\xi} \times \mathbf{B}_0)] \cdot \mathbf{B}_1 \, dV + \int \mathbf{j}_0 \cdot \mathbf{B}_1 \times \boldsymbol{\xi} \, dV,$$

or, using the Gauss divergence theorem on the first integral and Equation (7.15) for $\mathbf{B}_1$,

$$= -\mu^{-1}\int[(\boldsymbol{\xi}\times\mathbf{B}_0)\times\mathbf{B}_1]\cdot d\mathbf{S} - \mu^{-1}\int B_1^2\, dV + \int \mathbf{j}_0\cdot\mathbf{B}_1\times\boldsymbol{\xi}\, dV.$$

If there is no displacement ($\boldsymbol{\xi}$) on the boundary of the volume, the surface integral vanishes and similarly

$$\int \boldsymbol{\xi}\cdot\nabla p_1\, dV = -\int p_1(\nabla\cdot\boldsymbol{\xi})\, dV.$$

Then, after substituting for ρ_1 from Equation (7.16) and p_1 from Equation (7.13), the change in potential energy given by Equation (7.38) becomes

$$\delta W = \tfrac{1}{2}\int \bigl[B_1^2/\mu - \mathbf{j}_0\cdot(\mathbf{B}_1\times\boldsymbol{\xi}) + \gamma(p_0/\rho_0)\nabla\cdot(\rho_0\boldsymbol{\xi})(\nabla\cdot\boldsymbol{\xi})$$

$$+ (\boldsymbol{\xi}\cdot\mathbf{g})\nabla\cdot(\rho_0\boldsymbol{\xi})\bigr]\, dV, \qquad (7.39)$$

where $\mathbf{B}_1$ is given by Equation (7.15).

7.4.1. EXAMPLE: KINK INSTABILITY

Consider a cylindrically symmetric magnetic flux tube, whose equilibrium magnetic field in cylindrical polar coordinates (R, ϕ, z) is

$$\mathbf{B}_0 = B_{0\phi}(R)\hat{\boldsymbol{\phi}} + B_{0z}(R)\hat{\mathbf{z}}, \qquad (7.40)$$

and whose corresponding electric current density is

$$\mathbf{j}_0 = -\frac{dB_{0z}}{dR}\hat{\boldsymbol{\phi}} + \frac{1}{R}\frac{d}{dR}(RB_{0\phi})\hat{\mathbf{z}}.$$

Suppose now the field is force-free, with $\mathbf{j}_0\times\mathbf{B}_0 = 0$, or equivalently

$$\frac{d}{dR}(B_{0\phi}^2 + B_{0z}^2) = -\frac{2B_{0\phi}^2}{R}; \qquad (7.41)$$

this determines $\mathbf{B}_0$ after either $B_{0\phi}(R)$ or $B_{0z}(R)$ has been prescribed.

An intuitive explanation for the instability of the flux tube to a lateral kink-like perturbation (Figure 7.6) may be given as follows. Suppose that the magnetic field lines outside the equilibrium flux tube have only a ϕ-component, as shown in Figure 7.6(a). Then, after making the kink-like displacement in Figure 7.6(b), the field lines at A are closer together than those at B. The magnetic pressure is therefore stronger at A than B, and the resulting force from A to B is such as to increase the perturbation further.

This simple argument is modified by the presence of certain stabilising features, such as a magnetic field component along the axis of the tube, a gas pressure gradient, or line-tying at the ends of a tube anchored in the photosphere. Bernstein's energy method was applied to a magnetic flux tube by Newcomb (1960) and later, to a force-free tube in particular, by Callebaut and Voslamber (1962) and Anzer (1968).

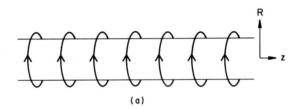

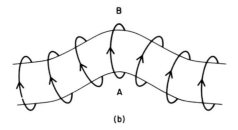

Fig. 7.6. (a) An equilibrium plasma tube surrounded by azimuthal field lines. (b) A lateral kink-like perturbation of the tube.

The effect of line-tying was included by Raadu (1972), that of pressure gradients by Giachetti et al. (1977) and that of both together by Hood and Priest (1979b), as outlined in Section 10.2.3. Here the treatment of Raadu is presented, the main object being to illustrate the method rather than obtain particular results.

In the force-free approximation, the gravitational and pressure forces are negligible in the equilibrium. Omitting them also in the perturbed state reduces the general expression (7.39) for the second-order potential energy to

$$\delta W = (2\mu)^{-1} \int B_1^2 - \mathbf{B}_1 \cdot \boldsymbol{\xi} \times (\nabla \times \mathbf{B}_0) \, dV, \tag{7.42}$$

where

$$\mathbf{B}_1 = \nabla \times (\boldsymbol{\xi} \times \mathbf{B}_0). \tag{7.43}$$

Here, only a certain class of perturbation ($\boldsymbol{\xi}$) will be considered for simplicity. If the resulting smallest value of δW is negative, the system is certainly unstable; but, if it is positive, the system is stable only to that class of perturbation.

The form adopted for the perturbation is

$$\boldsymbol{\xi} = \boldsymbol{\xi}^* f(z), \tag{7.44}$$

where

$$\boldsymbol{\xi}^* = \left[\xi^R(R)\hat{\mathbf{R}} - i\frac{B_{0z}}{B_0}\xi^0(R)\hat{\boldsymbol{\phi}} + i\frac{B_{0\phi}}{B_0}\xi^0(R)\hat{\mathbf{z}} \right] e^{i(m\phi + kz)}$$

and $f(0) = f(2L) = 0$. This has several important properties. It vanishes at the ends ($z = 0$ and $z = 2L$) of the flux rope and so satisfies rigid boundary conditions there;

this is needed if the ends are located in the photosphere with the bulk of the rope in the corona. If the functions $\xi^0(R)$ and $\xi^R(R)$ in (7.44) are purely real, the radial component is out of phase by $\tfrac{1}{2}\pi$ with the other components. Also, the perturbation has been constrained to lie in a direction normal to $\mathbf{B}_0$ (i.e., $\boldsymbol{\xi}\cdot\mathbf{B}_0 = 0$), since, by Equations (7.42) and (7.43), perturbations parallel to $\mathbf{B}_0$ have no effect on the energy. Furthermore, the presence of the factor $e^{i(m\phi + kz)}$ means that the kinking is helical, screwing out of the plane of Figure 7.6 rather than just being lateral (which would be represented by a factor such as $\cos kz \cos \phi$). The object of the analysis is to choose the arbitrary functions $f(z)$, $\xi^0(R)$ and $\xi^R(R)$ in turn in such a way as to minimise δW.

Consider first the choice of $f(z)$. Integration of Equation (7.42) with respect to R and ϕ gives

$$\delta W = \int_0^l A\left(\frac{df}{dz}\right)^2 + Cf^2\, dz, \tag{7.45}$$

where

$$A = \int_{R=0}^{\infty}\int_{\phi=0}^{2\pi} [\hat{\mathbf{z}} \times (\boldsymbol{\xi}^* \times \mathbf{B}_0)]^2 R\, dR\, d\phi \tag{7.46}$$

and

$$C = \int_{R=0}^{\infty}\int_{\phi=0}^{2\pi} [\nabla \times (\boldsymbol{\xi}^* \times \mathbf{B}_0)]\cdot[\nabla \times (\boldsymbol{\xi}^* \times \mathbf{B}_0)$$
$$- (\nabla \times \mathbf{B}_0) \times \boldsymbol{\xi}^*]R\, dR\, d\phi \tag{7.47}$$

are both constants. The aim is to minimise the integral in Equation (7.45) subject to a normalisation constraint

$$\int_0^{2L} f^2(z)\, dz = 1, \tag{7.48}$$

so that, after introducing a Lagrange multiplier (λ), we have

$$\delta W + \lambda = \int_0^{2L} \mathscr{F}(f, f')\, dz,$$

where $\mathscr{F}(f,f') = A f'^2 + (C + \lambda) f^2$. Then, from standard variational theory, the minimising function satisfies the *Euler–Lagrange equation*

$$\frac{\partial \mathscr{F}}{\partial f} = \frac{d}{dz}\left(\frac{\partial \mathscr{F}}{\partial f'}\right),$$

or $(C + \lambda)f = Af''$. The lowest-energy solution that satisfies the boundary conditions $f(0) = f(2L) = 0$ and the normalisation (7.48) is

$$f(z) = \left(\frac{1}{L}\right)^{1/2} \sin\frac{\pi z}{2L},$$

for which Equation (7.45) becomes

$$\delta W = \left(\frac{\pi}{l}\right)^2 A + C. \tag{7.49}$$

Forcing the ends of the flux tube to remain fixed has a stabilising effect, as can be seen from the presence of the positive-definite term $(\pi/l)^2 A$, which vanishes in the absence of such a constraint.

Next, perform the integrals over ϕ in the expressions (7.46) and (7.47) for A and C, with the result that the energy perturbation (δW) reduces to

$$\delta W = \int_0^\infty F\left(\frac{d\xi^R}{dR}\right)^2 - G\xi^{R2} + \frac{(k^2+h^2)R^2+1}{R}(B_0\xi^0 - H)^2 \, dR,$$

where

$$F(R) = \frac{R(B_{0\phi} + kRB_{0z})^2 + h^2 R^3 B_0^2}{1+(k^2+h^2)R^2}, \tag{7.50}$$

$$G(R) = -\frac{(B_{0\phi} + RkB_{0z})^2}{R} - \frac{(B_{0\phi} - kRB_{0z})^2}{R[1+(k^2+h^2)R^2]}$$
$$- \frac{h^2 R B_0^2}{1+(k^2+h^2)R^2} + h^2 R B_{0z}^2 - \frac{d}{dR}\left(\frac{B_0^2 + h^2 R^2 B_{0\phi}^2}{1+(k^2+h^2)R^2}\right), \tag{7.51}$$

$$H\left(R, \xi^R, \frac{d\xi^R}{dR}\right) = \frac{R}{1+(k^2+h^2)R^2}\left(\frac{d\xi^R}{dR}(kRB_{0\phi} - B_{0z}) - \frac{\xi^R}{R}(kRB_{0\phi} + B_{0z})\right),$$

$$h = \frac{\pi}{2L}.$$

Since the integrand contains $\xi^0(R)$ only in quadratic form, it may be minimised with respect to ξ^0 by simply choosing $\xi^0 = H/B_0$, for which δW becomes

$$\delta W = \int_0^\infty F\left(\frac{d\xi^R}{dR}\right)^2 - G\xi^{R2} \, dR. \tag{7.52}$$

Finally, the minimisation of this integral with respect to $\xi^R(R)$ is accomplished by solving the associated *Euler–Lagrange equation*

$$\frac{d}{dR}\left(F\frac{d\xi^R}{dR}\right) + G\xi^R = 0. \tag{7.53}$$

This is subject to the natural boundary condition for the variational problem, namely $d\xi^R/dR = 0$ at $R = 0$, together with (without loss of generality) $\xi^R = 1$ at $R = 0$; any other value of ξ^R at the origin (C, say) simply multiplies δW by C^2 without changing its sign. It can be seen from the expression (7.50) for $F(R)$ that the differential equation (7.53) has just one singular point, namely $R = 0$. Newcomb (1960) established that, if the solution $\xi^R(R)$ is positive for all R, then $\delta W > 0$ and so the perturbation is stable, whereas if $\xi^R(R)$ vanishes somewhere we have an instability. Once the components $(B_{0\phi}, B_{0z})$ of the equilibrium field are specified, the technique is therefore to test the solution for the presence of zeros for each wavenumber (k). Typical solutions are shown in Figure 7.7. As a particular example, consider the uniform-twist force-free field

$$B_{0z} = \frac{B_0}{1+(\Phi R/(2L))^2}, \quad B_{0\phi} = \frac{B_0(\Phi R/(2L))}{1+(\Phi R/(2L))^2},$$

INSTABILITY

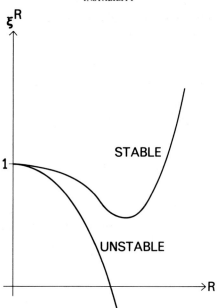

Fig. 7.7. Typical solutions to the Euler–Lagrange equation for the radial component (ξ^R) of the minimising perturbation for a flux tube.

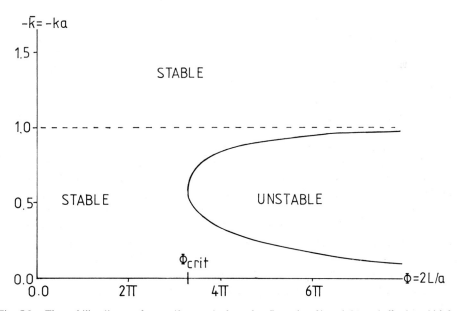

Fig. 7.8. The stability diagram for a uniform-twist force-free flux tube of length $2L$ and effective width $2a$, where k is the wavenumber of the perturbation along the tube and Φ is the twist (from Hood and Priest, 1979b).

for which the parameter Φ gives the angle of twist of a field line about the axis in going from one end of the tube to the other. The resulting stability diagram is shown in Figure 7.8. It indicates that the magnetic flux tube first goes unstable to kinking of the form (7.44) when the twist exceeds about 3.3π. This differs from the 2π of the *Kruskal–Shafranov limit* (Section 7.5.3.), because account has been taken of the flux-tube structure. It should be noted that only the form (7.44) was tested for stability. Thus, when $\Phi < 2\pi$ the tube is certainly stable, and when $\Phi > 3.3\pi$ it has certainly become unstable, but there may be some other form for the perturbation that goes unstable at a lower threshold. Indeed, a full solution of the partial differential equations of motion has since shown that the real threshold for instability lies at 2.5π (Hood and Priest, 1981b).

It should be noted that, at the present time, it is unclear what boundary condition to adopt at the ends of a loop to simulate photospheric line-tying, especially when plasmas pressure gradients are present. Possibilities under discussion are to impose either $\boldsymbol{\xi}_\perp = \mathbf{0}$, $\boldsymbol{\xi} = \mathbf{0}$, or $\mathbf{B}_1 = 0$, but the relevant condition will be rigorously established only when the stratification effects are modelled adequately.

7.4.2. Use of the Energy Method

There are several points to note about the use of the energy method.

(a) If some displacement ($\boldsymbol{\xi}$) is discovered that makes δW negative, the plasma is certainly unstable, but there may exist a more unstable mode. By contrast, it is difficult to prove stability, since one needs to demonstrate that δW is positive for *all* possible displacements. In practice, one often considers only a certain class of displacements; some skill is needed to choose the particular class when one is seeking an instability. There is no need for $\boldsymbol{\xi}$ to satisfy the equation of motion: Laval *et al.* (1965) established that, if a $\boldsymbol{\xi}$ makes δW negative, then there does exist a physical perturbation which grows exponentially.

(b) For a given displacement, Equation (7.37) may be used to determine the frequency of the oscillation ($\omega^2 > 0$) or the growth-rate of the instability ($\omega^2 < 0$). Note, however, that, if there exists a band of unstable wavenumbers, the most unstable mode (with the largest value of $|\omega^2|$) can be identified as the one with the largest value of $|\delta W|$ only if the normalisation $\int \frac{1}{2}\rho_0 \xi^2 \, dV = constant$ is employed.

(c) The minimisation of δW is often performed with a constraint in order to rule out a trivial solution. For instance, in the above example the constraint is $\int f^2 \, dz = 1$ (Equation (7.48)), so as to debar the solution $f(z) \equiv 0$, which would make $\delta W = 0$. The choice of such a normalisation is at our disposal, subject to the proviso that it does not change the sign of δW.

(d) The expression (7.39) for δW is derived under the assumption that $\boldsymbol{\xi}$ vanishes on the boundary; so, if some alternative boundary condition is employed, one must return to the more fundamental expression (7.38).

(e) If the equilibrium pressure (p_0) is uniform, the third term in the integral (7.39) for δW reduces to $\gamma p_0 (\nabla \cdot \boldsymbol{\xi})^2$, which is positive. This shows that a uniform plasma pressure has a stabilising influence and the perturbation that minimises this term alone (to zero) is an incompressible one.

(f) If $\delta W = 0$ to second order, then it is necessary to go to third or fourth order to investigate stability.

(g) If, as in modelling many laboratory situations, one wishes to investigate the stability of a region containing plasma and surrounded by a vacuum, two extra terms must be added to the expression (7.38) for δW. These are

$$\delta W_S = \frac{1}{2} \int_{S_0} (\mathbf{n} \cdot \boldsymbol{\xi})^2 \mathbf{n} \cdot \left\langle \nabla \left(p_0 + \frac{B_0^2}{2\mu} \right) \right\rangle dS,$$

which is the work done against the displacement of the interface S, and

$$\delta W_v = \frac{1}{2\mu} \int_{\text{vac}} B_v^2 \, dV,$$

the change of the vacuum magnetic energy resulting from the interface deformation; here $\mathbf{n}$ is the unit normal to the interface, $\langle X \rangle$ denotes the jump in X across the interface in the direction $\mathbf{n}$, and $\mathbf{B}_v$ is the perturbed vacuum magnetic field (Bernstein et al., 1958).

(h) Bernstein's energy principle has been extended by Frieman and Rotenberg (1960) to include steady flows in the equilibrium state. The result is the addition of extra terms to the expression for δW and the possibility of overstable modes. Another extension is due to Tasso (1977), who adds the effect of finite electrical conductivity.

(i) Here only *linear* stability has been considered. For a description of nonlinear methods, see, for instance, Stuart (1963) or Spies (1974). Also, in contrast to the macroscopic instabilities that are described by magnetohydrodynamics, a plasma may be subject to microscopic instabilities that lead to changes in the particle distribution functions according to a kinetic theory (e.g., Mikhailovsky, 1974).

7.5. Summary of Instabilities

The previous sections have described the normal mode and energy methods for testing stability. These are used in several other chapters, but here the properties of some of the basic instabilities that may be relevant to the Sun are summarised.

7.5.1. Interchange instability

Consider an interface between two plasmas with different pressures and containing magnetic fields of different strengths. A so-called *interchange* (or *exchange*) instability arises when two neighbouring bundles of field lines can be interchanged in such a way that the volumes occupied by the plasma remains constant but the magnetic energy decreases and so feeds the instability. The perturbed interface is rippled, with magnetic field lines parallel to the crests and troughs of the ripples. The instability is sometimes referred to as a *flute instability*, especially when the one plasma occupies a cylinder, so that its perturbed shape resembles a fluted column (Figure 7.5).

A special case of this instability occurs when a mass of plasma containing no magnetic field is confined by a magnetic field whose lines of force are concave towards the plasma (Figure 7.9(a)). (By contrast, plasma can be contained in stable equilibrium

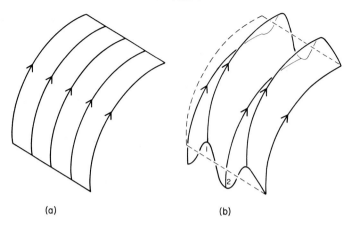

Fig. 7.9. (a) Part of the concave surface of a magnetic field that confines plasma in some region. (b) A flute-like displacement of the interface.

by field lines that are *convex* to it.) Consider the fluted displacement of such a concave interface shown in Figure 7.9(b). If the volume occupied by the plasma is unaltered, the pressure of the plasma does no work and its internal energy is unchanged. The volume (V) occupied by the magnetic field is also the same as in the equilibrium situation; so the curvature of the boundary means that the cross-sectional area of region 2 exceeds that of region 1; i.e., $A_2 > A_1$. But the net effect of the perturbation is to transfer magnetic flux from region 1 to region 2, so that the fluxes in the two regions are the same, namely $B_2 A_2 = B_1 A_1$.

The final magnetic field strength (B_2) must therefore be smaller than its initial value (B_1). In other words, during the course of the perturbation, the magnetic field lines have shortened and moved further apart, with a consequent fall in field strength (cf., Section 2.9.2). The change in magnetic energy produced by the displacement is

$$\frac{B_2^2}{2\mu}V - \frac{B_1^2}{2\mu}V = \frac{B_1^2 V}{2\mu}\left(\frac{A_1^2}{A_2^2} - 1\right).$$

This is negative, and so the magnetic energy is reduced and instability ensues. The perturbation grows like $e^{i\omega t}$, where, if the wrinkles have a wavenumber k and the radius of curvature of the interface is R_c, then approximately

$$\omega^2 = -\frac{2pk}{\rho R_c}. \tag{7.54}$$

The instability tends to be inhibited if the lines of force are sheared in such a way that it is difficult for them to be interchanged. Another stabilising effect is the anchoring of the field lines at points a distance L apart, say; this is relevant to coronal magnetic fields that are tied to relatively dense photospheric material. The influence of the line-tying propagates along the field at the Alfvén speed ($v_A \equiv B/(\mu\rho)^{1/2}$), so that the expression for the growth-rate in Equation (7.54) becomes modified to

$$\omega^2 = -\frac{2pk}{\rho R_c} + \frac{v_A^2}{L^2}, \text{ approximately.}$$

If the interface is limited in a direction normal to the magnetic field by a distance r_0, so that the minimum wavenumber (k) is roughly r_0^{-1}, it can be seen that line-tying stabilises the longest-wavelength perturbations when $(2\mu p/B^2) > (r_0 R_c/L^2)$; but sufficiently short waves are unstable in this approximation. Furthermore, Rosenbluth *et al.* (1962) have shown that, when the radius of curvature R_c is so large that $R_c > r_0^3/r_{iL}^2$, where r_{iL} is the ion Larmor radius, then finite ion Larmor radius effects stabilise an interchange mode (but not a kink mode).

There are a wide variety of interchange instabilities, other examples being the *Rayleigh–Taylor instability*, the *Kelvin–Helmholtz instability* and the *sausage instability* treated below. Indeed, when an interface is plane, an interchange of magnetic field lines in the two regions (with $\mathbf{k} \cdot \mathbf{B} = 0$) is often the most likely to give instability. This is because the field lines are not stretched during such a perturbation, and so no work is done against magnetic tension forces.

7.5.2. Rayleigh–Taylor Instability

Consider two incompressible, inviscid fluids of uniform densities ($\rho_0^{(-)}$ and $\rho_0^{(+)}$), separated by a horizontal boundary, with gravity acting vertically downwards. The fluid of density $\rho_0^{(+)}$ rests on top of the other. If the denser fluid is below ($\rho_0^{(-)} > \rho_0^{(+)}$), the system is stable, but, if it lies above ($\rho_0^{(+)} > \rho_0^{(-)}$), the system is unstable to a fluting of the boundary, as shown in Figure 7.3. Perturbations like $e^{i\omega t}$ grow at a rate $|\omega|$ given by

$$\omega^2 = -gk \frac{\rho_0^{(+)} - \rho_0^{(-)}}{\rho_0^{(+)} + \rho_0^{(-)}}. \tag{7.55}$$

The effect of a uniform vertical magnetic field ($B_0 \hat{z}$) is to modify the growth-rate but not to change the stability. For very long waves ($k \to 0$), the expression (7.55) is unaltered, but, for short waves ($k \to \infty$), the growth-rate is reduced to the value

$$i\omega \approx \frac{g\sqrt{\mu}}{B_0}(\sqrt{\rho_0^{(+)}} - \sqrt{\rho_0^{(-)}}),$$

which is independent of k.

A uniform horizontal field ($B_0 \hat{x}$) gives the dispersion relation

$$\omega^2 = -gk \frac{\rho_0^{(+)} - \rho_0^{(-)}}{\rho_0^{(+)} + \rho_0^{(-)}} + \frac{2B_0^2 k_x^2}{\mu(\rho_0^{(+)} + \rho_0^{(-)})},$$

in place of Equation (7.55). This implies that the magnetic field produces no additional effect when the wavenumber is normal to the field ($k_x = 0$); ripples along the field ($k = k_x$), however, produce a restoring force through the magnetic tension, which allows instability only when $0 < k < k_c$, where

$$k_c = \frac{(\rho_0^{(+)} - \rho_0^{(-)})g\mu}{2B_0^2}.$$

The *hydromagnetic Rayleigh–Taylor instability* (due to Kruskal and Schwarzschild, 1954) occurs when plasma of density $\rho_0^{(+)}$ is supported against gravity by a magnetic field ($B_0^{(-)}\hat{x}$), as shown in Figure 7.4. The dispersion relation is

$$\omega^2 = -gk + \frac{k_x^2 B_0^{(-)2}}{\mu \rho_0^{(+)}}; \tag{7.56}$$

it implies that the most unstable mode has a fluting of the boundary with $k_x = 0$ and a growth-rate

$$i\omega = (gk)^{1/2}. \tag{7.57}$$

Disturbances with $k_x \neq 0$ distort the field lines, so the magnetic tension acts to try and restore equilibrium, as indicated by the second term on the right-hand side of Equation (7.56). If the ends of the lines of force are fixed a distance L apart, the lines will bend when displaced; an Alfvén wave then propagates at the speed $v_A = B_0^{(-)}/(\mu \rho_0^{(+)})^{1/2}$ and Equation (7.57) is modified to $\omega^2 \approx -gk + (v_A^2/L^2)$, so that the long waves (with $k < v_A^2/(L^2 g)$) are stabilised.

Other analogues of the Rayleigh–Taylor instability occur when: either plasma is forced into a vacuum with acceleration f; or when its boundary is curved (with radius of curvature R_c), so that the containing field lines have a tension of magnitude $2p/R_c$ (see Equation (7.54)); or when the plasma is rotating with angular speed Ω, so that the centripetal acceleration at a distance r_0 is $\Omega^2 r_0$. In each of these cases, the growth-rate is given by Equation (7.57) with g replaced by f, $2p/(\rho R_c)$ or $\Omega^2 r_0$, respectively.

7.5.3. PINCHED DISCHARGE

A *linear pinched discharge* in the laboratory is a cylindrical plasma column (of radius a, say,) that is confined (or 'pinched') by the azimuthal magnetic field due to a current ($J\hat{z}$) flowing along its surface or through its interior, as shown in Figure 7.10(a). Since this configuration is similar to magnetic flux tubes present in the solar atmosphere, it is of interest to summarise its stability properties here. The radially inwards $\mathbf{j} \times \mathbf{B}$ force is balanced by an outwards pressure gradient. When the plasma (at pressure p_0 and density ρ_0) contains no magnetic field, the pinch is unstable to the interchange mode, since the confining field is concave to the plasma. The radius of curvature of the azimuthal field lines is denoted by a, and so from Equation (7.54) the growth-rate for this *sausage instability* (Figure (7.10)) is $i\omega = (2p_0 k/(\rho_0 a))^{1/2}$. For disturbances that destroy the column as a whole $k \approx a^{-1}$, which gives a growth-rate

$$i\omega = \frac{(2p_0/\rho_0)^{1/2}}{a}. \tag{7.58}$$

The linear pinch can be stabilised against the sausage mode by the presence of a large enough axial field (B_{0z}) inside the plasma. If the value of the azimuthal field at the interface is B_ϕ, the force balance on the interface gives

$$p_0 + \frac{B_{0z}^2}{2\mu} = \frac{B_\phi^2}{2\mu},$$

and the effect of an Alfvén wave propagating along the axis with speed $B_{0z}/(\mu\rho_0)^{1/2}$

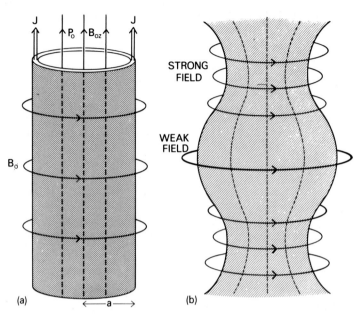

Fig. 7.10. (a) A linear pinch containing plasma at pressure p_0 and magnetic field of strength $B_{0z}\hat{z}$. A current $(J\hat{z})$ flows on the surface and produces a field $(B_\phi \hat{\phi})$ in the surrounding region. (b) A sausage perturbation of the interface.

is to modify the dispersion relation from Equation (7.58) to

$$\omega^2 = -\frac{2p_0}{\rho_0 a^2} + \frac{B_{0z}^2}{\mu \rho_0 a^2}, \text{ approximately.}$$

Substitution for p_0 from the above force-balance in this expression then gives stability ($\omega^2 > 0$) when $B_{0z}^2 > \frac{1}{2}B_\phi^2$.

The linear pinch (with a purely azimuthal field (B_ϕ) outside the plasma column) is also subject to the *kink instability* (Lundquist, 1951; Kruskal and Schwarzschild, 1954). A *helical kink* perturbation of the form $\boldsymbol{\xi} = \boldsymbol{\xi}(R)\exp(i(\phi + kz) + i\omega t)$ gives instability for all axial wavenumbers. The same is true for the *lateral kink* shown in Figure 7.6; its perturbation is proportional to $e^{i\phi}\cos kz$ and may be obtained by superposing two oppositely twisted helical perturbations like $e^{i(\phi + kz)}$ and $e^{i(\phi - kz)}$. For such displacements the cross-section of the column remains circular, but other modes have been considered that behave like $e^{i(m\phi + kz)}$ with $m \geq 2$, and they lead to a deformation of the section.

The kink instability cannot be stabilised by superposing an axial field outside the plasma column, since a *helical perturbation* of the above form is unstable provided the wavelength ($-2\pi/k$) is long enough that $B_\phi/R + kB_z \geq 0$. At equality, the wavenumber vector $(0, R^{-1}, k)$ is perpendicular to the equilibrium field ($\mathbf{k} \cdot \mathbf{B} = 0$), so that the crests and troughs of the perturbation follow the field as it spirals around the axis. In other words, the perturbation and the vacuum field have the same pitch, namely $2\pi a(B_z/B_\phi)$. (The *pitch* is simply the length along the z-axis of a field line that encircles the axis once.) Another consequence of $\mathbf{k} \cdot \mathbf{B} = 0$ is that it makes the perturbed Lorentz force ($\mathbf{j}_1 \times \mathbf{B}_0$) vanish. The above criterion may be written in terms of the amount $\Phi = 2LB_\phi/(RB_z)$ by which a field line is twisted about the tube axis in going from one end of the tube to the other, a distance $2L$. Thus, *for any twist*, the flux tube is unstable to a helical kink with wavenumber

$$k \geq -\frac{\Phi}{2L}. \tag{7.59}$$

For the particular case of a laboratory *torus* of length $2L$, two points that are located an axial distance $2L$ apart refer to the same location on the torus, and so $-k$ must equal π/L (or a multiple thereof). Helical kink instability is therefore present in such a torus when $\Phi \geq 2\pi$, a result due to Kruskal (1954), Shafranov (1957), and Tayler (1957), who also considered the stabilising effect of a conducting wall. A case of interest in a solar context is a *coronal flux loop*, but the Kruskal–Shafranov analysis is inapplicable as it stands, since it omits the stabilising effect of the photosphere; this effectively anchors the ends of magnetic lines and so forces the perturbations to vanish at the ends of the loop. It may be noted that such *line-tying* cannot be modelled with a helical kink perturbation of the above form, since there is no location (z) along the loop which makes it vanish for all ϕ.

In general, the electric current may flow through the whole of the pinched plasma and not just through a sheet current on its surface. Then the plasma pressure ($p(R)$) and reciprocal pitch

$$\frac{\tilde{\mu}(R)}{2\pi} = \frac{B_\phi(R)}{2\pi R B_z(R)}$$

both vary with radius. A *necessary* condition for kink perturbations of such an infinitely long 'diffuse' pinch to be *stable* is that

$$\frac{R}{4}\left(\frac{\tilde{\mu}'}{\tilde{\mu}}\right)^2 + \frac{2\mu p'}{B_z^2} \geq 0 \tag{7.60}$$

at every point in the plasma, where a prime denotes differentiation with respect to R. This is known as *Suydam's criterion* (Suydam, 1959). If the pinch is stable to kinking, then the criterion must be satisfied, but there may be examples satisfying the criterion which are unstable. If Equation (7.60) is not satisfied, the infinitely long pinch is unstable (without line-tying). A general method for treating the stability of a diffuse pinch has been presented by Newcomb (1960) (see Section 7.4). It has been applied by Hood and Priest (1979b, 1981b) to a coronal loop of finite length, including the effect of line-tying. They determine the amount of twist that is needed to give instability for a variety of loop structures; for instance, a uniform-twist force-free loop requires a twist of 2.5π before it becomes kink unstable.

Even if a flux tube were not subject to sausage and kink instabilities, Parker (1974c) has pointed out that, when twisted sufficiently, it is no longer under tension and becomes unstable to *buckling*. For a flux rope with only azimuthal and longitudinal components ($B_\phi(R)$ and $B_z(R)$), the condition for buckling is that the total stress across a section $(\pi/\mu)\int_0^\infty (B_z^2 - B_\phi^2) R \, dR$ be negative. In other words, the magnetic pressure of B_ϕ needs to exceed the magnetic tension of B_z, which is true when $B_\phi > B_z$ across enough of the section.

7.5.4. FLOW INSTABILITY

Laminar viscous flow between rigid boundaries such as the walls of a laboratory channel or the banks of a river is known to become unstable and develop into a turbulent flow when the *Reynolds number* (Re) exceeds a critical value (Re*). Re is defined by $Re = LV_0/v$, where $2L$ is the distance between the boundaries, V_0 is the fluid speed at the centre of the channel and v is the kinematic viscosity. The effect of a magnetic field on such a flow (when the fluid is electrically conducting) is often to raise the value of the critical Reynolds number. For a magnetic field directed across the channel, Lock (1955) showed theoretically that $Re^* \approx 50\,000$ Ha when the Hartmann number, $Ha = B_0 L_0 (\sigma/(\rho v))^{1/2}$, is much larger than unity; experiments exhibit a much smaller value for Re*. A magnetic field parallel to the flow direction yields instead $Re^* \approx 500$ Ha, according to the theory of Stuart (1954).

The steady circular flow of an incompressible liquid (with angular speed $\Omega(R)$) between two coaxial rotating cylinders is known as *Couette flow*. According to *Rayleigh's criterion*, such a flow is stable if and only if $d/dR(R^4\Omega^2) > 0$ everywhere, and so it is unstable if the same quantity is negative anywhere. The condition for stability is therefore that the angular momentum per unit mass $(R^2|\Omega|)$ increase with distance from the axis; it can be shown (Chandrasekhar, 1961) that this condition ensures that an (angular momentum conserving) interchange of (equally massive) elementary fluid rings produces an increase of kinetic energy, which requires a source of energy. The effect of viscosity is to postpone the onset of instability to a point beyond that predicted by Rayleigh (1916). A magnetic field also has a stabilising influence (Chandrasekhar, 1961). An arbitrarily small axial field (B_z) can stabilise the flow if $d/dR(|\Omega|) > 0$, while an azimuthal field $(B_\phi(R))$ makes the motion stable if and only if

$$\frac{d}{dR}(R^4\Omega^2) - \frac{\Omega_2}{\Omega_1}\frac{R^4}{\mu\rho}\frac{d}{dR}\left(\frac{B_\phi^2}{R^2}\right) > 0,$$

where Ω_1, Ω_2 are the angular speeds of the inner and outer cylinders.

When a uniform, inviscid fluid rests on top of another fluid and the two are in relative horizontal motion, they are subject to the *Kelvin–Helmholtz instability*. Suppose the density $(\rho^{(+)})$ of the upper fluid is less than that of the lower fluid $(\rho^{(-)})$, so that the interface is (Rayleigh–Taylor) stable in the absence of streaming; and let the velocities of the fluids be $U^{(+)}\hat{x}$ and $U^{(-)}\hat{x}$ (Figure 7.11). Then interchanges with wavenumber $k\hat{x}$ are unstable if

$$k > \frac{g(\rho^{(-)2} - \rho^{(+)2})}{\rho^{(-)}\rho^{(+)}(U^{(-)} - U^{(+)})^2}.$$

Thus, no matter how small the relative speed $(U^{(-)} - U^{(+)})$, instability will occur at a small enough wavelength. (If the density $(\rho(z))$ and horizontal flow speed $(U(z))$ are instead continuous functions of z such that

$$Ri \equiv -\frac{g\,d\rho/dz}{\rho(dU/dz)^2} > \tfrac{1}{4},$$

then the stratified flow is stable. Ri is known as the *Richardson number*, and this

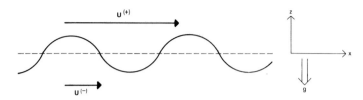

Fig. 7.11. Kelvin–Helmholtz instability for superposed fluids in relative motion.

condition ensures that the kinetic energy released during the vertical interchange of neighbouring elements is insufficient to do the necessary work against gravity. The role of shear-flow instabilities (when $\text{Ri} < \tfrac{1}{4}$) has not yet been considered in a solar context (e.g., in the photosphere).) The effect on the Kelvin–Helmholtz instability of uniform horizontal magnetic fields ($\mathbf{B}^{(-)}$ and $\mathbf{B}^{(+)}$) depends on their orientations relative to the flow direction. If they are both in the y-direction, perturbations with wavenumber $k\hat{\mathbf{x}}$ simply exchange lines of force without stretching them and the instability is unaffected. On the other hand, if the fields are parallel to the flow velocity, such a perturbation stretches field lines; their magnetic tension produces a restoring force, and the instability is suppressed provided the fields are so strong that

$$\frac{B^{(-)2} + B^{(+)2}}{\mu \rho^{(-)} \rho^{(+)}} (\rho^{(-)} + \rho^{(+)}) \geqslant (U^{(-)} - U^{(+)})^2.$$

(See Chandrasekhar (1961) or Cowling (1976), who also treats inclined fields.)

7.5.5. Resistive instability

The magnetic field in a current sheet of width l diffuses through the plasma on a time-scale $\tau_d = l^2/\eta$, where $\eta = (\mu\sigma)^{-1}$ is the magnetic diffusivity. During the process of magnetic diffusion, magnetic energy is converted ohmically into heat at the same rate, but solar values of τ_d are enormous unless l is tiny (Section 2.6.1). However, Furth et al. (1963) showed how the diffusion can drive three distinct instabilities and hence convert magnetic energy into heat and kinetic energy at a much faster rate. These instabilities occur provided the sheet is wide enough that $\tau_d \gg \tau_A$, where $\tau_A = l/v_A$ is the time taken to traverse the sheet at the Alfvén speed ($v_A = B_0/(\mu\rho_0)^{1/2}$). They occur on time-scales $\tau_d(\tau_A/\tau_d)^\delta$, where $0 < \delta < 1$ (see below), and have the effect of creating in the sheet many small-scale magnetic loops, which subsequently diffuse away and release magnetic energy.

During the instability, the magnetic field slips through the plasma in a region of width εl about the centre of the sheet, where the magnetic field vanishes. At the edge of this diffusion region the magnetic field is εB_0, namely a fraction (ε) of the value (B_0) at the edge of the sheet (Figure 7.12). If the velocity with which plasma enters the diffusion region is $-v_x\hat{\mathbf{x}}$, the resulting electric current is $\mathbf{j} \approx \sigma(\mathbf{v} \times \mathbf{B}) = \sigma v_x(\varepsilon B_0)\hat{\mathbf{z}}$ and the Lorentz force ($\mathbf{F}_L$) which opposes the flow is roughly

$$\mathbf{F}_L \equiv \mathbf{j} \times \mathbf{B} \approx -\sigma v_x(\varepsilon B_0)^2 \hat{\mathbf{x}}, \tag{7.61}$$

(where $v_x < 0$ for $x > 0$). Now, if an instability is to take place, this restoring force

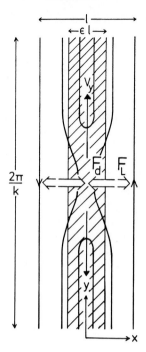

Fig. 7.12. Resistive instability in a current sheet for which the driving force (F_d) exceeds the restoring force (F_L). Significant diffusion takes place over a fraction (ε) of the width (l) of the sheet. Here just one wavelength ($2\pi/k$) is shown, but, in practice, there may be many such features end to end.

must be exceeded by a driving force ($\mathbf{F}_d$, say) which has roughly the same magnitude as $\mathbf{F}_L$. The rate of working of $\mathbf{F}_d$ during the instability is therefore

$$-\mathbf{v} \cdot \mathbf{F}_L \approx \sigma v_x^2 (\varepsilon B_0)^2. \tag{7.62}$$

The effect of the driving force is to accelerate plasma a distance $2\pi/k$ along the sheet to a speed v_y, given that the instability has a wavenumber $k(\gg l^{-1})$ in this direction. For an incompressible plasma, the equation $\nabla \cdot \mathbf{v} = 0$ implies that v_y is given in terms of the slow inflow speed ($-v_x$) by $kv_y + (\varepsilon l)^{-1} v_x \approx 0$, or $v_y \approx -(v_x/(k\varepsilon l))$. The rate of increase in kinetic energy over a time $(i\omega)^{-1}$ is therefore

$$i\omega \rho_0 v_y^2 \approx \frac{\omega \rho_0 v_x^2}{(k\varepsilon l)^2},$$

which, when equated to the work done driving the instability (Equation (7.62)), gives

$$(\varepsilon l)^4 = \frac{i\omega \rho_0 l^2}{\sigma k^2 B_0^2},$$

or, in terms of the time-scales τ_d and τ_A,

$$\varepsilon^4 = \frac{i\omega}{(kl)^2} \frac{\tau_A^2}{\tau_d}. \tag{7.63}$$

For two of the three resistive instabilities one can equate (in order of magnitude) the particular driving force ($\mathbf{F}_d$) to $-\mathbf{F}_L$ given by Equation (7.61), and so produce a relationship between ε, ω and k, which, together with Equation (7.63), determines the dispersion relation for ω as a function of k. (A more rigorous analysis produces the same result.)

The *gravitational mode* exists when a gravitational (or equivalent) force ($\rho g \hat{\mathbf{x}}$) transverse to the current sheet acts to produce a density stratification ($\rho_0(x)$). Plasma is transported without compression, so that density changes satisfy

$$\frac{\partial \rho_1}{\partial t} + v_x \frac{d\rho_0}{dx} = 0$$

from mass continuity; the driving force ($\rho_1 g \hat{\mathbf{x}}$) then becomes

$$\mathbf{F}_d = -\frac{v_x}{\omega} \frac{d\rho_0}{dx} g \hat{\mathbf{x}}, \tag{7.64}$$

and it pushes plasma into the sheet if $d\rho_0/dx > 0$. Equating Equation (7.64) to Equation (7.61) and using Equation (7.63) to eliminate ε gives

$$i\omega = \left(\frac{(kl)^2 \tau_A^2}{\tau_d \tau_G^4} \right)^{1/3},$$

where

$$\tau_G = \left(-\frac{g}{\rho_0} \frac{d\rho_0}{dx} \right)^{-1/2}$$

is a gravitational time-scale.

The *rippling-mode* instability occurs when there is a spatial variation across the sheet in magnetic diffusivity ($\eta_0(x)$). This may arise, for instance, from a temperature structure in the basic state. If the diffusivity at a point varies due to advection with the plasma, we have $(\partial \eta / \partial t) + \mathbf{v} \cdot \nabla \eta = 0$, and so linear departures (η_1) on a time-scale $(i\omega)^{-1}$ from the original diffusivity (η_0) are given by

$$\eta_1 = -\frac{v_x}{i\omega} \frac{d\eta_0}{dx}. \tag{7.65}$$

This change in diffusivity produces an extra current

$$\mathbf{j}_1 = -\frac{\eta_1}{\eta_0} \mathbf{j}_0 \tag{7.66}$$

in Ohm's Law; associated with it is an extra Lorentz force ($\mathbf{j}_1 \times \varepsilon B_0 \hat{\mathbf{y}}$) at the $(x > 0)$ edge of the diffusion region, where the magnetic field is $\varepsilon B_0 \hat{\mathbf{y}}$. Since the original current is, in order of magnitude, $\mathbf{j}_0 = B_0/(\mu l) \hat{\mathbf{z}}$, this force may be written (using Equations (7.65) and (7.66)) as

$$\mathbf{F}_d = -\frac{v_x}{\omega \eta_0} \frac{d\eta_0}{dx} \frac{\varepsilon B_0^2}{\mu l} \hat{\mathbf{x}}. \tag{7.67}$$

It represents a driving force with the sign shown in Figure (7.12) for $x > 0$ provided

$d\eta_0/dx < 0$. Similarly, on the left-hand side of the current sheet ($x < 0$), where the magnetic field is $-\varepsilon B_0 \hat{y}$ and $v_x > 0$, a driving force of the necessary direction is present provided $d\eta_0/dx > 0$ there. Thus, if the diffusivity (η_0) takes a *maximum* value in the centre of the sheet, the whole of the sheet is subject to the rippling mode. For a classical diffusivity, η_0 varies with temperature (T_0) like $T_0^{-3/2}$, so this would necessitate a temperature minimum at $x = 0$. If, on the other hand, the diffusivity increases (or decreases) monotonically through the sheet, the instability takes place only in the left- (or right-) hand side of the sheet. Equating (7.67) to (7.61) and using Equation (7.63) for ε determines the growth-rate as

$$i\omega = \left[\left(\frac{d\eta_0}{dx} \frac{l}{\eta_0} \right)^4 \frac{(kl)^2}{\tau_d^3 \tau_A^2} \right]^{1/5}.$$

It should be noted that, if conduction is efficient enough, diffusivity changes are not advected, so that Equation (7.65) fails and the rippling mode is stabilised.

The third type of resistive instability is the *tearing mode*, which, in contrast to the other two, occurs with a wavelength greater than the width of the sheet ($kl < 1$). Its growth-rate is given by $i\omega = [\tau_d^3 \tau_A^2 (kl)^2]^{-1/5}$, for wavenumbers in the approximate range $(\tau_A/\tau_d)^{1/4} < kl < 1$. Thus, the mode with longest wavelength has the fastest growth-rate, namely $i\omega = (\tau_d \tau_A)^{-1/2}$, a value that is the geometric mean of the diffusion and Alfvén frequencies. The tearing mode is possibly the most important of the three resistive instabilities, since it requires neither a gravitational force nor a resistivity gradient to be excited. It can occur not just in a current sheet but also in thin

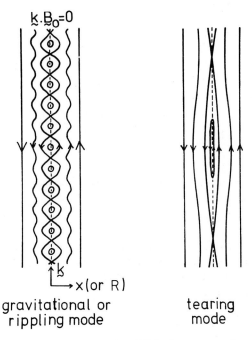

Fig. 7.13. Small- and long- wavelength resistive instabilities in a current sheet or a sheared magnetic field.

sheaths throughout a sheared magnetic structure. It may well be the cause of *coronal heating* (Section 6.4) since the growth-time for typical coronal conditions is about a week. The tearing mode offers one way in which reconnection can grow (Section 10.1), and it is possibly relevant to the *solar flare* (Chapter 10).

Resistive instabilities can occur whenever the magnetic field is sheared. A neutral current sheet is just one example of such a structure; other examples have fields of the form $B_{0y}(x)\hat{\mathbf{y}} + B_{0z}(x)\hat{\mathbf{z}}$ in plane geometry, or $B_{0\phi}(R)\hat{\boldsymbol{\phi}} + B_{0z}(R)\hat{\mathbf{z}}$ in cylindrical geometry. Sheared fields are in general resistively unstable at *many* thin sheaths throughout the structure. At *any* particular location specified by the value of x (or R), the instability has a wavenumber in a direction normal to the field, $\mathbf{k} \cdot \mathbf{B}_0 = 0$, and in Figure 7.12 the x-y axes refer to the x-k (or R-k) plane, as indicated in Figure 7.13. The effect of a transverse field has been included by Janicke (1982), *Solar Phys.* **76**, 29.

7.5.6. CONVECTIVE INSTABILITY

A horizontal layer of viscous, thermally conducting fluid that is heated from below can become unstable when the temperature difference between the upper and lower surfaces becomes too large (Section 1.3.1C); this was demonstrated by Bénard (1900), who conducted the first quantitative experiments on the instability. The temperature difference is measured by the *Rayleigh number* $\text{Ra} = g\alpha\theta d^3/\kappa v$, where α, κ, v are the coefficients of volume expansion, thermometric conductivity and kinematic viscosity, while θ is the temperature difference that is maintained across the layer and d is its thickness. In a fundamental theoretical paper, Rayleigh (1916) showed that instability sets in when Ra exceeds a critical value (Ra*). He assumed a Boussinesq fluid (Section 8.1.2) and found the value $\text{Ra}^* = \frac{27}{4}\pi^4 \approx 658$ when the boundaries are both free (Section 8.1.2). As Ra is increased, instability appears first at a wavenumber $k^* = (\pi d/\sqrt{2}) \approx 2.2d$, and it can develop into *convection* with a stationary pattern of motions (Figure 7.14). If the cells are infinitely elongated in one direction (x, say) they are called *rolls*, with a vertical velocity component of the form $v_z = V(z)\cos 2\pi y/L$. Three rolls may be combined to give *hexagonal (Christopherson) cells* with

$$v_z = \tfrac{1}{3}V(z)\left\{\cos\frac{2\pi}{3L}(\sqrt{3}x + y) + \cos\frac{2\pi}{3L}(\sqrt{3}x - y) + \cos\frac{4\pi}{3L}y\right\}.$$

The effects of rotation and magnetic field on the onset of convection have been summarised by Chandrasekhar (1961), Cowling (1976) and Roberts and Soward (1978). Suppose first that the fluid is rotating about the vertical (z-axis) with angular speed Ω, measured nondimensionally by the value of the *Taylor number* $\mathcal{T} = 4\Omega^2 d^4/v^2$. Rotation is found to inhibit convection somewhat, as indicated by the fact that the critical Rayleigh number (Ra*) for convection onset increases with $\mathcal{T}$ and the size of the cells decreases. For instance, with free boundaries when $\mathcal{T} = 100$ one has $\text{Ra}^* \simeq 830$ and $k^* \approx 2.6d$, while in the large-$\mathcal{T}$ limit $\text{Ra}^* \approx 8.7\,\mathcal{T}^{2/3}$ and $k^* \approx 1.3\,\mathcal{T}^{1/6}$. Usually, the instability sets in first as *stationary convection*; but, when the *Prandtl number* $\text{Pr} = v/\kappa$ is less than unity and $\mathcal{T}$ is large enough, instability shows itself first as *overstability*. For instance, when $\text{Pr} = 0.1$, overstability occurs first if $\mathcal{T}$ exceeds 730.

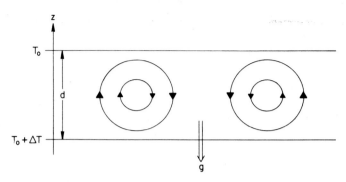

Fig. 7.14. Streamlines for convection that is generated when the temperature difference (ΔT) between the upper and lower surface is large enough.

The effect of a vertical magnetic field (parallel to **g**) is described in Section 8.1.2. It, too, inhibits the onset of convection, because the energy released by buoyancy goes into Joule heating as well as into viscous dissipation; thus, a higher adverse temperature gradient is needed than in the absence of a magnetic field. The type of instability depends on the ratio of the thermal diffusivity (κ) to magnetic diffusivity (η). If $\kappa < \eta$, the onset occurs as a *leak instability* and stationary convection results. The inhibiting effect of the magnetic field is shown by the fact that the critical Rayleigh number (Ra*) increases with the *Hartmann number* Ha $= B_0 d/(\mu\rho\eta\nu)^{1/2}$. For example, when Ha $= 10$ one finds Ra* ≈ 2650 and $k^* \approx 3.7\,d$, while as Ha increases so Ra* $\sim \pi^2 \text{Ha}^2$ and $k^* \sim (\pi^2\,\text{Ha}/\sqrt{2})^{1/3}$. If $\kappa > \eta$, the instability sets in as stationary convection when Ha is less than a critical value (Ha*), but as *overstability* when Ha exceeds Ha*. Ha* depends on the Prandtl number (Pr) and the *magnetic Prandtl number* $P_m = \nu/\eta$.

When the magnetic field is not parallel to the gravitational acceleration (**g**), convection sets in first as long horizontal rolls with their axes parallel to the horizontal component of the magnetic field. When rotation and a magnetic field are both present, their combined effect is to inhibit the onset of instability and to elongate the cells that appear at marginal stability; the critical Rayleigh number depends on both Ha and $\mathcal{T}$.

7.5.7. Radiatively-driven thermal instability

It was Parker (1953) who first pointed out that, if thermal conduction were ineffective, thermal instabilities would occur in the corona and upper atmosphere, because of the form of the radiative loss term in the energy equation. To illustrate this effect, suppose that the plasma is initially in equilibrium with temperature T_0 and density ρ_0, under a balance between mechanical heating of amount $h\rho$ per unit volume (where h is constant) and optically thin radiation; the latter may take the form $\chi \rho^2 T^\alpha$ (where χ and α are constants depending on the temperature range (see Equation (2.35c, d)), so that per unit mass

$$0 = h - \chi \rho_0 T_0^\alpha. \tag{7.68}$$

278   CHAPTER 7

For a perturbation at constant pressure (p_0), the energy equation is

$$c_p \frac{\partial T}{\partial t} = h - \chi \rho T^\alpha, \tag{7.69}$$

where

$$\rho = \frac{m p_0}{k_B T}, \tag{7.70}$$

or, substituting for h from Equation (7.68) and ρ from Equation (7.70),

$$c_p \frac{\partial T}{\partial t} = \chi \rho_0 T_0^\alpha \left(1 - \frac{T^{\alpha-1}}{T_0^{\alpha-1}}\right).$$

Thus, if $\alpha < 1$, a small decrease in temperature ($T < T_0$) makes the right-hand side negative, so that $\partial T/\partial t < 0$ and the perturbation continues:– we have a *thermal instability* with a time-scale $\tau_{rad} = c_p/(\chi \rho_0 T_0^{\alpha-1})$. Figure 2.2 implies that $\alpha < 1$ for temperatures roughly in excess of 10^5 K. Typical values for the variation of the growth-time with number density and temperature are shown in Table 7.1.

Usually, the thermal instability is prevented from taking place by the efficiency of heat conduction along magnetic field lines; this is represented by the extra term $\rho^{-1} \nabla \cdot (\kappa_\parallel \nabla T)$ (per unit mass) which needs to be added to the right-hand side of Equation (7.69), where $\kappa_\parallel = \kappa_0 T^{5/2}$ is the coefficient of thermal conduction parallel to the magnetic field. Thus, if the length of a field line is L, the conduction time is $\tau_c = L^2 \rho_0 c_p/(\kappa_0 T_0^{5/2})$, in order of magnitude. When L is so small that $\tau_c < \tau_{rad}$, the plasma is thermally stable; but, when L exceeds a value

$$L_{max} = \left(\frac{\kappa_0 T_0^{7/2-\alpha}}{\chi \rho_0^2}\right)^{1/2},$$

obtained by equating τ_c and τ_{rad}, the thermal instability takes place. It may be important in both prominence formation (Section 11.1) and coronal loop structure (Section 6.5.1).

The standard work on linear thermal instability theory in the absence of magnetic field and thermal conduction is by Field (1965), who considers a more general energy equation than Equation (7.69). The analysis has been generalised by several authors.

TABLE 7.1

The growth-time in seconds for radiative cooling in a plasma of number density $n_0(m^{-3})$ and temperature $T_0(K)$.

n_0	T_0				
	10^5	5×10^5	10^6	2×10^6	10^7
10^{14}	440	2200	3.2×10^4	1.3×10^5	3.2×10^6
10^{15}	44	220	3.2×10^3	1.3×10^4	3.2×10^5
10^{16}	4.4	22	320	1.3×10^3	3.2×10^4

Hunter (1970) includes both flows and thermal conduction, while Heyvaerts (1974) includes a magnetic field (Section 6.4.4C), and Hildner (1974) follows the nonlinear development numerically.

7.5.8. OTHER INSTABILITIES

In addition to the above instabilities, some other types are mentioned elsewhere in the book. Chapter 8 discusses those that are relevant to sunspots and photospheric flux tubes. More details about *convective instability* are presented in Section 8.1, and the *magnetic buoyancy instability* is described in Section 8.2.

Chapter 10 discusses instabilities that may give rise to solar flares. A *simple-loop* (or compact) *flare* (Section 10.2) may be produced by either a *thermal nonequilibrium* (when the cool core of an active-region loop loses equilibrium) or a (resistive) *kink instability* (when the loop is twisted too much) or an *emerging flux instability* (when the current sheet between new and old flux heats up). A *two-ribbon flare* (Section 10.3) is initiated when a magnetic arcade erupts outwards; this may be due to an *eruptive MHD instability* or the presence of another force-free equilibrium of lower energy when the shear is too great.

In Section 11.1 the *thermal instability* that causes coronal plasma to cool and form a *prominence* is discussed. This may take place either in a neutral sheet or a coronal arcade.

CHAPTER 8

SUNSPOTS

The existence of sunspots has been known since ancient times (Section 1.1), but it was only at the beginning of this century that they were found to be the sites of very strong magnetic fields, and it was realised that they represent the places where huge magnetic flux tubes burst through the solar surface. A theoretical understanding of sunspots has had to await the development of magnetohydrodynamics; however, even now, there is some controversy about answers to fundamental questions, such as: why is a sunspot cool (Section 8.3), what is its equilibrium structure (Section 8.4), and how is it formed (Section 8.6)? Other topics that are discussed in the present chapter include *magnetoconvection* (Section 8.1) and the process of *magnetic buoyancy* (Section 8.2) whereby a flux tube deep within the Sun tends to rise towards the surface because it is lighter than its surroundings. Outside active regions the magnetic flux is not spread out uniformly to a weak field of a few Gauss, but instead it is mainly concentrated at supergranulation boundaries (Section 1.3.2) into *intense flux tubes*, whose properties are discussed in Section 8.7. As a background to this chapter, the reader is referred to Section 2.9 for a summary of flux tube behaviour and to Sections 1.4.2 and 1.3.2B for an account of the observed properties of sunspots and photospheric magnetic fields in general.

8.1. Magnetoconvection

Before discussing the behaviour of sunspots, it is necessary to summarise briefly the extremely important topic of 'thermal convection in a magnetic field' or 'magnetoconvection' for short. To describe motions in the solar convection zone is a formidable task, since the convection is nonlinear, compressible and unsteady, and it is probably influenced by the intense flux tubes that thread the region. But great advances have recently been taken in the numerical simulation of magnetoconvection, reinforced by analytical work. Vigorous convection may possibly explain the concentration of flux tubes (Section 8.7), while its suppression within a flux tube may account for the coolness of sunspots (Section 8.3).

8.1.1. Physical effects

In the absence of a magnetic field, a fluid heated from below is unstable to convective motions when the temperature gradient exceeds its adiabatic value, as established by a parcel argument in Section 1.3.1C. For the solar convection zone (Section 1.3.1C), it is probably the increase in ionisation of H and He at certain depths which lowers the adiabatic gradient and so drives the convection. In some theoretical simulations, such convective cells eventually settle down into a nonlinear steady state when the viscous heating has grown to balance the work done against buoyancy. The effect of viscosity in changing the criterion for convective onset may be estimated by a modification of the parcel argument, as follows. As an element of plasma rises, the buoyancy force $(g \Delta \rho)$ is balanced by a viscous force $(\pi^2 \rho v v / d^2)$, where v is the vertical speed, d the distance moved and $\Delta \rho$ the density change. The latter may be written as $(\rho \alpha \Delta T)$ in terms of the *volume expansion coefficient* (α), so that the force balance becomes $\rho g \alpha \Delta T \approx \pi^2 \rho v v / d^2$.

Now, suppose that the parcel is not thermally isolated (as in Section 1.3.1C) but allows heat to leak in with *diffusivity* κ. Then the condition for instability becomes $d/v < d^2/\pi^2 \kappa$, so that the parcel can move a distance d (in a time d/v) faster than thermal diffusion smooths out the perturbation (on a time-scale $d^2/(\pi^2 \kappa)$). Eliminating v gives the criterion for convective onset as

$$Ra \equiv \frac{g \alpha \Delta T d^3}{\kappa v} > \pi^4, \tag{8.1}$$

where Ra is the *Rayleigh number*.

The difficulties in extrapolating results from both laboratory and theory to the Sun must not be underestimated. In the laboratory one has values for (v/κ) of 10^{-2} to 10^6 and for Ra of 10^3 to 10^8, whereas in the Sun the corresponding values are typically 10^{-14} and 10^{25}, respectively! For the theory, one generally makes the *Boussinesq approximation* (Section 8.1.2), valid over much less than a scale-height, but the Sun presents us with (nonstationary) convection extending over many scale-heights.

The *influence of a magnetic field* on convective motions is largely stabilising due to the restoring effect of the magnetic tension force. Its physical effects can be described as follows (Cowling, 1976). Consider a horizontal layer, of thickness d, heated from below, as shown in Figure 8.1. Gravity acts in the negative z-direction and the temperature difference is ΔT. The equilibrium is characterised by uniform density (ρ_0), uniform magnetic field $(\mathbf{B}_0)$ in the x–z plane, zero velocity, and a temperature stratification given by $T_0(z) = T_0 + \Delta T(1 - (z/d))$.

Now, suppose the magnetic field lines and plasma are displaced by a small amount (ξ) vertically, such that the field lines form sine curves of wave-number k. The radius of curvature is $(k^2 \xi)^{-1}$, and so the magnetic tension (B_0^2/μ) produces a restoring force $k^2 \xi B_0^2/\mu$. At the same time, since the temperature gradient is $\Delta T/d$, the displacement makes the temperature decrease by $\xi \Delta T/d$ and the density decrease by $\rho_0 \alpha (\xi \Delta T/d)$. This fall in density in turn gives rise to a buoyancy force $g(\rho_0 \alpha \xi \Delta T/d)$,

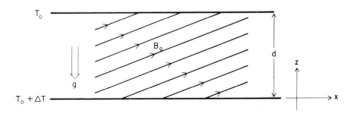

Fig. 8.1. The nomenclature for a horizontal layer of plasma, heated from below and containing a uniform magnetic field $(\mathbf{B}_0)$.

which tries to increase the displacement further and is able to overcome the tension force if

$$\frac{\rho_0 g \alpha \Delta T}{d} > \frac{k^2 B_0^2}{\mu}. \tag{8.2}$$

In other words, *overturning convection* can take place provided the magnetic field is weak enough or the wavelength long enough. Moreover, convection can always occur in the form of rolls parallel to the field (if the ends of the field lines are not tied), since the field lines can remain straight as they move and so produce no tension force.

When the condition for the above rapid overturning is not met, dissipative effects can still sometimes allow much gentler convection due to either a *leak instability* or *overstability*. The effect of a non-zero magnetic diffusivity (η) and thermal diffusivity (κ) is to allow the magnetic field and heat to diffuse through the plasma. For disturbances so slow that convection of heat and magnetic flux balances these diffusion effects, the greater we make η and κ the smaller both the tension and buoyancy forces become. The result is that the buoyancy and tension terms on the left and right sides, respectively, of Equation (8.2), need to be multiplied by κ^{-1} and η^{-1} to give the following condition for *leak instability*:

$$\frac{\rho_0 g \alpha \Delta T \eta}{d} > \frac{k^2 B_0^2 \kappa}{\mu}, \quad \text{for } \kappa < \eta. \tag{8.3}$$

Thus, if $\kappa < \eta$ leak instability occurs for smaller temperature differences than overturning (Equation (8.2)).

If, on the other hand, $\kappa > \eta$ and Equation (8.2) is not satisfied (so that overturning is prevented), the plasma can still be *overstable*. In this case, tension dominates buoyancy, and so vertical oscillations occur. During such an oscillation, thermal conduction reduces the horizontal temperature differences (and so the buoyancy force) by a factor proportional to κ, while the tension force falls by an amount proportional to η. Thus, as the plasma returns towards its equilibrium position both the destabilising and restoring forces have decreased (Figure 8.2). The *resultant* force

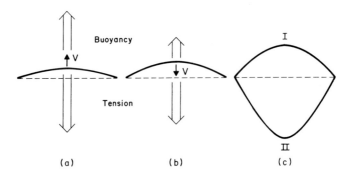

Fig. 8.2. An overstable oscillation. (a) As the plasma moves up, the restoring force of tension exceeds the destabilising force of buoyancy. (b) As the plasma moves down, both tension and buoyancy have been decreased by diffusion, but the resultant force has increased. (c) The amplitude of the second half-oscillation (II) therefore exceeds that of the first (I), and the process continues.

is therefore larger during the return than during the outgoing motion if

$$\frac{\rho_0 g \alpha \Delta T \kappa}{d} > \frac{k^2 B_0^2 \eta}{\mu}, \quad \text{for } \kappa > \eta, \tag{8.4}$$

and the result is *overstability*, with the amplitude of the oscillations increasing in time.

8.1.2. LINEAR STABILITY ANALYSIS

The standard equations for an incompressible plasma acted upon by forces due to pressure, magnetic field and gravity, and subject to both viscous, magnetic and thermal diffusion, are (Section 2.4)

$$\rho \frac{D\mathbf{v}}{Dt} = -\nabla p + \mathbf{j} \times \mathbf{B} + \rho \nu \nabla^2 \mathbf{v} - \rho g \hat{\mathbf{z}},$$

$$\frac{\partial \mathbf{B}}{\partial t} = \nabla \times (\mathbf{v} \times \mathbf{B}) + \eta \nabla^2 \mathbf{B},$$

$$\frac{DT}{Dt} = \kappa \nabla^2 T,$$

where $\nabla \cdot \mathbf{v} = \nabla \cdot \mathbf{B} = 0$. The energy equation has been taken in the form (2.30b), with a uniform coefficient of thermal diffusivity (κ) due to radiative diffusion.

Departures from an equilibrium plasma at rest, with a linear temperature profile ($T_0(z)$) and a uniform magnetic field ($\mathbf{B}_0$), are studied by writing

$$\mathbf{B} = \mathbf{B}_0 + \mathbf{B}_1, \quad \mathbf{v} = \mathbf{v}_1, \quad T = T_0(z) + T_1,$$

and linearising the equations. For simplicity, sound waves are filtered out and compressibility is incorporated only in the buoyancy force by making the *Boussinesq approximation*: in other words, variations in density are included only in the gravitational term, where it is written $\rho = \rho_0(1 + \alpha T_1)$. After eliminating all but one variable (v_{1z}, say), the equations reduce to

$$\left(\frac{\partial}{\partial t} - \kappa \nabla^2\right)\left(\frac{\partial}{\partial t} - \eta \nabla^2\right)\left(\frac{\partial}{\partial t} - \nu \nabla^2\right)\nabla^2 v_{1z}$$
$$= \frac{(\mathbf{B}_0 \cdot \nabla)^2}{\mu \rho_0}\left(\frac{\partial}{\partial t} - \kappa \nabla^2\right)\nabla^2 v_{1z} + \frac{g \alpha \Delta T}{d}\left(\frac{\partial}{\partial t} - \eta \nabla^2\right)\left(\frac{\partial^2 v_{1z}}{\partial x^2} + \frac{\partial^2 v_{1z}}{\partial y^2}\right). \tag{8.5}$$

A solution of the form

$$v_{1z} \sim e^{\omega t} e^{i(k_x x + k_y y)} \sin k_z z, \tag{8.6}$$

with $k_z = \pi/d$, vanishes at the upper and lower boundaries ($z = 0, d$), and it makes Equation (8.5) reduce to

$$(\omega + \kappa k^2)(\omega + \eta k^2)(\omega + \nu k^2)k^2$$
$$= -\frac{(\mathbf{B}_0 \cdot \mathbf{k})^2}{\mu \rho_0}(\omega + \kappa k^2)k^2 + \frac{g \alpha \Delta T}{d}(\omega + \eta k^2)(k_x^2 + k_y^2). \tag{8.7}$$

Consider first a *horizontal magnetic field* ($\mathbf{B}_0 = B_0\hat{\mathbf{x}}$), and neglect dissipative effects ($\kappa = \eta = \nu = 0$), so that the dispersion relation (8.7) becomes

$$\omega^2 = -\frac{B_0^2}{\mu\rho_0}k_x^2 + \frac{g\alpha\Delta T}{\alpha}\frac{k_x^2 + k_y^2}{k^2}.$$

It implies that rolls parallel to the field (with $k_x = 0$) are unstable for all wavenumbers $k_y \neq 0$. Also, when $k_x \neq 0$, the plasma is stable ($\omega^2 < 0$) provided

$$\frac{B_0^2}{\mu\rho_0} > \frac{g\alpha\Delta T}{d}\frac{k_x^2 + k_y^2}{k^2 k_x^2};$$

thus, for a given depth (d) of plasma and a non-zero k_x, convection is inhibited for all k_y if the magnetic field is so strong that

$$\frac{B_0^2}{\mu\rho_0} > \frac{g\alpha\,\Delta T}{dk_z^2},$$

which is equivalent to Equation (8.2).

Next, consider a *vertical magnetic field* ($\mathbf{B}_0 = B_0\hat{\mathbf{z}}$), and include the dissipative terms, so that Equation (8.7) becomes

$$(\omega + \kappa k^2)(\omega + \eta k^2)(\omega + \nu k^2)k^2$$
$$= -\frac{B_0^2}{\mu\rho_0}k^2 k_z^2(\omega + \kappa k^2) + \frac{g\alpha\,\Delta T}{d}(\omega + \eta k^2)(k_x^2 + k_y^2). \tag{8.8}$$

This allows the possibility of *instability* ($\omega^2 > 0$) or *overstability* (a complex frequency $\omega = \omega_r + i\omega_i$ with $\omega_r > 0$). The marginal condition for the onset of instability is $\omega = 0$, for which Equation (8.8) yields

$$\mathrm{Ra}(d^2 k^2 - \pi^2) = \pi^2\,\mathrm{Ha}^2\,d^2 k^2 + d^6 k^6, \tag{8.9}$$

where $\mathrm{Ha} = B_0 d/(\mu\rho\eta\nu)^{1/2}$ is the *Hartmann number* ($\gg 1$ for the Sun). A sketch of k against Ra is shown in Figure 8.3. The smallest value of Ra(Ra*, say) for which there exists a real k^2 is determined by the condition that the three roots of the cubic (8.9) in k^2 be real, namely $(\mathrm{Ra}^* - \pi^2\,\mathrm{Ha}^2)^3 = (27/4)\pi^4 \mathrm{Ra}^{*2}$.

Thus, when the magnetic field vanishes, Ha vanishes and the classical result Ra* $= 27\pi^4/4$ of Rayleigh (1916) is recovered (Section 7.5.6), which gives the precise version of the order-of-magnitude criterion (8.1). For the solar application Ha $\gg 1$, so that instead

$$\mathrm{Ra}^* \approx \pi^2\,\mathrm{Ha}^2, \tag{8.10}$$

and, as Ha increases, so Ra* and k^* increase; this is similar in form to Equation (8.3). Thus, the effect of raising the magnetic field strength (and therefore Ha) is to make the plasma more stable and any unstable cells smaller in size.

The marginal condition for the onset of growing oscillations (i.e., overstability) in a vertical field is $\omega_r = 0$, for which Equation (8.8) gives

$$\mathrm{Ra}\,(d^2 k^2 - \pi^2) = \pi^2\,\mathrm{Ha}^2\,d^2 k^2 \frac{\eta(\nu + \eta)}{\kappa(\nu + \kappa)} + d^6 k^6 \frac{(\eta + \kappa)(\eta + \nu)}{\kappa\nu}.$$

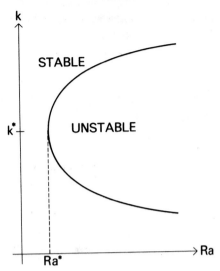

Fig. 8.3. The form of the marginal stability curve for convection in a uniform magnetic field, where k is the wavenumber and Ra the Rayleigh number.

Again the $k - $ Ra curve has a form similar to Figure 8.3, and this time overstability first begins (for Ha $\gg$ 1) at a Rayleigh number of

$$\text{Ra} \approx \pi^2 \text{ Ha}^2 \frac{\eta^2}{\kappa^2}, \tag{8.11}$$

which is essentially the same condition as Equation (8.4). The overstability is exhibited as a standing Alfvén wave, with the amplitude growing and the energy being gained from gravity. Roberts (1976) has included in this analysis a transverse structure, with the result that Equation (8.5) reduces to an ordinary differential equation for which a solution exists only for certain frequencies (ω); i.e., he has an eigenvalue problem.

8.1.3. MAGNETIC FLUX EXPULSION AND CONCENTRATION

Consider a weak magnetic field threading a convecting cell. If the magnetic energy density $(B^2/(2\mu))$ is much less than the kinetic energy density $(\frac{1}{2}\rho v^2)$, the convective motion is unhindered by the magnetic field. If, further, the magnetic Reynolds number is much larger than unity, the magnetic field is simply carried round and is wound up inexorably by the flow. This process stretches the field lines and increases the magnetic field strength until *either* the magnetic energy becomes comparable with the kinetic energy (and the flow is slowed down) *or* the local magnetic Reynolds number becomes of order unity (and the field lines slip through the plasma). In the latter case, the flux is expelled from the centre of the cell and accumulates round its boundaries.

Parker (1963a) has given a simple model for *kinematic concentration* of flux by, for instance, a supergranule cell. He considers the initial effect of an incompressible flow

$$\mathbf{v} = v_0 \sin kx \, \hat{\mathbf{x}} - v_0 kz \cos kx \, \hat{\mathbf{z}}$$

on a uniform vertical field ($B_0 \hat{z}$). The z-component of the induction equation is

$$\frac{\partial B_z}{\partial t} = -\frac{\partial}{\partial x}(v_0 \sin kx \, B_z),$$

which may be solved by the method of characteristics to give

$$B_z = \frac{B_0 \, e^{-kv_0 t}}{\cos^2 \tfrac{1}{2} kx + \sin^2 \tfrac{1}{2} kx \, e^{-2kv_0 t}}.$$

At $x = 0$ there is an upwelling and $B_z = B_0 \, e^{-kv_0 t}$, so the field is dispersed, but at $x = \pi/k$ there is a downdraft which concentrates the field like $B_z = B_0 \, e^{kv_0 t}$. A similar analysis for an initially horizontal field ($B_0 \hat{x}$) gives

$$B_x = B_0 (\cos^2 \tfrac{1}{2} kx + \sin^2 \tfrac{1}{2} kx \, e^{-2kv_0 t}) e^{kv_0 t},$$

so that the field is concentrated at the upwelling ($x = 0$).

In a classic paper, Weiss (1966) has simulated this kinematic process of flux expulsion numerically, neglecting the reaction of the field back on the flow. The two-dimensional, incompressible flow velocity of an eddy can be imposed with a stream function

$$\psi = \frac{UL}{\pi} \cos \frac{\pi x}{L} \cos \frac{\pi z}{L},$$

where U and L are a characteristic speed and length (Galloway and Weiss, 1981). The time-development of an initially uniform magnetic field ($B_0 \hat{z}$) is then deduced from the induction equation

$$\frac{\partial \mathbf{B}}{\partial t} = \nabla \times (\mathbf{v} \times \mathbf{B}) + \eta \nabla^2 \mathbf{B}, \tag{8.12}$$

subject to symmetry conditions ($B_x = 0$) on all boundaries. A *magnetic Reynolds number* is defined to be $R_m = UL/\eta$, and results for $R_m = 250$ are shown in Figure 8.4. The clockwise motion winds up the field and concentrates the flux at the sides of the cell. Near the cell centre the field is first amplified, but then reconnection takes place and it begins to decay. After a time of $6L/U$, where L/U is the *turnover time*, nearly all the flux has been expelled from the centre of the eddy and a steady state established.

Order-of-magnitude estimates for the *maximum field* (B_1) and the steady-state field (B_m) in the boundary layers at the side-walls of the cell may be obtained as follows. The time-scale for the initial amplification to a field strength B before diffusion sets in is

$$\tau \approx \frac{L}{U}, \tag{8.13}$$

obtained by equating the first and second terms in Equation (8.12). The time-scale for diffusion is

$$\tau_d \approx \frac{l^2}{\eta} \tag{8.14}$$

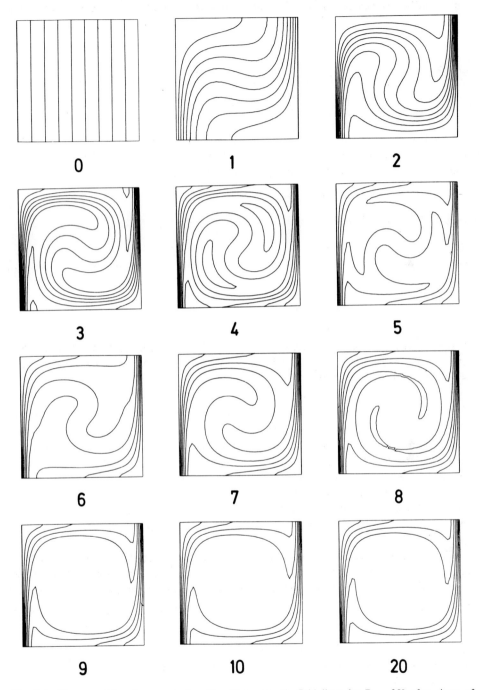

Fig. 8.4. Flux expulsion from a single eddy. The magnetic field lines for $R_m = 250$ after times of $t/\tau = 0 - 20$, where $\tau = 5L/(8U)$ (from Galloway and Weiss, 1981).

(by equating the first and third terms in Equation (8.12)), where the transverse length-scale for the wound-up field is

$$l \approx \frac{B_0}{B} L \qquad (8.15)$$

from flux conservation. Thus, equating τ and τ_d gives a maximum field strength of

$$B_1 \approx R_m^{1/2} B_0, \qquad (8.16)$$

as stressed by Moffatt (1978).

In the lateral boundary layers (of thickness d, say), there is a balance between convection and diffusion, and so, in a steady state, the flow speed (dU/L) into the layer must balance the diffusion speed (η/d), which gives

$$d \approx R_m^{-1/2} L. \qquad (8.17)$$

The flux is therefore concentrated to a field strength B_m given by flux conservation,

$$B_m d \approx B_0 L, \qquad (8.18)$$

as

$$B_m \approx R_m^{1/2} B_0. \qquad (8.19)$$

For two-dimensional cells (i.e., rolls) the flux is concentrated equally at both sides, but for a three-dimensional axisymmetric cell the flux is concentrated preferentially at the centre, which could represent the vertex between several cells. Also, the field strength is much higher than the above two-dimensional value (8.19), since flux conservation in a tube of radius d replaces Equation (8.18) by $B_m d^2 \approx B_0 L^2$, and so $B_m \approx R_m B_0$.

A further point concerns the estimate (8.16) for the maximum field strength (B_1). It is based on the condition that simple diffusion begins to occur, but, in practice, the *tearing mode instability* (Sections 2.10.2 and 7.5.5) may well take place, although it is not allowed in the numerical simulation; this would restrict the field strength to much lower values, namely the maximum of B_1^* and B_1^{**}, which are estimated as follows. Equating τ to the tearing mode time-scale

$$\tau_{tmi} \approx \left(\frac{l^2}{\eta} \frac{l}{v_A} \right)^{1/2}$$

gives

$$B_1^* \approx \left(\frac{R_m U}{v_{A0}} \right)^{1/4} B_0, \qquad (8.20)$$

where v_A and v_{A0} are the Alfvén speeds based on B and B_0, respectively. However, it is likely (Priest, 1981a) that the tearing mode can occur only if the normal field component B_0 is so small that $B_0 < (\eta/(v_A L))^{1/2} B$, which gives instead a maximum value for the amplified field strength of $B_1^{**} = (v_{A0}/U) R_m B_0$. This is the relevant value when

$$\left(\frac{v_{A0}}{U} \right)^5 > \frac{1}{R_m^3},$$

and otherwise Equation (8.20) should be adopted.

More recently, Weiss and his collaborators have conducted a comprehensive program of numerical simulations of magnetoconvection for a Boussinesq fluid, by coupling Equation (8.12) with equations of motion and energy, and so incorporating the reaction of the magnetic field on the motion. For example, Galloway and Moore (1979) have considered a three-dimensional axisymmetric cell and found several regimes of asymptotic large-time behaviour, depending on the initial field strength. If

$$\frac{v_{A0}}{U} < \frac{1}{R_m}, \tag{8.21}$$

(or $v_{A0}/U < R_m^{-1/2}$ in the two-dimensional case) the concentrated field (B_m) is weaker than the equipartition value (B_e), which satisfies

$$\frac{B_e^2}{2\mu} = \tfrac{1}{2}\rho U^2; \tag{8.22}$$

in this regime the kinematic solutions are valid and the flow is hardly affected by the field. At higher field strengths, an asymptotic regime is obtained in which the field is concentrated into a flux rope on the axis of the cell, as before, but there is no motion inside the rope, and the external motion is only slightly affected by the presence of the rope (Figure 8.5); thus, *two separate regions are created*, a stationary flux-rope region and a field-free convecting region. As the field strength increases still further, Galloway and Moore found one regime with overstability inside the flux rope and then another with overstability everywhere. The overstability is exhibited as a convective cell which periodically reverses its direction of motion. Naively, one may imagine that the field can be concentrated only to the equipartition

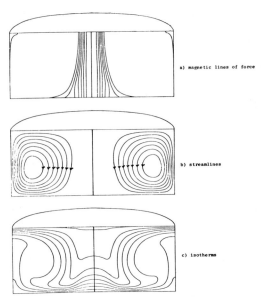

Fig. 8.5. (a) Magnetic field lines, (b) streamlines and (c) isotherms for a nonlinear steady state of axisymmetric magnetoconvection. Magnetic flux has been expelled from most of the cell and concentrated as a flux tube on the axis (from Galloway and Moore, 1979).

value (B_e), but Galloway and Moore found values that may be six times higher in their numerical experiments. The point is that, even if the Lorentz force is not balanced by the inertial term $\rho(\mathbf{v}\cdot\nabla)\mathbf{v}$ in the equation of motion (which would give Equation (8.22)), it can still be balanced by a pressure gradient. Thus, in the photosphere, for instance, the maximum field strength is limited by the photospheric pressure (p_e) to be at most B_p such that

$$\frac{B_p^2}{2\mu} = p_e, \tag{8.23}$$

when the pressure within the flux tube becomes negligible. Galloway and Moore found that, as the initial field (B_0) increases, so the flux rope field strength (B_m) increases in the *kinematic regime* like $B_m \approx R_m B_0$, until it reaches a maximum, and then it decreases in the *dynamic regime*. At first, the work done by bouyancy is balanced by viscous heating and

$$B_m \approx \left(\frac{\mathrm{Ra}}{\mathrm{Ha}^3}\right)^{2/3} \frac{\kappa}{\eta \log(B_m/B_0)} B_0.$$

Later, ohmic dissipation dominates and

$$B_m \approx \left(\frac{\mathrm{Ra}}{\mathrm{Ha}^2}\right)^{2/3} \frac{\kappa}{\eta} B_0.$$

In an idealised calculation, Galloway *et al.* (1978) found that the maximum value ($B_{\max}$) of B_m behaves like $B_{\max} \approx (\nu/\eta)^{1/2} B_e$, so that turbulent diffusivities ($\tilde{\nu} \approx \tilde{\eta}$) give $B_{\max} \approx B_e$. This would imply fields of a few hundred Gauss at the photosphere and 10^4 G near the base of the convective zone.

Other important advances in magnetoconvection theory include a study of linear stability when all three diffusivities (ν, κ, η) are present, as well as a magnetic field and rotation (Acheson, 1978). Also, Weiss (1981) has investigated, in nonlinear numerical simulations of two-dimensional convection, the bifurcations that are present in the transitions between oscillatory and steady convection as the Rayleigh number is increased; and this has been complemented by Knobloch *et al.* (1981), who analyse the ordinary differential equations that result from truncating the full system. One important aim of the nonlinear calculations is to establish the lowest value for B_0 for which vigorous steady convection is possible in the case ($\eta < \kappa$) when feeble convection has already set in as overstable oscillations (Busse, 1975b). Weiss (1981) and Galloway and Weiss (1981) show that, when B_0 is large enough, this criterion for the onset of steady convection reduces to the simple condition for overturning convection given in Equation (8.2). In future, it will be interesting to discover the effects of relaxing the Boussinesq approximation, of taking the initial field horizontal (as at the base of the convection zone), of adopting free-surface conditions (so as to allow downflow in intense tubes and penetration into stable regions), and of modelling three-dimensional convection. As a start, Latour *et al.* (1976) and Glatzmaier (1980) have relaxed the Boussinesq condition slightly by considering the *anelastic equations*, valid for highly subsonic flows, while Graham (1975) has presented some fully compressible simulations, and Hathaway (1980) has developed a code for three-dimensional flow.

8.2. Magnetic Buoyancy

The term 'magnetic buoyancy' refers to two distinct situations. Firstly, a horizontal isothermal flux tube cannot be in equilibrium and must rise, since it is lighter than its surroundings (Section 8.2.1). Secondly, a stratified magnetic field can be in equilibrium, but it may become unstable to the formation of rising flux tubes (Section 8.2.2).

8.2.1. Qualitative Effect

In the convection zone, magnetic fields tend to be concentrated into tubes (Section 8.1.3) and pushed around by turbulent motions. Any field with a smaller energy content than the turbulence is stretched until the Lorentz force is strong enough to react back (or diffusion sets in). At equipartition with a typical kinetic energy density of 10^5 J m^{-3} (10^6 erg cm^{-3}), one finds from Equation (8.22) a field strength of about 5000 G, but magnetoconvection simulations suggest that this may well be exceeded (Section 8.1.3).

The fundamental paper by Parker (1955a) suggests that, once such a flux tube is formed, it rises by *magnetic buoyancy* and produces a pair of sunspots where it breaks the photospheric surface (Figure 8.6). Suppose the gas pressure and magnetic field strength of the tube are p_i and B_i and the ambient external pressure at the same height is p_e. Then *lateral* total pressure balance implies $p_e = p_i + (B_i^2/(2\mu))$. If the temperature ($T$) is uniform and the corresponding densities are ρ_e, ρ_i, this becomes

$$\frac{k_B T \rho_e}{m} = \frac{k_B T \rho_i}{m} + \frac{B_i^2}{2\mu}, \tag{8.24}$$

so that ρ_e must exceed ρ_i. The plasma in the tube therefore feels a resultant buoyancy force of $(\rho_e - \rho_i)g$ per unit volume, which tends to make the tube rise. When the tube becomes curved it will also experience a restoring force due to magnetic tension; this is not large enough to completely counteract buoyancy as long as

$$(\rho_e - \rho_i)g > \frac{B_i^2}{\mu L},$$

or, using Equation (8.24),

$$L > \frac{2k_B T}{mg} \equiv 2\Lambda. \tag{8.25}$$

In other words, a tube that is longer than twice the local scale-height will be buoyed up. When it reaches the surface, its magnetic field strength will have fallen much below 5000 G in view of the smaller kinetic energy density at the photosphere (Section 8.7).

The importance of the magnetic buoyancy effect can be estimated from the size of the density deficit

$$\frac{\rho_e - \rho_i}{\rho_e} = \frac{B_i^2 m}{2\mu \rho_e k_B T}$$

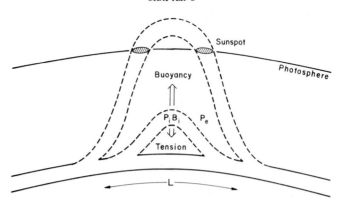

Fig. 8.6. Successive positions in the upper convective zone for a magnetic flux rope, part of which rises through magnetic buoyancy and produces a sunspot pair where it breaks the photosphere.

from Equation (8.24). At a distance 20 000 km below the photosphere, values of $\rho_e \approx 0.25$ kg m^{-3}, $T \approx 2.5 \times 10^5$ K and $B_i \approx 1000$ G make $(\rho_e - \rho_i)/\rho_e \approx 10^{-5}$, so that magnetic buoyancy is small (but still comparable with thermal buoyancy). On the other hand, at only 1000 km below the photosphere one may take $\rho_e \approx 0.8 \times 10^{-5}$ kg m^{-3}, $T \approx 1.5 \times 10^4$ K and again $B_i \approx 1000$ G, so that $(\rho_e - \rho_i)/\rho_e \approx 0.004$. The buoyancy force is therefore much stronger in the *upper* part of the convection zone. Similar arguments suggest that the cooling associated with a sunspot is a relatively shallow phenomenon extending, say, only 2000 km below the photosphere, although the flux tube itself may originate deep in the convective zone.

If the footpoints of a flux tube are separated by less than a few scale-heights (Λ), the tube can form an arch and remain in equilibrium under a balance between the forces of magnetic buoyancy and magnetic tension. Parker (1979a, Section 8.6) estimates the shape ($z = z(y)$) of such an arch in a vertical $y - z$ plane as follows. With z directed vertically upwards, the gradient of the arch is given in terms of its inclination ($\theta(z)$) to the horizontal by $dz/dy = \tan\theta$. Since the tube is in horizontal equilibrium, $(B^2/\mu)\cos\theta = B_s^2/\mu$, where B^2/μ and B_s^2/μ are the tensions in the tube at height z and at the summit ($y = 0$, $z = z_s$). However, the field strength ($B(z)$) for a tube in an isothermal medium is given from Equation (8.86) by

$$B(z)^2 = B_s^2\, e^{-(z_s - z)/\Lambda}.$$

Thus, after eliminating θ and B between the above equations and integrating, the equation of the arch may be shown to be

$$\tan^2 \frac{y}{2\Lambda} = e^{(z_s - z)/\Lambda} - 1.$$

As $z \to -\infty$, so $y \to \pm \pi\Lambda$, and the *maximum footpoint separation* is therefore $2\pi\Lambda$. If the flux tube is twisted, it must obey an extra constraint to be in equilibrium, namely that it extend vertically over no more than a few scale-heights (Section 9.6 of Parker, 1979a).

8.2.2. Magnetic Buoyancy Instability

Consider a horizontal magnetic field $(B_0(z)\hat{\mathbf{x}})$ in equilibrium, and suppose gravity is acting downwards along the negative z-axis (Figure 8.7(a)). Such a system is unstable if the field strength decreases fast enough with height, but the exact criterion depends on what effects are incorporated. Suppose, first of all, that the field lines remain straight. Then instability ensues if the *field strength falls off faster than the density*, so that

$$\frac{d}{dz}\left(\frac{B_0}{\rho_0}\right) < 0, \qquad (8.26)$$

which may be established by the following parcel argument (e.g., Moffatt, 1978; Acheson, 1979a). Suppose a flux tube rises by an amount δz, such that its internal field strength, density and pressure change by amounts δB, $\delta \rho$ and δp, while the corresponding changes in ambient values are $\delta B_0, \delta \rho_0, \delta p_0$. Conservation of mass and magnetic flux imply that B/ρ remains constant in the flux tube, so that

$$\frac{\delta B}{B_0} = \frac{\delta \rho}{\rho_0}. \qquad (8.27)$$

Also, horizontal equilibrium means a balance in total pressure $(p + B^2/(2\mu))$, so that

$$\delta p + \frac{B_0 \, \delta B}{\mu} = \delta p_0 + \frac{B_0 \, \delta B_0}{\mu},$$

or, since $\delta p = (k_B/m)T_0 \, \delta \rho$ and $\delta p_0 = (k_B/m)T_0 \, \delta \rho_0$ for an isothermal medium,

$$\frac{k_B T_0 \, \delta \rho}{m} + \frac{B_0 \, \delta B}{\mu} = \frac{k_B T_0 \, \delta \rho_0}{m} + \frac{B_0 \, \delta B_0}{\mu}. \qquad (8.28)$$

The condition $(\delta \rho < \delta \rho_0)$ for the tube to continue rising implies that $\delta B_0 < \delta B$ from Equation (8.28), and so Equation (8.27) gives $(\delta B_0/B_0) < (\delta \rho_0/\rho_0)$, which establishes

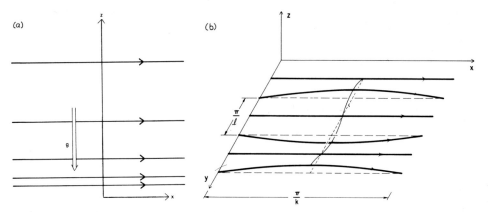

Fig. 8.7. Field lines for the magnetic buoyancy effect: (a) equilibrium field in a vertical plane; (b) perturbed field lines at one particular height.

Equation (8.26). When the additional stabilising influence of normal buoyancy is incorporated, the criterion for instability becomes modified to

$$\frac{1}{\Lambda}\frac{\rho_0}{B_0}\frac{d}{dz}\left(\frac{B_0}{\rho_0}\right) < -\frac{\gamma N^2}{v_A^2}, \qquad (8.29)$$

where N is the *Brunt frequency* (Section 4.4).

There are many extra effects that can be included, as summarised in the rest of this section, following the clear review by Acheson (1979a). For instance, if the field is allowed to bend, it is potentially more unstable. Parker (1966, 1979a) and Gilman (1970) then found that *'magnetic buoyancy drives an instability if the basic field strength decreases with height'*, so that

$$\frac{dB_0}{dz} < 0. \qquad (8.30)$$

The unstable perturbations form slightly curved flux tubes, which allow plasma to drain from the summit, and so they enhance the buoyancy effect by making the summit lighter. This *magnetic buoyancy instability* is really only a form of *hydromagnetic Rayleigh–Taylor instability* (Sections 7.3.1A and 7.5.2), since a negative value for dB_0/dz means that the field is holding up extra mass against gravity and the density is falling off with height more slowly than normal (see Equation (8.32)).

The basic stability analysis of Gilman (1970) and Schatzman (1963) may be summarised as: the equilibrium variables $(p_0(z), \rho_0(z), T_0(z), B_0(z))$ satisfy the perfect gas law

$$p_0 = \frac{k_B}{m}\rho_0 T_0 \qquad (8.31)$$

and the equation of magnetostatic balance (Section 3.2)

$$\frac{d}{dz}\left(p_0 + \frac{B_0^2}{2\mu}\right) + \rho_0 g = 0. \qquad (8.32)$$

In general, a discussion of the stability of such a basic state would involve the solution of a complicated set of ordinary differential equations for the perturbed variables subject to certain boundary conditions – an eigenvalue problem. But the two degrees of freedom present allow one to consider the particular case when the (isothermal) sound and Alfvén speeds,

$$c_{s0} = \left(\frac{p_0}{\rho_0}\right)^{1/2}, \quad v_{A0} = \frac{B_0}{(\mu\rho_0)^{1/2}},$$

are constants with height, so that the resulting dispersion relation has constant coefficients. The solutions to Equations (8.31) and (8.32) are then

$$p_0(z) = p^* e^{-z/\Lambda_B}, \quad \rho_0(z) = \rho^* e^{-z/\Lambda_B}, \quad B_0(z) = B^* e^{-(1/2)z/\Lambda_B}, \qquad (8.33)$$

where p^*, ρ^*, B^* are the values at the reference height $z = 0$ and

$$\Lambda_B = \frac{p^* + B^{*2}/(2\mu)}{\rho^* g}$$

is the scale-height ($\Lambda = p^*(\rho^* g)^{-1}$) increased by the presence of the magnetic field. Perturbations from equilibrium are taken of the form $f(z)\,e^{i(kx+ny-\omega t)}$, on a timescale large enough for any temperature variations to be smoothed out by radiation and conduction. The linearised magnetohydrodynamic equations yield the dispersion relation

$$(c_{s0}^2 + v_{A0}^2)\omega^4 - v_{A0}^2[(2c_{s0}^2 + v_{A0}^2)k^2 + c_{s0}^2/(2\Lambda\Lambda_B)]\omega^2$$
$$+ k^2 v_{A0}^4 c_{s0}^2(k^2 - (2\Lambda\Lambda_B)^{-1}) = 0 \qquad (8.34)$$

in the limit $n^{-1} \ll k^{-1}, n^{-1} \ll \Lambda$, of a small wavelength normal to the magnetic field (which gives the most unstable modes). The perturbed field lines in a horizontal plane therefore possess a long-wavelength structure along their length and a short-wavelength structure in the y-direction, reminiscent of neighbouring flux loops rising and sinking from a unidirectional field (Figure 8.7(b)). As a loop rises, its field strength tends to increase due to horizontal stretching but also to decrease due to vertical expansion. If the resulting field strength exceeds the neighbouring field strength at the new height, instability results by Parker's argument. When $dB_0/dz < 0$ and the wavelength ($2\pi/k$) along the field is so large that

$$0 < k^2\Lambda < \frac{1}{2\Lambda_B} \equiv -\frac{1}{B_0}\frac{dB_0}{dz}, \qquad (8.35)$$

it can be seen that the last term in Equation (8.34) is negative, and so the configuration is unstable ($\omega^2 < 0$). Equation (8.35) is thus an updated version of Parker's earlier condition (8.25), with magnetic tension forces being overcome by magnetic buoyancy for small enough wavenumbers k. For $c_{s0}^2 \gtrsim v_{A0}^2$ the most unstable wave has a wavelength along the field ($2\pi k^{-1}$) of about $16\,\Lambda$. By contrast, when there are no variations along the field ($k = 0$), Equation (8.34) gives stability, and, if the initial field is uniform with height ($\Lambda_B^{-1} = 0$), Equation (8.34) yields the two stable solutions $\omega^2 = k^2 v_{A0}^2$ and

$$\omega^2 = \frac{k^2 v_{A0}^2 c_{s0}^2}{v_{A0}^2 + c_{s0}^2},$$

representing Alfvén waves and slow magnetoacoustic waves (at almost perpendicular propagation to the magnetic field (Section 4.6)). The additional stabilising influence of stratification modifies the criterion (8.35) for instability to

$$\frac{1}{\Lambda B_0}\frac{dB_0}{dz} < -k^2 - \frac{\gamma N^2}{v_A^2}. \qquad (8.36)$$

The above simple analysis of magnetic buoyancy is complicated considerably by the extra effects of dissipation (v, η, κ) and rotation (Ω) inclined at θ to the vertical. Magnetic and thermal diffusion reduce the stabilising effect of stratification, and they modify Equations (8.29) and (8.36) to

$$\frac{1}{\Lambda}\frac{\rho_0}{B_0}\frac{d}{dz}\left(\frac{B_0}{\rho_0}\right) < -\frac{\eta\gamma N^2}{\kappa\, v_A^2} \quad \text{and} \quad \frac{1}{\Lambda B_0}\frac{dB_0}{dz} < -k^2 - \frac{\eta\gamma N^2}{\kappa\, v_A^2},$$

for instability of modes with $k = 0$ and $k \neq 0$, respectively.

Gilman (1970) showed that *rapid rotation alone stabilises the equilibrium against magnetic buoyancy*, since it counteracts buoyancy by inhibiting vertical motions. He finds stability if $\gamma \beta_0 = 2c_{s0}^2/v_{A0}^2$ is large enough or if rotation is so fast that $2\Omega \sin \theta (\Lambda/c_{s0}) > 1$. However, this result is itself modified by the effect of diffusion. Roberts and Stewartson (1977) found that dissipation can counteract rotation and still allow the magnetic buoyancy instability to occur in the large-β_0 limit; growth is on the ohmic diffusion time-scale, which makes it important in the solar interior if a turbulent diffusivity is adopted. Acheson (1978, 1979a) and Acheson and Gibbons (1978) have also considered diffusion and rotation together. They make the approximations $c_s \gg v_A$, so that sound waves are filtered out, and $\Omega \Lambda \gg v_A$, so that the plasma is rapidly rotating, which is valid in most of the convection zone. In the absence of normal buoyancy, this gives (for adiabatic variations) a low-frequency mode ($\omega \ll kv_A$) with

$$\omega = \frac{kv_A^2}{2\Omega}\left[\frac{1}{\gamma\Lambda} \pm i\left\{-\frac{1}{\gamma\Lambda}\frac{d}{dz}\left(\log\frac{B_0}{\rho_0}\right) - \left(1 + \frac{l^2}{n^2}\right)k^2\right\}^{1/2}\right] - i\eta(l^2 + n^2), \tag{8.37}$$

which represents the effects of magnetic buoyancy and diffusion on the *hydromagnetic inertial wave* (see Equation (4.49′)). The mode consists of a wave propagating along the magnetic field and driven unstable by magnetic buoyancy if B_0/ρ_0 decreases so rapidly with height that

$$\frac{\Lambda \rho_0}{\gamma B_0}\frac{d}{dz}\left(\frac{B_0}{\rho_0}\right) < -\frac{3\pi^2\sqrt{3}}{c_*}, \tag{8.38}$$

where $c_* = v_A^2/(2\Omega\eta)$. The unstable modes have $k\Lambda \lesssim 1$ when $c_* > 1$ and growth-rates of typically $v_A^2/(\Omega\Lambda^2)$, so that the effect of rotation has been to make the instability grow much more slowly than the Alfvén rate (v_A/Λ). The effect of diffusion is essentially to suppress the instability if

$$\frac{v_A^2}{2\Omega\eta} < 1. \tag{8.39}$$

Modifications due to curvature are treated by Acheson (1979a), who considers a toroidal equilibrium field $(B_0(r)\hat{\phi})$ in cylindrical geometry. Also, Spruit and van Ballegooijen (1982) show that a slender isolated flux tube in the convection zone is unstable, so that any dynamo should be located at the base of the convection zone. Magnetic buoyancy instability then occurs only outside a critical radius of $2\gamma\Lambda$, and the growing waves propagate in an easterly direction. When $c_* \gg 1$ the critical mode for Equation (8.37) has a long wavelength ($\approx c_*^{1/2}\Lambda$) along the field, which could not fit in to a cylindrical geometry. The corresponding modified instability criterion is therefore

$$\left(\frac{1}{\gamma\Lambda} - \frac{2}{r}\right)r\frac{d}{dr}\log\left(\frac{B_0}{\rho_0 r}\right) < -1. \tag{8.40}$$

8.2.3. THE RISE OF FLUX TUBES IN THE SUN

Parker (1955a) suggested that toroidal flux tubes rise in sections a few scale-heights (Λ) long by magnetic buoyancy and create sunspots where they break through the solar surface along lines of latitude in the sunspot zones. The slight inclination (about 10° typically) of a sunspot pair (Section 1.4.2E) is probably caused by the action of the *Coriolis force* as the flux tube rises. The time-scale for the rise, according to Parker (1975), is just the *Alfvén time*

$$\tau = \frac{\Lambda}{v_A}, \tag{8.41}$$

which varies considerably with depth. According to Equation (8.41), a 100 G field would rise from a depth of 20 000 km in only 2 months and from the middle of the convection zone in a year. In order to obtain a time-scale of 10 yr appropriate to dynamo generation (Chapter 9), such a field would need to start from the base of the convection zone (200 000 km in depth). Moreover, fields stronger than this (which are needed to account for the flux emergence as active regions) would rise much too quickly. Similar estimates have been given by Parker (1979a, Section 8.7), who suggests that the speed of rise is determined by a balance between magnetic buoyancy and aerodynamic drag to be typically the Alfvén speed; also, the pressure of the surrounding plasma flattens the tube into a horizontal ribbon.

These estimates have been challenged by Schüssler (1977), who suggests that the expansion of a tube as it rises can slow it down considerably, and by Acheson (1979a, b), who points out that the extra effect of rotation is to make the buoyancy instability grow on the much longer time-scale of

$$\tau = \frac{4\Omega\Lambda^2}{v_A^2}. \tag{8.42}$$

The depth at which this becomes 10 yr is 30 000 km for a field of 100 G, or 100 000 km for 1000 G; at greater depths the time-scale is much longer. Furthermore, according to Equation (8.39) the effect of magnetic diffusion is to allow the instability only if $v_A^2/(2\Omega\eta) > 1$. With typical turbulent values for η of $10^8 - 10^9$ m² s⁻¹ (Section 2.1.5), this means that field strengths in excess of 1000 G are needed to overcome the stabilising effect of rotation. Thus, Acheson concludes that *fields of about* 1000 G *or more are needed* and that they are generated in the *middle of the convective zone or above*. (Moreover, Galloway *et al.* (1977) expect the fields to be concentrated by convection to strengths of about 10 000 G (Section 8.1.3), and this is the subsurface value that Zwaan (1978) feels is necessary to account for active-region fluxes.)

Reservations about the above estimate are that it is based on the *linear* growth-time and that perhaps tearing modes (Sections 2.10.2 and 7.5.5) can operate faster than normal diffusion. Also, convection will make the flux tubes rise in a few turn-over-times, which is much faster than the above rates. Galloway and Weiss (1981) have suggested therefore that magnetic flux can be held down only in regions of downflow and that the dynamo region may be located at the interface between the radiative and convective zones. Finally, it should be mentioned that Meyer *et al.* (1979) have modelled the motion of filamentary flux tubes under the combined action of magnetic buoyancy, Lorentz forces and aerodynamic drag.

8.3. Cooling of Sunspots

The most widely held explanation for the cooling of sunspots is the *inhibition of convection* by their magnetic field. Consider first part of the convection zone just below the photosphere and free of magnetic flux. In the absence of motions the temperature would decrease outwards faster than adiabatic. But the effect of convection is to mix up the different layers, trying to equalise their temperatures and thus reducing the temperature gradient below the static value. Biermann (1941) suggested that the sunspot's magnetic field inhibits this convection completely in the top few thousand kilometers of the convection zone and so removes the mixing-up process. The temperature therefore decreases outwards faster inside the spot than in the surrounding convecting region, and this leads to a lower temperature in the sunspot, as indicated schematically in Figure 8.8. In other words, the magnetic field reduces the heat flow, damming it back and so cooling the spot. Cowling (1953) modified this idea by suggesting that there is *some* convection inside the spot but that it is less vigorous than outside and so transports less heat. He points out that, from linear theory (Section 8.1.2), as the magnetic field increases (with $\kappa > \eta$) one proceeds from steady convection to weaker, oscillating (overstable) convection and then to a state where convection is suppressed completely. Furthermore, the observations (Section 1.4.2B) do indeed indicate the presence of weak convective motions inside umbrae and an energy flux of about one-fifth the normal photospheric value.

A possible difficulty with the above explanation for sunspot coolness is that the missing energy flux would be diverted around the sunspot and should show up at the solar surface as a *bright ring* around it. Since no such bright ring is observed, its absence has presented a problem for proponents of the inhibition theory. Recently, however, this problem has been overcome by Spruit (1977), who has presented a

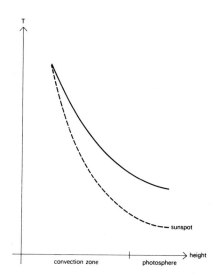

Fig. 8.8. The variation of temperature with height inside a sunspot (– – –) and in a neighbouring region that has the same temperature at some depth below the photosphere.

series of model calculations for heat flow below a sunspot. These indicate that, if the sunspot depth exceeds 10^4 km, the convection is so efficient at transferring heat that there is only a small pile-up of heat below the spot, and it spreads out over such a large region around the spot that no easily observable bright ring appears. However, the Solar Maximum Mission has observed variations of up to 10^{-3} in the solar irradiance at the same time as the appearance of sunspots, suggesting that the convection zone can store the missing energy.

Another difficulty with the inhibition theory is that the pile-up of heat below a spot should raise the temperature and pressure, and so cause the field to disperse (Parker, 1974a). Two ways of resolving this problem have been pointed out in a useful summary of sunspot theory by Spruit (1981a). Either the field strength is high enough everywhere along the flux tube to stop the pressure enhancement from destroying the spot, or there is an extra force to confine the flux tube at the blocking level. Meyer et al. (1974) suggested that such a force would be produced by a *moat cell* (Sections 1.4.2A and 8.6.2); this dynamic confinement is, however, only possible in the later stages of long-lived spots, and it appears to provide a strong enough force only if the spot is deep (10^4 km) and small (radius 2000 km).

As an alternative to inhibition of convection, Parker (1974a) and Roberts (1976) suggest that sunspots are cool because *convective over-stability converts the heat into Alfvén waves*, which are emitted preferentially downwards (Parker, 1979g). Parker estimates that an Alfvén wave flux of 2.5×10^7 W m^{-2} should be emitted upwards from the overstable region below the sunspot. At the photosphere, where the Alfvén speed is 15 km s^{-1} and the field strength 3000 G, the corresponding wave amplitude would be 3 km s^{-1}. However, strong reflection in the photosphere (Thomas, 1978) reduces the amplitude to 2.1 km s^{-1}, in agreement with the observations of Beckers (1976a). Parker (1979h) also estimates that a downdraft of 1 to 2 km s^{-1} at a depth of 1000 to 4000 km is sufficient to reduce the heat flux to the value found in a sunspot umbra.

8.4. Equilibrium Structure of Sunspots

This section describes the equilibrium of a sunspot, concentrating on the magnetohydrostatic aspects rather than the problems of radiative transfer. It leads naturally to questions of stability, both convective (Section 8.1) and interchange (Section 8.4.2). Comments on umbral oscillations can be found in Section 4.9.4.

8.4.1. MAGNETOHYDROSTATIC EQUILIBRIUM

For simplicity, consider first the effect of a purely vertical magnetic field $(B(R)\hat{z})$ on the equilibrium of a stratified atmosphere (Figure 8.9). Suppose $B(R)$ possesses a maximum value of B_i on the axis at $R = 0$, and that it approaches zero at large values of R, the corresponding pressures being $p_i(z)$ and $p_e(z)$. Then horizontal pressure balance and vertical equilibrium imply

$$p(R, z) + \frac{B^2(R)}{2\mu} = p_e(z) \tag{8.43}$$

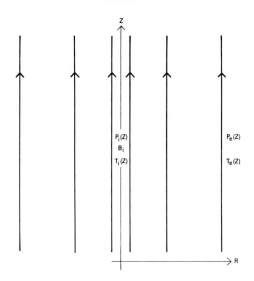

Fig. 8.9. A vertical magnetic field threading a vertically stratified medium.

and

$$\frac{\partial p}{\partial z} = -\rho g. \tag{8.44}$$

At large distances from the axis, Equation (8.44) becomes

$$\frac{dp_e}{dz} = -\rho_e g, \tag{8.45}$$

which needs to be supplemented by an energy equation before one can determine the temperature stratification ($T_e(z)$). On the axis, Equation (8.43) gives

$$p_i(z) + \frac{B_i^2}{2\mu} = p_e(z), \tag{8.46}$$

which determines p_i once p_e is known, and Equation (8.44) becomes

$$\frac{dp_i}{dz} = -\rho_i g. \tag{8.47}$$

Now, the derivative of Equation (8.46) with respect to z implies that the pressure gradients dp_e/dz and dp_i/dz are equal, and so from Equations (8.45) and (8.47)

$$\rho_i = \rho_e. \tag{8.48}$$

Furthermore, Equation (8.46) implies that the plasma pressure (p_i) inside the spot is smaller than the ambient pressure (p_e), and the corresponding temperature deficit is

$$\frac{T_i(z)}{T_e(z)} = 1 - \frac{B_i^2}{2\mu p_e(z)}.$$

Thus, the presence of the vertical magnetic field has no effect on the density, but it produces a pressure deficit (and hence a temperature deficit) in order to maintain horizontal pressure balance. Furthermore, the result implies that only if the thermal equilibrium of the plasma produces this particular form for the temperature difference will the magnetic field lines be straight; in general, one expects them to be curved.

According to Equation (8.46), the magnetic pressure $(B_i^2/(2\mu))$ in the vertical flux rope is smaller than the external plasma pressure $(p_e(z))$. This may be true below the photosphere, but at the photospheric level the magnetic pressure 2.4×10^4 N m^{-2} (2.4×10^5 dyne cm^{-2}) due to a field of 3000 G, say, exceeds the ambient gas pressure (at $\tau_{5000} = 1$) of 1.4×10^4 N m^{-2}. The magnetic field lines have a pressure that is too strong for them to be contained, and so they spread out, as indicated in Figure 1.26. The resulting decrease in B_i^2 means from Equation (8.46) that the spot pressure gradient (dp_i/dz) exceeds the ambient value, and so from Equations (8.45) and (8.47) $\rho_i < \rho_e$. This density deficit within the sunspot produces the Wilson effect (Section 1.4.2D).

Below a few thousand kilometers the plasma pressure exceeds 10^7 N m^{-2} and, if there is no comparable increase of magnetic pressure, the flux rope may be shredded by turbulent motions. For this reason some authors believe that the flux tube beneath the sunspot is a rather shallow phenomenon extending to only 10 000 km in depth. Others suggest that a flux tube continues to be cooler than its surroundings deep in the convection zone.

A model for the *magnetostatic equilibrium* of a sunspot was proposed by Schlüter and Temesvary (1958); it is based on the force balance $\mathbf{0} = -\nabla p + \mathbf{j} \times \mathbf{B} - \rho g \hat{\mathbf{z}}$, with gravity acting along the negative z-axis. For a cylindrically symmetric $(\partial/\partial\phi = 0)$, untwisted $(B_\phi = 0)$ magnetic field $(B_R(R, z), 0, B_z(R, z))$, the equation has components

$$0 = -\frac{\partial p}{\partial R} + \frac{B_z}{\mu}\left(\frac{\partial B_R}{\partial z} - \frac{\partial B_z}{\partial R}\right) \tag{8.49}$$

and

$$0 = -\frac{\partial p}{\partial z} - \frac{B_R}{\mu}\left(\frac{\partial B_R}{\partial z} - \frac{\partial B_z}{\partial R}\right) - \rho g. \tag{8.50}$$

They seek a *similarity solution* that automatically satisfies $\text{div } \mathbf{B} = 0$ and that has the form

$$B_R = -\frac{R}{2}f(\zeta)\frac{dB_i}{dz}, \qquad B_z = f(\zeta)B_i(z),$$

where $\zeta = RB_i^{1/2}(z)$ and $f(0) = 1$, so that $B_i(z)$ represents the field on the axis of symmetry ($R = 0$). The flux (F) through the spot is $F = 2\pi \int_0^\infty \zeta f(\zeta)\, d\zeta$. Equation (8.49) may then be integrated with respect to R from 0 to ∞ at constant z, to give, after some manipulation

$$0 = 2(p_e(z) - p_i(z)) + 4a_2 B_i^{1/2}\frac{d^2 B_i^{1/2}}{dz^2} - B_i^2, \tag{8.51}$$

where the subscripts e and i denote values at $R = \infty$ and $R = 0$, respectively, and $a_2 = \int_0^\infty \frac{1}{2}\zeta f^2(\zeta)\, d\zeta$. On the other hand, Equation (8.50) gives

$$\frac{dp_0}{dz} = -\rho_0 g \tag{8.52}$$

at $R=0$ (where B_R vanishes), and

$$\frac{dp_e}{dz} = -\rho_e g, \tag{8.53}$$

provided f tends to zero fast enough at infinity. Thus, if the temperature structure, the flux (F) and the shape factor $f(\zeta)$ ($=e^{-\zeta^2}$, for instance) are prescribed, together with the values of B_i and dB_i/dz at $z=0$, Equations (8.51) to (8.53) determine the functions $B_i(z)$, $p_i(z)$ and $p_e(z)$. The analysis has been extended by Deinzer (1965), who uses an energy equation to determine the temperature, taking into account the influence of convection inhibition on the heat flux. Further work on this model can be found in Jakimiec (1965), Yun (1968) and Landman and Finn (1979). However, there is still the need to solve Equation (8.49), Equation (8.50) and an energy balance equation (including the effects of convection in a magnetic field) without invoking the (unrealistic) self-similar assumption; in particular, it would be interesting to determine the shape of the free surface of a sunspot. Osherovich (1979, 1982) has improved the Schlüter–Temesvary theory in two ways. He first included an azimuthal field component (B_ϕ) and next overcame a limitation of the theory (namely that it implies a maximum value (of typically 67°) for the inclination of field lines to the vertical). This was accomplished by replacing the magnetic vector potential $A = -\frac{1}{2} e^{-\zeta^2}$ by

$$A = -\tfrac{1}{2} e^{-\zeta^2} \left(1 + a_3 \zeta^2 \exp - \int_0^z A(z)^{-1} dz \right),$$

and thereby incorporating field lines which return to the photosphere. After choosing the constant a_3 as -5.6, excellent agreement was obtained with the observed values of $B_R(R)$ and $B_z(R)$ at the surface of a typical spot.

Meyer et al. (1977) have set up some simple magnetohydrostatic models, in which a sunspot is represented by a flux tube with pressure (p_i), density (ρ_i), field strength (B_i). It occupies a volume (V_i) and is surrounded by a field-free region (V_e) with pressure (p_e) and density (ρ_e), as indicated in Figure 8.10. The equilibrium equations in the two regions are

$$\nabla \left(p_i + \frac{B_i^2}{2\mu} \right) = \rho_i \mathbf{g} + (\mathbf{B}_i \cdot \nabla) \frac{\mathbf{B}_i}{\mu} \tag{8.54}$$

and

$$\nabla p_e = \rho_e \mathbf{g}. \tag{8.55}$$

These are supplemented by a boundary condition at the interface S (with unit normal $\mathbf{n}$), namely that the total pressure be continuous

$$p_i + \frac{B_i^2}{2\mu} = p_e. \tag{8.56}$$

Near the photosphere, where the umbral pressure is much less than the magnetic pressure, some simpler *vacuum models* (Meyer et al., 1977) are often useful. For them the internal pressure and gravitational force are negligible, so that the magnetic

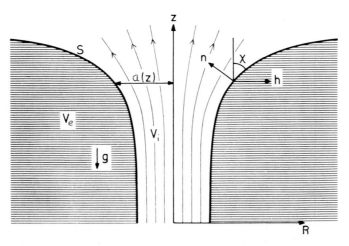

Fig. 8.10. A magnetic flux rope surrounded by field-free plasma (after Meyer et al., 1977).

field (B_i) is potential. It may be written

$$\mathbf{B}_i = \frac{1}{R}\left(-\frac{\partial \psi}{\partial z}, 0, \frac{\partial \psi}{\partial R}\right)$$

in cylindrical polars, where $\psi(R, z)$ satisfies

$$\frac{\partial^2 \psi}{\partial R^2} - \frac{1}{R}\frac{\partial \psi}{\partial R} + \frac{\partial^2 \psi}{\partial z^2} = 0, \tag{8.57}$$

subject to the boundary condition that the total pressure be continuous

$$\frac{B_i^2}{2\mu} = p_e(z) \tag{8.58}$$

on the interface ($a = a(z)$). Upper and lower boundary conditions are also imposed. As $z \to \infty$ the field is assumed to be monopolar, so that on the interface the field is nearly horizontal with

$$B_R \approx B_i \approx \frac{F}{2\pi a^2}. \tag{8.59}$$

As $z \to -\infty$, the field becomes vertical with

$$B_z \approx B_i \approx \frac{F}{\pi a^2} \tag{8.60}$$

on S. The object should be to find the radius ($a(z)$) of the interface and the flux function ($\psi(R, z)$) for a given plasma pressure ($p_e(z)$) in the field-free region and a given flux $F = 2\pi(\psi)_s$. However, it is much easier to solve the inverse problem, namely of choosing analytic solutions to Equation (8.57) and then deducing $p_e(z)$. There are several simple examples of such vacuum solutions. The *Bessel-function model* has a potential $\psi = ARJ_1(kR)e^{-kz}$ (for $J_1(kR) > 0$) and a radial field

$$B_R = AkJ_1(kR)e^{-kz}, \tag{8.61}$$

which becomes singular as $z \to -\infty$. (The corresponding solution for a twisted spot is given in Equation (3.45).) The *dipole model* has $\psi = A \sin^2 \theta / r$ (for $\cos^2 \theta > \frac{1}{3}$), with

$$B_R = \frac{3\psi^3 \cos\theta}{A^2 \sin^3 \theta}, \qquad (8.62)$$

(which is singular at $\theta = \frac{1}{2}\pi$). Also, the *perforated current sheet model* has field lines $v = $ constant and a radial field

$$B_R = \frac{\tanh u \sin v}{\cosh^2 u - \sin^2 v}, \qquad (8.63)$$

where (u, v, ϕ) are oblate spheroidal coordinates such that $R = \cosh u \sin v$, $z = \sinh u \cos v$.

Unfortunately, none of these models give pressures ($p_e(z)$) which agree well with the solar stratification, so Meyer et al. also use a *mean model*, by assuming that the mean field ($\bar{B}_i(z)$) across the flux tube is approximately equal to the field at the boundary, which is valid for a *slender flux tube* (Section 8.7.1). Then Equation (8.58) gives

$$\bar{B}_i(z) \approx (2\mu p_e(z))^{1/2}, \qquad (8.64)$$

where $p_e(z)$ is imposed from a standard model of the convection zone based on mixing-length theory; the radius of the tube for a given flux (F) follows from

$$\pi a(z)^2 \bar{B}_i(z) = F. \qquad (8.65)$$

Recently, Parker (1979b) has challenged the conventional picture of a sunspot as a single large flux tube (Figure 1.26). He has suggested that it divides into many separate tubes of diameters down to 300 km at about 1000 km beneath the photosphere (Figure 8.11) because of interchange or (overturning) convective instability. In this *spaghetti model* the loose cluster of flux tubes (or strands) is held together at the solar surface by magnetic buoyancy and by a downdraft beneath the sunspot (Section 8.6). Below a depth of 1000 km the flux tubes (within which the heat transport is greatly inhibited) have fields of 5000 G and are separated by field-free gas that is convecting heat normally; the net effect is that only one-fifth of the normal heat flux is presented to the underside of the umbra. The *sub-surface down-draft* is assumed to flow between the individual flux strands towards the sunspot axis and to provide a confining force (Section 8.3) by exerting an *aero-dynamic drag* on the strands (Parker, 1979c); it also removes some of the excess heat that has piled up below the spot (Parker, 1979h) and so reduces the force necessary for confinement (Spruit, 1981a). How does the downdraft originate? Parker points out that a downdraft is associated with a rising tube and is amplified by convective forces to give the tube a convective propulsion (Parker, 1979f); also, individual flux tubes are observed to be located probably in downdrafts between granules and supergranules. However, it is likely that such a convective cell exists only in the early formative stage of a sunspot, since a moat cell appears after a few days with its circulation in the opposite direction. Spruit (1981d) suggests that the strands may be held together instead by being anchored at the base of the convection zone.

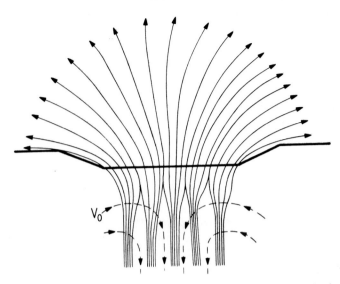

Fig. 8.11. Parker's model for a sunspot as a cluster of magnetic flux tubes, held together by a convective downdraft indicated by dashed arrows (from Parker, 1979b).

The individual flux tubes in Parker's model are identified with *magnetic knots*, which have fluxes of 10^{11} Wb (10^{19} Mx), corresponding to fields of 1500 G across a diameter of 1000 km (Section 1.4.2A). When 3 or 4 knots coalesce they form a *pore* with diameter of 1500 to 2000 km and a field strength of 1500 G. Occasionally, the strong subsurface downdraft is established and the accumulation continues to form a sunspot from 30 or so knots; such a small spot has a field strength of 3000 G over a diameter of 4000 km; it represents a flux of 4×10^{12} Wb and may later grow much bigger. The bright *umbral dots* (Section 1.4.2B) that are often present in the umbra are taken as evidence for photospheric gas between the flux strands (Parker, 1979i). Another piece of evidence which Parker cites in favour of his theory is the fact that the surface temperature of a sunspot umbra appears to be independent of its diameter, which is difficult to explain on a single flux tube theory; regardless of whether the umbral diameter is 4000 km or 40 000 km, the surface temperature is about 3900 K and the field strength close to 3000 G. Although Parker's theory has many attractive features and our understanding of the behaviour of individual intense flux tubes is rapidly developing (Section 8.7), the theory for the interactions between them is only in its infancy. The elements of a spaghetti (or cluster) model have been presented by Spruit (1981d), who balances the attractive force of buoyancy and the repulsive magnetic force between tubes above the photosphere.

A firm supporter of the concept of discrete flux tubes is Piddington (1978), who discusses qualitative ideas for a twisted kilogauss flux tube (called a *flux rope*) with a typical flux of 10^{13} Wb (10^{21} Mx). It consists of a few hundred *flux fibres*, which each have fluxes of about 3×10^{10} Wb and are themselves twisted for stability against fluting, although such twists would probably disappear by local reconnections, and twists are not found in many observations (Zwaan, 1978). Small bundles of flux fibres are termed *flux strands*, which may account for X-ray bright points and join together below the photosphere to form a flux rope. Individual fibres can themselves fray at the photosphere into many *threads*, which correspond to filigree. Piddington suggests that sunspots are not concentrated from weak uniform (100 G) fields, such as those imposed initially in magnetoconvection simulations (Section 8.1.3), since this would imply that black sunspots form from much larger grey areas. instead, he claims that sunspots (and intense flux tubes (Section 8.7)) are already concentrated to kilogauss strengths before they emerge (see also Zwaan, 1978).

8.4.2. SUNSPOT STABILITY

Convective instability in the presence of a magnetic field has already been described in Section 8.1. On linear theory one finds that the onset of steady (overturning) convection is unaltered by the presence of a weak field, whereas moderate fields allow only a weaker form of overstable (oscillatory) convection (when $\kappa > \eta$), and strong fields suppress convection altogether. The importance of this instability for the structure of sunspots is described in Sections 8.3 and 8.6, while its role in the formation of intense tubes is discussed in Section 8.7.

The equilibrium structure of a sunspot is quite complicated, since it involves a balance between the three forces due to gravity, the magnetic field, and the plasma pressure. Its stability has therefore not yet been analysed completely, but much progress has been made by Meyer *et al.* (1977), who treat local stability near the photosphere. There are two competing effects. The first is destabilising and occurs because the fanning of the field lines makes the boundary of the sunspot curve around the external region where the plasma pressure is higher; such a configuration would be unstable to the interchange or flute instability (Section 7.5.1) if gravity were absent. The second effect is stabilising and arises because the magnetic field of the outer part of the spot lies on top of the denser external plasma, a situation which would be stable to the hydromagnetic Rayleigh–Taylor mode if there were no curvature (Section 7.5.2). Meyer *et al.* (1977) determine the resulting criteria for stability of magnetohydrostatic sunspot models given by Equations (8.54) to (8.56) as follows.

The energy principle of Bernstein *et al.* (1958) provides the second-order change in energy due to an adiabatic perturbation (Section 7.4) as $\delta W = \delta W_i + \delta W_e + \delta W_s$. Here the surface contribution is

$$\delta W_s = \tfrac{1}{2} \int_S (\mathbf{n} \cdot \boldsymbol{\xi})^2 \mathbf{n} \cdot \left[\nabla \left(p_i + \frac{B_i^2}{2\mu} \right) - \nabla p_e \right] dS, \qquad (8.66)$$

and the volume terms (δW_i and δW_e) are assumed to be both positive, since convective or (volume) magnetic instabilities are not being considered. Then a necessary and sufficient condition for stability is that the integrand of Equation (8.66) be positive, so that

$$\mathbf{n} \cdot \left[\nabla \left(p_i + \frac{B_i^2}{2\mu} \right) - \nabla p_e \right] > 0, \qquad (8.67)$$

which is just the condition that the net restoring force on the interface act in opposisition to the displacement. After using Equations (8.54) to (8.56) for an axisymmetric field $(B_R, 0, B_z)$, the stability condition reduces to

$$\frac{dB_R^2}{dz} < 0, \tag{8.68}$$

taken along a field line in the surface S. Thus the net effect is *stability if the magnitude of the radial magnetic field component decreases upwards on the boundary of the flux tube*. An alternative form for this condition for stability at uniform temperature is

$$R_c \sin \chi > 2\Lambda_e \tag{8.69}$$

in terms of the scale-height

$$\Lambda_e = \frac{p_e}{\rho_e g}, \tag{8.70}$$

as well as the radius of curvature (R_c) of a field line in S and the inclination (χ) of S to the vertical. This shows that the boundary is stable if its radius of curvature or inclination are large enough. Furthermore, R_c and χ can be written in terms of the tube radius (a) and its derivatives $(a' \equiv da/dz, a'' \equiv d^2a/dz^2)$, so that (8.69) reduces to

$$\frac{a'(1 + a'^2)}{a''} > 2\Lambda_e. \tag{8.71}$$

Parker (1979b) has given a simple physical derivation of Equation (8.69). Suppose an elementary flux tube becomes separated from the surface S. It will experience a tension force $(B^2/(\mu R_c))$ normal to S, pulling it out of the surface. In addition, there will be a buoyancy force $(g \Delta\rho)$ acting vertically upwards, where $\Delta\rho$ is the density deficit inside the tube. However, at uniform temperature $\Delta\rho/\rho = \Delta p/p$, where the pressure deficit is $\Delta p = B^2/(2\mu)$. Thus, the component of the buoyancy force normal to S becomes $B^2 \sin \chi/(2\mu\Lambda)$, and the condition that this exceed the tension force gives the stability criterion (8.69).

Meyer *et al.* (1977) apply the stability condition (8.68) to the vacuum models of Section 8.4.1. Near the top of the sunspot, Equations (8.58) and (8.59) imply that

$$B_R^2 \approx 2\mu p_e(z),$$

which decreases with height, and so the interface is stable. Near the base of the spot Equation (8.60) gives

$$B_R \equiv \frac{da}{dz} B_z \sim \frac{da}{dz} \frac{1}{a^2},$$

while Equations (8.58) and (8.60) give the radius as $a \sim p_e(z)^{-1/4}$. Thus

$$B_R \sim -\frac{1}{p_e^{3/4}} \frac{dp_e}{dz},$$

and Equation (8.68) implies that $p_e(z)$ must increase downwards faster than z^4 for a stable configuration. At all heights (z), the Bessel-function model is stable, since Equation (8.61) gives $B_R = k\psi/R$, which decreases upwards on S (where $\psi = F/(2\pi)$).

The dipole model too is stable, since, as one moves upwards along the interface (ψ = constant), so θ increases and, from Equation (8.62), B_R decreases. At the plane $z = 0$ ($u = 0, v < \frac{1}{2}\pi$), the perforated current-sheet model possesses a throat near which $\partial B_R/\partial u > 0$ from Equation (8.63), and so it is unstable.

More realistically, Meyer *et al.* (1977) consider *mean models* for which the mean field ($\bar{B}_i(z)$) and radius ($a(z)$) are determined by Equations (8.64) and (8.65) in terms of F and $p_e(z)$. The derivatives of a follow from Equations (8.55), (8.64), (8.65) and (8.70) as

$$\frac{da}{dz} = \frac{a}{4\Lambda_e}, \qquad \frac{d^2a}{dz^2} = \frac{a}{16\Lambda_e^2}\left(1 - 4\frac{d\Lambda_e}{dz}\right).$$

The condition (8.71) for stability can then be written as

$$F > 8\pi(2\mu p_e)^{1/2}\Lambda_e^2\left(-4\frac{d\Lambda_e}{dz} - 1\right). \tag{8.72}$$

For a given $p_e(z)$, $\Lambda_e(z)$ and $d\Lambda_e/dz < -\frac{1}{4}$, flux tubes with a small enough flux are therefore unstable. In other words, *slender flux tubes are locally unstable to fluting* when $d\Lambda_e/dz < -\frac{1}{4}$, which holds for the standard model atmosphere over the height range -700 km $< z < 30$ km. At the opposite extreme, large sunspots have $R_c \approx a \gg \Lambda_e$ and $\sin \chi > \frac{1}{2}$, so the stability condition is easily satisfied. Meyer *et al.* use Equation (8.72) and their standard model for $p_e(z)$ to estimate the critical flux separating unstable from stable flux tubes; they conclude that a vacuum model of a *flux tube that converges monotonically with depth is stable when its flux exceeds about* 10^{11} Wb (10^{19} Mx). However, below a certain depth the flux-tube pressure exceeds its magnetic pressure, and the vacuum model fails. Also, the sunspot models considered by Parker (1979b) possess a shallow 'throat' or 'wasp waist', below which the field diverges; although buoyancy can stabilise the upper part of such a tube where the field is fanning out, the lower part of the boundary at (and below) the throat is still locally unstable to the interchange mode; it is this which, according to Parker, breaks up a single flux tube into many strands (Section 8.4.1).

Leading spots in a sunspot group tend to outlive following spots (Section 1.4.1), so that many active regions possess just one (leading) spot. Meyer *et al.* suggest that this is because, as a stitch of field rises from deep in the convection zone, the decreasing angular velocity causes the leading part of the flux tube to become nearly vertical, and so it is stable. The following part is inclined to the vertical, which makes its inner edge subject to fluting, and it is this which breaks up the following spot.

It will be interesting to see in future the effects on flux-tube stability characteristics of adopting full vacuum models, of including the thermodynamics, and of allowing for the surrounding convective cell.

8.5. The Sunspot Penumbra

The classical interpretation of penumbral structure as consisting of *convective rolls* was given by Danielson (1961), who suggests that the light (hot) material is rising and the dark (cool) material is falling. The dispersion relation (8.7) for the linear stability analysis in an inclined uniform magnetic field (B_x, B_z) is third-order for ω in terms of k_x and k_y. As shown in Section 8.1.2, a horizontal field is unstable to rolls with $k_x = 0$. Also, for an inclined field the most unstable perturbation is often a roll aligned with the horizontal field direction, again such that $k_x = 0$. Danielson determines ω for a roll of given length (k_x^{-1}) and width (k_y^{-1}), for parameters appropriate to sunspot penumbrae. He finds that, as the inclination of the magnetic field increases, so convection sets in when the field is nearly horizontal.

Meyer et al. (1977) have suggested that in addition convective elements beneath the spreading penumbral field may drive local interchanges and may penetrate through the field to give a bright filament at the photosphere. Moreover, Parker (1979b) regards penumbral filaments as part of the loose cluster of flux tubes which make up the sunspot in his model. Spruit (1981d) has suggested that the rising of a bright filament should tend to drive an inflow, whereas the falling of a dark filament should cause plasma to flow out again.

Clearly, more work is needed on these ideas to understand the properties of a penumbra. The nonlinear numerical simulations of magnetoconvection (Section 8.1.3) have been carried out mainly for a vertical field, so it is important to see what happens in an inclined field. Even static models for the penumbral part of a sunspot are much more difficult to set up than the vacuum models of the umbra (Section 8.4.1), since the plasma pressure cannot be neglected in the penumbra. Also, a satisfactory penumbral model cannot be said to exist until it explains Evershed flow in a self-consistent manner. The outward flow (at an altitude of, say, 500 km) may be due to the effect of geometry or nonlinear interactions in roll-like convection or due to siphon flow towards a low-pressure magnetic knot (Section 6.5.2), whereas the inward flow along loops that are, say, 500 km high may represent a siphon flow driven into the low-pressure sunspot (Section 6.5.2). Finally, an interesting feature of the penumbra is the observation of running penumbral waves that are continually propagating outwards; they have been interpreted as trapped fast magnetoacoustic modes (Sections 4.8 and 4.9.4).

8.6. Evolution of a Sunspot

8.6.1. Formation

In an important paper, Meyer et al. (1974) outlined a model for the growth and decay of sunspots. To begin with, a magnetic flux tube is buoyed up to the photosphere, where its axis is approximately vertical. Then the *growth phase* represents a compression of the flux in two stages. During the *first stage*, supergranulation flow down to a depth of at least 10 000 km concentrates the field (Figure 8.12). Equating the kinetic and magnetic energy densities (8.22) at the photosphere gives a field strength for $\rho = 3 \times 10^{-4}$ kg m^{-3} and $v = 0.3$ km s^{-1} of only 50 G. Locally, a flux tube may also be concentrated by granulation with higher speeds of typically 2 km s^{-1} to higher equipartition field strengths of about 400 G. Inside the flux rope this magnetic field is strong enough to halt overturning convection, and so during the *second stage* the plasma cools and falls. The magnetic field strength therefore increases still further to a value at which the external gas pressure roughly balances the internal magnetic pressure (8.23). This gives a field strength of 2200 G

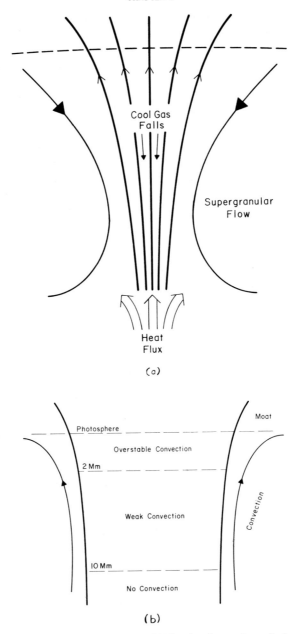

Fig. 8.12. (a) The growth phase of a sunspot. (b) The slow decay phase of a long-lived sunspot.

for a photospheric pressure of 2×10^4 N m^{-2}, or 3500 G for a pressure of 5×10^4 N m^{-2} at a depth of 200 km. The build-up of heat below the spot then halts the supergranulation flow, and spot growth ceases. At this point many spots break apart, especially the following ones, for which the flux rope is inclined somewhat to the vertical. If the flux rope is, however, vertical down to a depth of at least 10 000 km, an annular convection cell (or *moat*) is formed around the spot with a sense of motion opposite to that of the previous supergranule. It is driven by the leakage of heat below the spot and allows the spot to survive for many weeks while it slowly decays away.

The above ideas for the growth of a sunspot are similar to those of Parker (1979a, Chapter 10). However, in his spaghetti model (Section 8.4.1), Parker (1979b) suggests that a cluster of small flux tubes is assembled by *hydrodynamic attraction* during the formation process. As they rise through the convection zone, the flux tubes are attracted towards one another by the *Bernouilli effect*. Consider, for example, two horizontal cylinders (of radius a) that are rising side-by-side with constant vertical speed (u) and separation x (Parker, 1979a, Section 8.9), as indicated in Figure 8.14(a). A force of attraction is produced because the fluid is accelerated between the gap between the cylinders, and so the pressure is reduced there. If the motion is irrotational and $x \gg a$, the total kinetic energy of cylinders plus moving fluid can be written

$$K(x) = \tfrac{1}{2}\pi a^2 u^2 [\rho_c + \rho(1 + \tfrac{1}{2}a^2/x^2 + \ldots)], \tag{8.73}$$

where ρ_c and ρ are the density of the cylinder and fluid, respectively. The force of attraction for a length L is therefore $F_a = -(dK/dx)L$, or, after differentiating Equation (8.73),

$$F_a = \tfrac{1}{2}\pi a u^2 \rho(a/x)^3 L. \tag{8.74}$$

However, the magnetic repulsion between the two tubes with field strength B_0 is $F_r = \tfrac{1}{8}\pi a^2 \rho v_A^2 (a/x)^2$, in terms of the Alfvén speed ($v_A = B_0/(\mu\rho)^{1/2}$). Thus the hydrodynamic attraction is able to overcome the repulsion provided the length of the cylinders exceeds

$$L = \frac{x v_A^2}{4u^2}. \tag{8.75}$$

Parker also suggests that one flux tube rising behind another (Figure 8.14b) is attracted in the wake of the leading tube (Parker, 1979d, e). Moreover, after the flux tubes begin to cluster at the solar surface, a subsurface downdraft (Figure 8.11) is set up and holds them together by aerodynamic drag (Section 8.4.1). If the tubes are held together like this at some depth, the effect of magnetic buoyancy is to keep them together in the shape of a circle at the surface, rather like a group of tethered balloons.

8.6.2. Decay

For the *slow decay phase*, Meyer et al. (1974) propose the structure shown in Figure 8.12(b). Outside the spot, small-scale convection is present in the annular moat flow and can be represented by eddy diffusion coefficients. Inside the flux rope, convection is affected by the magnetic field in a manner that depends on the ratio κ/η. Equation (8.10) in dimensional variables shows that, if $\kappa < \eta$, convection occurs first as a leak instability at a temperature difference of

$$\Delta T = \frac{\pi^2}{\alpha} \frac{\kappa}{\eta} \frac{B^2}{\mu\rho g d}, \tag{8.76a}$$

where $B^2/(\mu\rho gd)$ is the ratio of magnetic tension to buoyancy forces. If $\kappa > \eta$, according to Equation (8.11) convection begins as an overstable oscillation when

$$\Delta T = \frac{\pi^2}{\alpha}\frac{\eta}{\kappa}\frac{B^2}{\mu\rho gd}. \tag{8.76b}$$

In each case, overturning convection, which is more efficient at transporting heat, sets in when

$$\Delta T = \frac{\pi^2}{\alpha}\frac{B^2}{\mu\rho gd}. \tag{8.77}$$

The variations with depth of η and κ are shown in Figure 8.13, which leads one to expect overstable oscillations in the top 2000 km of the spot, with motions in columns parallel to the magnetic field. Between a depth of 2000 and 10 000 km, if ΔT lies between the values given by Equation (8.76a) and (8.77), there is small-scale (leak) convection, with a vigour limited by the magnetic field. At a depth of 6000 km, for instance, a cell-size (d) of about 2000 km, a temperature difference (ΔT) of 50 K and a vertical plasma speed of 20 m s^{-1} give a heat flux (10^7 W m^{-2}) comparable with the observed umbral value. However, if ΔT exceeds the value given by Equation (8.77) overturning occurs, and it may concentrate the field into tubes at cell boundaries (Section 8.1.3) to give Parker's spaghetti model (Figure 8.11), as pointed out by Spruit (1981a). Below 10 000 km the field is strong enough to suppress convection altogether.

It has long been recognised (Cowling, 1953) that the slow decay of a sunspot cannot

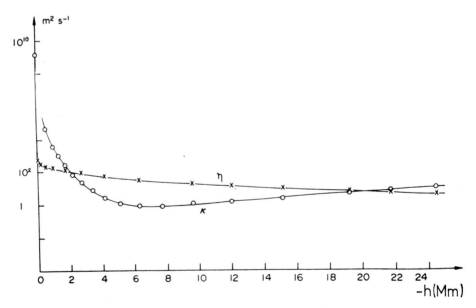

Fig. 8.13. Variation of magnetic diffusivity (η) and thermal conductivity (κ) (in m^2 s^{-1}) with depth ($-h$ in Mm) below the photosphere (1 Mm $\equiv$ 1000 km). By comparision, the eddy magnetic diffusivity at the photosphere is 10^7 to 10^9 m^2 s^{-1} (from Meyer et al., 1974).

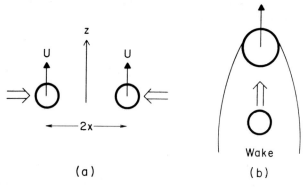

Fig. 8.14. The hydrodynamic attraction of two flux tubes moving (a) upwards side-by-side with speed u and (b) one behind the other.

be due to simple *ohmic diffusion*, since the diffusion-time $\tau_d = l^2/\eta$ is far too long. For instance, a scale-length of 3000 km and a classical magnetic diffusivity (η) of 300 m² s⁻¹ give $\tau_d \approx 1000$ yr, by comparison with the observed lifetime of a few months. Meyer et al. (1974) suggest that magnetic flux leaks slowly from the spot due to an *eddy diffusivity* ($\tilde{\eta}$), and it is then carried rapidly across the moat by the surrounding annular circulation (see also Meyer et al., 1979). However, Wallenhorst and Howard (*Solar Phys.* (1982) **76**, 203) find that the magnetic flux of an active region decreases as a sunspot disappears. They suggest that flux tubes sink below the surface or reconnect rather than simply spreading out. Figure 8.5 simulates this effect if it is tipped upside down so that the sense of circulation is correct. In order to model diffusion in the central region (2000 to 10 000 km), where the field is roughly vertical $(B(R,t)\hat{\mathbf{z}})$, consider the vertical component of the induction Equation (2.50) in cylindrical polars and with no flow, namely

$$\frac{\partial B}{\partial t} = \frac{\tilde{\eta}}{R}\frac{\partial}{\partial R}\left(R\frac{\partial B}{\partial R}\right).$$

A self-similar solution to this (in terms of the total flux (ϕ_0) out to infinity) is

$$B = \frac{\phi_0}{4\pi\tilde{\eta}t}\exp\left(-\frac{R^2}{4\tilde{\eta}t}\right). \tag{8.78}$$

If the edge of the spot ($R = a(t)$, say) is defined to be such that the magnetic field reaches a prescribed value (B_s, say), the sunspot flux becomes $F = \int_0^a 2\pi BR\, dR = F_0 - 4\pi\tilde{\eta}B_s t$, after substituting for B from Equation (8.78). Thus, the rate of decrease of the flux is

$$-\frac{dF}{dt} = 4\pi\tilde{\eta}B_s. \tag{8.79}$$

The fact that this is constant agrees with observations (Figure 1.25), which give

$$-\frac{dF}{dt} = 1.2 \times 10^8 B_{max}\, \text{m}^2\, \text{s}^{-1}, \tag{8.80}$$

in terms of the *maximum field strength* (B_{max}) in tesla. A value for $\tilde{\eta}$ may then be deduced by equating (8.79) and (8.80) and adopting values for B_s and B_{max} of, say, 1500 G and 3000 G, respectively. The result is $\tilde{\eta} \approx 2 \times 10^7 \text{ m}^2 \text{ s}^{-1}$, which is consistent with an estimate for the eddy diffusivity due to small-scale convection at a depth of 6000 km.

8.7. Intense Flux Tubes

Photospheric observations indicate that most of the photospheric flux is concentrated into intense flux tubes, with a field strength of 1500 to 2000 G and diameters of only 100 to 300 km (Section 1.3.2B), representing a flux of typically 5×10^9 Wb (5×10^{17} Mx). They are probably located between granules at supergranulation boundaries and are associated with filigree and small-scale faculae. However, their extremely high field strengths cannot be explained on the basis of equipartition with granular kinetic energy since this gives fields of only about 200 G at the photosphere or 600 G at a depth of 1000 km. One possibility is that *magnetoconvection* can concentrate fields to strengths considerably in excess of this equipartition value, as in the numerical experiments of Weiss, Galloway and coworkers (Section 8.1.3). These suggest that turbulent convection can concentrate the field into flux ropes, which are then shuffled around as the convection pattern changes, with a tendency for them to remain between the cells. The flux tubes may also be continually shredding apart due to the *flute instability* (Section 8.4.2). Indeed, it is likely that the whole convection zone consists of a mixture of convecting cells interspersed with a tangled web of intense flux tubes. Another explanation for the high field strengths is given below in Section 8.7.2.

To begin with, it is necessary to describe the properties of a slender flux tube anywhere in the solar atmosphere (Section 8.7.1), not just near the photosphere, and then one can discuss photospheric intense tubes in particular (Sections 8.7.2–8.7.4). The term *slender* flux tube has been used with several (somewhat related) meanings as follows:

(i) a flux tube that is so narrow that its magnetic field is approximately uniform across its width; this means its flux may be written as $F = \pi a^2 B(s)$, in terms of its radius (a) and field strength (B) at a distance s along the tube; also, the field strength at the surface of the tube (which is used in the surface boundary condition) does not therefore differ greatly from the value at the centre of the tube;

(ii) a tube whose width (a) is much smaller than the scale-height (Λ_e) in the external plasma;

(iii) a tube whose width is much smaller than the wavelength of waves being considered; this would be better referred to as a long wavelength approximation.

8.7.1. Equilibrium of a Slender Flux Tube

Consider a flux tube that is slender in the sense of (i) above, with gravity acting along the negative z-direction (Figure 8.15). Suppose with Parker (1979a, Chapter 8) that the ambient medium is in hydrostatic equilibrium. The pressure (p_e) is then given from Equation (3.6) by

$$p_e(z) = p_e(0) e^{-\bar{m}_e(z)}, \tag{8.81}$$

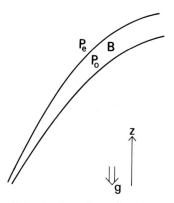

Fig. 8.15. An isolated (slender) flux tube confined by an external pressure (p_e).

where $\bar{m}_e(z) = \int_0^z \Lambda_e(z)^{-1}\,dz$ is the *number of scale-heights* above some reference level ($z = 0$), and Λ_e is the *external scale-height*. Suppose also that the tube is in magnetohydrostatic equilibrium under a balance between pressure gradients, gravity and the Lorentz force; the component of this force balance along the tube gives the internal pressure (p_i) as

$$p_i(z) = p_i(0)\, e^{-\bar{m}_i(z)}, \tag{8.82}$$

where $\bar{m}_i(z) = \int_0^z \Lambda_i(z)^{-1}dz$, in terms of the *internal scale-height* (Λ_i). Now, across the surface of the flux tube, the total pressure is conserved, and so

$$p_i + \frac{B_i^2}{2\mu} = p_e. \tag{8.83}$$

Substitution for p_i and p_e from Equations (8.81) and (8.82) therefore gives the magnetic pressure as

$$\frac{B(z)^2}{2\mu} = p_e(0)\, e^{-\bar{m}_e(z)} - p_i(0)\, e^{-\bar{m}_i(z)}. \tag{8.84}$$

In other words, if the pressures inside and outside the tube are specified at some level and the temperature is specified everywhere, the field strength along the whole length of the tube is determined by Equation (8.84). The resulting *tube radius* ($a(z)$) follows by flux conservation from

$$a(z)^2 = \frac{a(0)^2 B(0)}{B(z)}. \tag{8.58}$$

For the special case when the internal and external temperatures are the same, so that $\bar{m}_e(z) = \bar{m}_i(z) \equiv \bar{m}(z)$, Equations (8.84) and (8.85) reduce to

$$B(z) = B(0)\, e^{-(1/2)\bar{m}(z)} \tag{8.86}$$

and $a(z) = a(0) e^{(1/4)\bar{m}(z)}$. (In particular, a uniform temperature makes $\bar{m}(z) = z/\Lambda$.) Thus *the tube widens and the field weakens with height*.

When the tube interior is cooler than the exterior, by an amount ΔT, say, $p_i(z)$ declines more rapidly than $p_e(z)$. For a cooling that extends over large heights, this has the effect of draining most of the material from the upper parts of the tube. Above essentially $T/\Delta T$ scale-heights, p_i is much smaller than p_e, and so $B^2/(2\mu)$ becomes approximately equal to p_e.

For a tube whose plasma pressure is negligible, the internal field is potential, as in the vacuum models of Section 8.4.1. The magnetic field is given by

$$\frac{B(z)^2}{2\mu} = p_e(z), \tag{8.87}$$

in terms of the external pressure, which in turn is determined from

$$\frac{dp_e}{dz} = -\frac{p_e}{\Lambda_e}, \tag{8.88}$$

after the external scale-height ($\Lambda_e(z)$) has been prescribed; the radius follows from

$$\pi a(z)^2 B(z) = F, \tag{8.89}$$

in terms of the flux (F). In order to allow for departures from slenderness, one needs to remember that $B(z)$ refers to the surface value in Equation (8.87) and to the mean value in Equation (8.89). They may be treated by an expansion in Λ_e/a (Meyer et al., 1977, Appendix A).

8.7.2. Intense Magnetic Field Instability

A general picture for the way the Sun may assemble intense tubes from a diffuse magnetic field has been put forward by Parker (1978) and Spruit (1979), and mathematical aspects of it have been developed by Webb and Roberts (1978) and Spruit and Zweibel (1979). It is sketched in Figure 8.16 and may be described as follows. A diffuse vertical field is first swept into the corners and boundaries of supergranules, as discussed by Leighton (1963) and Parker (1963a) and described in Section 8.1.3. This forms a flux tube of moderate strength (several hundred Gauss), which is subject to convective collapse by the *intense magnetic field instability* (sometimes called the superadiabatic effect or convective instability). The tube plasma cools and falls, which may explain the *downdrafts* of 0.5 to 2.0 km s^{-1} that are found in supergranule boundaries provided new unstable tubes are continually being created. (Section 1.3.2A). At the same time, the field strength increases and the flux tube narrows until the field is so strong (typically 1 to 2 kG) that it can overcome the convective instability of the superadiabatic atmosphere and suppress the motion. The resulting convectively stable intense tube is shuffled around by the flow and at the same time tends to disperse due to interchange instabilities (Section 8.4.2).

Vertical motions in a slender vertical flux tube that is slowly diverging with height may be described by the (nonlinear) *slender flux tube equations* of Roberts and Webb (1978, 1979), as an extension to those of Defouw (1976). The density ($\rho(z,t)$), pressure ($p(z,t)$), vertical speed ($v(z,t)$) and field ($B(z,t)$) on the tube axis are given by the following equations of continuity, vertical momentum, transverse momentum and

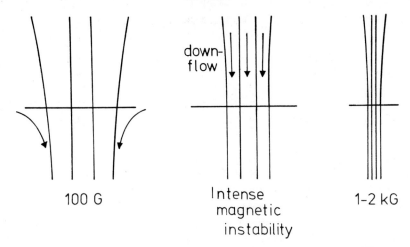

Fig. 8.16. The formation of kilogauss flux tubes by the intense magnetic field instability. Convective intensification is followed by instability accompanied by a downflow, and the final state is convectively stable.

isentropic energy:

$$\frac{\partial}{\partial t}\left(\frac{\rho}{B}\right) + \frac{\partial}{\partial z}\left(\frac{\rho v}{B}\right) = 0, \tag{8.90}$$

$$\rho\left(\frac{\partial v}{\partial t} + v\frac{\partial v}{\partial z}\right) = -\frac{\partial p}{\partial z} - \rho g, \tag{8.91}$$

$$p + \frac{B^2}{2\mu} = p_e, \tag{8.92}$$

$$\frac{\partial p}{\partial t} + v\frac{\partial p}{\partial z} = \frac{\gamma p}{\rho}\left(\frac{\partial \rho}{\partial t} + v\frac{\partial \rho}{\partial z}\right), \tag{8.93}$$

where p_e is the external pressure. Equation (8.90) is obtained by eliminating div **v** between the equations of continuity and induction for a perfectly conducting plasma. The equations $\nabla \cdot \mathbf{B} = 0$ and $\partial \rho / \partial t + \nabla \cdot (\rho \mathbf{v}) = 0$ then determine the resulting values of $\partial B_R / \partial R$ and $\partial v_R / \partial R$ on the tube axis.

For linear perturbations $v(z, t) = \bar{v}(z) \, e^{i\omega t}$ about a basic state given by Equations (8.81) to (8.83), with $T_i(z) = T_e(z)$ and no external pressure perturbation, Equations (8.90) to (8.93) imply that the velocity amplitude ($\bar{v}(z)$) satisfies

$$\frac{d^2\bar{v}}{dz^2} - \frac{1}{2\Lambda_i}\frac{d\bar{v}}{dz} + \left(\frac{\omega^2 - N_i^2}{c_T^2} + (1 - \tfrac{1}{2}\gamma)\frac{N_i^2}{c_{si}^2}\right)\bar{v} = 0. \tag{8.94}$$

Here

$$c_T(z) = \frac{c_{si} v_A}{(c_{si}^2 + v_A^2)^{1/2}} \tag{8.95}$$

is the *tube speed* (Section 4.10.1), and

$$c_{si}(z) = \left(\frac{\gamma p_i}{\rho_i}\right)^{1/2}, \quad v_A(z) = \frac{B}{(\mu\rho_i)^{1/2}}$$

are the *sound* and *Alfvén speeds*, while

$$N_i(z)^2 = \frac{g}{\Lambda_i}\left(\frac{\gamma-1}{\gamma} + \frac{d\Lambda_i}{dz}\right)$$

is the square of the *Brunt-Väisälä frequency*. When Equation (8.94) is supplemented by the two boundary conditions that $\bar{v}(z)$ vanish at both $z = 0$ and $z = -d$, say, it represents an eigenvalue problem for the frequency (ω). It is then of interest to determine the conditions on $\Lambda_i(z)$ and d for stability ($\omega^2 > 0$) and instability ($\omega^2 < 0$).

A local approximation to Equation (8.94) gives $\bar{v}(z) \sim e^{z/(4\Lambda_i)} \sin \pi z/d$ and

$$\omega^2 = \left(\frac{\gamma}{2c_{si}^2} + \frac{1}{v_A^2}\right)c_T^2 N_i^2 + \left(\frac{\pi^2}{d^2} + \frac{1}{16\Lambda_i^2}\right)c_T^2.$$

This dispersion relation shows that the plasma is stable for $N_i^2 > 0$. But, if $N_i^2 < 0$, unstable modes arise when the depth d is large enough. For conditions typical of an intense tube ($c_{si} = v_A$, $\gamma = 1.2$, $\Lambda_i = 150$ km, $d\Lambda_i/dz = -0.25$), the time-scale for this convective instability is typically 100 s, and the tube is stable if the field is so large (1 to 2 kG) that $\beta_i \equiv (2\mu p_i/B_i^2) < \frac{1}{2}$. A full non-local solution for the convection zone yields $\beta_i = 1.8$ (Spruit and Zweibel, 1979).

Solutions to the nonlinear Equations (8.90) to (8.93) have not yet been investigated in detail, although Roberts has discovered soliton solutions and Spruit (1979) has constructed the possible equilibrium solutions that are derived from the original equilibrium by an adiabatic displacement. Also, Hollweg (1982) has used them to model a spicule. Assuming $T_i(z) = T_e(z)$ initially, Spruit finds that the vertical displacement ($\xi(z_0)$) from a reference level (z_0) is governed by a nonlinear equation of the form

$$\frac{d^2\xi}{dz^2} = F\left(z_0, \xi, \frac{d\xi}{dz}\right).$$

This equation is solved numerically for the possible equilibria, subject to the vanishing of ξ at two levels. It possesses solutions only for $\beta_i > 2$. In the collapsed state, the tube is cooler than its surroundings at low levels and hotter at high levels with a typical surface field strength of 1800 G.

The above explanation for field intensification is not really distinct from concentration by magnetoconvection, in the sense that they are complementary attempts to model the effect of convective instability on a magnetic field. The numerical simulations follow the nonlinear development, but they are essentially incompressible, and the steady intense tube occurs after many turnover times when buoyancy balances dissipation. The intense magnetic field instability includes compressibility and stratification, but it neglects dissipation and the external convection; it has been analysed only in the linear regime, and the intense tube occurs when magnetic tension overcomes buoyancy.

8.7.3. Spicule Generation

Intense flux tubes provide a natural channel for fluid motions between the photosphere and the overlying atmosphere. They may be a source for spicules, representing paths along which the plasma is ejected upwards by the squeezing and *buffeting of granules* on their sides (Parker, 1974b; Roberts, 1979). Also, they may act as a sink for returning spicular material and other plasma falling from the corona. At any one time, spicules and intense tubes are comparable in number (10^5 spicules, 4×10^4 tubes), and both are concentrated in the network above supergranule boundaries; also, the lifetimes of spicules (8 to 15 min) and granules (5 to 10 min) are similar, so it is natural to relate them.

Roberts (1979) shows that granular buffeting can increase the tube pressure and drive a large-amplitude flow upwards along an intense tube: the narrower the tube, the larger the upflow. This is a *resonant effect*: it occurs when the phase speed of the external driving oscillation approaches the natural speed c_T (Section 4.10.1) for longitudinal wave propagation in the tube. The resonance may be illustrated by considering the linear equation for isentropic wave propagation in a uniform field $(B_0 \hat{z})$ threading a uniform atmosphere. For two-dimensional motions of the form $\mathbf{v} = (v_x(x), 0, v_z(x)) e^{i(\omega t + kz)}$, the transverse velocity amplitude satisfies $(d^2 v_x/dx^2) - m^2 v_x = 0$, where

$$m^2 = \frac{(k^2 c_s^2 - \omega^2)(k^2 v_A^2 - \omega^2)}{(k^2 c_T^2 - \omega^2)(c_s^2 + v_A^2)}. \tag{8.96}$$

In an unbounded medium, this gives the usual magnetoacoustic wave solutions (Section 4.6), but, for a column of field confined to $|x| < x_0$, Equation (8.96) has a solution

$$v_x = A \sinh mx, \quad |x| < x_0. \tag{8.97}$$

Now suppose that granular buffeting of a turbulent flow (v_e) with density (ρ_e) is represented by an oscillating pressure force $\delta p_e = \frac{1}{2} \rho_e v_e^2 \sin \omega t \cos kz$ on the tube boundary. Then the value of A is determined by the constancy of total pressure at the boundary, and the longitudinal speed (v_z) follows from the wave equation. On the axis $(x = 0)$ of the tube it has an amplitude

$$v_z(0) = \frac{\rho_e}{2\rho_0} \left(\frac{v_e}{v_A}\right)^2 c_T f(k) \cos \omega t \sin kz,$$

where

$$f(k) = \frac{\omega k c_T}{(k^2 c_T^2 - \omega^2) \cosh m x_0}.$$

For frequencies $\omega < k c_T$, $f(k)$ has a maximum that grows as the width of the tube decreases. An alternative way of generating spicules is by the kink (or transversal) tube waves described in the next section (Spruit, private correspondence).

8.7.4. Tube Waves

Intense flux tubes provide an efficient means of magnetic communication between

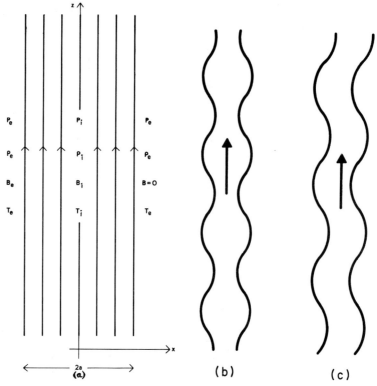

Fig. 8.17. (a) A magnetic slab surrounded by a field-free medium. (b) A sausage mode disturbance travelling along the tube. (c) A kink mode.

the photosphere and chromosphere. In particular, they form a natural channel for wave energy, but there are many more types of waves that can propagate in such non-uniform structures than in a simple uniform medium (Chapter 4). Roberts (1981b) has categorised these *tube waves* by considering for simplicity a *uniform slab* of width $2a$ and surrounded by a uniform field-free plasma (Figure 8.17); equilibrium values within and outside the slab are denoted by subscripts i and e, respectively. The dispersion relation may be derived in a similar manner to Equation (4.62), and it takes the form

$$(k^2 v_A^2 - \omega^2) m_e = (\rho_e/\rho_i) \omega^2 m_i \tanh m_i a \qquad (8.98\text{a})$$

for the *sausage mode* (v_x an odd function of x) or

$$(k^2 v_A^2 - \omega^2) m_e = (\rho_e/\rho_i) \omega^2 m_i \coth m_i a \qquad (8.98\text{b})$$

for the *kink mode* (v_x an even function of x), where

$$m_i^2 = \frac{(k^2 c_{si}^2 - \omega^2)(k^2 v_A^2 - \omega^2)}{(c_{si}^2 + v_A^2)(k^2 c_T^2 - \omega^2)}, \qquad m_e^2 = \frac{k^2 c_{se}^2 - \omega^2}{c_{se}^2},$$

in terms of the usual sound speed (c_s), Alfvén speed (v_A) and *tube speed*

$(c_T = c_{si}v_A/(c_{si}^2 + v_A^2)^{1/2})$. In view of the constancy of total pressure, the speeds c_{si}, c_{se} and v_A are related by

$$c_{si}^2 + \frac{\gamma}{2}v_A^2 = \frac{\rho_e}{\rho_i}c_{se}^2.$$

The fact that m_i and m_e are functions of ω and k makes Equations (8.98a) and (8.98b) complicated transcendental equations for $\omega = \omega(k)$. The results for the case when v_A is smaller than both c_{se} and c_{si} are shown in Figure 8.18. It is assumed that $m_e^2 > 0$, so that the waves are evanescent outside the slab and decay away to zero as $x \to \pm \infty$ like $e^{\mp m_e x}$. (The alternative case (when $m_e^2 < 0$) gives waves approaching (or receding from) the slab, and it makes the slab disturbance grow (or decay) with time as the energy accumulates (or declines).) When $m_i^2 = -n_i^2 < 0$, the perturbations behave like $e^{i(\omega t + kz \pm n_i x)}$ inside the tube, and so they represent *body waves* which may account for umbral oscillations (Roberts, 1981c); but when $m_i^2 > 0$ they behave like $e^{i(\omega t + kz) \pm m_i x}$ inside the tube, and so they represent *surface waves*. The mode $\omega = kv_A$ represents an Alfvén wave propagating along the tube, but the interfaces are undisturbed, and so there is no motion outside; for a flux tube it would correspond to a torsional Alfvén wave (Section 4.3.1). All the other modes are magnetoacoustic in nature, so that there are no truly Alfvén surface waves in a compressible plasma: they arise only in the incompressible limit from the slow modes.

In the *long wavelength limit* ($ka \ll 1$), some of the sausage modes may be investigated by approximating $\tanh m_i a$ in Equation (8.98a) by $m_i a$, so that

$$(k^2 v_A^2 - \omega^2)m_e = (\rho_e/\rho_i)\omega^2 m_i^2 a, \tag{8.99}$$

which may incidently be derived directly from the linearised *slender flux tube equations* ((8.90) to (8.93)). As ka approaches zero, there are two solutions to Equation (8.99)): either m_i approaches infinity and one finds

$$\omega = kc_T \tag{8.100}$$

for the *slow surface wave* (sometimes called simply the *tube wave*), or m_e tends to zero and so

$$\omega = kc_{se} \tag{8.101}$$

for a *fast wave*, which may be of surface or body type depending on the sign of m_i^2 and represents a sound wave in the external medium. Two other modes are apparent in Figure 8.18 for $ka \ll 1$: *slow body waves* (both sausage and kink) have $\omega = kc_T$ and cause rapid oscillations across the slab; also, there is a *slow surface wave* that is kink in nature and has

$$\omega = kv_A(ka\rho_i/\rho_e)^{1/2}. \tag{8.102}$$

When $v_A > c_{se}$ the diagram is similar to Figure 8.18, but with no waves above v_A and fast body waves between c_{se} and c_{si}.

In the *incompressible limit* ($c_{si}^2/v_A^2 \to \infty$), $c_T = v_A$, $m_i = m_e = k$ and Equation (8.98) reduces to $k^2 v_A^2 - \omega^2 = (\rho_e/\rho_i)\omega^2 \tanh ka$ and $k^2 v_A^2 - \omega^2 = (\rho_e/\rho_i)\omega^2 \coth ka$. Thus, in Figure 8.18 c_T collapses on to v_A and c_{si} disappears to infinity, so that the body

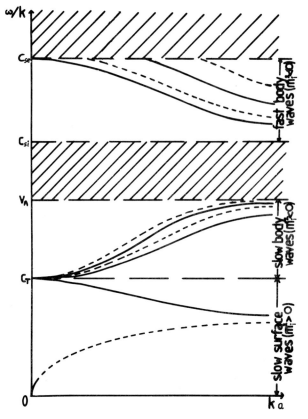

Fig. 8.18. The phase speed (ω/k) as a function of wavenumber (k) for 'tube' waves in a slab of width $2a$ when $c_{se} > c_{si} > v_A$. Sausage modes are shown by solid curves and kink modes by dashed ones. In the shaded region, there are no modes evanescent outside the slab (from Roberts, 1981b).

waves vanish and only the lower part of the figure remains. For $ka \ll 1$, this gives slow surface waves with

$$\omega = kv_A \quad \text{(sausage)}, \tag{8.103}$$

corresponding to Equation (8.100), and also

$$\omega = kv_A(ka\rho_i/\rho_e)^{1/2} \quad \text{(kink)}, \tag{8.104}$$

which are sometimes called 'Alfvén' surface waves.

Parker (1979a, Section 8.10) has given a derivation of the slow kink mode (for $\rho_i = \rho_e$) from first principles as follows. Consider a small transverse sinusoidal displacement of the slab $\xi(z, t) = \varepsilon \exp i(\omega t - kz)$, with $ka \ll 1$. For an incompressible medium surrounding the slab, the velocity is potential and its longitudinal component may be written

$$v_x(x, z, t) = \begin{cases} i\omega\varepsilon \exp(-k(x-a) + i(\omega t - kz)), & x \geq a, \\ i\omega\varepsilon \exp(k(x+a) + i(\omega t - kz)), & x \leq -a, \end{cases} \tag{8.105}$$

so that $v_x(\pm a, z, t) = \partial \xi/\partial t$. The resulting pressure fluctuations are given by the linear equation of motion in the ambient medium, namely

$$\frac{\partial p}{\partial x} = -\rho \frac{\partial v_x}{\partial t},$$

and so

$$p(x, z, t) = -\frac{\rho \omega^2 \varepsilon}{k} \exp(-k(x-a) + i(\omega t - kz)),$$

$$p(-x, z, t) = \frac{\rho \omega^2 \varepsilon}{k} \exp(k(x+a) + i(\omega t - kz)).$$
(8.106)

Now, the slab has a curvature $\partial^2 \xi/\partial z^2$, which means that the tension (B^2/μ per unit area) produces a restoring force of $2a(B^2/\mu)\partial^2\xi/\partial z^2$ for a width of $2a$. The slab is in rough equilibrium at each point between this tension and the pressures on either side, so that

$$2a \frac{B^2}{\mu} \frac{\partial^2 \xi}{\partial z^2} = p(a, z, t) - p(-a, z, t),$$

which reduces to $\omega = k v_A(ka)^{1/2}$, as required. The frequency has been reduced below the value (kv_A) for a normal Alfvén wave by the restoring pressure forces in the ambient medium. For a slender flux tube rather than a slab, the above analysis can be repeated in cylindrical coordinates. The result is that the radial velocity decreases like R^{-2} rather than e^{-kx} (in Equation (8.105)), and so the pressure at the boundary is proportional to ω^2 rather than ω^2/k (in Equation (8.106)). This in turn gives a final dispersion relation of $\omega = kv_A/\sqrt{2}$, for the kink surface wave in a tube: thus, the ambient medium reduces the frequency by only a factor of $1/\sqrt{2}$, rather than $(ka)^{1/2}$. When $\rho_i \neq \rho_e$ and the external Alfvén speed (v_{Ae}) is non-zero, Spruit (1981b) finds that the kink surface wave has

$$\frac{\omega}{k} = \left(\frac{\rho_i v_{Ai}^2 + \rho_e v_{Ae}^2}{\rho_i + \rho_e}\right)^{1/2},$$

which lies between v_{Ai} and v_{Ae}; he refers to this wave and the slow surface wave ($\omega = kc_T$) as *transversal* and *longitudinal tube waves*, respectively.

It is clear that there are many ways of generalising the above basic analysis. Propagating modes outside the slab can be considered by allowing $m_e^2 < 0$, and reflection and transmission coefficients can be calculated for a wave incident from one side. The presence of a non-zero field outside the slab can be incorporated. The whole analysis can be performed in a cylindrical rather than a Cartesian geometry, which leads to few changes in the quantitative results (Spruit, 1981b). The effect of gravity can be included in the basic state. For instance, Roberts and Webb (1978) have considered this for the slow (sausage) surface wave by using the slender flux tube equations (8.90) to (8.93): the result is that the wave becomes evanescent when the wave period exceeds a critical value that decreases with height. Also, in a clear review of waves in a strongly structured atmosphere, Roberts (1981c) has pointed out that the effect of gravity may be described for many modes by the Klein–Gordon equation, which implies that an oscillatory wake may trail behind a wavefront and trapped modes may exist. The nonlinear development of such a wake may show up as a spicule (Hollweg; 1982). Another effect is the departure from an adiabatic variation: Webb and Roberts (1980) find that, above the photosphere, tube waves are subject to strong radiative damping, which may create the high temperature and extra brightening that are observed in the network.

CHAPTER 9

DYNAMO THEORY

9.1. Introduction

At the photosphere the Sun's magnetic field is concentrated into intense flux tubes and also into sunspots (Sections 1.3.2B, 1.4.2, and Chapter 8), whose detailed evolution is highly complex, but whose overall behaviour is remarkably ordered as the solar cycle proceeds (Section 1.4.2E). This underlying pattern shows up in several ways:

(1) the 11-yr oscillation in sunspot number;
(2) the restriction of sunspots to two belts of latitude;
(3) the spread and drift towards the equator of the sunspot belts (Sporer's Law);
(4) the inclination of sunspot groups (by typically $10°$) to the equator;
(5) the laws of polarity (Figure 1.29);
(6) the reversal of the polar fields near sunspot maximum.

All these features are commonly thought to be caused by some kind of dynamo mechanism operating in the largely unobservable depths of the convective zone, but the details of such an interaction between plasma and magnetic field are not yet fully understood. The problems are first of all to show that a magnetic field may, in fact, be maintained and then to reproduce the above features of the solar cycle.

Since the classical diffusion time $(R_\odot^2/\eta)$ for the decay of a global solar magnetic field is about 10^{10} yr, and therefore comparable with the age of the Sun, it might at first may be thought that the Sun's magnetic field is primordial. This is, however, unlikely to be the case because the above estimate for the decay-time of the field may well be a gross overestimate. Resistive instabilities might operate much faster by creating much smaller length-scales than $R_\odot$ (Section 7.5.5). Also, magnetic buoyancy would tend to expel magnetic flux at some fraction of the Alfvén speed (Section 8.2), and finally an eddy magnetic diffusivity $(\tilde{\eta})$ in the convection zone of typically 10^9 m^2 s^{-1} would destroy flux there after only about 10 yr, (i.e., 10 $L^2/\tilde{\eta}$ where L is the depth of the convection zone). Even if a primordial field were present in the core, it would probably be decoupled from the solar cycle. We are, therefore, impelled to seek some kind of dynamo to explain the maintenance of the solar magnetic field. By comparison, the oscillator theories favoured by some authors (Piddington, 1977; Layzer et al., 1979; Dicke, 1979) are at a rudimentary stage of development (Cowling, 1981). They offer no means of maintaining a toroidal field component against resistive decay, and there is no observational evidence for the ambient field about which such oscillations are claimed to occur or for variations in angular velocity with a 22-year period.

In dynamo theories a magnetic field is maintained by currents induced in a plasma by its motion across lines of force (Figure 9.1). Thus, a motion (**v**) across a magnetic field (**B**) leads to an induced electric field (**v** × **B**), which drives an electric current by Ohm's Law (**j** = σ(**E** + **v** × **B**)) and gives a magnetic field from Ampère's Law (**j** = curl **B**/μ). The magnetic field then creates both an electric field through Faraday's Law (curl **E** = − ∂**B**/∂t) and also a Lorentz force (**j** × **B**), which can oppose the force that drives the motion and so completes the circuit of cause and effect. For a complete solution to this highly nonlinear dynamo problem, it is necessary to solve the full magnetohydrodynamic equations and demonstrate that:

(i) there is a motion (**v**) which can maintain an oscillating magnetic field;
(ii) this motion is itself maintained by the available forces.

These two steps involve solving the induction equation

$$\frac{\partial \mathbf{B}}{\partial t} = \text{curl}\,(\mathbf{v} \times \mathbf{B}) + \eta \nabla^2 \mathbf{B} \tag{9.1}$$

and the equation of motion, respectively. The complete problem is so difficult that most attention has been restricted to the first step alone, which is referred to as the *kinematic dynamo problem* (Section 9.4). In other words, is it possible to construct a velocity field which produces a growing (monotonic or oscillating) magnetic field? However, the full problem of constructing a realistic *magnetohydrodynamic dynamo* (Section 9.5) has barely begun.

The present chapter offers no more than a brief glimpse at a huge topic. Further details may be found in the excellent books by Moffatt (1978), Krause and Rädler (1980), and Parker (1979a), as well as the clear reviews by Weiss (1974, 1982), which we have followed in part, by Cowling (1981) and by Stix (1981). We shall here restrict attention mostly to the *solar* dynamo, although rather different dynamos may be operating in the Earth (where the field is relatively steady, with occasional reversals), in Jupiter (which is rapidly rotating), in other stars (Bonnet and Dupree, 1981), and even in galaxies. Slowly-rotating stars with convective envelopes are found to possess a CaK intensity and angular velocity which both decrease with age (*as* $t^{-1/2}$). Since CaK is a good indicator of magnetic flux (see Figure 1.3(c)), this suggests that the efficiency of a dynamo is proportional to the rotation rate. Furthermore,

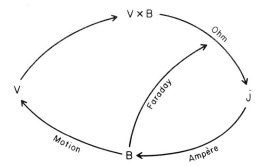

Fig. 9.1. The interaction between plasma and magnetic field, as described by the equation of motion and by the laws of Ohm, Faraday and Ampère, which are normally combined as the induction equation.

another important parameter for stellar dynamos is probably the depth of the convection zone.

9.2. Cowling's Theorem

Many years ago Cowling discussed the dynamo problem and essentially posed it, by showing that the simplest configuration will not work. There are several extensions to the theorem, but its essence is that *a steady axisymmetric magnetic field cannot be maintained* (Cowling, 1934). It may be established as follows.

Write a steady, axisymmetric field as the sum of a toroidal (i.e., azimuthal) component (B_φ) and a poloidal component $\mathbf{B}_p$ (which itself represents the sum of the radial and axial components in cylindrical polars)

$$\mathbf{B} = B_\varphi \mathbf{i}_\varphi + \mathbf{B}_p. \tag{9.2}$$

Because of the axisymmetry, the magnetic configuration in all meridional planes (through the axis of symmetry) is the same and must consist of closed field lines. In each meridional plane there must therefore exist at least one *0-type neutral point* (N), where $\mathbf{B}_p$ vanishes so that the field is purely azimuthal (Figure 9.2).

Now, Ohm's Law in the form $\mathbf{j}/\sigma = \mathbf{E} + \mathbf{v} \times \mathbf{B}$ may be integrated around the closed line of force (C) through the neutral points (N) to give

$$\oint_C \mathbf{j}/\sigma \cdot \mathbf{ds} = \oint_C \mathbf{E} \cdot \mathbf{ds} + \oint_C \mathbf{v} \times \mathbf{B} \cdot \mathbf{ds}$$

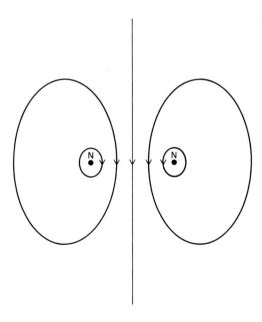

Fig. 9.2. The magnetic field lines in a meridional plane for an axisymmetric field.

or

$$\oint_c j_\varphi ds/\sigma = \int_s \text{curl } \mathbf{E} \cdot \mathbf{dS} + \oint_c \mathbf{v} \times \mathbf{B} \cdot \mathbf{ds},$$

where Stokes' theorem has been used to transform the first term on the right-hand side. Since the magnetic field is steady by assumption, curl $\mathbf{E}$ vanishes due to Faraday's Law. Also, at N, $\mathbf{B}$ is parallel to the path element $\mathbf{ds}$, so that the triple scalar product $\mathbf{v} \times \mathbf{B} \cdot \mathbf{ds}$ vanishes, and the integral of Ohm's Law reduces to $\oint_c j_\varphi ds = 0$. Since j_φ does not vanish at N, this cannot be satisfied, and so a steady field cannot be axisymmetric.

Cowling's argument led to fears that no dynamo could possibly work, but, after 25 yr the air of pessimism was dispelled when some particular kinematic dynamo models were rigorously established (Backus, 1958; Herzenberg, 1958). Since then a swarm of models has appeared and the emphasis has shifted towards seeking those with a realistic behaviour. For instance, dynamo action in an infinite region has been proved by Childress (1969) and G. Roberts (1972) for both

$$\mathbf{v} = (\sin y - \cos z, \sin z - \cos x, \sin x - \cos y) \tag{9.3a}$$

and

$$\mathbf{v} = (\cos y - \cos z, \sin z, \sin y). \tag{9.3b}$$

Similar periodic motions inside a sphere also work, and G. Roberts (1970) has even shown that 'almost all' spatially periodic motions will do! Cowling's argument holds only for exact axisymmetry. Slight departures from axisymmetry may allow a dynamo to work, but only with great difficulty. For instance, the Braginsky dynamo needs very high velocities (Section 9.4.1).

9.3. Qualitative Dynamo Action

9.3.1. Generation of toroidal and poloidal fields

The magnetic field may be split into toroidal and poloidal components (Equation (9.2)), and so one needs to establish that both components can be generated by a flow. On the Sun the equatorial regions are seen to rotate faster than the polar regions (Section 1.3.1C), and such *differential rotation* tends to pull out a purely poloidal field and create toroidal flux (Figure 9.3(a)). This effect may be demonstrated mathematically by considering an *axisymmetric* magnetic field and a flow $\mathbf{v} = v_\varphi \mathbf{i}_\varphi + \mathbf{v}_p$, so that the φ-component of the induction equation (9.1) becomes

$$\frac{\partial B_\varphi}{\partial t} + R[\mathbf{v}_p \cdot \mathbf{V}]\left[\frac{B_\varphi}{R}\right] = R\mathbf{B}_p \cdot \mathbf{V}\left[\frac{v_\varphi}{R}\right] + \eta[\nabla^2 - R^{-2}]B_\varphi. \tag{9.4}$$

The first term gives the time rate of change of the toroidal field, while the second term represents its advection with the flow. On the right-hand side, the first term shows how a shear in angular velocity (v_φ/R) can produce toroidal flux from a poloidal field $(\mathbf{B}_p)$. Such a stretching of the field lines would continue until it is balanced by the ohmic diffusion represented by the second term on the right.

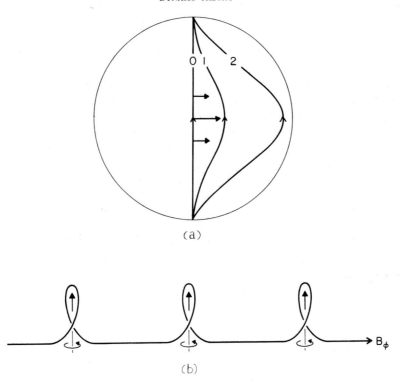

Fig. 9.3. (a) The stretching out of an initially poloidal field line (labelled 0) to subsequent positions (1 and 2) by differential rotation (solid-headed arrows). (b) The creation of poloidal flux from toroidal flux by rising and twisting motion.

The difficulty with an axisymmetric field is that the *poloidal* field component cannot be maintained (Cowling's theorem). After writing $\mathbf{B}_p = \operatorname{curl}(A_p \mathbf{i}_\varphi)$, the poloidal component of Equation (9.1) may be integrated to give

$$\frac{\partial A_p}{\partial t} + \frac{\mathbf{v}_p}{R} \cdot \nabla(R A_p) = \eta(\nabla^2 - R^{-2}) A_p, \tag{9.5}$$

which (for an incompressible flow) does not allow the generation of $\mathbf{B}_p$ from B_φ, since it implies that A_p (and therefore both $\mathbf{B}_p$ and, by (9.4), B_φ) decays away.

A suggestion for generating $\mathbf{B}_p$ from B_φ and so resolving this difficulty was put forward in a fundamental paper by Parker (1955b). He pointed out that as rising blobs of plasma expand they also tend to rotate because of the Coriolis force (Figure 9.3b). Such anticylonic motions are clockwise in the northern hemisphere and anti-clockwise in the southern hemisphere. If they carry flux tubes up with them, the twist converts toroidal fields into poloidal ones. The rate of generation of $\mathbf{B}_p$ is proportional to B_φ, and so Parker modelled the net effect of many convection cells by adding an electric field,

$$E_\varphi = \alpha B_\varphi, \tag{9.6}$$

to Equation (9.5), which therefore becomes

$$\frac{\partial A_p}{\partial t} + \frac{\mathbf{v}_p}{R} \cdot \nabla(RA_p) = \alpha B_\varphi + \eta(\nabla^2 - R^{-2})A_p, \tag{9.7}$$

so that dynamo action is now possible. The constant of proportionality (α) in the mean e.m.f. (E_φ) over many eddies gives this so-called *α-effect* its name (although Parker himself uses a different notation, Γ). It has the units of a velocity and is a measure of the mean rotational speed of eddies. Its value may be estimated from mean-field electrodynamics (Section 9.4.2). Of course, there are also falling motions, which rotate the field in the opposite direction, and so there needs to be some asymmetry between up and down motions to create a net effect. The main cause of such asymmetry is the stratification, since the rising plasma expands while the falling plasma contracts. Other possible causes are the geometry (since plasma tends to rise at the centre of a cell and fall at its boundary) and magnetic buoyancy (Section 8.2), which aids the rising motions. Alternative ways of creating an *α-effect* include *hydromagnetic inertial waves* (Section 4.5) and *magnetic buoyancy* motions, both of which possess a non-zero helicity $\mathbf{v} \cdot \nabla \times \mathbf{v}$ (Moffatt, 1978, Chapter 10).

9.3.2. Phenomenological Model

Babcock (1961) and Leighton (1964, 1969) were stimulated by magnetograph observations of the photosphere to develop a qualitative model of the solar cycle. They suggest that at sunspot minimum there exists below the surface a weak poloidal field that is subsequently stretched by differential rotation to form a strong toroidal field. Kinks in the toroidal field rise to the surface at low latitudes to form sunspots, but in the process of rising they are rotated by Coriolis forces, so that a poloidal field is generated in the opposite direction to the original one. The reversed poloidal flux is then able to diffuse to the poles due to a supergranular eddy diffusion in time for the next sunspot minimum.

The dynamo equations for the evolution of B_φ and B_r are averaged over r and φ, and written in the form

$$\frac{\partial B_\varphi}{\partial t} = r \sin\theta \, B_r \frac{\partial \omega}{\partial r} - C' \delta |B_\varphi| B_\varphi, \tag{9.8}$$

$$\frac{\partial B_r}{\partial t} = \delta \frac{C}{\sin\theta} \frac{\partial}{\partial \theta}[B_\varphi \sin\gamma] + \frac{1}{\tau_D \sin\theta} \frac{\partial}{\partial \theta}\left[\sin\theta \frac{\partial B_r}{\partial \theta}\right], \tag{9.9}$$

where $\omega(r)$ is the angular velocity, C and C' are constants, δ is a constant which vanishes when $|B_\varphi| < B_c$ and equals unity when $|B_\varphi| > B_c$, τ_D is the characteristic supergranular eddy diffusion time (assumed equal to 22 years). In Equation (9.8) the first term on the right represents the production of B_φ from B_r by a radial shear in angular velocity. When the toroidal field strength exceeds B_c, it is assumed that the toroidal flux tubes develop kinks and rise through the solar surface, so enhancing B_r and decreasing B_φ, as indicated by the terms containing δ. Finally, the last term in (9.9) represents the diffusion over the solar surface of the radial field.

A numerical integration of the above equations was very successful in reproducing the main features of the solar cycle, although the lack of an r-dependence means that the generation of the poloidal field is not incorporated adequately. It was found that a negative value for $d\omega/dr$ produces dynamo action most easily. The turbulent diffusion plays two roles; it connects together all the small-scale flux loops into a smooth field and also spreads the large-scale field throughout the convection zone from its generation site.

9.4. Kinematic Dynamos

The dynamo equations may be derived rigorously in two ways. One is for a laminar flow (Section 9.4.1), while the other is for a turbulent flow (Section 9.4.2) and is more relevant to the Sun. The equations possess simple wave-like solutions (Section 9.4.3) and have been used to attempt a more realistic modelling of the solar cycle (Section 9.4.4).

9.4.1. Nearly-symmetric dynamo

In view of the limitations imposed on a workable dynamo by Cowling's theorem, Braginsky (1965) and Soward (1972) have considered the effect of a small amount of magnetic diffusion in creating small departures from an axisymmetric toroidal magnetic field $(B_\varphi(R, z)\hat{\varphi})$ and fluid velocity $(V_\varphi(R, z)\hat{\varphi})$ of order $U_0\hat{\varphi}$. They expand **B** and **v** in powers of $\varepsilon(\ll 1)$, which is related to the magnetic Reynolds number $(R_m = U_0 a/\eta)$ for a sphere of radius a by $\varepsilon = R_m^{-1/2}$.

The fluid velocity and magnetic field are written

$$\mathbf{v} = \mathbf{v}_0(R, z, t) + \varepsilon \mathbf{v}_1(R, z, \varphi, t), \qquad \mathbf{B} = \mathbf{B}_0(R, z, t) + \varepsilon \mathbf{B}_1(R, z, \varphi, t),$$

where the azimuthal averages

$$\mathbf{v}_0 = V_\varphi \hat{\varphi} + \varepsilon^2 \mathbf{v}_p, \qquad \mathbf{B}_0 = B_\varphi \hat{\varphi} + \varepsilon^2 \mathbf{B}_p,$$

are dominated by their toroidal components.

An azimuthal average of the induction equation then yields

$$\frac{\partial \mathbf{B}_0}{\partial t} = \nabla \times (\mathbf{v}_0 \times \mathbf{B}_0) + \nabla \times \mathscr{E} + \eta \nabla^2 \mathbf{B}_0, \tag{9.10}$$

with a new electric field $\mathscr{E} = \varepsilon^2 \langle \mathbf{v}_1 \times \mathbf{B}_1 \rangle$, which is the aximuthal mean of $\varepsilon^2 \mathbf{v}_1 \times \mathbf{B}_1$ and may produce dynamo action. The toroidal and uncurled poloidal components of Equation (9.10) are

$$\frac{\partial B_\varphi}{\partial t} + \varepsilon^2 R(\mathbf{v}_p \cdot \nabla)\frac{B_\varphi}{R} = \varepsilon^2 R(\mathbf{B}_p \cdot \nabla)\frac{V_\varphi}{R} + (\nabla \times \mathscr{E})_\varphi + \eta(\nabla^2 - R^{-2})B_\varphi,$$

$$\frac{\partial A_p}{\partial t} + \varepsilon^2 R(\mathbf{v}_p \cdot \nabla)\frac{A_p}{R} = \varepsilon^{-2}\mathscr{E}_\varphi + \eta(\nabla^2 - R^{-2})A_p,$$

and so are similar in form to the dynamo Equations (9.4) and (9.7).

After some manipulation, it may be shown that $\mathbf{v}_p$ and $\mathbf{B}_p$ may be replaced by 'effective' values and that the regenerative term $\varepsilon^{-2}\mathscr{E}_\varphi$ takes the form $\langle \mathbf{v}_1 \times \mathbf{B}_1 \rangle = \alpha B_\varphi$, where

$$\alpha = \frac{2\eta}{R}\left\langle \frac{\partial s_z}{\partial R}\frac{\partial^2 s_R}{\partial R \partial \varphi} + \frac{1}{R^2}\frac{\partial s_z}{\partial \varphi}\frac{\partial^2 s_R}{\partial \varphi^2} + \frac{\partial s_z}{\partial z}\frac{\partial^2 s_R}{\partial z \partial \varphi}\right\rangle + \frac{2\eta}{R^2}\left\langle \frac{s_R}{R}\frac{\partial s_z}{\partial \varphi} + \frac{\partial s_z}{\partial z}\frac{\partial s_z}{\partial \varphi}\right\rangle,$$

in terms of a poloidal vector (**s**) which is related to $\mathbf{B}_1$ and $\mathbf{v}_1$ by

$$\mathbf{B}_1 = \nabla \times (\mathbf{s} \times B_\varphi \hat{\boldsymbol{\varphi}}), \qquad \mathbf{v}_{1p} = \left[\frac{\partial}{\partial t} + \frac{V_\varphi}{R}\frac{\partial}{\partial \varphi} - \hat{\mathbf{z}} \times \right]\mathbf{s}.$$

It should be stressed that the Braginsky dynamo is diffusive in origin and operates only for large azimuthal velocities (i.e. $R_m \gg 1$).

9.4.2. Turbulent Dynamo: Mean-Field Electrodynamics

Parker suggested that the net effect of averaging many small-scale convective motions would be to produce the large-scale electric field (αB_φ) in Equation (9.7) and so allow regeneration of the poloidal magnetic field. The idea has been given a formal basis and investigated in detail by Krause and Rädler (1980) and others, starting with the fundamental paper by Steenbeck, Krause and Rädler (1966). They consider a small-scale turbulent motion (**v**) which is statistically steady and homogeneous but not isotropic. It produces a fluctuating magnetic field (**b**) on a small scale (l) and maintains a field ($\mathbf{B}_0$) on a much larger scale (L), so that the total field is $\mathbf{B} = \mathbf{B}_0 + \mathbf{b}$, and the induction equation becomes

$$\frac{\partial}{\partial t}(\mathbf{B}_0 + \mathbf{b}) = \operatorname{curl} \mathbf{v} \times (\mathbf{B}_0 + \mathbf{b}) + \eta \nabla^2 (\mathbf{B}_0 + \mathbf{b}). \tag{9.11}$$

Averages over some scale intermediate between l and L are denoted by an over-bar, so that the means of the fluctuating velocity and magnetic field must vanish ($\bar{\mathbf{v}} = \bar{\mathbf{b}} = 0$). The average of Equation (9.11) then produces an induction equation for $\mathbf{B}_0$,

$$\frac{\partial \mathbf{B}_0}{\partial t} = \operatorname{curl} \overline{\mathbf{v} \times \mathbf{b}} + \eta \nabla^2 \mathbf{B}_0, \tag{9.12}$$

which may be subtracted from Equation (9.11) to give an equation for **b** in terms of $\mathbf{B}_0$, namely,

$$\frac{\partial \mathbf{b}}{\partial t} = \operatorname{curl}(\mathbf{v} \times \mathbf{B}_0 + \mathbf{v} \times \mathbf{b} - \overline{\mathbf{v} \times \mathbf{b}}) + \eta \nabla^2 \mathbf{b}. \tag{9.13}$$

In order to close this pair of equations, however, it is necessary to make some assumption about the form of $\overline{\mathbf{v} \times \mathbf{b}}$. Usually, one considers *pseudo-isotropic* turbulence, such that the flow is not invariant under reflections about the origin. The lack of symmetry may be produced by, for instance, fast rotation or stratification. If one supposes that

$$\langle \mathbf{v} \times \mathbf{b} \rangle = \alpha \mathbf{B}_0 - \tilde{\eta} \nabla \times \mathbf{B}_0, \tag{9.14}$$

Equation (9.12) becomes

$$\frac{\partial \mathbf{B}_0}{\partial t} = \operatorname{curl}(\alpha \mathbf{B}_0) + (\eta + \tilde{\eta})\nabla^2 \mathbf{B}_0, \qquad (9.15)$$

and it can be seen that the effect of the turbulence is to provide the extra emf $(\alpha \mathbf{B}_0)$ and also to enhance the large-scale diffusion through the term $\tilde{\eta}\nabla^2 \mathbf{B}_0$.

The evaluation of the coefficients α and $\tilde{\eta}$ may be performed under various circumstances by means of transform or Greens-function techniques (e.g., Moffatt, 1978, Chapter 7). In general, Equation (9.13) is difficult to solve because of the presence of the nonlinear terms $\operatorname{curl}(\mathbf{v} \times \mathbf{b} - \overline{\mathbf{v} \times \mathbf{b}})$, but when the small-scale (turbulent) magnetic Reynolds number (vl/η) is small these terms are negligible. For such a *quasi-linear* (or *first-order smoothing*) *approximation*, (9.13) reduces to $0 = \operatorname{curl}(\mathbf{v} \times \mathbf{B}_0) + \eta\nabla^2 \mathbf{b}$ or

$$0 = (\mathbf{B}_0 \cdot \nabla)\mathbf{v} + \eta\nabla^2 \mathbf{b}. \qquad (9.16)$$

Since this is linear in $\mathbf{v}$ and $\mathbf{b}$, Fourier transforms may be used as follows to solve for $\mathbf{b}$ and hence obtain $\overline{\mathbf{v} \times \mathbf{b}}$ in terms of $\mathbf{v}$ and $\mathbf{B}_0$.

After writing

$$\mathbf{v}(\mathbf{r}, t) = \int e^{i\mathbf{k}\cdot\mathbf{r}} \, d\mathbf{Z}(\mathbf{k}, t), \qquad \mathbf{b}(\mathbf{r}, t) = \int e^{i\mathbf{k}\cdot\mathbf{r}} \, d\mathbf{Y}(\mathbf{k}, t),$$

and, assuming $\mathbf{B}_0$ is uniform, Equation (9.16) may be transformed to give

$$d\mathbf{Y} = \frac{i\mathbf{B}_0 \cdot \mathbf{k}}{\eta k^2} \, d\mathbf{Z}. \qquad (9.17)$$

Substitution of Equation (9.17) into the expression $\overline{\mathbf{v} \times \mathbf{b}} = \int d\mathbf{Z}^* \times d\mathbf{Y}$ therefore gives

$$(\overline{\mathbf{v} \times \mathbf{b}})_i = \alpha_{ij} B_{0j}, \qquad (9.18)$$

where

$$\alpha_{ij} = \frac{i}{\eta} \int \frac{k_j}{k^2} \overline{[d\mathbf{Z}^* \times d\mathbf{Z}]}_i.$$

Equivalently, α_{ij} may be written as

$$\alpha_{ij} = \frac{i}{\eta} \int \frac{k_j}{k^2} \varepsilon_{ikl} \Phi_{kl} \, d\mathbf{k} \qquad (9.19)$$

in terms of the *spectrum tensor* $(\Phi_{ij}(\mathbf{k}))$, which is related to the velocity correlation tensor $\overline{v_i v_j}$ by $\overline{v_i v_j} = \int \Phi_{ij} \, d\mathbf{k}$.

If the turbulence is *isotropic*, the spectrum tensor is

$$\Phi_{ij} = \Phi_{ij}^0 \equiv \frac{E(k)}{4\pi k^4}(k^2 \delta_{ij} - k_i k_j) \qquad (9.20)$$

in terms of the *energy spectrum function* $(E(k))$ such that $\tfrac{1}{2}\overline{\mathbf{v}\cdot\mathbf{v}} = \int E(k) \, dk$. According to Equation (9.19) this makes α_{ij} identically zero, and so *dynamo action is not possible*.

However, when turbulence is *pseudo-isotropic*

$$\Phi_{ij} = \Phi_{ij}^0 + \frac{iF(k)}{8\pi k^4}\varepsilon_{ijk}k_k, \qquad (9.21)$$

which is invariant with respect to a rotation of axes but not with respect to rotation in the origin. In this case Equation (9.19) becomes $\alpha_{ij} = \alpha\delta_{ij}$, so that $\overline{\mathbf{v} \times \mathbf{b}} = \alpha\mathbf{B}$, where

$$\alpha = -\frac{1}{3\eta}\int_0^\infty \frac{F(k)}{k^2}\,dk. \qquad (9.22)$$

The function $F(k)$ introduced in Equation (9.21) is the *helicity spectrum function*, which is related to the *helicity*

$$\overline{\mathbf{v}\cdot\nabla\times\mathbf{v}} = i\int \overline{d\mathbf{Z}^*\cdot\mathbf{k}\times d\mathbf{Z}} = -i\int k_i\varepsilon_{ijk}\Phi_{jk}\,dk,$$

since for the form (9.21) we find $\overline{\mathbf{v}\cdot\nabla\times\mathbf{v}} = \int_0^\infty F(k)\,dk$.

The effect of slow variations in the large-scale field may be included as well by writing $\mathbf{B}_0(\mathbf{r}) = \mathbf{B}_0 + (\mathbf{r}\cdot\nabla)\mathbf{B}_0$. Repeating the above analysis then yields an expression for the *eddy diffusion coefficient* ($\tilde{\eta}$) as

$$\tilde{\eta} = \frac{2}{2\eta}\int_0^\infty \frac{E(k)}{k^2}\,dk \qquad (9.23)$$

in terms of the energy spectrum function ($E(k)$).

Alternatively, if the *correlation time* (τ) is much smaller than the turnover time one may write $\alpha = -\tfrac{1}{3}\tau\overline{\mathbf{v}\cdot\nabla\times\mathbf{v}}$ and $\tilde{\eta} = \tfrac{1}{3}\tau\overline{\mathbf{v}\cdot\mathbf{v}}$.

A useful approximation to α is

$$\alpha \approx -\frac{\tau^2 v^2 \omega\cos\theta}{\rho v}\frac{\partial(\rho v)}{\partial r}$$

or, in terms of the *density scale-height* (Λ),

$$\alpha \approx -\frac{\tau^2 v^2 \omega\cos\theta}{\Lambda}. \qquad (9.24)$$

The helicity is a measure of the lack of symmetry (the right- or left-handedness) of the small-scale flow. It can be seen from Equations (9.22) and (9.24) that the α-effect depends on a non-zero helicity, corresponding to a twisting motion with a preferred sense of rotation due to, for instance, a coriolis force. Just such a cyclonic twist was required to generate poloidal flux from toroidal flux in Parker's analysis. Also, the laminar flows (9.3) of Childress and Roberts have curl $\mathbf{v} = \mathbf{v}$, which implies that in this case the helicity $\mathbf{v}\cdot\text{curl }\mathbf{v} = v^2$ is positive. Thus, the presence of a non-vanishing helicity seems highly desirable for a successful dynamo.

9.4.3. SIMPLE SOLUTIONS: DYNAMO WAVES

The kinematic dynamo equations (9.4) and (9.7) are less tractable in a spherical geometry than a plane one, which is strictly relevant to the Sun only in the limit of a thin spherical shell. Parker (1955b, 1979a) writes

$$\mathbf{B} = \left[-\frac{\partial A}{\partial z}, B_y, \frac{\partial A}{\partial x} \right]$$

in rectangular cartesian coordinates, with z normal to the solar surface locally and y in the eastward (toroidal) direction. He considers a toroidal velocity $\mathbf{v} = v_y(z)\hat{\mathbf{y}}$ with a vertical shear and supposes everything is uniform in the y-direction ($\partial/\partial y = 0$). Equations (9.4) and (9.7) then reduce to

$$\left[\frac{\partial}{\partial t} - \eta \nabla^2 \right] B_y = \frac{dv_y}{dz} \frac{\partial A}{\partial x}, \qquad \left[\frac{\partial}{\partial t} - \eta \nabla^2 \right] A = \alpha B_y. \tag{9.25}$$

When dv_y/dz is constant, these linear dynamo equations possess plane-wave solutions of the form

$$B_y = B_0 \exp\left[pt + i(k_x x + k_y y) \right], \tag{9.26}$$

where

$$p = -\eta k^2 \pm (1+i) \left[\frac{\alpha k_x}{2} \frac{dv_y}{dz} \right]^{1/2}.$$

The second term can make $\mathbb{R}(p)$ (the real and imaginary parts of p are denoted by $\mathbb{R}(p)$ and $\mathbb{I}(p)$, respectively) positive, and so it represents field generation by the α-effect, whereas the first term gives simple ohmic decay. The square of the ratio of these two terms is

$$N_D = \frac{\alpha k_x}{2\eta^2 k^4} \frac{dv_y}{dz}, \tag{9.27}$$

which is known as the *dynamo number*. It can be seen that when $|N_D| > 1$ there are growing solutions ($\mathbb{R}(p) > 0$) in the form of *migrating dynamo waves*, which travel northwards (negative x-direction) when $N_D > 0$ and southwards when $N_D < 0$. For the Sun the anticyclonic motions have a different sense in the north and south hemispheres, so that α changes its sign at the equator and the waves approach or leave the equator depending on the sign of dv_y/dz. In particular, the waves migrate equatorwards like sunspots if the *toroidal velocity increases with depth* ($dv_y/dz < 0$). The wave period of $(\mathbb{R}(p))^{-1} \approx R_\odot^2/\eta$ agrees with the solar-cycle duration if an *eddy value* of 10^9 m^2 s^{-1} is adopted for η.

Parker has also solved the equations in a bounded layer, and, by varying the value of D, he claims to describe qualitatively both cyclic and steady ($D = 1$) generation of terrestrial, solar and galactic fields

9.4.4. Solar cycle models: the α–ω dynamo

Roberts (1972) tackled the important problem of solving the dynamo equations in a sphere (of radius $R_\odot$), including both an α-effect and a differential rotation due to a shear in angular velocity ($\omega = v_\varphi/R$). The assumed forms for α and ω are, for example,

$$\alpha = \alpha_0 \cos\theta, \tag{9.28}$$

which reverses sign at the equator ($\theta = \frac{1}{2}\pi$), and $\omega = \omega'_0 r$, so that the angular speed is a linearly increasing or decreasing function of radius depending on the sign of the constant ω'_0.

With a large-scale flow ($\mathbf{v}_0$) and both a turbulent diffusivity ($\tilde{\eta} \gg \eta$) and an α-effect due to the statistical properties of the small-scale motions ($\mathbf{v}$), the induction equation becomes

$$\frac{\partial \mathbf{B}_0}{\partial t} = \nabla \times (\mathbf{v}_0 \times \mathbf{B}_0) + \nabla \times (\alpha \mathbf{B}_0 - \tilde{\eta} \nabla \times \mathbf{B}_0).$$

In particular, for an axisymmetric flow $\mathbf{v}_0 = R\omega(R, z)\hat{\boldsymbol{\varphi}} + \mathbf{v}_p$ and a field

$$\mathbf{B} = B_0(R, z)\hat{\boldsymbol{\varphi}} + \nabla \times (A_p(R, z)\hat{\boldsymbol{\varphi}}),$$

this may be replaced by the two scalar dynamo equations

$$\frac{\partial B_\varphi}{\partial t} + R(\mathbf{v}_p \cdot \nabla)\left[\frac{B_\varphi}{R}\right] = R(\mathbf{B}_p \cdot \nabla)\omega + \nabla \times (\alpha \mathbf{B}_p) + \tilde{\eta}(\nabla^2 - R^{-2})B_\varphi, \quad (9.29)$$

$$\frac{\partial A_p}{\partial t} + \frac{1}{R}(\mathbf{v}_p \cdot \nabla)(RA_p) = \alpha B_\varphi + \tilde{\eta}(\nabla^2 - R^{-2})A_p. \quad (9.30)$$

These equations are similar to Equations (9.4) and (9.7), except that the diffusivity is turbulent and also the α-effect has been included in Equation (9.4).

The two source terms on the right-hand side of Equation (9.29) are both capable of regenerating B_ϕ from $\mathbf{B}_p$. When the rotation is weak ($|\alpha_0| \ll |L^2 \omega'_0|$) the first term on the right is negligible and so the α-effect alone regenerates both the toroidal and poloidal fields. For these so-called 'α^2-dynamos' the fields that are produced tend to be steady rather than oscillatory, and so they are more relevant to the terrestrial dynamo.

For the Sun the rotation is probably so strong ($|\alpha_0| \gg |L^2 \omega'_0|$) that the second term on the right is negligible and we have an 'α-ω dynamo', with the toroidal and poloidal components maintained by differential rotation and the α-effect, respectively. The structure of the fields and the existence of dynamo action depends on the value of a *dynamo number* $X = \alpha_0 \omega'_0 R_\odot^3 \tilde{\eta}^2$, which (like N_D) gives the ratio of field generation to dissipation. Since Equations (9.29) and (9.30) are linear in B_ϕ and A_p, they possess solutions behaving like e^{pt}, so that the object is to determine p as a function of X. In particular, what is the smallest value of X which allows the α-effect to overcome the diffusion and so produce a growing mode ($\mathbb{R}(p) > 0$)? Also, is the mode oscillatory ($\mathbb{I}(p) \neq 0$) or not ($\mathbb{I}(p) = 0$)? When $\alpha_0 \, d\omega/dr < 0$, Roberts (1972) finds that the most easily excited dynamo is oscillatory and of dipole type (A_p even, B_φ odd in z) with -X between 74 and 206, depending on the particular forms for α and ω. Its activity progresses from pole to equator (Figure 9.4), just like Parker's dynamo waves. As the toroidal (or poloidal) fields from the two hemispheres reach the equator they disappear by diffusing into one another. When $\alpha_0 \, d\omega/dr > 0$, the preferred mode (i.e., at the lowest X-value) is oscillatory and quadrupolar (A_p odd, B_φ even), with a critical X between 76 and 212 and a migration towards the poles. Adding a small meridional circulation ($\mathbf{v}_p$) is found to halve the marginal value of X and to change the character of the preferred mode to that of a dipole.

DYNAMO THEORY

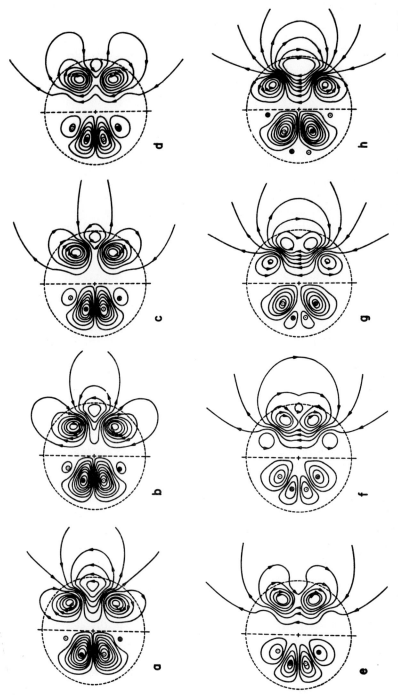

Fig. 9.4 A half-cycle for an α–ω dynamo with $\alpha = \alpha(R, \theta)$ and $\omega = \omega(R)$. The dynamo number for this marginal oscillation is $X = -206$, and $\Omega = (a^2/\eta)\Omega(p)$ is 47.4. $\mathbf{B}_p$-lines are shown in each right-hand hemisphere and lines of constant B_φ on the left at intervals of $\Omega t = \pi/8$. (From Roberts, 1972.)

The α–ω dynamo models of Roberts (1972) possess a period of $2\pi R_\odot^2/(100\tilde{\eta})$, which is smaller than the crude estimate of $R_\odot^2/\tilde{\eta}$ (Section 9.4.3) because diffusion is taking place on scales smaller than $R_\odot$. The period evidently depends on the value adopted for the turbulent eddy diffusivity $\tilde{\eta}$ (assuming $\tilde{\eta} \gg \eta$). For granules $\tilde{\eta} \approx 10^9$ m² s⁻¹ (Section 2.1.5), which gives a period of only 1 yr, although granules do not last long enough to acquire much helicity from Coriolis forces. Supergranules, which probably do give rise to an α-effect, have a slightly larger eddy diffusivity, and so they give an even shorter period. In order to obtain a period comparable with that of the solar cycle (22 yr) it is necessary to take $\tilde{\eta} \approx 10^8$ m² s⁻¹ (although nonlinear effects may lengthen the period). Assuming that $\alpha_0 > 0$ for the Sun, the other suggestion from the models is that ω should increase with depth ($d\omega/dr < 0$), which is certainly consistent with the tentative observation that sunspots rotate more rapidly than the photospheric plasma (Section 1.3.1C), since sunspots are presumably anchored in the interior.

In an attempt to reproduce detailed features of the solar cycle such as the butterfly diagram, various authors have studied more sophisticated α–ω dynamos. Deinzer and Stix (1971) and Steenbeck and Krause (1969) have considered an $\alpha(R, \theta)$ such that the α-effect is present near one or two radial shells, whereas Stix (1973) has studied its localisation in two rings. Roberts and Stix (1972) have incorporated a θ-variation in ω. Yoshimura (1975, 1977) allows for R- and θ-variations in both ω and α, including a reversal in the sign of α at some depth. Also, he parametrises several physical effects and describes the evolution of the resulting potential coronal field. Stix (1976) has considered a tensorial α-effect and the phase relation between poloidal and toroidal fields.

9.5. Magnetohydrodynamic Dynamos

9.5.1. Modified kinematic dynamos

Solving both the equations of induction and motion presents a formidable problem, which has not yet been completed. Instead, several authors have modified the kinematic dynamos to include a feedback of the Lorentz force on the motions; they solve the induction equation alone and assume a variation of α with the magnetic field. For example, Stix (1972) has taken

$$\alpha = \frac{\alpha_0}{2}\left[1 - erf\frac{B - B_c}{c}\right],$$

such that α approaches zero when B greatly exceeds a critical value (B_c). He finds nonlinear dynamo waves in a plane layer, and Jepps (1975) has repeated the analysis for a spherical geometry. Other authors have put $\alpha \approx 1 - cB^2$ for weak fields and $\alpha \approx B^{-3}$ for strong ones. Yoshimura (1975, 1978, 1979) has energetically developed much further this idea of parametrising extra physical effects. He adopts various functional forms for both α and ω that include parameters to describe flux eruption by magnetic buoyancy and a time-delay for the feedback of the magnetic field on the dynamo process. By varying the parameters he is able to reproduce many observed features of the solar cycle, such as: a realistic butterfly diagram; the poleward migration of mid-latitude poloidal flux in a secondary dynamo wave; a 55-year modulation of the 11-yr cycle: a weakening of the magnetic field to describe the Maunder Minimum.

The α^2-dynamo in a sphere of radius a with $\alpha = \alpha_0 \cos\theta$ gives a growing dipolar magnetic field when α_0 exceeds $\alpha_c = 7.64(\eta/a)$. Malkus and Proctor (1975) and Proctor (1977) have therefore modelled the nonlinear saturation of such a field due to a mean flow driven by the Lorentz force. They make progress by expanding the equations of motion and induction (with an α-effect) in powers of $(\alpha_0/\alpha_c - 1)$.

9.5.2. Strange Attractors

Possible nonlinear effects which may limit the growth of a linear $\alpha - \omega$ dynamo are the removal of flux by magnetic buoyancy (e.g., Leighton, 1969; Yoshimura, 1975), the reduction of α due to the inhibition of convection when the fields become large (e.g., Stix, 1972; Jepps, 1975), and the limitation of differential rotation by the Lorentz force (Gilman, 1969). Jones (1982) has considered some simple model equations for such nonlinear $\alpha - \omega$ solar dynamos. In spherical polars he writes the $\alpha - \omega$ dynamo equations as

$$\frac{\partial A_p}{\partial t} = \alpha R B_\varphi + \eta R(\nabla^2 - R^{-2})\frac{A}{R}, \tag{9.31}$$

$$\frac{\partial B_\varphi}{\partial t} = \frac{1}{r}\frac{\partial(\omega, A)}{\partial(r, \theta)} + \eta(\nabla^2 - R^{-2})B_\varphi - bB_\varphi^3, \tag{9.32}$$

together with the azimuthal component of the equation of motion (omitting viscous drag) in the form

$$\rho\frac{\partial}{\partial t}(\omega R) = (\mathbf{j} \times \mathbf{B})_\varphi + GR^{-1}, \tag{9.33}$$

where $R = r\sin\theta$. In Equation (9.32) a buoyancy term is included having the form bB_φ^3 (where b is constant), and in Equation (9.33) G represents the couple that produces differential rotation and so drives the dynamo. Taking the average of these equations over a hemisphere and neglecting buoyancy gives a simple third-order system of ordinary differential equations which is a special case of the Lorentz equations (1963) and has been studied by Robbins (1976):

$$\frac{d\bar{B}_p}{dt} = -\frac{\bar{B}_p}{\tau_d} + \alpha\frac{\bar{B}_\varphi}{L}, \tag{9.34}$$

$$\frac{d\bar{B}_\varphi}{dt} = -\frac{\bar{B}_\varphi}{\tau_d} + \omega\bar{B}_p, \tag{9.35}$$

$$\frac{d\bar{\omega}}{dt} = \frac{G}{I} - \frac{\bar{B}_p\bar{B}_\varphi}{\mu\rho L^2}, \tag{9.36}$$

where τ_d is the ohmic diffusion-time, L is the width of the convection zone, and I is the moment of inertia. At large values of α the solutions are oscillatory, but at low values they are aperiodic. A similar model for a coupled-disc dynamo has been studied by Cook and Roberts (1970) and Ito (1980), who finds period doubling bifurcations.

Jones (1982) considers less severe truncations of the full equations, (9.31) to (9.33), and so is able to model dynamo waves. He allows the mean values of RB_r and RB_φ at latitude 30° (namely $\bar{B}_{p1}$ and $\bar{B}_{\varphi 1}$) to differ from the corresponding values ($\bar{B}_{p2}$ and $\bar{B}_{\varphi 2}$) at latitude 60°, and supposes the radial shear is dominant, so that the mean induction equation gives

$$\frac{d\bar{B}_{p1}}{dt} = -\frac{\bar{B}_{p1}}{\tau_d} + \alpha\frac{\bar{B}_{\varphi 2}}{L}, \qquad \frac{d\bar{B}_{\varphi 1}}{dt} = -\frac{\bar{B}_{\varphi 1}}{\tau_d} + \bar{\omega}\bar{B}_{p1} - b\bar{B}_{\varphi 1}^3,$$

$$\frac{d\bar{B}_{p2}}{dt} = -\frac{\bar{B}_{p2}}{\tau_d} - \alpha\frac{\bar{B}_{\varphi 1}}{L}, \qquad \frac{d\bar{B}_{\varphi 2}}{dt} = -\frac{\bar{B}_{\varphi 2}}{\tau_d} + \bar{\omega}\bar{B}_{p2} - b\bar{B}_{\varphi 2}^3.$$

In one case Jones assumes the shear is the same at both latitudes, so that

$$\frac{d\bar{\omega}}{dt} = \frac{G}{I} - \frac{\bar{B}_{p1}\bar{B}_{\varphi 1} + \bar{B}_{p2}\bar{B}_{\varphi 2}}{\mu\rho L^2}.$$

The resulting solutions oscillate in time with a phase lag between the fields at 30° and those at 60°, and so they represent nonlinear dynamo waves that migrate towards the equator. The effect of the Lorentz force is to determine the field strength, while magnetic buoyancy shortens the period of the cycle. In another model Jones supposes that the equation of motion provides no coupling between the rotation of the two latitudes, which have angular speeds $\bar{\omega}_1$ and $\bar{\omega}_2$, so that

$$\frac{d\bar{\omega}_1}{dt} = \frac{G}{I} - \frac{\bar{B}_{p1}\bar{B}_{\varphi 1}}{\mu\rho L^2} \quad \text{and} \quad \frac{d\bar{\omega}_2}{dt} = \frac{G}{I} - \frac{\bar{B}_{p2}\bar{B}_{\varphi 2}}{\mu\rho L^2}.$$

The solutions are irregular (with a rough periodicity) when the buoyancy is weak. The lack of a dynamical coupling between the polar and equatorial regions produces a magnetic field behaviour that is different in the two regions, with poloidal flux concentrated at high latitudes and toroidal flux at low latitudes.

The low level of activity during the Maunder Minimum (Section 1.4.2E) may have been caused by the dynamo temporarily working less efficiently, since a weaker field strength would inhibit the eruption of toroidal flux by magnetic buoyancy. Such nonlinear behaviour cannot be explained by standard (linear) kinematic dynamo theory (Section 9.4), but it is typical of nonlinear systems. Zeldovich and Ruzmaikin (1980) have presented a simple third-order nonlinear system of ordinary differential equations which has properties reminiscent of the Maunder Minimum phenomenon. The system is similar to Equations (9.34) to (9.36) and takes the form

$$\frac{d\bar{A}_p}{dt} = -\frac{\bar{A}_p}{\tau_p} + \alpha\bar{B}_\varphi - C\bar{B}_\varphi, \tag{9.37}$$

$$\frac{d\bar{B}_\varphi}{dt} = -\frac{\bar{B}_\varphi}{\tau_\varphi} + \frac{\bar{A}_p}{L\tau_\varphi}, \tag{9.38}$$

$$\frac{dC}{dt} = -\frac{C}{\tau_\alpha} + \frac{\bar{A}_p\bar{B}_\varphi}{\mu\rho L^2}, \tag{9.39}$$

where L is the width of the convection zone, τ_α is the diffusion time for helicity, τ_p and τ_φ are ohmic diffusion times for $\bar{A}_p$ and $\bar{B}_\varphi$, and C represents the deviation of the α-effect from the kinematic value, so that when $C \approx 0$ we have a pair of equations similar in form to Equations (9.7) and (9.4). The set of Equations (9.37) to (9.39) possesses three critical points for $(\bar{A}_p, \bar{B}_\varphi, C)$, namely $(\pm L[(\alpha - L/\tau_p)\mu\rho L/\tau_a]^{1/2}$, $\pm [(\alpha - L/\tau_p)\mu\rho L/\tau_\alpha]^{1/2}, \alpha - L/\tau_p)$ and $(0, 0, 0)$. It has *strange attractor* solutions which remain near one of the critical points for a long time and then rapidly move over to another one.

These strange attractors exhibit fascinating behaviour and are attractive in themselves to mathematicians. However, it should be borne in mind that, although they may be able to model properties of the solar cycle, they do not necessarily explain them. Any success in reproducing solar behaviour does not prove the model is sound.

9.5.3. Convective dynamos

Several convective dynamo models have been proposed, with buoyancy as the driving force in the Boussinesq approximation. Soward (1974) has considered a Rayleigh number (Ra) that is just above the critical value (Ra*), and so he uses (Ra/Ra* − 1) as a small expansion parameter. He considers three-dimensional motions in the shape of square or hexagonal cells and obtains nonlinear oscillatory solutions. The motions possess helicity, which is able to maintain a steady, periodic dynamo. Busse (1973, 1975a) has obtained dynamo action for a motion which consists of a shear flow parallel to the axis of a horizontal two-dimensional roll. He too makes analytic progress by taking a marginally supercritical Rayleigh number, and has also considered an annular geometry.

Gilman (1977, 1978) has modelled Boussinesq convection in a rotating spherical shell with a Taylor number ($\mathcal{T}$) of 10^5, a Prandtl number (Pr) of 1 and a Rayleigh number (Ra) of 2×10^4. The resulting differential rotation is driven by Reynolds stresses rather than meridional circulation and possesses equatorial acceleration. The helicity of the convection suggests that the system may be able to drive a dynamo, although the magnitude of the helicity necessary to give the observed differential rotation is too large to produce the value of α that is typical for α–ω dynamos.

Several dynamo models have parametrised the physics of small-scale processes and of features not present in the induction equation itself. As a complementary approach, Gilman and Miller (1981) have extended the work of Gilman (1977) by numerically solving the coupled equations of induction, motion, energy and continuity for a convectively-driven dynamo. A rotating and stratified shell of Boussinesq plasma is heated uniformly from the base. No α-effect term is included in the induction equation, but turbulent values for the diffusion coefficients are adopted. The dimensionless parameters $\mathcal{T}(=10^5)$ and Pr$(=1)$ are prescribed, and then Ra is chosen so that the differential rotation driven by convection is similar to the observed form (Gilman, 1977). In particular, equatorial acceleration is produced only if the effect of rotation on convection is strong and the angular velocity *decreases* with depth ($d\omega/dr > 0$). The initial magnetic field is purely toroidal, and most of the subsequent energy is in the toroidal flow and magnetic field. As boundary conditions the magnetic field is taken to be radial at the surface and tangential at the base. The results show that dynamo action is indeed produced, as evidenced by a growing magnetic field when the magnetic Prandtl number (η/κ) exceeds 0.2 (with a magnetic Reynolds number of 150). New effects are the tendency for the motion to inhibit dynamo action and for magnetic fields with energies much smaller than the kinetic energy to create instability and so modify the flow substantially. However, the magnetic field that is generated in this dynamo differs markedly in character from the solar magnetic field, since there is no evidence for field-reversals, equatorial migration or a preferred magnetic symmetry about the equator. The main reason is that the helicity (and therefore the effective strength of the α-effect) present in the model is much larger (by three orders of magnitude) than in α–ω-dynamos, and so the model behaves much more like an α^2-dynamo. It will be interesting to see in the future whether a better simulation of the solar cycle can be produced by allowing both for compressibility and for the inclusion of important additional physics such as flux concentration into intense tubes and magnetic buoyancy. In particular, compressibility may lower the effective α, and flux concentration may allow most of the magnetic flux to be immune from the twisting associated with the strong helicity.

9.6. Difficulties with Dynamo Theory

Dynamo theory has proceeded a long way towards providing an explanation for the existence of the solar magnetic field and its variation with the solar cycle. Nevertheless, several problems remain to be overcome, as pointed out by, for instance, Moffatt (1978) and Stix (1981).

(1) The effect of the concentration of magnetic flux tubes and the probable filamentary nature of magnetic fields in the convective zone needs to be incorporated. It can probably only be done by a parametrisation, but, as a start, Childress (1979) has added a toroidal velocity to the magnetoconvection problem of Galloway *et al.* (1978) (Figure 8.5) and finds that $\alpha \approx U_0 R_m^{-1/2}$. The behaviour of α at high magnetic Reynolds numbers (R_m) is rather uncertain, but Kraichnan (1976) finds instead $\alpha \approx v_0$, where v_0 is the rms velocity.

(2) The first-order smoothing approximation (Section 9.4.2) has been used to close the turbulent dynamo equations crudely and so calculate α. It requires either that the small-scale magnetic Reynolds number be small or that the lifetime of an eddy be much shorter than its circulation time, so that $\tau \ll l/v$, where v is the turbulent velocity while τ and l are the correlation time and length. However, these conditions do not hold on the Sun, where $\tau \approx l/v$.

(3) The value of α that is believed necessary for the Sun can be estimated from the ratio of B_p to B_φ to be 1–10 cm s^{-1}, but the value of α calculated from, for example, mixing-length theory is much too high (typically 100 m s^{-1}). In other words, the present theory gives too large an effect of the turbulence on the magnetic field, but this may be reduced by incorporating a feedback from the Lorentz force and a filamentation into flux tubes. Furthermore, the tensor properties of α need to be calculated in detail and may produce magnetic pumping.

(4) The value of $\bar{\eta}$ estimated from mixing-length theory or from the observed dispersal of active regions is typically $10^9 \text{m}^2 \text{s}^{-1}$ or larger, but this is a factor of ten higher than required in some kinematic dynamo models (Section 9.4.4), so it is important to predict its value rigorously. The effect of velocity correlations higher than second order has been considered by Knobloch (1977), while Kraichnan (1973) has treated the effect of helicity fluctuations and shown that $\bar{\eta}$ can even be negative. Numerical simulations of MHD turbulence by Pouquet *et al.* (1976, 1978) have shown a transfer from kinetic energy to magnetic energy, with a cascade of kinetic energy and an inverse cascade of magnetic energy, but there is a need to extend the simulations to include compressibility and a spherical geometry.

(5) In view of the importance of the sign of $d\omega/dr$ in dynamo theories, it is important to determine its value reliably from observations (e.g., of solar p-modes (Section 4.9.1)). The apparent observation that magnetic elements rotate faster by a few percent than photospheric plasma is in agreement with the requirement that $d\omega/dr < 0$ for some dynamo models (in order to give a magnetic drift towards the equator). By contrast, differential rotation models tend to have $d\omega/dr > 0$. It is also important to measure the surface velocity accurately and its variation with the solar cycle. Howard (1976) finds that the equator is rotating faster by about 3 to 4% at sunspot minimum, whereas Howard and Labonte (1980) have detected a small torsional oscillation at 3 m s^{-1}, and Topka *et al.* (1982) have presented evidence that magnetic flux is carried polewards by a meridional flow of 10 m s^{-1} rather than by diffusion.

(6) The whole concept of the turbulent diffusion of a magnetic field needs to be put on a firmer foundation (Piddington, 1973; Cowling, 1981), since there may not be enough time to expel all the flux from eddies and annihilate it at cell boundaries by simple diffusion. Clearly, the effect of fast magnetic reconnection needs to be incorporated (Section 10.1).

Looking to the future, there are several other features that need to be studied.

(1) The apparent rigid-body rotation of some coronal holes (Section 1.3.4C) is not well understood. It may represent a dynamo wave travelling along lines of latitude, or it may reflect a large-scale banana-like convection pattern.

(2) Dynamo theory needs to take account of the large amount of magnetic flux that emerges as ephemeral active regions (i.e., X-ray bright points) rather than normal active regions (Section 1.3.4B). Also, the fact that the number of bright points is greatest at solar minimum in such a way that the total rate of flux emergence appears roughly constant in time needs to be explained.

(3) Other observations which would shed light on the solar cycle are changes in the solar radius, luminosity and surface temperature, and also the properties of stellar cycles (Wilson *et al.*, 1981).

In conclusion, despite the above difficulties, it must be stressed that dynamo theory remains one of the most highly developed and successful branches of solar magnetohydrodynamics. Clearly, it will remain a topic of great interest in the future.

CHAPTER 10

SOLAR FLARES

The *observations* of flares have been briefly summarised in Section 1.4.4, whereas a comprehensive account can be found in Svestka (1976b), and results from the Skylab satellite observations have been compiled by Sturrock (1980). The *theories* for solar flares have been reviewed by many people, such as Sweet (1969), Van Hoven (1976), Priest (1976), Brown and Smith (1980), Sturrock (1980, Chapter 9), Spicer and Brown (1981), and Priest (1982a, b), while an analysis of the magnetohydrodynamics of the basic flare mechanism is described in Priest (1981a). This latter book includes reviews by Craig and Van Hoven of *simple-loop flares* (their thermal evolution and magnetic instability), as well as chapters by Birn and Schindler and Pneuman on *two-ribbon flares* (their magnetostatic equilibria and 'post'-flare loop evolution). In addition, Svestka describes new observations that have been made since his book was published, and Heyvaerts discusses particle acceleration mechanisms.

It has been said that there are as many flare theories as there are flare theorists! Reviewing them all is a vast undertaking, and it has already been well-accomplished in the above references. This chapter does not therefore attempt to repeat the process, but instead gives a rather personal view about the most likely mechanisms for the basic flare instability. It focuses on the key problems as to why the flare occurs and how the energy release is initiated. Although important, the secondary effects of the energy release are not addressed.

The source of the flare energy is generally believed to be the magnetic field, and the main way of accomplishing its release is by magnetic reconnection, as described in Section 10.1. A *simple-loop flare* shows up as the brightening and decay of a single stationary loop. Several mechanisms at present provide possible explanations for such a flare. It may result when a new magnetic flux tube emerges from below the photosphere and reconnects with the overlying magnetic field, as in the *emerging flux model* (Section 10.2.1). Alternatively, it may be an example of *thermal nonequilibrium*, when a cool loop ceases to be in thermal equilibrium and heats up (Section 10.2.2); this is the most speculative mechanism and therefore more work is needed to determine its viability. A third possibility is that the loop becomes *kink* (or *resistive kink*) *unstable* when it is twisted too much (Section 10.2.3).

A *two-ribbon flare* (Figure 10.1) begins with the eruption and opening of a magnetic arcade containing a filament. During the main phase the field then reconnects back down and forms 'post'-flare loops (Section 10.3.3). The onset of the eruptive instability has been modelled indirectly by seeking multiple equilibria (Section 10.3.1), or directly by a magnetohydrodynamic stability analysis (Section 10.3.2). It may occur spontaneously when the arcade shear or height become too great. Or else it may be

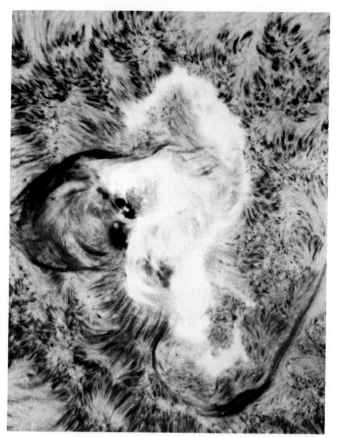

Fig. 10.1. The main phase of the two-ribbon flare of August 7, 1972 in Hα. The Hα ribbons are well separated and are joined by 'post'-flare loops that can just be seen in projection. Part of the dark active-region filament that originally snaked its way through the complex sunspot group between the present location of the bright ribbons is still visible (courtesy H. Zirin, Big Bear Observatory).

triggered by emerging flux, thermal nonequilibrium or a tearing mode. Models for the accompanying *coronal transient* can be found in Section 11.4.

10.1. Magnetic Reconnection

The ways in which a current sheet may *form* have been summarised in Section 2.10.1, and a list of its properties, including reconnection, has been presented in Section 2.10.2. For example, the creation of such a sheet may be due to magnetic instability or nonequilibrium (Section 6.4.4B). Reconnection itself can develop in several distinct ways:

(i) it may be generated spontaneously by a resistive instability such as the *tearing mode* (Section 7.5.5), either in a current sheet or throughout a sheared structure;
(ii) it may be driven from outside when separate flux systems are pushed together, with the result that the region near the neutral line collapses (Section 2.7) and a thin reconnecting sheet is created (Section 6.4.4B);

(iii) it may be driven locally by a sudden enhancement of the resistivity at some location.

The *nonlinear results* of a growth of reconnection (by, for instance, the tearing mode) depend crucially on the assumed boundary conditions at large distances. If the flow velocity (and field strength) are held constant on the inflowing boundary, the configuration evolves to a state of steady reconnection based on that inflow speed. This is relevant when the reconnection is being driven by distant magnetic evolution (such as emerging flux). If the boundary conditions are free, in the sense that the inflow speed may adopt any value it likes, then the nonlinear state is likely to approach one of steady reconnection based on the maximum allowable possible rate. If the boundary is rigid (but conducting), no new flux can enter the region and the field ultimately decays to a potential one as the excess magnetic energy is converted into heat; in this case the tearing mode may speed up the decay by creating small-scale structure, and for some part of the decay a quasi-steady state of reconnection may be set up.

A state of *steady* reconnection can develop in several ways, as mentioned above, and the main realisation of recent years is that, if oppositely directed fields are being forced together at a prescribed rate, the *plasma will just respond by allowing reconnection at that speed*. This applies up to a certain *maximum rate*, which is typically 0.01 to 0.1 v_A and depends weakly on the magnetic Reynolds number. If, on the other hand, the system is free to reconnect at any speed, it will tend to choose the maximum rate. The object of this section is simply to describe some of the attempts to model the steady reconnection process, governed by Ohm's Law

$$\mathbf{E} + \mathbf{v} \times \mathbf{B} = \eta \, \text{curl} \, \mathbf{B}, \tag{10.1}$$

the equation of motion

$$\rho(\mathbf{v} \cdot \nabla)\mathbf{v} = -\nabla p + \text{curl} \, \mathbf{B} \times \frac{\mathbf{B}}{\mu}, \qquad \text{div} \, \mathbf{B} = 0, \tag{10.2}$$

and the continuity equation

$$\text{div}\,(\rho \mathbf{v}) = 0. \tag{10.3}$$

10.1.1. Unidirectional Field

A current sheet tends to broaden in time as it diffuses away (Section 2.6.1), but a steady state can be set up if new magnetic flux is brought in from the sides, so that the outwards diffusion is balanced by inwards convection. However, plasma will be carried in with the flux, and this must be turned and escape along the sheet if there is to be no accumulation of mass near the neutral line. Such a process may be modelled simply by the incompressible stagnation point flow

$$v_x = -V_0 \frac{x}{a}, \qquad v_y = V_0 \frac{y}{a} \tag{10.4}$$

in a unidirectional field

$$\mathbf{B} = B(x)\hat{\mathbf{y}}, \tag{10.5}$$

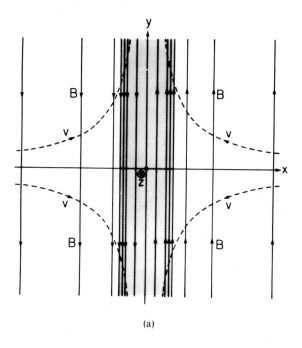

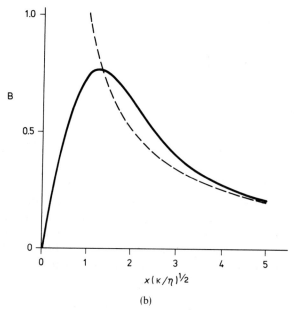

Fig. 10.2. Magnetic annihilation in a current sheet. (a) Oppositely directed magnetic field lines (———) are carried in from two sides by a stagnation point flow with streamlines (– – – – –). In the diffusion region (shaded) the field is no longer frozen to the plasma and magnetic energy is converted into heat by ohmic dissipation. (b) The field strength (B) as a function of x with $k = V_0/a$, expressed in units such that dB/dx is unity at $x = 0$ (from Priest and Sonnerup, 1975).

where V_0 and a are constants. The equation of motion determines the pressure as

$$p = \text{const} - \tfrac{1}{2}\rho v^2 - \frac{B^2}{2\mu}, \tag{10.6}$$

while Ohm's Law

$$E - \frac{V_0 x}{a} B = \eta \frac{dB}{dx} \tag{10.7}$$

may be solved for the magnetic field with E constant (Parker, 1973). The result is shown in Figure 10.2.

If diffusion is absent ($\eta = 0$), the solution of Equation (10.7) is $B = Ea/(V_0 x)$, which is represented by the dashed curve in Figure 10.2(b); it shows how the field strength increases to infinity as the inflow speed declines to zero. If diffusion is present this process of field intensification is halted when the current density ($\mu^{-1} dB/dx$) reaches such a high value that the right-hand side of Equation (10.7) becomes important and the field is able to slip through the plasma. The width (l) of the diffusion region is given by equating the diffusion and convection terms in Equation (10.7) as $l = (\eta/(aV_0))^{1/2}a$, or, in terms of the *magnetic Reynolds number* ($R_m = LV_0/\eta$), $l = L/R_m^{1/2}$.

The above solution has been generalised in a straightforward manner by Sonnerup and Priest (1975) and Priest and Sonnerup (1975) to include z-components ($v_z(x)$ and $B_z(x)$) of both the flow velocity and magnetic field. It is possible to bring together fields which are not just antiparallel but may be inclined at any angle in the y–z plane. As the fields approach one another, their directions rotate and their magnitudes change, giving (in general) a non-zero field at the centre of the sheet and annihilating part of the magnetic energy in the process.

10.1.2. Diffusion Region

Most of the solar atmosphere can be considered to be perfectly conducting so that $\mathbf{E} + \mathbf{v} \times \mathbf{B} \simeq \mathbf{0}$, and the plasma is frozen to the field lines. In a current sheet or *diffusion region*, the current density is so large that the right-hand side of Equation (10.1) is no longer small, which means that diffusion is important and the field lines can slip through the plasma. In practice the diffusion region possesses a finite length (L), and so the one-dimensional model of the previous section fails near the ends of the region, where there are significant transverse field components (in the x-direction).

The configuration of interest (Figure 10.3) is one in which field lines are carried into the diffusion region from the sides (with speed v_i) and out through the top and bottom (with speed v_0). In the process, the field lines are said to have been 'reconnected' at the neutral point (N), where the magnetic field strength in the plane of the figure vanishes. The outflow field strength (B_0) is less than the inflow strength (B_i). Thus for a steady-state, two-dimensional situation, in which the electric field (E) is constant, the inflow of magnetic energy (EB_i) exceeds the outflow value (EB_0). Some of the energy is converted into heat and some into kinetic energy (since v_0 exceeds v_i in value).

For steady, compressible flow Sweet (1958) and Parker (1963a) derived the following order-of-magnitude relationships between the input and output parameters:

$$v_0 = v_A^* \equiv \frac{B_i}{(\mu \rho_c)^{1/2}}, \tag{10.8}$$

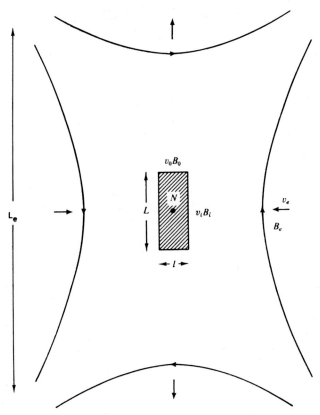

Fig. 10.3. The configuration for steady magnetic reconnection. Oppositely-directed field lines of strength B_e, frozen to the plasma, are carried towards one another at a speed v_e by a converging flow. They enter a diffusion region (shaded) with dimensions l and L, are reconnected at the neutral point (N), and are finally ejected from the ends.

$$v_i = \frac{\eta}{l}, \tag{10.9}$$

$$\rho_i v_i L = \rho_c v_0 l. \tag{10.10}$$

Equation (10.8) arises because plasma is accelerated away from N by an excess gas pressure of amount $B_i^2/(2\mu)$, so that the outflow speed is a hybrid Alfvén speed, based on the inflow field and the current sheet density; Equation (10.9) expresses a balance between inwards convection and outwards diffusion; Equation (10.10) is a consequence of mass conservation. It can be seen from Equations (10.8) and (10.10) that $l \ll L$ provided $v_i \ll v_A^*(\rho_c/\rho_i)$.

In most applications the incompressible limit ($\rho_c = \rho_i$) has been adopted, on grounds of simplicity, so that Equations (10.8) to (10.10) simply determine v_0 and the dimensions of the diffusion region in terms of prescribed input values (v_i and B_i). However, when there is an appreciable temperature difference between the diffusion region and its surroundings, the density ratio (ρ_c/ρ_i) differs from unity. For a thin diffusion region ($l \ll L$), it is determined from the constancy of total pressure across the region, namely

$p_c = p_i + B_i^2/(2\mu)$, as

$$\frac{\rho_c}{\rho_i} = \frac{T_i}{T_c}(1 + \beta_i), \qquad (10.11)$$

in terms of the inflow beta $\beta_i = p_i 2\mu/B_i^2$. Thus, for given inflow parameters, Equations (10.8) to (10.11) determine the outflow values (v_0, l, L, ρ_c) in terms of the current

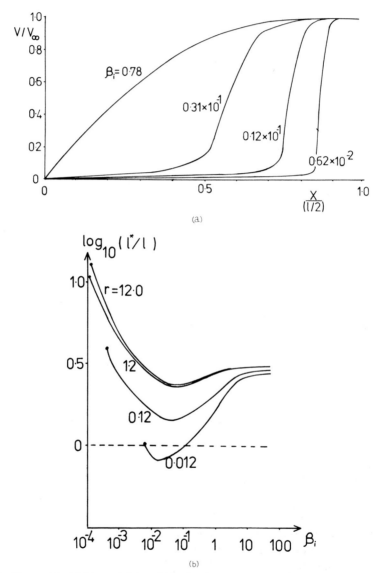

Fig. 10.4. The model of Milne and Priest (1981) for the central diffusion region. (a) The inflow speed as a function of distance (x) from the sheet centre for various values of β_i and $r = 0.012$ (β_i is the ratio of plasma to magnetic pressure for the inflow and r is the ratio of radiation to convection.) (b) The width (l^*) of the region in terms of its order-of-magnitude value ($l = \eta/v_i$).

sheet temperature (T_c), which is in turn governed by an energy equation (Tur and Priest, 1978) of the form

$$E = J + H + K - R. \tag{10.12}$$

Here E represents the convective transfer of heat through the sheet; J, H and K represent the Joule heating, mechanical heating and conduction into the sheet, respectively, while R denotes the radiative loss.

The above order-of-magnitude treatment of the diffusion region has been improved upon by Milne and Priest (1981), who solve the flow equations approximately within the region. The solutions depend on β_i and

$$r = \frac{\text{radiation}}{\text{convection}}.$$

The resulting inflow velocity profiles and sheet width (l^*) are plotted in Figure 10.4 in terms of the order-of-magnitude value $l = \eta/v_i$. It can be seen that the sheet may be a factor of 10 thicker than previously predicted. Also, the curves are terminated at dots, so that there is a β-limitation, with no solution if β is too small; it is caused by the maximum in the radiative loss function. Another feature is the presence at low values of β of steep gradients within the sheet.

10.1.3. The Petschek Mechanism

The objects of steady reconnection theory have been to determine the details of the flow in the external region (which surrounds the diffusion region) and to find the maximum allowable value for the speed of approach (v_e) at a distance L_e in terms of the Alfvén speed ($v_{Ae} \equiv B_e/(\mu\rho)^{1/2}$) there. Petschek (1964) was the first to propose a fast reconnection model (Figure 2.11(c)) for which $L \ll L_e$ so that the diffusion region occupies a small part of the area of consideration. One quadrant of his model is shown in Figure 10.5(a). A *slow magnetohydrodynamic shock wave* (or finite Alfvén wave in the incompressible limit) remains stationary at the position OA, while plasma carries magnetic flux through it. The direction of the magnetic field vector rotates towards the normal, and the strength of the field decreases in the process. Pairs of shock waves propagate away from the diffusion region, and they may be regarded loosely as current sheets extending from the central current sheet. In a sense, this central sheet has bifurcated to give pairs of sheets far from the neutral point.

When the inflow speed is much less than the Alfvén speed, the angle between OA and the vertical axis becomes very small indeed, which makes the external flow almost uniform and $v_i \simeq v_e$. However, as the flow speed increases, so the inclination of the waves increases, which in turn decreases the field strength at the diffusion region (i.e., $B_i < B_e$). This speeds up the flow ($v_i > v_e$), since the electric field is uniform for this steady, two-dimensional configuration (i.e. $v_i B_i = v_e B_e$). If the inflow speed (v_e) increases beyond a maximum value, the field strength (B_i) is reduced to zero and the flow chokes off completely.

Petschek estimates the *maximum flow speed* (v_e^*) by assuming that the field in the inflow region (to the right of OA in Figure 10.5(a)) is potential. If the field strength is uniform ($B_e \hat{\mathbf{y}}$) at large distances and has a uniform horizontal component ($B_x \hat{\mathbf{x}}$)

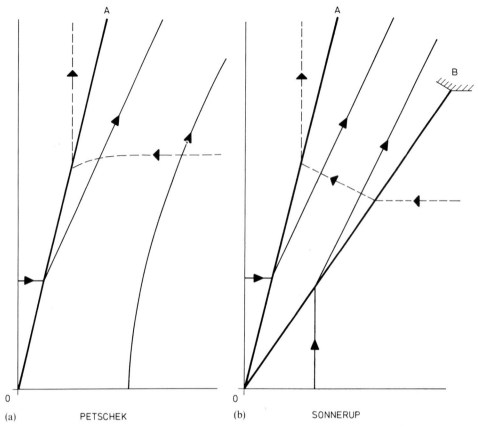

Fig. 10.5. (a) Magnetic field lines (———) and streamlines (- - - - -) for the first quadrant of Petschek's mechanism. A finite Alfvén wave (or slow shock) is situated at OA. It is generated at the diffusion region, which is here regarded as a point O, and serves to turn the flow and magnetic field through large angles. (b) The first quadrant for Sonnerup's model. A second discontinuity is present at OB.

along a length L_e of OA, this implies that the field at O outside the diffusion region (of length L) is

$$B_i = B_e - \frac{4B_x}{\pi} \log_e \frac{L_e}{L}. \tag{10.13}$$

But the fact that OA is an Alfvén wave means that $v_e = B_x/(\mu\rho)^{1/2}$, and hence B_i has fallen to $\tfrac{1}{2} B_e$ when

$$v_e \equiv v_e^* = \frac{\pi B_e/(\mu\rho)^{1/2}}{8 \log_e (L_e/L)}. \tag{10.14}$$

By using Equations (10.8) to (10.10) for L and $v_i B_i = v_e B_e$, this may be written in terms of the *Alfvén speed* $(v_{Ae} = B_e/(\mu\rho)^{1/2})$ and the *magnetic Reynolds number* $(R_{me} = L_e v_{Ae}/\eta)$ as

$$v_e^* = \frac{\pi v_{Ae}}{8 \log_e (8 v_e^{*2} R_{me}/v_{Ae}^2)}. \tag{10.15}$$

It can be seen that the maximum reconnection rate depends weakly on R_{me} and is typically 1 to 10% of the external Alfvén speed (v_{Ae}). A more precise repetition of Petschek's analysis was carried out by Roberts and Priest (1975), who found somewhat lower reconnection rates than Petschek.

But Petschek's mechanism was not generally accepted, partly because of the semi-quantitative nature of the analysis, and also because Green and Sweet (1966) proposed that the field strength should decrease outwards along OA. This implies that the field lines to the left of OA are bowed away from O, rather than being straight, which presents a problem for matching to the diffusion region. Petschek and Thorne (1967) suggested that, if the field does decrease along OA, so that the tangential field component reverses sign across OA, then (in the compressible case) OA should split into a slow shock and an intermediate wave standing ahead of the shock. Even more doubt was cast on the validity of Petschek's analysis when alternative solutions for the external region were put forward by Sonnerup (1970) and Yeh and Axford (1970).

By comparison with Figure 10.5(a), Sonnerup's model (Figure 10.5(b)) possesses an extra discontinuity (OB) to the right of OA. The discontinuities separate the quadrant into three regions, in each of which the magnetic field and plasma velocity are uniform. This enables one to relate the values of variables in each region analytically. However, the discontinuity OB needs to be generated at a corner B in the flow, and so the model is not applicable in detail to a current sheet in the solar atmosphere. However, this model does have the great virtue of simplicity, and the discontinuity OB may be regarded as a mathematical lumping together of expansion effects in the inflow region. The elegance of the model has meant that it has been possible to generalise it in several ways: Cowley (1974a) has included magnetic field and velocity components normal to the plane of the figure; Yang and Sonnerup (1976) have included compressibility effects, for which OB spreads into a slow expansion fan centred at the external corner in the flow and produces a complicated interaction region ahead of the diffusion region; asymmetry in the opposing inflow regions has been treated by Cowley (1974b) and by Mitchell and Kan (1978).

10.1.4. EXTERNAL REGION

Yeh and Axford (1970) were the first to seek *mathematical* solutions for the flow in the external region. They argue that the magnetohydrodynamic variables should be of *self-similar* form on some scale intermediate between the size of the diffusion region and the distance between the magnetic field sources. They simply look for *incompressible* solutions to Equations (10.1) to (10.3) with $\eta = 0$, namely

$$\mathbf{E} + \mathbf{v} \times \mathbf{B} = \mathbf{0}, \tag{10.16}$$

$$\rho(\mathbf{v} \cdot \nabla)\mathbf{v} = -\nabla p + \operatorname{curl} \mathbf{B} \times \frac{\mathbf{B}}{\mu}, \tag{10.17}$$

$$\nabla \cdot \mathbf{v} = 0. \tag{10.18}$$

They adopt the forms $\mathbf{B} = \mathbf{B}(\theta)$, $\mathbf{v} = \mathbf{v}(\theta)$, for the magnetic field and fluid velocity. In other words, they assume $\psi = rg(\theta)$, $A = rf(\theta)$, where the stream function (ψ) and vector potential ($A\hat{z}$) are related to **v** and **B** by

$$(v_r, v_\theta) = \left(\frac{1}{r}\frac{\partial \psi}{\partial \theta}, -\frac{\partial \psi}{\partial r}\right), \quad (B_r, B_\theta) = \left(\frac{1}{r}\frac{\partial A}{\partial \theta}, -\frac{\partial A}{\partial r}\right).$$

The resulting ordinary differential equations for $g(\theta)$ and $f(\theta)$ are solved subject to the boundary conditions $g = df/d\theta = 0$ on $\theta = 0$ and $\theta = \frac{1}{2}\pi$, which imply that the x- and y-axes form streamlines on which $B_r = 0$. Yeh and Axford's model contains the extra discontinuity OB, which is unacceptable physically. But unfortunately OB is not acceptable mathematically either, since it proves to be impossible to join the flow on both sides (Vasyliunas, 1975).

Soward and Priest (1977) have tackled the fast reconnection problem in the same spirit, by seeking incompressible, two-dimensional solutions of Equations (10.16) to (10.18) which are asymptotic in form and so are valid at large distances from the diffusion region. But they stipulate that there should be no second discontinuity like OB, and seek solutions that to lowest order behave like

$$\mathbf{B} = (\log r)^{1/2}\,\mathbf{f}(\theta), \qquad \mathbf{v} = (\log r)^{-1/2}\,\mathbf{g}(\theta),$$

so that B is increasing slowly with r while v decreases and their product remains of order unity. The analysis is rather complex and has been summarised by Priest and Soward (1976). They define $R = \log_e (r/L) + R_0$, where R_0 is a constant, which is chosen for convenience as $(\pi/(8M_i) - \log_e M_i)$, and $M_i \equiv v_i(\mu\rho)^{1/2}/B_i$. Then they seek series solutions in powers of R for which the Lorentz force dominates the inertial term in Equation (10.17). It is necessary to expand ψ and A in the forms

$$\psi = rg(R, \theta) \equiv r\bigl[R^{-1/2}g_0(\theta) + R^{-3/2}(g_{11}(\theta)\log_e R + g_1(\theta)) + \ldots\bigr],$$
$$A = rf(R, \theta) \equiv r\bigl[R^{1/2}f_0(\theta) + R^{-1/2}(f_{11}(\theta)\log_e R + f_1(\theta)) + \ldots\bigr], \quad (10.19)$$

where θ is measured from the x-axis, shown vertical in Figure 10.6(a). These expansions are substituted into Equations (10.16) and (10.17), and the coefficients of powers of R are equated to zero. The resulting ordinary differential equations for the unknown functions $(g_0, f_0, g_{11}, f_{11}, g_1, f_1, \ldots)$ are solved for $0 < \theta < \frac{1}{2}\pi$, subject to the boundary conditions $g = \partial f/\partial\theta = 0$ at $\theta = \frac{1}{2}\pi$, to all orders. The expansion (10.19) represents only an 'outer' solution, since it fails to satisfy the required boundary condition at $\theta = 0$. The 'inner' solution, which is valid near $\theta = 0$, takes the form

$$\psi = rG(R, \xi) \equiv r\bigl[R^{-1/2}G_0(\xi) + R^{-3/2}(G_{11}(\xi)\log_e R + G_1(\xi)) + \ldots\bigr],$$
$$A = rF(R, \xi) \equiv r\bigl[R^{-1/2}F_0(\xi) + R^{-3/2}(F_{11}(\xi)\log_e R + F_1(\xi)) + \ldots\bigr], \quad (10.20)$$

where $\xi = \theta/\Theta(R)$, and

$$\theta = \Theta(R) \equiv \Theta_0 R^{-1} + (\Theta_{11}\log_e R + \Theta_1)R^{-2} + \ldots$$

is the position of the discontinuity OA (Figure 10.6(a)). Again, they derive equations for the unknown functions in Equation (10.20) and solve them for $0 \leq \xi < \infty$, subject to the boundary conditions

$$G = \frac{\partial F}{\partial \xi} = 0 \quad \text{at } \xi = 0, \text{ (i.e., } \theta = 0\text{)}.$$

Finally, the inner and outer expansions (10.19) and (10.20) are matched by means of the conditions

$$\lim_{\theta \to 0} g = \lim_{\xi \to \infty} G \quad \text{and} \quad \lim_{\theta \to 0} f = \lim_{\xi \to \infty} F,$$

to all orders. The inner solution exhibits a discontinuity in **v** and **B** at $\xi = 1$ (i.e., at OA), but it may be chosen to be continuous throughout the remainder of the domain. It transpires that, to the left of OA, the magnetic field behaves like $R^{-1/2}$, while to the right it behaves like $R^{1/2}$. Thus, as one approaches the diffusion region along the line $\theta = \frac{1}{2}\pi$ or moves away along $\theta = 0$, the magnetic field decreases in strength while the plasma speeds up. Furthermore, the position of OA (namely, $\theta = \Theta(R)$) is found to be to lowest order $\theta = \pi/(8R)$, so that the discontinuity bends round slightly, as indicated diagrammatically in Figure 10.6(a). The streamlines and magnetic field lines are found to have curvature in the senses shown. In particular, the field lines to the left of OA are convex towards O, which allays the previous fears of Green and Sweet

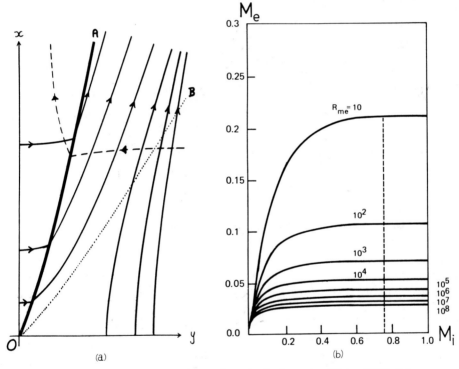

Fig. 10.6. (a) The modification of Petschek's configuration by Soward and Priest (1977). OA now curves slightly towards the x-axis. OB is a second 'Alfvén line', at which the normal components of the plasma velocity and magnetic field are equal, but the plasma flows smoothly through it without jumps in the tangential components. Any discontinuity at OB would need to be generated externally, and so it is most unlikely in the solar atmosphere. By comparison with Figure 10.5a, extra field lines have been sketched to indicate that, as one approaches O, the field strength decreases along Oy and increases along Ox. (b) The Alfvén Mach number ($M_e \equiv v_e/v_{Ae}$) for fast reconnection (as given by Soward and Priest (1977)) is plotted as a function of $M_i \equiv v_i/v_{Ai}$ for a series of values of $R_{me} \equiv v_{Ae}L_e/\eta$. v_A is the Alfvén speed, and subscripts e and i refer to values at large distances from the diffusion region and close to it, respectively, as indicated in Figure 10.3. Regardless of the value of R_{me}, M_e has a maximum at $M_i = \pi/4$, as shown by the dashed line.

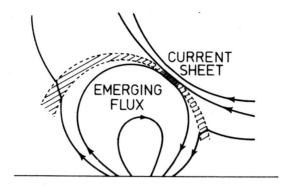

(a) Preflare Heating

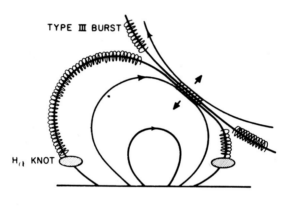

(b) Impulsive Phase

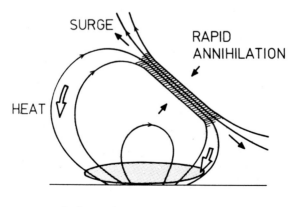

(c) Main Phase

(1966). The figure is exaggerated in the sense that, in practice, OA and OB lie very close to the x-axis.

The significance of the above analysis is that *it puts Petschek's mechanism on a sound mathematical basis as the only workable model for fast steady-state reconnection* in the solar atmosphere. Furthermore, when the solution is linked to the diffusion region by means of Equations (10.8) to (10.10), it determines a relation between the Alfvén Mach numbers $M_e \equiv v_e(\mu\rho)^{1/2}/B_e \equiv v_e/v_{Ae}$ and $M_i \equiv v_i(\mu\rho)^{1/2}/B_i$. (They are characteristic of the flow at large distances and at the diffusion region inflow, respectively, as indicated in Figure 10.3.) Figure 10.6(b) shows the resulting graphs of M_e as a function of M_i, for various values of the magnetic Reynolds number (R_{me}). It can be seen that M_e does indeed possess a maximum value, as suggested by Petschek, which is typically a half of what Petschek gave. For instance, the maximum inflow speed v_e (or, loosely, 'reconnection rate') varies from $0.2 v_{Ae}$ when $R_{me} = 10$ to $0.02\ v_{Ae}$ when $R_{me} = 10^6$. The value of the reconnection rate required for the main phase of a large flare is likely to be below this maximum.

The above analysis has been extended by Soward and Priest (1982) to fast reconnection in a *compressible* plasma. There is no longer a logarithmic singularity at OA, because the slow-mode characteristics intersect OA from both sides. OA becomes a slow shock, with no need for the intermediate wave envisaged by Petschek and Thorne. Furthermore, the analysis on reconnection has been complemented by some remarkable numerical simulations (e.g., Ugai and Tsuda, 1977) and laboratory experiments (e.g., Baum and Bratenahl, 1977); for a summary see Priest (1981b).

10.2. Simple-loop Flare

10.2.1. Emerging (or evolving) flux model

Heyvaerts *et al.* (1977) and Canfield *et al.* (1974) have outlined a flare model based on high-resolution ground-based observations suggesting that flares occur after the emergence of new flux from below the photosphere (Section 1.4.4B). They propose that the type of flare depends on the magnetic environment into which the new flux emerges. Usually, just an *X-ray bright point* is produced. But emergence near a unipolar sunspot or into a unipolar area near the edge of an active region may give rise to a *simple-loop flare*. Furthermore, if the new flux appears near the sheared field around an active-region filament, then a *two-ribbon flare* may result (Figure 1.37), with the emerging flux triggering the release of stored energy in the much more extensive overlying field.

Fig. 10.7. The emerging flux mechanism of Heyvaerts *et al.* (1977) for a *small flare* (a simple-loop flare).

(a) *Preflare phase*. The emerging flux reconnects with the overlying field. Shock waves (dashed) radiate from a small current sheet and heat the plasma as it passes through them into the shaded region.

(b) *Impulsive phase*. The onset of turbulence in the current sheet (when it has reached a critical height) causes a rapid expansion. The resulting electric field accelerates particles, which then escape along field lines and produce an impulsive microwave burst as they spiral. Those that move downwards give rise to hard X-rays by collisional excitation (and possibly Hα knots), while those that escape upwards onto open field lines produce type III radio bursts.

(c) *Flash and main phases*. The current sheet reaches a new steady state, with reconnection based on a marginally turbulent resistivity. It is much bigger than before, and both heat and particles are conducted down to the lower chromosphere, where they produce the Hα flare.

It should be noted that the model also applies to a situation of *evolving flux* rather than emerging flux. When distinct flux systems evolve, current sheets form at the interfaces between the systems; thus, horizontal motions of photospheric flux will produce current sheets in just the same way as vertical motions.

It should also be stressed that not all emerging flux can produce a flare. Simple-loop flares are triggered only if sufficient flux emerges that the current sheet occupying the interface between new and old flux reaches a critical height that depends on emergence speed and field strength. (Two-ribbon flares also need the large-scale field to be highly sheared, so that it contains the necessary stored energy.)

The model suggests that *simple-loop flares* occur in three stages when the emerging magnetic flux interacts with the overlying field (Figure 10.7). As the new flux eats its way into the ambient field, so the current sheet moves up in the atmosphere. If the sheet reaches a critical height (h_{crit}), its current density exceeds the threshold value for the onset of microturbulence, and the flare is triggered. This condition may be rewritten very approximately in terms of the current-sheet temperature (T_c) as

$$T_c^2 > T_{turb}^2 \equiv 1.8 \times 10^{16} \frac{B_i}{v_i},$$

where the inflow field (B_i) and speed (v_i) are in tesla and m s^{-1}, respectively.

In order to determine T_c, however, it is necessary to investigate the energy balance (10.12) inside the sheet, including the effects of mechanical heating, Joule heating, radiative loss and thermal conduction (Heyvaerts and Priest, 1976; Tur and Priest,

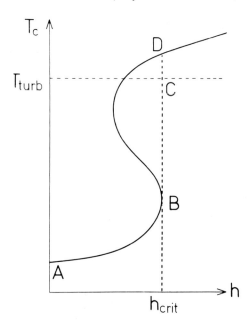

Fig. 10.8. The thermal equilibrium temperature (T_c) in a reconnecting current sheet is shown schematically as a function of height (h) in the solar atmosphere. As the sheet gains in height, the equilibrium solution moves along *AB*. When the critical height (h_{crit}) is attained, there are no neighbouring equilibria, and the sheet heats up dynamically along the path *BD*. But, as the temperature T_{turb} is exceeded, the critical current density for turbulence onset is surpassed.

1976). Each term is approximated and the equation is solved for the internal temperature (T_c) of the sheet as a function of the height (h) above the photosphere. The result is that, low down in the atmosphere, there exists a unique solution to the energy equation. The main balance is between radiation and Joule heating and T_c is lower than T_{turb} (Figure 10.8). Eventually, a critical height is reached beyond which there is no thermal equilibrium with such a balance! The sheet is therefore in a metastable state. It heats up rapidly, and if T_{turb} is exceeded the flare is triggered. The current sheet broadens very rapidly due to the onset of a turbulent resistivity, so that strong electric fields are generated which accelerate particles to high energies and produce the impulsive phase.

Tur and Priest (1978) investigate the energy balance of the current sheet in more detail by solving the order-of-magnitude form of Equation (10.12), for a wide range of ambient magnetic field strengths (B_i) and velocities (v_i), in order to determine their effect on the critical height for the current sheet and to find the temperature at which the sheet becomes turbulent; also, one can distinguish the different ways in which the critical conditions may be reached. The results have been summarised in Priest (1981b). Milne and Priest (1981) have repeated the analysis with their improved model for the diffusion region (Section 10.1.2). The resulting values for the critical height (h_{crit}) are shown in Figure 10.9, from which it can be seen that h_{crit} decreases as B_i increases. This is because of the low-β limitation, which makes equilibrium solutions impossible when β is too low, as indicated by the dots in Figure 10.4(b).

Further work, both theoretical and observational, is needed to set the emerging

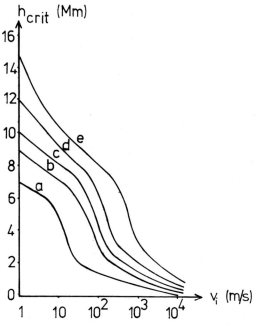

Fig. 10.9. The variation of the critical height (h_{crit}) in Mm (1Mm = 10^6m) for simple-loop flare onset with the emergence speed (v_i) for several values of the ambient magnetic field (B_i) in tesla. a, b, c, d, e refer to $B_i = 10^{-1}$, $10^{-1.5}$, 10^{-2}, $10^{-2.5}$, 10^{-3}, respectively. (A weakly turbulent diffusivity that is a 100 times larger than the classical value has been adopted) (from Milne and Priest, 1981).

flux model on a firmer footing. Theoretically, the dynamics of the current sheet need to be followed during its search for a new equilibrium state. Mercier and Heyvaerts (1980) suggest evolution to an unsteady state of reconnection or to a steady state with gravity important. Observationally, more measurements of emerging flux magnetic and velocity fields prior to flares are urgently required.

10.2.2. THERMAL NONEQUILIBRIUM

Usually, it is the magnetic field that is considered to be the source of the flare energy, but Hood and Priest (1981a) have suggested that a simple-loop flare may occur when the cool core of an active-region loop no longer exhibits thermal equilibrium. The possibility of such a *thermal nonequilibrium* is demonstrated by solving approximately the energy balance equation. It is found that, if one starts with a cool equilibrium at a few times 10^4 K and gradually increases the heating or decreases the loop pressure, then critical *metastable* conditions are ultimately reached beyond which no cool equilibrium exists. The plasma heats up *explosively* to a new quasi-equilibrium at typically 10^7 K. During such a thermal flaring the strong magnetic tube in which the plasma is located maintains its position, and any particle acceleration is secondary in nature.

It is suggested that thermal nonequilibrium is the essence of some simple-loop flares. Also, it may act as the *trigger* for the magnetic eruption of a two-ribbon event by creating a thermal 'preflare' temperature rise in an active-region filament; the plasma expands along the filament, which slowly rises until, at a critical height, the magnetic configuration becomes magnetohydrodynamically unstable (Section 10.3.2) and erupts violently outwards.

Thermal instability (Section 7.5.7), driven by radiation, had previously been suggested as a trigger for a flare by Sweet (1969), Kahler and Kreplin (1970), and Spicer (1976, 1977). They followed Field (1965) in assuming an isothermal basic state under a balance between radiation and heating. Hood and Priest extended their work by including the effects of thermal conduction in the non-uniform basic state of a coronal loop. The result is that rapid heating may occur through nonequilibrium rather than instability. The difference between the two is that in the latter case an equilibrium becomes unstable, but in the former case the equilibrium ceases to exist at all and the subsequent evolution may be more violent. (For the role of thermal nonequilibrium in prominence formation, see Section 11.1.)

Sturrock (1966) and Sweet (1969) have stressed that the preflare configuration must be *metastable* just prior to the energy release so that the flaring is explosive. In other words, the system is marginally stable on a linear analysis, but, after a finite perturbation, it is unable to find a neighbouring equilibrium. Demonstrating linear instability is not sufficient to explain a flare, because the nonlinear response of the system may be simply to evolve into another nearby equilibrium with no violent consequences. Thermal non-equilibrium certainly meets this requirement of providing a metastable state.

Hood and Priest consider a simple static equilibrium governed by

$$\frac{d}{ds}\left(\kappa_0 T^{5/2} \frac{dT}{ds}\right) = n_e^2 Q(T) - H, \qquad (10.21)$$

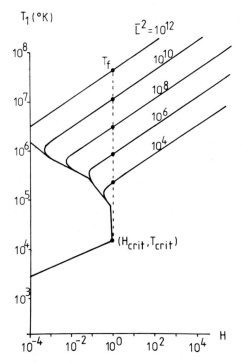

Fig. 10.10. The equilibrium summit temperature (T_1) of a loop as a function of the heating H (expressed in units of the radiative loss at $T = 2 \times 10^4$ K and $n_c = 5 \times 10^{14}$ m^{-3}), for various values of the dimensionless half-length $(\bar{L} \equiv (Q/(\kappa_0 T^{7/2}))^{1/2} n_c L)$ (from Hood and Priest, 1981a).

where $n_e = p/(2k_B T)$, and the pressure is uniform. Here s is the distance measured along a field line, $d/ds(\kappa_0 T^{5/2} dT/ds)$ represents conduction along the field (see (2.33)), and $n_e^2 Q(T)$ is the optically thin radiation (2.35c). Also, H is the heating term, which is assumed to be uniform per unit volume, although the qualitative features of the solution are not strongly dependent on its form (Section 6.5.1). The summit temperature (T_1) of the loop depends on the pressure (p), the heating (H) and the loop length $(2L)$. The way in which T_1 depends on H and L for a fixed p is shown in Figure 10.10 when the conduction term is replaced by an order-of-magnitude approximation. As the heating rate slowly increases, so the summit temperature increases through equilibria on the lower branch. But, when the heating exceeds a critical value (H_{crit}), there is no longer a neighbouring equilibrium, and so the plasma rapidly heats up along the dashed line towards a quasi-steady flaring value (T_f) above 10^6K, denoted by a dot.

Similar results are obtained in several other cases: with different forms of heating; with full (rather than order-of-magnitude) solutions to (10.21); and with the effect of gravity incorporated. A lack of equilibrium can also be produced if the pressure is reduced. Finally, it must be pointed out that T_f is only a rough estimate of the flaring temperature; its value and the transition from the low- to the high-temperature state depend on the detailed plasma dynamics (Craig, 1981), which need to be studied in more detail.

10.2.3. KINK INSTABILITY

Solar *coronal loops* are observed to be remarkably stable structures, but occasionally they can lose stability and produce a flare. For many years this was puzzling, since a simple magnetohydrodynamic stability analysis (Section 7.5.3) shows that flux tubes are unstable to kinking whenever they are twisted. But then Raadu (1972), Giachetti et al. (1977) and Hood and Priest (1979b) included in the analysis the dominant stabilising effect, namely *line-tying* of the ends of the loop in the dense photosphere. Due to the inertia of the photospheric plasma, any perturbation that is initiated in the corona must vanish at the loop footpoints, which behave as if they had lead weights attached to them. The result is that a loop is stable when only slightly twisted, but it becomes kink unstable when the twist is too great.

The *critical twist* (Φ) for instability depends on the aspect ratio of the loop (L/a), where $2L$ is its length and $2a$ is its width. It also varies with the ratio of plasma to magnetic pressure and the detailed transverse magnetic structure, but it is typically between 2π and 6π.

Pressure gradients can sometimes be stabilising. They were incorporated without line-tying by Giachetti et al. (1977) and with line-tying by Hood and Priest (1979b). The latter analysis follows closely the energy method of Section 7.4.1 and is outlined below.

Consider a cylindrical flux tube (of length $2L$) with equilibrium pressure $p(R)$ and magnetic field components $(0, B_\phi(R), B_z(R))$ satisfying the force balance

$$0 = \frac{dp}{dr} + \frac{d}{dr}\left(\frac{B_\phi^2 + B_z^2}{2\mu}\right) + \frac{B_\phi^2}{\mu r}. \tag{10.22}$$

Suppose the plasma is displaced by an amount

$$\xi = \left[\xi^R(R), -i\frac{B_z}{B}\xi^0(R), i\frac{B_\phi}{B}\xi^0(R)\right]\cos\frac{\pi z}{2L} e^{i(m\phi + kz)}, \tag{10.23}$$

which vanishes at the ends ($z = \pm L$) of the loop. Then the change in energy (δW) is given by Equation (7.39) with $g = 0$, and, by generalising the treatment of Section 7.4.1, it may be minimised with respect to ξ^0 to give

$$\delta W = \frac{1}{\mu}\int_0^\infty F\left(\frac{d\xi^R}{dR}\right)^2 - G\xi_R^2 \, dR. \tag{10.24}$$

Here F and G are functions of R, p, B_ϕ, B_z, k and m, similar to Equations (7.50) and (7.51) but containing extra terms from p. In order to minimise Equation (10.24) with respect to the radial perturbation (ξ^R) one solves the *Euler–Lagrange equation*

$$\frac{d}{dR}\left(F\frac{d\xi^R}{dR}\right) + G\xi^R = 0, \tag{10.25}$$

subject to

$$\xi^R = 1, \frac{d\xi^R}{dR} = 0 \text{ at } R = 0, \quad \text{for } m = 1,$$

$$\xi^R = 0, \frac{d\xi^R}{dR} = 1 \text{ at } R = 0, \quad \text{for } m \neq 1.$$

The boundary between stability and instability for the perturbation (10.23) is then obtained by varying the parameters until $\xi^R(R)$ vanishes somewhere (Section 7.4.1).

The detailed results depend on the forms that are adopted for $p(R)$, $B_\phi(R)$, $B_z(R)$ satisfying Equation (10.22). For example, a force-free field of uniform twist gives the stability diagram in Figure 7.8. Another field considered by Hood and Priest has a uniform axial field $B_z = B_0$ and a twist

$$\Phi(R) \equiv \frac{2LB_\phi}{RB_z} = \frac{\Phi_0}{1 + R^2/a^2}.$$

The result is a pressure profile of

$$p(R) = p_\infty + \left(\frac{\Phi a}{2L}\right)^2 \frac{B_0^2}{2\mu},$$

which decreases from a maximum on the axis (Section 3.3.4). This *variable-twist field* may model the effect of a localised twisting on a uniform field. Its stability characteristics depend on five parameters, namely k, m, L/a, Φ_0 and β (the ratio $(p_\infty 2\mu/B_0^2)$ of plasma to magnetic pressure at large distances from the axis). The computed stability curves for $\beta = 0$ are given in Figure 10.11(a), and are similar to the force-free case of Figure 7.8. The dashed lines represent the *Kruskal–Shafranov limit* (7.59), which may be written

$$\frac{1}{2}\Phi(0) = -\frac{2L}{a}\bar{k}$$

at radius $r = a$; it gives the critical twist for instability in the absence of pressure gradients and line-tying. It can be seen that at large k the loop is less stable than the Kruskal–Shafranov limit (because of the destablising effect of the pressure gradient), whereas at small k it is more stable (because of the influence of line-tying).

For each value of L/a, Figure 10.11a gives a critical twist Φ_crit (at $r = a$), beyond which the loop is unstable to some range of k. Φ_crit increases with L/a because the equilibrium pressure decreases and so becomes more stabilising. The variation of Φ_crit with L/a and β is shown in Figure 10.11(b). As β rises in value, the pressure rises without affecting the pressure gradient, which increases the value of a positive-definite term in δW and so makes the equilibrium more stable. For an aspect ratio of $L/a = 10$, typical of active-region loops, it can be seen that the smallest twist to give instability is 2.8π at $\beta = 0$ and about 3.3π at $\beta = 1$.

Later Hood and Priest (1981b) conducted a full stability analysis to all possible line-tied perturbations, which required the solution of partial differential equations. For example, the force-free field of uniform twist becomes unstable at a twist of 2.5π, whereas their previous analysis had given instability to perturbations of the form (10.23) at a twist of 3.3π.

In the future, there are three aspects of kink instability theory that need to be investigated further, namely the nonlinear development (e.g., Sakurai, 1976), the effect of the curvature of the loop, and departures from infinite conductivity. The effect of non-zero resistivity on the kink mode gives the so-called *resistive kink* or *cylindrical tearing-mode* (Sections 7.5.5 and 10.2.4), which has been summarised by Van Hoven (1981) and is described in the next section. Spicer (1976, 1977, 1981) in particular has stressed the importance of cylindrical tearing modes for flares; he has

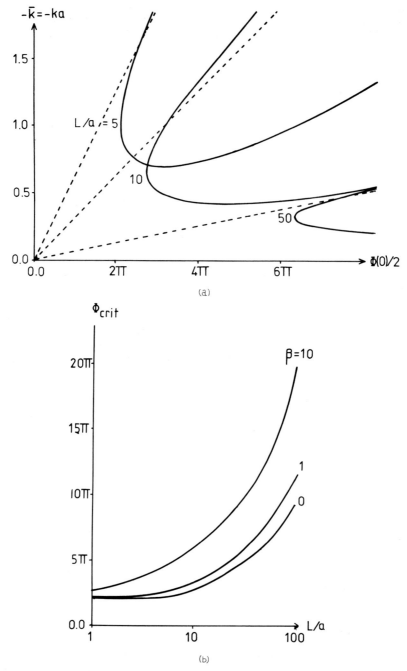

Fig. 10.11. Kink instability ($m = 1$) of a variable-twist field, with $\beta \equiv 2\mu p_\infty / B_0^2$, where p_∞ is the coronal pressure and B_0 is the uniform axial field. (a) The stability diagram for $\beta = 0$ and for different values of the aspect ratio (L/a). The equilibrium is stable to the perturbation (10.23) to the left of each curve and unstable to the right. $\tfrac{1}{2}\Phi(0)$ is the twist at $R = a$, the 'edge' of the loop. (b) The variation of the critical twist (Φ_{crit}) at $R = a$ with L/a and β (from Hood and Priest, 1979b).

pointed out that double tearing occurs faster than ordinary tearing and has considered nonlinear effects for increasing dissipation. However, the crucial influence of line-tying has not yet been incorporated.

10.2.4. RESISTIVE KINK INSTABILITY

In the previous section the effect of line-tying on the *ideal* kink instability was considered. During its nonlinear development the flux tube may be highly contorted locally, so that the effects of finite electrical conductivity become important in local current sheets. However, it is possible instead that a loop reaches resistive instability (Section 7.5.5) before the onset of the ideal kink instability.

Throughout most of the configuration the resistive term in the linearised induction equation

$$\frac{\partial \mathbf{B}_1}{\partial t} = \nabla \times (\mathbf{v}_1 \times \mathbf{B}_0) - \nabla \times (\eta \nabla \times \mathbf{B}_1)$$

is negligible, since the magnetic Reynolds number is very large. However, it does become important in singular layers where

$$\nabla \times (\mathbf{v}_1 \times \mathbf{B}_0) \equiv (\mathbf{B}_0 \cdot \nabla)\mathbf{v}_1 - \mathbf{B}_0(\nabla \cdot \mathbf{v}_1) = 0,$$

so that $\partial \mathbf{B}/\partial t$ can no longer be balanced by $\nabla \times (\mathbf{v}_1 \times \mathbf{B}_0)$. In particular, for a wave-like perturbation ($\mathbf{v}_1 \sim e^{i\mathbf{k}\cdot\mathbf{r}}$) in an incompressible medium such that $\nabla \cdot \mathbf{v}_1 = 0$, the locations of the singular layers are given by $\mathbf{k} \cdot \mathbf{B}_0 = 0$. For example, in a cylindrical geometry where $\mathbf{v}_1 \sim e^{i(m\phi + kz)}$, the locations ($R = R_s$) for each k and m are determined by

$$kB_{0z}(R_s) + \frac{m}{R_s}B_{0\phi}(R_s) = 0.$$

The simplest force-free equilibrium in cylindrical geometry is the constant-α field

$$B_z = B_0 J_0(\alpha R), \qquad B_\phi = B_0 J_1(\alpha R).$$

Its stability may be investigated by seeking solutions of the form

$$\mathbf{B}_1 = \mathbf{B}_1(R) \exp(\omega t + i(m\phi + kz))$$

to the linearised resistive MHD equations (Furth *et al.*, 1973; Coppi *et al.*, 1976). Assuming ideal kink stability and no line-tying, it is found that a flux tube is unstable to the so-called *resistive internal kink mode* (or *cylindrical tearing mode*). The fastest growing perturbations have long wavelengths ($kR_s \ll 1$) and $m = 1$, with a growth-rate

$$\omega \approx [R_s^2 |(\mathbf{k} \cdot \mathbf{B})'_s|/B_0]^{2/3} \tau_d^{-1/3} \tau_A^{-2/3},$$

in terms of the resistive diffusion-time ($\tau_d = R_s^2/\eta$) and the Alfvén travel-time ($\tau_A = R_s/v_A$). In other words, $\omega \sim \tau_d^{-1/3} \tau_A^{-2/3}$, so that for a large magnetic Reynolds number ($\tau_d/\tau_A \gg 1$) the growth-rate lies between τ_d^{-1} and τ_A^{-1}, and is larger than the planar growth-rate ($\tau_d^{-1/2} \tau_A^{-1/2}$).

The above result is obtained by matching the value of $\xi_R^{-1} d\xi_R/dR$ at the edge of a singular layer, whose width is $(\tau_A/\tau_d)^{1/3} R_s$. Outside it, diffusion is negligible and the

R-component of the equation of motion reduces to the form $(f\xi'_R)' - g\xi_R = 0$. The layer is located where $f(R)$ vanishes, and ξ'_R/ξ_R behaves like $(R - R_s)^{-2}$ there. It is so thin that a planar approximation may be adopted (with $x = z + \theta/k$ and $y = R$). Inside the layer, diffusion is important, and the equations of motion and induction normal to the layer reduce to

$$\omega B_{1y} = ikB_{0x}v_{1y} + \eta B''_{1y},$$
$$\omega\mu\rho_0 v''_{1y} = ikB_{0x}B''_{1y},$$

whose solution for ξ'_R/ξ_R again behaves like $(R - R_s)^{-2}$ at the outer edge.

Several features of the nonlinear development of the tearing mode have been discovered. Van Hoven and Cross (1973) find that 20 to 30% of the stored magnetic energy is released in 3 to 4 linear growth-times, after which the instability saturates with an equipartition between kinetic energy and small-scale magnetic energy. Similar results are established for cylindrical tearing, both in the tokamak regime, where $B_z \gg B_\phi$ and $\beta \ll 1$ (Waddell et al., 1976), and also in the reversed-field pinch regime, where $B_z \lesssim B_\phi$ and $\beta \sim 1$ (Schnack and Killeen, 1979). Furth et al. (1973) and Schnack and Killeen (1978) discuss the *double-tearing mode*, which occurs when there are two singular surfaces where $\mathbf{k} \cdot \mathbf{B}_0$ vanishes for the same k; it has a fast linear growth-rate ($\omega \approx \tau_d^{-1/4}\tau_A^{-3/4}$). One nonlinear effect is the *overlapping of resonances* (Finn, 1975), for which magnetic islands from neighbouring layers interact. It may well be important for some loops (Spicer, 1977), especially long ones (Van Hoven, 1981). Another nonlinear effect is *mode coupling* of tearing modes with different helicities (Waddell et al., 1979), but the resulting growth is much slower than the fundamental resistive kink mode, which Van Hoven (1981) suggests is adequate by itself to explain the time-scale for loop flares. He also points out that *island coalescence* of small wavelengths into longer ones is not relevant in the solar context ($\tau_d/\tau_A \gg 1$), where the longest available wavelength grows fastest.

An important limitation of the above linear and nonlinear analyses is that linetying has not been incorporated. In order of magnitude, it would imply $2\pi/k \lesssim 2L$, so that the wavelength ($2\pi/k$) can fit into a loop of length $2L$. But k satisfies $\mathbf{k} \cdot \mathbf{B}_0 = 0$, and so this condition reduces to $\Phi \geq 2\pi$ in terms of the twist ($\Phi = 2LB_\phi/(RB_z)$). However, a more precise derivation of the threshold for cylindrical tearing has not been completed, so that one does not yet know whether it is lower or higher than the threshold for the much faster growing ideal kink mode.

Finally, it must be remembered that a study of microinstabilities (which is outside the scope of this book) is essential to determine the anomalous resistivity and deduce the particle acceleration (e.g., Spicer, 1981). Moreover, in order to maintain the current at the level necessary for marginal stability, it is essential that the configuration is driven by external motions, such as are envisaged in, for instance, the emerging flux model (Section 10.2.1).

10.3. Two-ribbon Flare

The generally accepted scenario for the events leading up to a two-ribbon flare is that a magnetic arcade (supporting a plage filament) responds to the slow photospheric motions of its foot-points by evolving passively through a series of largely force-free equilibria. At some critical amount of shear, the configuration becomes unstable and erupts outwards (Figure 10.12). Subsequently, the field closes back down, as described in Section 10.3.3, and produces loops and Hα ribbons (Figures 1.37, 1.38, 10.1), but the main problem has been to explain the initial *eruptive instability*. Two approaches have been tried. The first is to seek the existence and multiplicity of

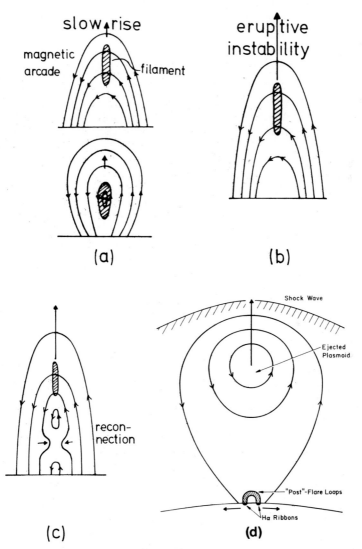

Fig. 10.12. The overall magnetic behaviour in a two-ribbon flare, as seen in a section through the magnetic arcade. (a) Slow preflare rise of a filament, possibly due to thermal nonequilibrium, emerging flux or the initial stages of an MHD instability. The surrounding field may just be sheared (upper diagram) or it may contain a magnetic island so that the filament lies along a flux tube (lower diagram). (b) Eruptive instability of a magnetic arcade and filament. (c) Field lines below the filament are stretched out until reconnection can start. (d) As reconnection proceeds, 'post'-flare loops rise and Hα ribbons move apart.

equilibrium solutions to the force-free equations (Section 10.3.1), subject to the relevant boundary conditions. If the field is evolving slowly through one set of equilibria, it may erupt if a neighbouring equilibrium ceases to exist or if a new equilibrium with lower energy becomes possible. The second approach is to study the magnetohydrodynamic stability of model arcades directly (Section 10.3.2).

The two approaches are complementary. To demonstrate instability directly by the second method is in practice only easy for relatively simple equilibrium configurations. By contrast, the first method can deal with more complex equilibria, and then it makes the hypothesis that a fast evolution occurs from one equilibrium to another. The advantage of the second approach is that it may demonstrate instability at a lower threshold, but the first approach has an advantage too: it may reveal nonequilibrium or imply the possibility of a *nonlinear* instability for an equilibrium that is linearly stable by the second method.

As described in Chapter 1 of Priest (1981a), it should be noted that the *eruptive instability* may occur spontaneously or it may be triggered by some other mechanism. Possibilities are thermal nonequilibrium (Section 10.2.2), the tearing mode, or emerging flux (Section 10.2.1); recent evidence for the latter cause of some flares has been presented by Martin (private correspondence), who finds that filaments tend to erupt when there is rapidly emerging flux nearby with a strong magnetic field.

10.3.1. Existence and Multiplicity of Force-Free Equilibria

Since a coronal arcade is longer than it is wide, it is natural to neglect the fine structure and consider *two-dimensional* force-free fields, independent of the longitudinal (horizontal) coordinate, y say, with components

$$B_x(x, z) = \frac{\partial A}{\partial z}, \quad B_z(x, z) = -\frac{\partial A}{\partial x}, \quad B_y(A), \tag{10.26}$$

which automatically satisfy div $\mathbf{B} = 0$. Here $A(x, z)\hat{\mathbf{y}}$ is the magnetic vector potential or flux function for the transverse components (B_x, B_z) in the vertical plane perpendicular to the filament axis. As shown in Section 3.5.4, it satisfies

$$\nabla^2 A + \frac{d}{dA}(\tfrac{1}{2}B_y^2) = 0. \tag{10.27}$$

Much effort has been expended on trying to solve Equation (10.27) above the plane $z = 0$, taken to represent the photosphere, subject to two boundary conditions on $z = 0$. The first is that the normal component be prescribed:

$$(B_z)_{z=0} = B_n(x), \tag{10.28}$$

say. For the second condition two distinct types have been considered, giving problems that may be labelled (I) and (II). The simpler is

PROBLEM I:

$$B_y = f(A), \tag{10.29}$$

for which the functional form $(f(A))$ of the axial field is prescribed. The more relevant problem, however, is to specify the photospheric connections of each field line, namely,

PROBLEM II:

$$d = d(x), \qquad (10.30)$$

which means that the displacement $(d(x))$ of the footpoints from the x-axis is imposed (Figure 3.10). The aim is then to follow the evolution of the force-free field through a series of equilibria determined by the footpoint motion. Some of the early attempts to solve problems I to II with different forms for $B_n(x)$ and $f(A)$ are described in Section 3.5.6, and more recent work is reviewed by Birn and Schindler (1981).

The first numerical attempts at Problem II were by Sturrock and Woodbury (1967) and Barnes and Sturrock (1972), but they find only one solution. Several authors have tackled Problem I numerically (Section 3.5.6). Low and Nakagawa (1975) seek only one solution. Jockers (1978) found two solutions for $f(A) = \lambda A^{(n+1)/2}$, with the parameter λ less than a maximum value (λ_{max}); however, as the shear increases, so λ increases to λ_{max} through one class of solution and then it decreases through the other class (Figure 10.13(a)). Also, Heyvaerts et al. (1980) have discovered a third solution with a partially open field (Figure 10.13(b) – see below). More recently they have found up to 7 equilibria for the same λ! Problem I has also been considered analytically. Low (1977) found two solutions for $f(A) = e^{-A}$ with circular field lines and a maximum λ of 2. Birn et al. (1978) establish some general results when B_y^2 is

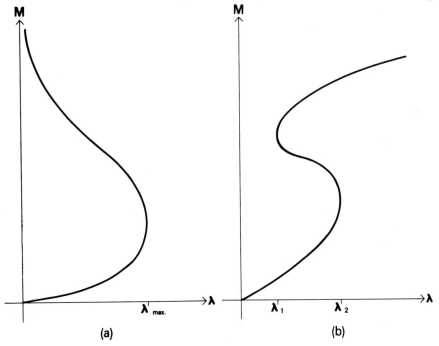

Fig. 10.13. The multiplicity of force-free solutions, as represented by some parameter M (such as the shear gradient (m) or the maximum value of $|A|$) as a function of the parameter λ, for the solutions of (a) Birn et al. (1978) and (b) Heyvaerts et al. (1980).

proportional to λ: they find that, subject to certain conditions on the form of $B_y(A)$, there exists a λ_{max} such that when $\lambda < \lambda_{max}$ there is at least one solution but when $\lambda > \lambda_{max}$ there are no solutions. They then put forward the hypothesis that, if the system is forced into a state where λ_{max} is exceeded, then the magnetic field will erupt violently. Their general ideas may hold for other applications such as increasing the pressure (to give a surge) or the length-scale, but it is clear from the results of Jockers that, for the present problem (where λ is changed by shearing motions), λ_{max} may not be exceeded.

The approach of Heyvaerts et al. (1980) is to solve Problem I numerically within a finite region for the particular current density

$$J(A) \equiv \frac{d}{dA}(\tfrac{1}{2}B_y^2) = \lambda A^4(1-A)^2 \text{ for } 0 \leqslant A \leqslant 1$$

and 0 elsewhere. The assumed boundary conditions are of Dirichlet type on the

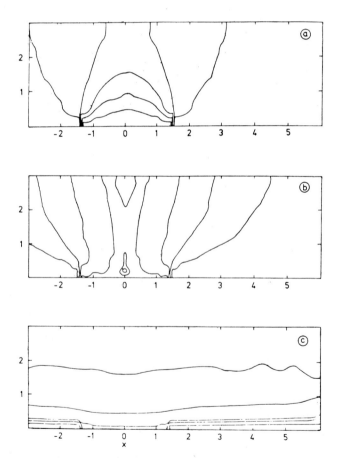

Fig. 10.14. Magnetic field pattern in the (x, z) plane for three force-free solutions obtained by Heyvaerts et al. (1980), corresponding to a solution curve of the type shown in Figure 10.13 (b): (a) lower branch, (b) middle branch, (c) upper branch.

photospheric boundary, namely

$$A(x, 0) = \begin{cases} 0.2 & \text{for } |x| \leq 1.3, \\ 1.5 - |x| & \text{for } 1.3 \leq |x| \leq 1.5, \\ 0 & \text{for } |x| \geq 1.5, \end{cases}$$

and they are of Neumann type on the open boundaries:

$$\frac{\partial A}{\partial x} = 0 \quad \text{for } |x| = 6, \qquad \frac{\partial A}{\partial z} = 0 \quad \text{for } z = 3.$$

For a given λ between λ_1 and λ_2, three topologically different solutions are discovered (Figure 10.14), corresponding to the different branches of Figure 10.13(b). Clearly, in general one may expect to find an even greater multiplicity of solutions as more bend-backs and bifurcations occur. Also, Heyvaerts et al. (1982, Astron. Astrophys.) show that a critical λ exists for a wide range of conditions ($J(A)$ bounded and with compact support).

Birn et al. (1978) establish the following two solutions to Problem I in cylindrical polars with the Z-axis horizontal and located in the photospheric boundary:

$$B_R = \frac{4b}{c\pi R} \tanh\left[\frac{2b\phi}{\pi} - b\right], \qquad B_\phi = -\frac{2}{cR},$$

$$B_Z = \lambda \left[\frac{2}{c}\right]^{1/2} \frac{2a \cosh b}{R\pi \cosh(2b\phi/\pi - b)}, \tag{10.31}$$

where b is either of the two solutions of $b = (\tfrac{1}{2}c)^{1/2} a\lambda \cosh b$. The resulting shear is linear, $d(x) = x \sinh b$, and the maximum value of λ^2 is approximately $0.88/(a^2 c)$. The two solutions possess similar features to those of Jockers, since they merge as $\lambda_{\max}$ is approached. As the shear increases (by raising the value of b), λ just increases to $\lambda_{\max}$ and then it decreases without any irregular behaviour. Priest and Milne (1980) generalised these to obtain a family of force-free fields that are separable in R and ϕ. The result is that

$$B_R = -\frac{R^{-1-k}}{k}\frac{dF}{d\phi}, \qquad B_\phi = -R^{-1-k}F, \qquad B_Z = \lambda R^{-1-k}F^{1+1/k}, \tag{10.32}$$

where $F(\phi)$ satisfies

$$\frac{d^2 F}{d\phi^2} + k^2 F + \lambda^2 k(k+1)F^{1+2/k} = 0.$$

For example, when $k = -\tfrac{1}{2}$ the two solutions are $F = (1 \pm (1 - \lambda^2)^{1/2} \cos \phi)^{1/2}$, which merge at the maximum λ, namely 1, when the field lines are semi-circles (Figure 10.15). The shear takes the form $d(x) = mx$, where the shear-gradient (m) is a monotonically decreasing function of λ for the upper solution and an increasing function of λ for the lower solution.

Analytical progress has also been made by considering a pair of cylindrically symmetric force-free fields (Priest and Milne, 1980). The field components are $(0, B_\phi(R), B_Z^{(1)}(R))$ and $(0, B_\phi(R), B_Z^{(2)}(R))$, with their axes of symmetry (Z) horizontal and located

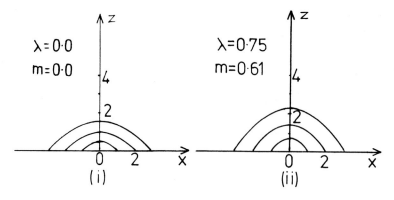
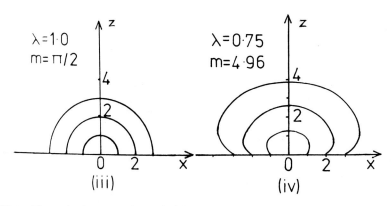

Fig. 10.15(a). The projections onto the vertical (x, z) plane of the field lines for the force-free solutions (10.32) of Priest and Milne (1980) with $k = -\frac{1}{2}$. As the shear gradient (m) increases, so λ increases to a maximum of 1 and then it decreases.

a distance h above and below the photosphere (Figure 10.16). For both of these fields the normal component at the photosphere is the same, namely

$$B_n(x) = \frac{x}{R_0} B_\phi(R_0), \tag{10.33}$$

where $R_0 = (h^2 + x^2)^{1/2}$. The photospheric shears, however, are different; they are defined here as the Z-separation of the two footpoints of a field line, and may be written

$$d^{(1)}(x) = 2\tilde{\Phi} R_0 \frac{B_z^{(1)}(R_0)}{B_\phi(R_0)} \tag{10.34}$$

and

$$d^{(2)}(x) = 2(\pi - \tilde{\Phi}) R_0 \frac{B_z^{(1)}(R_0)}{B_\phi(R_0)}, \tag{10.35}$$

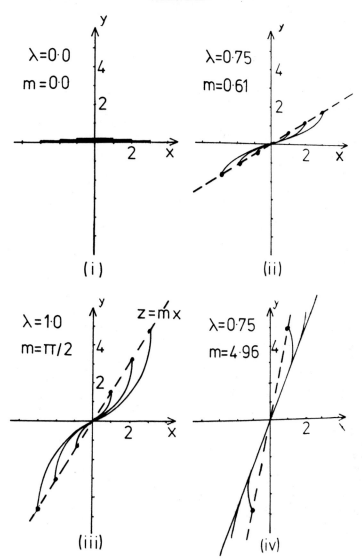

Fig. 10.15(b). The corresponding field-line projections onto the horizontal (x, y) plane. The dashed line $y = mx$ gives the location of footpoints initially lying on the x-axis.

where the inclination $\tilde{\Phi}(R_0)$ of the radius vector to a field-line footpoint is given by $\cos \tilde{\Phi}(R_0) = h/R_0$. Since the fields are force-free, both $B_z^{(1)}$ and $B_z^{(2)}$ satisfy

$$\frac{d}{dR}\left[\frac{B_\phi^2 + B_z^2}{2}\right] + \frac{B_\phi^2}{R} = 0. \tag{10.36}$$

Whole families of solutions to Problem I may now easily be constructed by just

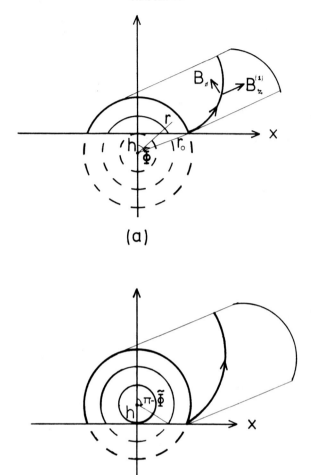

Fig. 10.16. Cylindrically symmetric magnetic arcades whose axis of symmetry (Z) is (a) a distance h below the photosphere ($z = 0$) and (b) a distance h above the photosphere. The projections of field lines onto the vertical x-z plane are arcs of circles (from Priest and Milne, 1980).

prescribing a form for $B_n(x)$, deducing $B_\phi(R)$ from Equation (10.33), and calculating $B_Z^{(1)}(R) \equiv B_Z^{(2)}(R)$ from Equation (10.36).

The above formalism may also be used to find a solution to Problem II with

$$B_Z^{(2)^2} = B_Z^{(1)^2} - c, \tag{10.37}$$

where c is a constant, so that both $B_Z^{(1)}$ and $B_Z^{(2)}$ satisfy Equation (10.36) for the same B_ϕ. The fact that the shears ($d^{(1)}$ and $d^{(2)}$) must be the same for this problem means from Equations (10.34) and (10.35) that

$$\tilde{\Phi} B_Z^{(1)} = (\pi - \tilde{\Phi}) B_Z^{(2)}. \tag{10.38}$$

Then, it is simply a matter of solving Equations (10.36) to (10.38) for $B_\phi(R)$, $B_Z^{(1)}(R)$,

$B_Z^{(2)}(R)$, where $\cos \tilde{\Phi}(R) = h/R$. The result is

$$B_Z^{(1)^2} = \frac{c(\pi - \tilde{\Phi})^2}{\pi(\pi - 2\tilde{\Phi})}, \qquad B_Z^{(2)^2} = \frac{c\tilde{\Phi}^2}{\pi(\pi - 2\tilde{\Phi})},$$

and

$$B_\phi^2 = (1 - cI(\tilde{\Phi})) \cos^2 \tilde{\Phi},$$

where

$$I(\tilde{\Phi}) = \int_0^{\tilde{\Phi}} \frac{2\tilde{\Phi}(\pi - \tilde{\Phi})}{\pi \cos^2 \tilde{\Phi}(\pi - 2\tilde{\Phi})^2} \, d\tilde{\Phi}.$$

Unfortunately, these solutions fail beyond the radius given by $I(\tilde{\Phi}) = c^{-1}$ where B_ϕ vanishes, but their importance is in demonstrating for the first time that it *is* possible to construct two solutions with the same shear (Problem II). Furthermore, as the parameter c increases then so does the shear, and, when c is large enough, the energy of the first solution can exceed that of the second.

In future, it is important to find numerical solutions to Problem II, but ultimately one would like to include pressure variations and even couple the magnetohydrostatics with an energy balance equation.

10.3.2. Eruptive instability

The magnetohydrodynamic stability of coronal arcades has been reviewed by Birn and Schindler (1981). They establish the general result that any equilibrium situated on the lower branch (Figure 10.13) is stable against all *two-dimensional* disturbances.

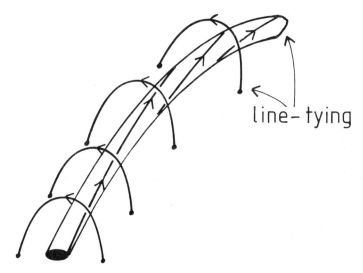

Fig. 10.17(a). A possible preflare configuration consisting of a weakly twisted flux tube anchored at its ends and located within a magnetic arcade. The plage filament is assumed to be located along the flux tube.

This suggests that the critical point (λ_{max}) may be reached by a quasi-static evolution along the lower equilibrium branch.

Hood and Priest (1980b) have incorporated the stabilising effect of photospheric line-tying in a stability analysis of some particular magnetic arcade configurations. The field structure near an active-region filament before it erupts is not yet known in detail, so two types are considered:

(I) a *simple sheared arcade*, with a topology similar to that of Figure 10.16(a) or the upper part of Figure 10.12(a), and with field lines crossing the filament;

(II) a *large flux tube contained within an arcade* (Figure 10.17(a)), with a topology in a plane transverse to the arcade similar to that of Figure 10.16(b) or the lower part of Figure 10.12(a), and with the filament located along the flux tube.

For both types, a cylindrically symmetric, force-free basic state of the form $\mathbf{B} = (0, B_\phi(R), B_z(R))$ is adopted. Then the effect of a perturbation which vanishes at the photosphere is analysed by the energy method (Sections 7.4.1 and 10.2.3). However, only particular forms for the perturbation are considered, and so, when an instability is found, it certainly exists, but it may well commence at a lower threshold. For a sheared arcade of Type (I) above, Hood and Priest were unable to find instability, even though a wide range of equilibria and perturbations were investigated. This strongly suggests that *equilibria on the lower branch are stable*.

Preflare configurations of Type (II) with an overlying external arcade that is line-tied were also tested for stability. The magnetic field is directed predominantly along the filament, and the fact that motions are frequently observed along plage filaments suggests that this may be the more realistic structure. It was modelled by a cylindrical field of length $2L$, line-tied at its ends and surrounded by a magnetic arcade; a vertical section through the structure has a similar topology to that of

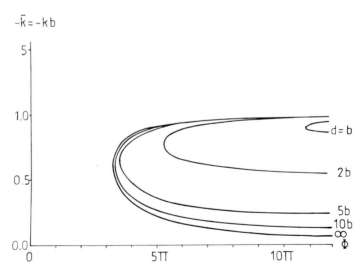

Fig. 10.17(b). Sufficient conditions for the amount of twist ($\Phi \equiv 2L/b$) required to produce instability of a flux tube embedded in an arcade. The flux tube has length $2L$, and its axis is situated at a height d above the photosphere. For Φ greater than some critical value (i.e., to the right of each curve), the equilibrium is unstable for a range of wavenumbers (k) (from Hood and Priest, 1980b).

Figure 10.16(b), and the weak curvature normal to the plane of the figure is neglected. The axis is located a distance d above the photosphere, and the form adopted for the magnetic field is simply the cylindrically symmetric field of uniform twist, namely

$$B_\theta = \frac{B_0(R/a)}{1 + R^2/a^2}, \quad B_z = \frac{B_0}{1 + R^2/a^2}, \quad \text{for } R \leq d, \tag{10.39}$$

with the distance R measured from the flux tube axis. Sufficient conditions for instability are obtained by integrating an Euler–Lagrange equation of the form (10.25). The critical twist (Φ) that is sufficient for instability depends on the axial wavenumber (k), the height (d) of the axis above the photosphere, and the constant a. The results are plotted in Figure 10.17(b). For Φ greater than a critical value (Φ_{crit}), there exists a range of values of k for which the tube is unstable. To the right of each curve the equilibrium is definitely unstable, to a perturbation of the form

$$\xi = \begin{cases} (\xi^R(R), -\dfrac{iB_z}{B}\xi^0(R), \dfrac{iB_\theta}{B}\xi^0(R))\sin\dfrac{\pi Z}{L} e^{i(m\theta + kZ)}, & R < d, \\ 0, & R > d; \end{cases}$$

thus, for example, when $d = 3b$, the configuration has certainly become unstable by the time $\Phi > 4.2\pi$. Moreover, when $\Phi \leq 2\pi$ the flux tube is stable to all kink perturbations, regardless of the value of d.

From this analysis, *simple sheared arcades of Type (I) appear so far to be stable, whereas configurations of Type (II) can become unstable if the length, twist or filament height are too great*. Two possible ways for this to happen are as follows. In the first place, if the length or twist of the flux tube is increased while its height d remains fixed, the effect of line-tying is gradually reduced until instability ultimately ensues; for example, when $d = 5b$, $\Phi_{\text{crit}} = 3.6\pi$. On the other hand, if the height increases, while the length and twist remain fixed, the stabilising effect of the overlying arcade is gradually reduced until again the structure becomes unstable; for example, with $\Phi = 5\pi$ we find $d_{\text{crit}} = 2.2b$. A means of making the height increase and so triggering the flare is for thermal nonequilibrium to occur first.

More recently, Hood (1983) have investigated the stability of coronal arcades further. Since the linear problem is therefore now well-understood, it is crucial in future to try and follow the nonlinear development of the instability.

It is highly likely that the same magnetic instability is responsible for the eruption of both *plage* filaments (to give two-ribbon flares) and *quiescent* filaments. Therefore much can probably be learnt about the cause of two-ribbon flares by observing eruptions of quiescent prominences, where the magnetic structure is much simpler and on a larger scale than for plage filaments. Some more comments on mechanisms for quiescent filament eruption can be found in Sections 11.2.2 and 11.3.1.

10.3.3. THE MAIN PHASE: 'POST'-FLARE LOOPS

After the eruption of a magnetic arcade containing a filament, the main phase of the flare is characterised by 'post'-flare loops which rise upward, at first rapidly (10 to 50 km s^{-1}) and later much more slowly (0 to 1 km s^{-1}) for a day or longer (Section

1.4.4C). Examples can be seen in Figures 1.38 and 10.18. Cool Hα loops (with strong down-flows) are located below hot X-ray loops, which may reach an altitude of 100 000 km. The loop feet are rooted in two ribbons of Hα emission, which separate as the loops rise. The summit temperature and density early in the event may be as much as 10^7 K $-$ 10^8 K and 10^{16} m^{-3}. Essentially the same phenomenon (but less energetic) occurs when a quiescent filament erupts without producing an Hα flare.

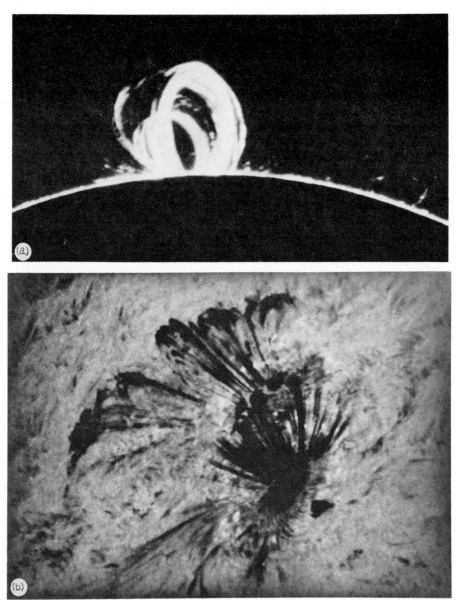

Fig. 10.18. Examples of cool 'post'-flare loops: (a) limb photograph in Hα; (b) disc photograph at Hα + 0.8 Å on 10 September. 1974 (© AURA Inc., Sacramento Peak Observatory).

Pneuman (1981) has summarised the model of Kopp and Pneuman (1976) for these flare loops in terms of magnetic reconnection as the configuration closes back down after being blown open by the initial eruption. He stresses that it is not one rising loop but a series of loops that become visible in turn as the reconnection proceeds and energy is continuously released. Also, he emphasises that the term 'post'-flare is misleading, since it implies that the loops occur after the flare and are a by-product of it, whereas in fact they begin near the onset of the event and may well produce the chromospheric ribbons.

Kopp and Pneuman derived the equations for isothermal flow of a perfectly conducting plasma along moving magnetic field lines. If v_s and v_n are the plasma speeds along and normal to the field, the equations of motion and continuity in a frame moving with a curved field line are

$$\frac{Dv_s}{Dt} + v_s \frac{Dv_s}{Ds} = -\frac{1}{\rho}\frac{Dp}{Ds} - \frac{GM_\odot}{r^2}\cos\alpha + v_n \frac{D\alpha}{Dt} + \frac{v_n^2}{r}\cos\alpha + \frac{v_n v_s}{R_c}, \quad (10.40)$$

$$\frac{D}{Dt}(\rho A) + \frac{D}{Ds}(\rho v_s A) - \frac{\rho v_n A}{R_c} = 0, \quad (10.41)$$

where D/Ds and D/Dt are derivatives in the moving frame, $GM_\odot/r^2$ is the gravitational force at a distance r from the solar centre and at an inclination α to the magnetic field, R_c is the radius of curvature of the field line, A is the cross-sectional area of the flux tube, and the pressure is given by the perfect gas law. These equations are solved (by the method of characteristics) for the density (ρ) and flow speed (v_s) in a prescribed evolving magnetic configuration (for which v_n, α, R_c and A are prescribed functions of space and time).

The adopted magnetic field is a dipole superimposed on a uniform radial field in such a way that the neutral point is located at a radial distance ($r_1(t)$) which increases with time (Figure 10.19(a)). In other words,

$$B_r = B_0 \frac{1 + 2(r_1/r)^3}{1 + 2(r_1/R_\odot)^3} \sin\theta, \quad B_\theta = -B_0 \frac{1 - (r_1/r)^3}{1 + 2(r_1/R_\odot)^3} \cos\theta,$$

for $R_\odot < r < r_1$, and

$$B_r = B_0 \frac{3(r_1/r)^2}{1 + 2(r_1/R_\odot)^3} \sin\theta, \quad B_\theta = 0$$

for $r \geq r_1$. The neutral point location is assumed to rise with time like

$$r_1(t) = (1.5 - 0.5\,e^{-\omega t})R_\odot,$$

where $\omega = 5.7 \times 10^{-5}\,\text{s}^{-1}$, so that initially $r_1 = R_\odot$ and $dr_1/dt = 20\,\text{km s}^{-1}$, and ultimately $(t \to \infty) r_1 \to 1.5 R_\odot$. The initial flow and pressure are taken from an isothermal solar wind solution ($T = 1.5 \times 10^6$ K), and the base density remains constant ($10^{16}\,\text{m}^{-3}$). The resulting flow along a single flux tube is found to be much higher than the solar wind flow, with speeds in excess of 100 km s^{-1} in the low corona (Figure 10.19(b)). When the tube closes, a shock propagates downwards and leaves behind it a stationary state (Figure 10.20). Subsequently, the plasma cools and falls to give the Hα loops.

CHAPTER 10

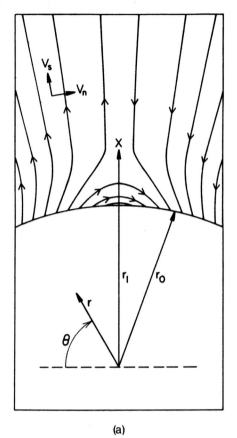

(a)

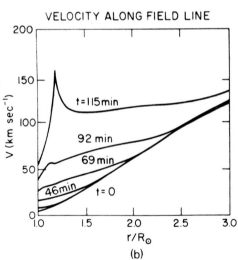

(b)

Fig. 10.19. The Kopp–Pneuman model of 'post'-flare loops: (a) field configuration and notation; (b) velocity versus radial distance along a moving flux tube that is swinging over from a vertical location towards the neutral point, which it reaches at $t = 115$ min at a height of $0.2\ R_\odot$ above the solar surface (from Pneuman, 1980).

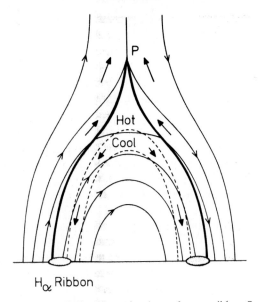

Fig. 10.20. The magnetic structure during the main phase of a two-ribbon flare. Plasma motions are indicated by dark-headed arrows, and the thick curve shows the position of the slow shock wave which trails from the rising neutral point P and brings the plasma to rest to form hot loops. When the plasma cools it falls and produces the cool loops outlined by dashed curves.

The shock is modelled for simplicity by Kopp and Pneuman as a *hydrodynamic shock*, but the resulting temperature rise (by a factor of 1.8) is not sufficient to explain the observed temperatures, especially early in the event. But Cargill and Priest (1982a) have shown that the extra release of energy from the magnetic field in a *slow magnetoacoustic shock* can provide the necessary heating. They find that the flare temperature depends on the value of the ambient plasma beta (β); for $\beta = 0.1$ the plasma is heated by the shock to typically 10^7 K, but for $\beta = 0.01$ the temperature can become even larger, as observed on the Solar Maximum Mission Satellite.

The mass flow is enhanced over the solar wind value by a factor of ten due to the field line motion. But the mass contained in a loop prominence system may be even larger than this flow would provide, and so the excess is probably supplied by chromospheric evaporation.

Since the above analysis is purely kinematic, it will be interesting in future to see the results of a completely magnetohydrodynamic simulation, including nonisothermality, shock waves and evaporation.

CHAPTER 11

PROMINENCES

The observations of the cool, dense sheets of plasma known as prominences have been summarised in Section 1.4.3 and in the books by Tandberg–Hanssen (1974) and Jensen et al. (1979). Here the main aim is to dicuss the *magnetic* properties of *quiescent* prominences. The important aspects of radiative transfer and energy balance are outside the scope of this book, and there is not room to discuss the many types of active prominence. An exception is the *loop prominence*, which occurs after a prominence eruption (Section 10.3.2) and has been described in Section 10.3.3. The eruption of an *active-region* (or plage) *prominence* is generally violent: it may show up as a spray, and usually it gives rise to a two-ribbon flare. Quiescent prominences are much larger and have weaker fields; they erupt more gently and do not usually produce an Hα flare, but the basic instability may well be the same.

This chapter treats the *formation* of quiescent prominences by thermal instability (Section 11.1) and their overall *magnetohydrostatics* (Sections 11.2 and 11.3), as well as the theory of *coronal transients*, which accompany prominence eruptions. As a background, the reader is referred to Sections 7.5.7 and 6.4.4C for the basic theory of radiatively-driven thermal instability, and to Chapter 3 for a general discussion of magnetohydrostatics. A distinction will be made here between the two possible locations for prominences, namely *magnetic arcades* (Sections 11.2) and *current sheets* (Section 11.3.1). Also, it is only the overall structure of a prominence that will be considered, since the detailed fine structure and associated downflow does not yet seem to be understood*: it may possibly be caused by the hydromagnetic Rayleigh–Taylor instability (Sections 7.3.1 and 7.5.2) or by a filamenting thermal instability (Sections 6.4.4C and 7.5.7).

11.1. Formation

Consider hot coronal plasma, with temperature T_0 and density ρ_0, and in thermal equilibrium under a balance between heating (h) and radiation ($\tilde{\chi}\rho_0$) per unit mass, so that

$$0 = h - \tilde{\chi}\rho_0. \tag{11.1}$$

Perturbations from this equilibrium at constant pressure (p_0) are given by

$$c_p \frac{\partial T}{\partial t} = h - \tilde{\chi}\rho + \frac{\kappa_\parallel}{\rho}\frac{\partial^2 T}{\partial s^2}, \tag{11.2}$$

where the last term represents the effect of heat conduction at a distance s along the magnetic field, and

$$\rho = \frac{m p_0}{k_B T}. \tag{11.3}$$

*See Low (1982) *Solar Phys.* **75**, 119 for a static model.

After putting $T = T_0 + T_1$ and $\rho = \rho_0 + \rho_1$, linearising, and eliminating ρ_1, Equations (11.2) and (11.3) give an equation for the temperature perturbation (T_1), namely,

$$c_p \frac{\partial T_1}{\partial t} = \frac{\tilde{\chi} \rho_0}{T_0} T_1 + \frac{\kappa_\parallel}{\rho_0} \frac{\partial^2 T_1}{\partial s^2}. \tag{11.4}$$

Now, suppose the plasma is contained in a magnetic structure of length L, and assume the perturbation vanishes at the ends of the structure, so that T_1 can be taken proportional to $\exp(\omega t + 2\pi i s/L)$ and Equation (11.4) reduces to

$$\omega = \frac{\tilde{\chi} \rho_0}{c_p T_0} - \frac{\kappa_\parallel 4\pi^2}{\rho_0 L^2}. \tag{11.5}$$

Thus, if conduction is absent, ω is positive and the plasma is thermally unstable. But the presence of conduction stabilises the plasma provided

$$L < L_m = 2\pi \left[\frac{c_p \kappa_\parallel T_0}{\tilde{\chi} \rho_0^2} \right]^{1/2}. \tag{11.6}$$

When L_m is exceeded, the plasma becomes *thermally unstable* and cools down until a new equilibrium is reached and a prominence-like condensation has formed. The above order-of-magnitude estimate for L_m would be modified by incorporating a temperature dependence in the radiative loss (Section 7.5.7). But, more importantly, the initial equilibrium is in general only uniform if the conduction completely dominates the energy balance. Near the point of instability, radiation is of roughly the same importance as conduction, and so the equilibrium is non-uniform. It then transpires that, when L slowly increases, a point is eventually reached at which *thermal nonequilibrium* sets in rather than instability (Section 10.2.2). A neighbouring equilibrium no longer exists, and the plasma cools down towards a new equilibrium. The rate of cooling may be much higher than the linear instability rate, which overcomes a difficulty with some of the early prominence-formation models, namely of obtaining a fully formed prominence within the day or so that observations require. The object of the sections below is to describe the onset of prominence formation in three different magnetic configurations, namely a loop, an arcade, and a current sheet. An account of the earlier models for prominence formation, mainly in a uniform medium, can be found in Tandberg-Hanssen (1974). A particularly interesting idea has been suggested by Pneuman (1972b), who models a large-scale coronal arcade (a helmet), whose base is at the temperature maximum. Each field line is assumed isothermal and has a global energy balance between mechanical heating and radiation, since there is no net conductive loss. The resulting density-temperature distribution throughout the arcade possesses a small region of enhanced density and a diminished temperature (a prominence) near its base, surrounded by a low-density cavity: these are natural consequences of the effect of the magnetic structure on the energy balance.

11.1.1. Formation in a Loop (Active-Region Prominences)

As described in Section 6.5.1, Hood and Priest (1979a) have solved the energy balance equation

$$\frac{d}{ds}\left(\kappa_0 T^{5/2} \frac{dT}{ds} \right) = \rho^2 \tilde{\chi} T^\alpha - h\rho, \tag{11.7}$$

with

$$\rho = \frac{mp}{k_B T}, \tag{11.8}$$

along a loop of uniform cross-sectional area and uniform pressure (p). They find that, if the pressure (p) or length ($2L$) of a hot loop is too great, or if the heating (h) is too small, then a state of *thermal nonequilibrium* ensues. There is no neighbouring equilibrium (Figures 6.10 and 6.11), so the loop cools down towards a new equilibrium at prominence temperatures. This mechanism of prominence formation works with a wide range of forms for the heating, and it is also present if gravity is incorporated in the basic state so that the loop pressure is not uniform (Section 6.5.1C).

The above analysis for obtaining the temperature-density structure along a single field line may be applied to a whole magnetic structure in which the pressure varies from one field line to another. For example, a coronal arcade is treated in Section 11.1.2, and a twisted coronal loop may be modelled as a structure with cylindrical symmetry, as follows. Suppose it is in equilibrium between a pressure gradient and a Lorentz force, such that

$$\frac{dp}{dR} = -\frac{d}{dR}\left(\frac{B_\phi^2 + B_z^2}{2\mu}\right) - \frac{B_\phi^2}{\mu R}, \tag{11.9}$$

where

$$\Phi(R) = \frac{2LB_\phi}{RB_z} \tag{11.10}$$

is the *twist* (see Section 3.3). As the loop is twisted up by footpoint motions, one would expect the field to bow out as well, but this is being neglected here, partly for simplicity and partly because coronal loops do possess rather uniform cross-sections and significant radial pressure variations (Section 1.3.4B).

For a given axial field (B_z) and twist (Φ), Equations (11.9) and (11.10) may be solved to give $p(R)$, and then Equations (11.7) and (11.8) determine $T(s)$ and $N(s)$ along the field lines at each radius. This procedure has been carried out for both the uniform-twist and variable-twist fields of Section 3.3.4, in order to determine the effect of twist on a loop's thermal structure. As the twist increases, so the axial pressure rises and it provides an outwards pressure gradient to balance the extra inwards tension force. When heating exceeds radiation, this in turn produces a hot sheath, but when radiation exceeds heating the radiative losses are enhanced; they therefore lower the temperature, ultimately to the point where thermal nonequilibrium sets in on the loop axis. If the twisting continues, the core of the loop, where the plasma has cooled down to prominence temperatures, will broaden, as indicated in Figure 11.1.

The above analysis may be relevant to the formation of an *active-region prominence* along a flux tube that has been stretched so much that thermal conduction is no longer able to stabilise the condensation process. Such a magnetic configuration is consistent with the fact that the ends of prominences are sometimes located in regions of opposite polarity, and it is also more likely to be able to account for prominence eruption (Section 10.3.2). The analysis may apply to quiescent prominences if they consist of a collection of twisted loops, but such prominences are more likely to be located in a simple coronal arcade, as modelled below.

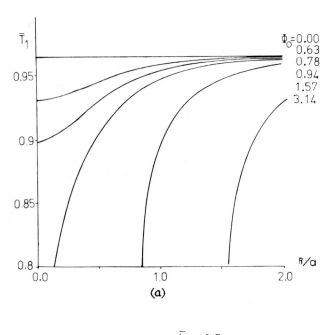

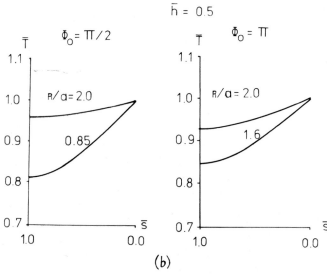

Fig. 11.1. Prominence formation in a twisted loop, showing the effect of increasing the twist (Φ_0) on the temperature structure. (a) The temperature ($\bar{T} = T/T_0$) as a function of radial distance (R). (b) $\bar{T}$ as a function of distance ($\bar{s} = s/L$) along the loop from the summit ($\bar{s} = 1$) to the footpoint ($\bar{s} = 0$). As Φ_0 increases, so the temperature falls and creates a cool core, which broadens to a radius $0.75a$ for $\Phi_0 = \frac{1}{2}\pi$ and $1.5a$ for $\Phi_0 = \pi$. In this particular loop $B_0 = 10$ G, $L = 5 \times 10^7$ m, $a = 10^7$ m, $T_0 = 10^6$ K, $n_0 = 5 \times 10^{14}$ m^{-3}, and the dimensionless heating is 0.5 (from Hood and Priest, 1979a).

11.1.2. Formation in a Coronal Arcade

The temperature and density have been obtained for coronal plasma in thermal and hydrostatic equilibrium and located in a force-free magnetic arcade (Priest and Smith, 1979). It is found that, when the coronal pressure becomes too great, the equilibrium ceases to exist and the plasma cools to form a quiescent prominence, as described below. The same process can be initiated at low heating rates when the width or shear of the arcade exceeds a critical value. Priest and Smith therefore suggest that a prominence should be modelled as a dynamic structure, with plasma continually draining downwards; new material is sucked up along field lines of the ambient arcade and into the region of nonequilibrium, where it cools to form new prominence material (Figure 11.2). This may explain transition region upflow of 6–10 km s^{-1} either side of a filament (Lites et al., 1976). However, the effect of such a circulation has not been analysed in detail except that Ribes and Unno (1980) have modelled siphon flow in an arcade.

Suppose the coronal arcade is in equilibrium under the force balance represented by Equation (3.1), namely,

$$\mathbf{0} = -\nabla p + \mathbf{j} \times \mathbf{B} - \rho g \hat{\mathbf{z}}. \tag{11.11}$$

If the magnetic force is dominant, the components of Equation (11.11) normal to and along the field may be approximated by

$$\mathbf{j} \times \mathbf{B} = \mathbf{0} \tag{11.12}$$

and

$$\frac{dp}{dz} = -\rho g, \tag{11.13}$$

respectively, where $p = (k_B/m)\rho T$. Thus, one may consider any solution to Equation

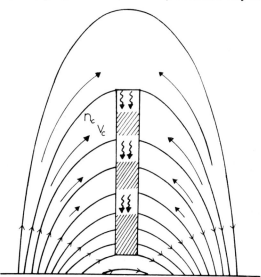

Fig. 11.2. A dynamic model for a prominence in which coronal plasma enters the prominence from the sides and then slowly dribbles through the magnetic field (from Priest and Smith, 1979).

(11.12) for the field, and then solve Equation (11.13) together with an energy equation such as

$$\frac{d}{ds}\left(\kappa_{\|}\frac{dT}{ds}\right) - \frac{\kappa_{\|}}{B}\frac{dB}{ds}\frac{dT}{ds} = \rho^2 \tilde{\chi} T^{\alpha} - h\rho \qquad (11.14)$$

for ρ and T along each field line, where $\kappa_{\|} = \kappa_0 T^{5/2}$.

In particular, a magnetic arcade of width L may be modelled by the linear field (3.44):

$$B_x = -\frac{L}{\pi a} B_0 \cos\frac{\pi x}{L} e^{-z/a},$$

$$B_y = \left(1 - \frac{L^2}{\pi^2 a^2}\right) B_0 \cos\frac{\pi x}{L} e^{-z/a}, \qquad (11.15)$$

$$B_z = B_0 \sin\frac{\pi x}{L} e^{-z/a},$$

for which the field lines are inclined at an angle $\gamma = \sec^{-1}(\pi a/L)$ to the horizontal y-axis (Figure 3.8). The boundary conditions that are adopted for solving Equations (11.13) and (11.14) are

$$\left.\begin{array}{c} n(\equiv \rho/(2m)) = n_0 = 5 \times 10^{14} \text{ m}^{-3} \\ T = T_0 = 10^6 \text{ K} \end{array}\right\} \text{ at the base } (z = 0)$$

$$\frac{dT}{ds} = 0 \quad \text{at the summit } (z = H).$$

The summit height $(z = H)$ of a field line with its feet a distance x_0 from the y-axis is given by $H = -a \log_e \cos(\pi x_0/L)$.

The resulting model coronal arcades depend on the five parameters $\rho_0, T_0, h, L, \gamma$. If the base heating exceeds the radiation, the temperature increases with height; otherwise, it falls initially, and it eventually starts to rise when the radiation has become smaller than the heating. The effect of increasing T_0 is to make conduction more important, and so to create a more isothermal plasma. Raising the base density (ρ_0) makes the radiation more important and so has the opposite effect. Increasing the heating tends to make the plasma hotter and less dense, with the central regions of the arcade being somewhat denser and cooler than those away from the axis. Widening the arcade enhances the relative density- and temperature-variations; at high altitudes the temperature rises and the density falls.

It is found that, if ρ_0 exceeds a critical value (corresponding to typically 10^{15} m^{-3}), a hot equilibrium is no longer possible and the plasma cools down to form a prominence, sucking new material up along the field lines in the process. The critical density decreases as either the arcade width (L) or shear (γ) are increased. At high heating rates this is the only way to create thermal nonequilibrium, but at low heating (smaller than radiation at the base) it occurs when the width (L) or shear (γ) becomes too large. The reason why shearing the field leads to a lack of equilibrium is that it causes the field lines to rise, which lengthens them and reduces the stabilising effect of conduction. As an example, Figure 11.3 shows the plasma structure just below the critical shear for nonequilibrium. A slightly cool, dense region has formed

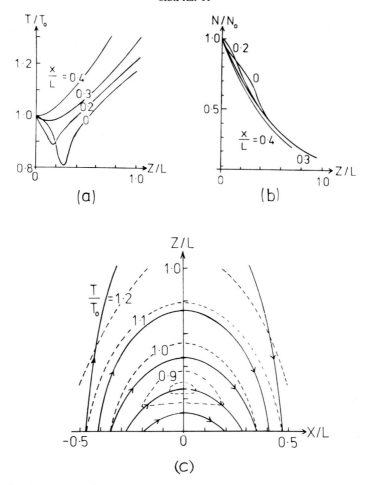

Fig. 11.3. The development of a cool central region just prior to prominence formation in an arcade with a width (L) of 100 000 km and a base heating equal to half the radiation: (a) temperature profiles, (b) density profiles, (c) isotherms (dashed) and field lines (solid) (from Priest and Smith, 1979).

at a height of about 20 000 km. If the shear is increased any further, it is not possible to find a hot equilibrium solution along field lines that thread the central part of the core. As one continues to increase the shear, so the range of altitudes over which equilibrium is impossible increases.

Consider the field lines in the arcade. As one goes to higher locations in the structure so the length of the field line increases, which tends to decrease the size of the conduction term in Equation (11.14). But, at the same time, the density declines, which tends to lower the value of the radiation term too. It is this competition between increasing a field line's length and decreasing its density which determines whether or not thermal nonequilibrium sets in at some height. If conduction exceeds radiation everywhere, the whole arcade will be filled with hot stable plasma in excess of 10^6 K. If radiation dominates for some range of altitude, the plasma will cool down to form a prominence.

Below the prominence, the field lines are short enough to prevent nonequilibrium, while above the prominence the *density* is low enough to do so. These features may be demonstrated in an order-of-magnitude way as follows.

By neglecting the heating term and approximating the conduction term, Equation (11.14) may be written as

$$\kappa_0 T_1^{5/2} \frac{T_0 - T_1}{H^2} = - \tilde{\chi} \rho_1^2 T_1^\alpha \tag{11.16}$$

for the density (ρ_1) and temperature (T_1) at the summit ($z = H$) of a field line. This determines T_1 as a function of H, with the density given roughly by Equation (11.13) as

$$\rho_1 = \frac{\rho_0 T_0}{T_1} \exp - \frac{H T_0}{\Lambda_0 T_1} \tag{11.17}$$

in terms of the scale-height $\Lambda_0 = k_B T_0/(mg)$. Equation (11.16) therefore has the form $f(T_1) = g(\xi)$, where $f(T_1) = T_1^{(5/2) - \alpha}(T_0 - T_1)$, $g(\xi) \sim \rho_0^2 \xi^2 \, e^{-2T_0\xi/\Lambda_0}$, $\xi = H/T_1$.

Three possible types of solution are sketched in Figure 11.4, where the graphs of f and g are plotted on the left and the resulting solution $T_1(H)$ on the right. For a given H the two possible solutions for T_1 are those which make $f = g$. When ρ_0 is so large that $n_0 > 1.8 \times 10^{15}$ m^{-3}, the maximum of g exceeds the maximum of f, and a region of H develops in which there is no hot equilibrium solution. It first occurs at a height of $H_I = 45\,000$ km.

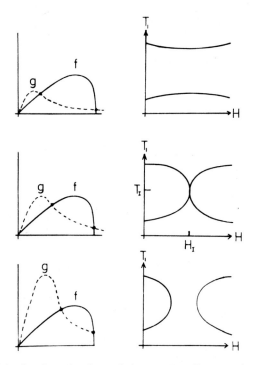

Fig. 11.4. Sketches of the functions f and g and the corresponding summit temperature (T_1) in an arcade as a function of altitude (H) (from Priest and Smith, 1979).

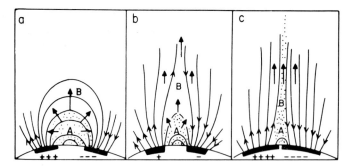

Fig. 11.5. Prominence formation in a current sheet: the magnetic field overlying an active region during: (a) the early phase of activity; (b) the main phase of activity; (c) the post-active stage (from Kuperus and Tandberg-Hanssen, 1967).

The inflow of plasma to the prominence along field lines is similar to the siphon flow of Pikel'ner (1971). But the driving mechanism is quite different here. Pikel'ner argued that a flow is initiated because heating is reduced in a region where field lines dip down. In the present case, a flow is driven during the condensation process when the prominence is forming due to the breakdown of coronal equilibrium. Then, once the prominence has formed, the downflow within the prominence sucks new material in from the sides, with typical upflow speeds of 5 km s^{-1}. Also, the feet of the field lines may be moving due to supergranules. A converging flow would make the field lines of Figure 11.2 fall slowly, while a diverging flow would make them rise.

11.1.3. Formation in a current sheet

Active-region (or plage) filaments are often located in the *middle* of an active region and snake their way between sunspots. In this case the magnetic configuration is likely to be either an extended flux tube (Section 11.1.1) or an arcade (Section 11.1.2). However, they are sometimes situated around the *edge* of an active region or at the boundary between the opposing fields of two active regions (e.g., Martin, 1973). In such cases the formation site may well be a large-scale current sheet.

Quiescent prominences may also form in large current sheets. They sometimes appear at the boundary between two weak magnetic regions of opposite polarity which are moving together (Martin, 1973). This occurs commonly when old remnant active regions push up against magnetic field of the polar region and form the polar crown. Quiescent prominences are often located also at the base of coronal streamers, which indicate low-lying closed field lines surmounted by an open structure (e.g., Figure 1.4). Kuperus and Tandberg-Hanssen (1967) have proposed a model for prominence formation in which the closed magnetic field of an active region is first blown open by flare activity; the prominence then condenses in the resulting current sheet, and, during the process, some field lines close over the top of the prominence to give the characteristic streamer configuration.

The theory for formation in a sheet has several advantages. As the plasma condenses and drags the magnetic field with it, the build-up of magnetic pressure (which would inhibit condensation) is relieved by the onset of a tearing-mode instability (Section 7.5.5). It destroys the field and creates magnetic loops, which may insulate the plasma. Also, the reconnection produces a closed field at the base of the sheet which may help to support the condensing plasma.

11.1.3A. Thermal Nonequilibrium

Equation (11.6) shows that, for a temperature and density characteristic of the lower corona, a neutral sheet becomes thermally unstable when its length L exceeds about 100 000 km. A more accurate estimate may be obtained by considering the sheet's energy balance as follows (Smith and Priest, 1977). Suppose equilibrium conditions inside the sheet (both magnetic and thermal) are characterised by a plasma pressure (p_{20}), density (ρ_{20}), temperature (T_{20}) and a vanishing magnetic field, while the corresponding external values are p_1, ρ_1, T_1, B_1 (Figure 11.6). For simplicity, the details of the sheet structure and the effect of gravity are neglected (see Weber (1979) for a numerical solution). Horizontal force-balance and thermal equilibrium give

$$p_{20} = p_1 + \frac{B^2}{2\mu}, \tag{11.18}$$

$$\frac{d}{dy}\left(\kappa_0 T^{5/2} \frac{dT}{dy}\right) - \rho^2 \tilde{\chi} T^\alpha + h\rho = 0, \tag{11.19}$$

where

$$p_{20} = \frac{k_B}{m} \rho_{20} T_{20}, \tag{11.20}$$

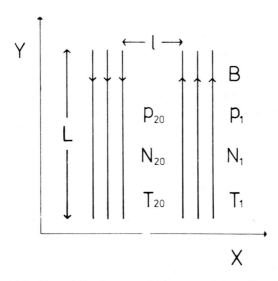

Fig. 11.6. The notation for an equilibrium neutral sheet of length L.

and, if heating balances radiation outside the sheet, $h = \rho_1 \tilde{\chi}_1 T_1^{\alpha_1}$. The conduction term in Equation (11.19) may be approximated by $\kappa_0 T_{20}^{5/2}(T_1 - T_{20})/L^2$, so that

$$\kappa_0 T_{20}^{5/2} \frac{T_1 - T_{20}}{\rho_{20} L^2} - \rho_{20} \tilde{\chi} T_{20}^{\alpha} + \rho_1 \tilde{\chi}_1 T_1^{\alpha_1} = 0. \qquad (11.21)$$

The three Equations (11.18), (11.20), (11.21) determine p_{20}, ρ_{20} and T_{20} in terms of L and B. The results for coronal conditions ($T_1 = 10^6$ K, number density $= 10^{14}$ m^{-3}) are plotted in Figure 11.7(a). It shows that, as the sheet length increases, so one moves along an equilibrium curve from the bottom right-hand corner, and the sheet temperature decreases slightly from the coronal value of 10^6 K. If the length exceeds a maximum value (L_{max}), a hot equilibrium no longer exists and the plasma cools down (along the dotted line for $B = 1$ G) to a new equilibrium at prominence temperatures (T_{prom}). It can be seen that, for example, a field strength of 1 G gives a maximum length of 50 000 km, approximately the height of a quiescent prominence.

The way the current sheet cools down after the onset of thermal nonequilibrium may be modelled very roughly as follows. Since the cooling occurs much more slowly than the propagation of magnetohydrodynamic waves, the sheet will remain in total pressure equilibrium with its surroundings. Its pressure ($p_2(t)$), density ($\rho_2(t)$) and temperature ($T_2(t)$) are therefore given by

$$p_2 = p_1 + \frac{B^2}{2\mu}, \qquad p_2 = \frac{k_B}{m}\rho_2 T_2,$$

together with a time-dependent energy equation

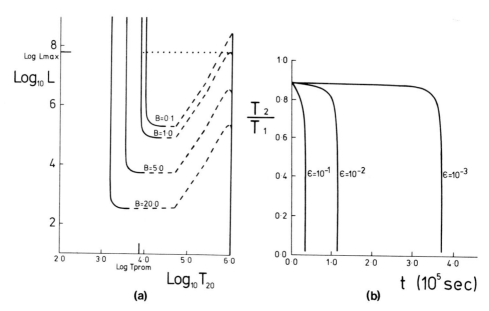

Fig. 11.7. Prominence formation in a neutral sheet. (a) The length L (in m) of an equilibrium sheet as a function of its temperature (T_{20} (K)) for several values of the external magnetic field (B(in G)). (b) The time-development of the sheet temperature (T_2) during prominence formation in a sheet of length $L_{max}(1 + \varepsilon)$ and magnetic field 0.8 G (from Smith and Priest, 1977).

$$c_p \frac{\partial T_2}{\partial t} = \rho_1 \tilde{\chi}_1 T_1^\alpha - \rho_2 \tilde{\chi} T_2^\alpha + \kappa_0 T_2^{5/2} \frac{T_1 - T_2}{\rho_2 L^2}, \tag{11.22}$$

where an order-of-magnitude approximation has been used in the conduction term for simplicity. These equations may be used to investigate the thermal stability of the equilibrium curves in Figure 11.7(a), with the result that the solid parts of the curves represent stable equilibria and the dashed parts are unstable. Also, just at the maximum length, the plasma is neutrally stable on a linear analysis, but is unstable to cooling according to a second-order treatment.

If the sheet is assumed to be gradually increasing in length, the actual development of the cooling depends on the length ($L_{\max}(1 + \varepsilon)$) which the sheet has attained before it gets under way. The time-development may therefore be estimated by integrating Equation (11.22) while keeping L constant at $L_{\max}(1 + \varepsilon)$. The result is shown in Figure 11.7(b). The time (τ) for the temperature to fall to prominence values depends on the size of ε. It is the same as the observed time of typically 10^5 s (1 day) when ε is chosen to be about 10^{-2}. For small values of ε it can be seen that the temperature decreases slowly at first and then suddenly plummets when t is close to τ. Clearly, the above analysis represents only a rough attempt at modelling the condensation, and in future there is a need to solve the full partial differential equations for the process. Somov and Syrovatsky (1980, 1982) have considered more formally the question of filament formation in a current sheet for arbitrary wavelengths along the sheet, while Chiuderi and Van Hoven (1979) have studied a related problem, namely the dynamic effect of the magnetic field on filament formation in a one-dimensional force-free field.

11.1.3B. Line-Tying

During the condensation of plasma in a vertical current sheet, Lorentz forces will tend to oppose the transverse motions because the magnetic field lines are anchored in the dense photosphere (Figure 11.8(a)). Raadu and Kuperus (1973) set up a simple model to take account of such line-tying. They assume horizontal motions of a perfectly conducting plasma, with the variables depending on θ and t alone. Also, the condensation is assumed to proceed so slowly that the magnetic field passes through a series of configurations with horizontal force-balance

$$-\frac{\cos \theta}{r} \frac{\partial p}{\partial \theta} + j_z B_y = 0, \tag{11.23}$$

where

$$j_z = -\frac{1}{\mu r \cos \theta} \frac{\partial B_y}{\partial \theta},$$

by Ampere's Law. Because the plasma is frozen to the field, they suppose its density follows from $\rho/\rho_0 = B_y/B_0$, where a zero subscript denotes initial equilibrium values (although vertical force-balance is possibly more appropriate). If the entropy, radiation ($\rho^2 \tilde{Q}(T)$) and heating ($h\rho$) are the only important terms in the energy balance, it may be written

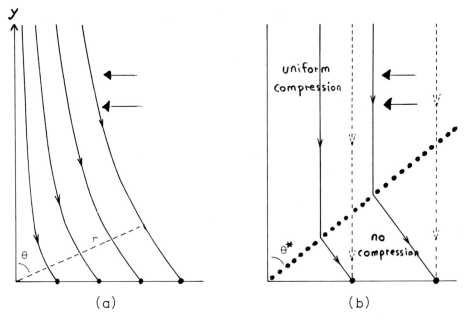

Fig. 11.8. The effect of line-tying on prominence formation in a neutral sheet, one half of which is shown. (a) The notation and the form of the field lines for a diffuse condensation. (b) Condensation in a wedge with an initially vertical field (dashed) (after Raadu and Kuperus, 1973).

$$\frac{Dp}{Dt} - \frac{\gamma p}{\rho}\frac{D\rho}{Dt} = (\gamma - 1)\rho[h - \tilde{Q}(T)\rho], \qquad (11.24)$$

where the equilibrium conditions ($\rho = \rho_0$, $T = T_0$) imply $h = \tilde{Q}(T_0)\rho_0$.

The initial growth of the instability is studied by writing $B_y = B_0(1 + \varepsilon(\theta)e^{\sigma t})$, $p = p_0 + p_1(\theta)e^{\sigma t}$, so that Equations (11.23) and (11.24) give two relations for ε and p_1. After eliminating p_1, they become

$$\left[\gamma p_0(\sigma - \sigma_p)\cos^2\theta + \frac{B_0^2}{\mu}(\sigma - \sigma_\rho)\right]\frac{d\varepsilon}{d\theta} = 0, \qquad (11.25)$$

where

$$\sigma_p = \frac{\rho}{c_p}\left(\frac{\tilde{Q}}{T} - \frac{d\tilde{Q}}{dT}\right)$$

and

$$\sigma_\rho = -\frac{\rho}{c_v}\frac{d\tilde{Q}}{dT}$$

are the values of the growth-rate for motions at constant pressure and constant density, respectively.

Equation (11.25) implies that ε is uniform, apart from a possible discontinuity at $\theta = \theta^*$, say, where the expression in square brackets vanishes; in other words, where

$$\sigma = \frac{\sigma_\rho + (\gamma p_0 \mu/B_0^2)\sigma_p \cos^2\theta^*}{1 + (\gamma p_0 \mu/B_0^2)\cos^2\theta^*}. \qquad (11.26)$$

Since no compression is required at the photosphere ($\theta = \frac{1}{2}\pi$), the required solution is

$$\varepsilon = \begin{cases} \varepsilon_0, & \theta < \theta^*, \\ 0, & \theta > \theta^*, \end{cases}$$

as indicated in Figure 11.8(b). θ^* is constant, so that as time progresses the field lines near the photosphere swing over more and more, while the plasma condenses in the form of a wedge. The growth-rate σ (from Equation (11.26)) depends on the value of the wedge-angle, and it can be shown that the maximum growth-rate is positive and occurs when $\theta^* = 0$, provided

$$-\frac{3}{2} < \frac{T}{\tilde{Q}} \frac{d\tilde{Q}}{dT} < 0.$$

In other words, if $\tilde{Q}$ decreases with T but not too rapidly, the effect of line-tying is to favour the formation of *thin* wedges.

Features that one would like to see included in the above analysis are thermal conduction, gravity and vertical motions, but that would be a difficult undertaking.

11.2. Magnetohydrostatics of Support in a Simple Arcade

11.2.1. Kippenhahn–Schlüter Model

Kippenhahn and Schlüter (1957) modelled a prominence as a thin isothermal sheet, which is one-dimensional in the sense that all the variables depend on one coordinate (x) alone. The field lines are bowed down by the dense prominence plasma, as shown in Figure 11.9, and they play two roles. The magnetic tension provides an upward force to balance gravity and support the prominence (as in Figure 2.6), whereas the magnetic pressure increases with distance from the z-axis and so it provides a transverse force to compress the plasma and balance the plasma pressure gradient.

The prominence equilibrium is governed by the force balance (Section 3.1)

$$0 = -\nabla p - \rho g \hat{\mathbf{z}} - \nabla(B^2/(2\mu)) + (\mathbf{B}\cdot\nabla)\mathbf{B}/\mu, \tag{11.27}$$

together with

$$\rho = \frac{mp}{k_B T}, \tag{11.28}$$

where T and the horizontal field components (B_x, B_y) are all assumed to be uniform. The pressure ($p(x)$), density ($\rho(x)$) and vertical field ($B_z(x)$) are assumed functions of x alone, so that div $\mathbf{B} = 0$ is satisfied identically, while the x- and z-components of Equation (11.27) reduce to

$$0 = -\frac{d}{dx}\left(p + \frac{B^2}{2\mu}\right), \tag{11.29}$$

$$0 = -\rho g + \frac{B_x}{\mu}\frac{dB_z}{dx}. \tag{11.30}$$

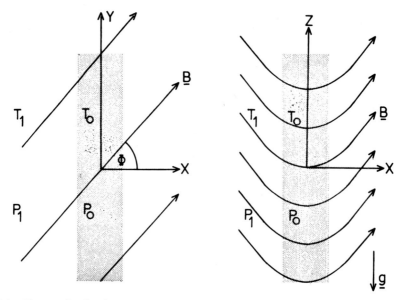

Fig. 11.9. The notation for the support and compression of a prominence represented as a sheet. The z-axis is directed normal to the solar surface, and the y-axis runs along the length of the prominence. For the basic Kippenhahn–Schlüter model the temperature is uniform ($T_1 = T_0$) and the shear angle Φ (between the prominence normal and the horizontal field) is zero.

The boundary conditions are
$$p \to 0, \quad B_z \to \pm B_{z\infty}, \quad \text{as } x \to \pm \infty, \qquad (11.31)$$
and by symmetry
$$B_z = 0 \quad \text{at } x = 0. \qquad (11.32)$$
Then the integral of Equation (11.29) together with Equation (11.31) gives
$$p = \frac{B_{z\infty}^2 - B_z^2}{2\mu}, \qquad (11.33)$$
and so, after using Equations (11.28) and (11.33), Equation (11.30) may be written
$$0 = -\frac{B_{z\infty}^2 - B_z^2}{2\Lambda} + B_x \frac{dB_z}{dx},$$
where $\Lambda = k_B T/(mg)$ is the scale-height (Section 3.1). Its solution, subject to Equation (11.32), is simply
$$B_z = B_{z\infty} \tanh \frac{B_{z\infty} x}{B_x \Lambda}, \qquad (11.34)$$
while Equation (11.33) gives
$$p = \frac{B_{z\infty}^2}{2\mu} \operatorname{sech}^2 \frac{B_{z\infty} x}{B_x \Lambda}, \qquad (11.35)$$

as sketched in Figure 11.10. Clearly, the central plasma pressure is equal to the external magnetic pressure associated with the vertical field component, and the half-width of the sheet is of order $B_x \Lambda / B_{z0}$, the distance over which the plasma pressure falls appreciably from its central value.

Several points may be noted. The axial field (B_y) plays a purely passive role and it doesn't affect the structure of the model at all, as represented by the solutions (11.34) and (11.35). One limitation of the model is that the ambient gas pressure must vanish (Equation (11.31)), since otherwise the corresponding mass would need a vertical force to support it; in other words, the assumption of one-dimensionality would fail. However, it should be possible to include this small effect by linearising about the present model. Other limitations are that the external field joining the prominence

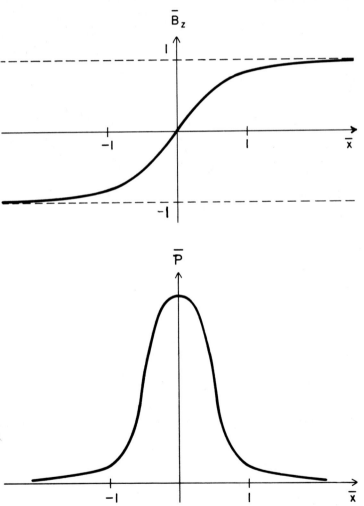

Fig. 11.10. The vertical magnetic field (B_z) and plasma pressure (p) for the Kippenhahn–Schlüter model, shown schematically as functions of distance (x) across the prominence. $\bar{B}_z$ is defined as $B_z/B_{z\infty}$, $\bar{p}$ as $p2\mu/B_{z\infty}^2$ and $\bar{x}$ as $B_{z\infty} x/(B_x \Lambda)$.

field to the photosphere is not incorporated (Section 11.2.3) and that the temperature is assumed uniform (Section 11.2.2).

11.2.2. GENERALISED KIPPENHAHN–SCHLÜTER MODEL

Many authors have considered the energy balance of prominences separately from the support mechanism (e.g., Orrall and Zirker, 1961), but recently there have been several attempts to combine the energetics and magnetostatics. For example, Low (1975) and Lerche and Low (1977) have achieved a notable advance by considering an energy equation in which thermal conduction is balanced by a heat-loss term proportional to the local density. The use of this particular form for the heat-loss function simplifies matters greatly, but it is only a first approximation to the sum of the expected heating and radiation. Also, it may be noted that they did not apply boundary conditions at the corona. Heasley and Mihalas (1976) considered the radiative transfer equations in detail, and in some cases they attempted to solve them simultaneously with the equations of magnetohydrostatic equilibrium. Their models apply primarily to the central, low-temperature region of prominences, and give a width much smaller than observed. Thermal conduction is neglected, so that the effect on the energy balance of shearing the magnetic field is absent.

Milne et al. (1979) have coupled the magnetohydrostatics and energy balance for a simple equilibrium model of a prominence embedded in the corona, as described below. The main aim is to determine the effect on the prominence structure of the horizontal field and shear, and so the energy balance is treated in a much less comprehensive way than Heasley and Mihalas. Equations (11.28) to (11.30) are solved together with

$$\frac{d}{dx}\left(\kappa_0 T^{5/2} \frac{dT}{dx} \frac{B_x^2}{B^2}\right) = \tilde{\chi}\rho^2 T^\alpha - h\rho, \tag{11.36}$$

where the presence of B_x^2/B^2 takes account of the fact that conduction is along the field lines rather than in the x-direction. The boundary conditions are

$$p = p_1, \quad T = T_1, \quad \text{at } x = \pm \Lambda_1, \tag{11.37}$$

and, by symmetry,

$$B_z = \frac{dT}{dx} = 0 \quad \text{at } x = 0, \tag{11.38}$$

where coronal conditions ($T_1 = 2 \times 10^6$ K, $p_1 = 2.76 \times 10^{-3}$ Nm^{-2}, corresponding to a number density of 10^{14} m^{-3}) are assumed to hold at $|x| = \Lambda_1 (= k_B T_1/(mg))$. The set of Equations (11.28) to (11.30) and (11.36) to (11.38) represent a two-point boundary-value problem, and the central values of pressure (p_0) and temperature (T_0) at $x = 0$ are determined by the solution. The solutions depend on two parameters, namely

$$\beta = \frac{2\mu p_1}{B_x^2} \quad \text{and} \quad \frac{B_y}{B_x}. \tag{11.39}$$

Increases in β may be achieved by decreasing the horizontal field strength (B_x), while varying B_y/B_x is equivalent to altering the shear angle

$$\Phi = \tan^{-1}(B_y/B_x) \tag{11.40}$$

indicated in Figure 11.9. The solutions possess some very interesting properties. For example, if the energy equation is first of all not included, the magnetostatic equations show that the solution has the form

$$B_z = (2\mu p_0)^{1/2} \tanh\left[\tfrac{1}{2}(\beta p_0/p_1)^{1/2} l(x)\right],$$
$$p = p_0 \operatorname{sech}^2\left[\tfrac{1}{2}(\beta p_0/p_1)^{1/2} l(x)\right],$$

where

$$l(x) = \frac{T_1}{\Lambda_1}\int_0^x \frac{dx}{T},$$

and the boundary condition $p = p_1$ at $x = \Lambda_1$ gives

$$p_0^{1/2} = p_1^{1/2} \cosh\left[\tfrac{1}{2}(\beta p_0/p_1)^{1/2} l_1\right] \tag{11.41}$$

with $l_1 = l(\Lambda_1)$. Equation (11.41) determines two values for p_0 when $\beta^{1/2} l_1$ is less than a critical value and none if it is greater. Thus, for given $T(x)$, there exists a *maximum allowable* β (approximately $1.7\, l_1^{-2}$) *for the existence of equilibrium solutions*. This same feature of *magnetostatic nonequilibrium* is present in the full numerical solutions, and it even shows up for isothermal solutions ($T \equiv T_1$) as follows. Integrating Equation (11.29) and evaluating it at $x = 0$ and $x = \Lambda_1$ gives $p_0 = p_1 + B_{z1}^2/(2\mu)$, whereas Equation (11.30) may be approximated in order of magnitude by $0 = -p_0 + B_x B_{z1}/\mu$. Eliminating B_{z1} then gives $\beta p_0^2 - 4p_1 p_0 + 4p_1^2 = 0$, which has two real solutions

$$p_0 = \frac{2p_1}{\beta}\left[1 \pm (1-\beta)^{1/2}\right] \tag{11.42}$$

for the central pressure (p_0) only if $\beta < 1$.

Another interesting special case is when the horizontal magnetic field (B_x) is so large that β and B_z approach zero, while the pressure becomes uniform. Then Equation (11.36) determines the temperature with $B^2 = B_x^2 + B_y^2$ and $\rho = mp/(k_B T)$. The resulting central temperature must exceed a minimum value which makes the right-hand side of Equation (11.36) vanish. Also, solutions to Equations (11.36) to (11.38) then exist *only if the magnetic field is sheared beyond a critical angle* of 82.5°, corresponding to $B_y/B_x = 7.5$.

From the full numerical solutions to Equations (11.28) to (11.30) and (11.36) to (11.38), it is found that the central temperature varies with β and B_y/B_x (or shear Φ) in the complex manner shown in Figure 11.11(a). The corresponding temperature and pressure profiles for zero shear and $\beta = 0.6$ are given in Figures 11.11(b) and 11.11(c). It can be seen from Figure 11.11(a) that for $\beta > \beta_{\max} = 1.70$ there are no prominence-like solutions. In general, T_0 is multiple-valued, and the number of solutions depends on β and Φ. For example, if $\Phi = 0$ one finds: only the almost-isothermal solution when $\beta < \beta_{\min} = 0.60$; four solutions when $0.60 < \beta < 0.68$, with only the lowest at

(a)

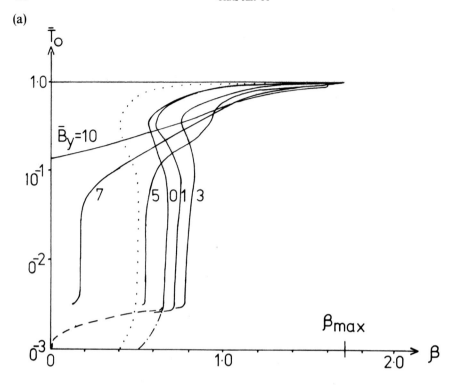

(b)

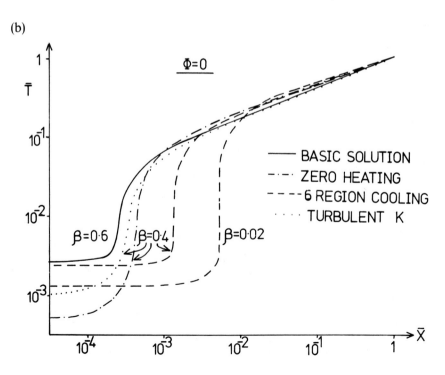

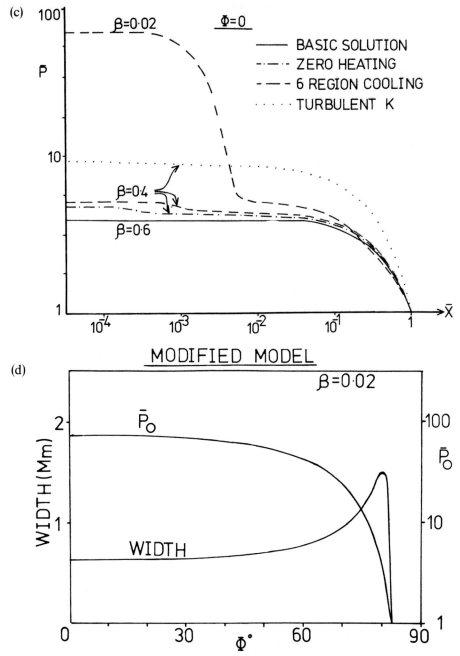

Fig. 11.11. The prominence solutions of Milne et al. (1979) in terms of the two parameters β and $\bar{B}_y =$ $= B_y/B_x (= \tan \Phi)$. Solid curves show the standard model, and several modified models are also given for $\Phi = 0$, namely ones with no heating (—·—·—), no heating and a modified radiation law (— — —), and an eddy thermal conductivity (· · · ·). (a) The variation of central temperature ($\bar{T}_0 = T_0/T_1$) with β and $\bar{B}_y$. (b) The temperature ($\bar{T} = T/T_1$) as a function of distance ($\bar{x} = x/\Lambda_1$) from the prominence centre for $\Phi = 0$. (c) The corresponding profile of pressure ($\bar{p} = p/p_1$). (d) The variation of prominence width (1 Mm $\equiv 10^3$ km) and central pressure (p_0/p_1) with shear angle (Φ) for $\beta = 0.02$ and the modified radiation law.

prominence temperatures; and two solutions when $0.68 < \beta < 1.7$, but neither is cool enough for prominences. As Φ increases, so β_{min} decreases, reaching zero when $\Phi > 83°$. The net result is that prominence-like solutions to the standard equations are possible for only a narrow range of β and Φ. Milne et al. (1979) therefore, modified the equations in several ways.

If the heating is by waves, they may dissipate their energy before reaching the densest part of the prominence, and so the first modification is to put $h = 0$ near the centre of the sheet for temperatures less than $0.1\ T_1$, say. The result is that there is no longer a lower limit on T_0 or a minimum β for prominence-like solutions, as indicated in Figure 11.11. However, for low values of β this produces prominences which are much too cool, because the radiative loss term in (11.36) is appropriate for an optically thin plasma, and so it is much too high for the central optically thick parts of a prominence. The second modification is therefore to reduce the radiation below 10^4 K by taking $\alpha = 17.4$, say, which produces much more reasonable central temperatures. As the shear (Φ) increases, so the width of the prominence increases, while its central pressure decreases, as shown in Figure 11. 11(d). For $\beta = 0.02$, the maximum width is 1500 km, which is of the same order as the observed widths. It occurs for a shear angle of $80°$, in agreement with the results of Tandberg-Hanssen and Anzer (1970) that prominences in general have fields with a shear angle of order $75°$. Also, Figure 11.11(d) shows the existence of a maximum shear angle of $83°$, beyond which no cool equilibrium exists. A final modification is to include an (enhanced) eddy thermal conductivity, which lowers the central temperature but does not change the general features of the model.

The maximum in β for equilibrium to exist is a result of the magnetohydrostatics, and it corresponds to a maximum in the coronal pressure (p_1). This implies that the prominence cannot exist below a certain height, which increases with field strength (B_x), so that active-region prominences can form lower in the atmosphere than can quiescent prominences. The maximum β also corresponds to a minimum in B_x for constant p_1. Thus, if B_x is smaller than this minimum, the plasma cannot be supported, and so it presumably sags down, pulling the magnetic field with it and giving rise to the 'feet' that are often observed in quiescent prominences.

The maximum in the shear (Φ) is essentially a result of the energetics. As the shear increases, the amount of heat conducted into the region declines, and so the temperature increases until no cool solution is possible. As the maximum shear is approached, the prominence rapidly heats up, while its pressure, density and width decrease. Simultaneously, the magnetic field lines spring up and possibly produce an erupting prominence. (For other explanations of eruptions, see Section 10.3.)

In the future, the model could be improved by adding a two- or three-dimensional structure, and matching to some external magnetic configuration such as a coronal arcade, rather than imposing boundary conditions at $x = \varLambda_1$; until that is done, it will not be known how realistic those boundary conditions are. Also, a more sophisticated treatment of radiation, conduction and heating is needed, but it is expected that the basic features of the solution will remain. Clearly, it would be fascinating to obtain more observations of the magnetic field structure inside and around prominences, so that the variation of prominence characteristics with shear and horizontal field strength could be compared with theory.

11.2.3. The external field

Anzer (1972) has modelled the magnetic field surrounding a prominence. The prominence is treated as an infinitely thin current sheet stretching a distance H along the (vertical) z-axis from the origin, as shown in Figure 11.12, while the surrounding field is assumed to be potential and two-dimensional in the x–z plane. The problem is then to solve

$$\nabla^2 \mathbf{B} = \mathbf{0} \tag{11.43}$$

in the first quadrant of the x–z plane, subject to the boundary conditions

$$B_z = \begin{cases} 0 & \text{on } x = 0, z > H, \\ 0 & \text{on } z = 0, x > \sqrt{a}, \\ f(x) & \text{on } z = 0, 0 \leqslant x \leqslant \sqrt{a}, \end{cases} \tag{11.44}$$

$$B_x = g(z) \quad \text{on } x = 0, 0 \leqslant z \leqslant H.$$

The function g may be determined in principle from observations of a particular prominence on the limb, whereas f may be deduced from observations of the photosphere beside the same filament when situated at the centre of the disc, assuming that conditions remain constant as the filament moves from limb to disc.

First of all, write

$$B_x = \frac{\partial u}{\partial z}, \quad B_z = -\frac{\partial u}{\partial x}, \tag{11.45}$$

so that the problem reduces to that of determining $(u + iv)$ as an analytic function of $x + iz$ in the first quadrant of the x–z plane, with v prescribed as zero on the z-axis between 0 and H, and u prescribed on the rest of the x- and z-axes. Next, apply the conformal transformation $x + iz = \sqrt{\zeta}$, where $\zeta = \xi + i\eta$, so that now it is necessary to find a function $u + iv$ that is analytic in the upper-half ζ-plane with the following conditions on the ξ-axis:

$$v = 0, \quad \text{for } \xi < h \equiv -H^2,$$
$$u = F(\xi), \quad \text{say, for } h < \xi < a,$$
$$u = 0, \quad \text{for } \xi > a.$$

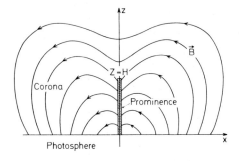

Fig. 11.12. A two-dimensional model of the magnetic field in the neighbourhood of a quiescent prominence (from Anzer, 1972).

The solution which is bounded at infinity is then simply

$$u + iv = \frac{\sqrt{\zeta - h}}{\pi i} \int_h^a \frac{F(t)}{\sqrt{t - h}(t - \zeta)} dt, \qquad (11.46)$$

and it may be transformed back to give the solution in the x–z plane.

Anzer applied his technique to a particular prominence with $H = 40\,000$ km, $\sqrt{a} = 100\,000$ km, $f(x) = 0$ for $x < 10\,000$ km, $f(x) = 4$ G for $x > 10\,000$ km, and g a linearly increasing function. In terms of the field component (B_{x0}) normal to the prominence sheet at $x < 0$, the Lorentz force averaged over the width of the prominence is

$$F_L = JB_{x0}, \qquad (11.47)$$

where $J = 2B_{zd}/\mu$ is the current flowing through the prominence and B_{zd} is the value of B_z at $x = \tfrac{1}{2}d$ (equal to that at $x = -\tfrac{1}{2}d$, by symmetry). B_{x0} is prescribed, whereas B_{zd} is calculated from the solution. The result is that F_L is found to be positive for z above a certain height ($\simeq 17\,000$ km), and it can support a reasonable mean plasma mass of $nd \simeq 1.8 \times 10^{24}$ m^{-2}.

A limitation of the model is that below 17 000 km F_L is negative, and so the Lorentz force is not capable of supporting prominence material against gravity. This could probably be avoided by generalising the model to include a height on the z-axis which marks the prominence base and below which B_z vanishes. It would also be interesting to see the effect of a more general force-free field outside the prominence.

11.2.4. MAGNETOHYDRODYNAMIC STABILITY

Anzer (1969) has investigated the stability of the Kippenhahn–Schlüter model using Bernstein's energy principle (Section 7.4). He models the prominence as a thin vertical plasma sheet of width d along the z-axis, containing a current $J = 2B_{zd}/\mu$, using the same notation as above. In the limit as $d \to 0$, the change (δW) in energy produced by a perturbation (ξ) of the plasma reduces to

$$\delta W = \tfrac{1}{2} \iint \left(J \frac{dB_{x0}}{dz} \xi_x^2 - B_{x0} \frac{dJ}{dz} \xi_z^2 \right) dy\, dz + G,$$

where G is a positive term (see Equation (7.38)). From this it can be seen that sufficient conditions for stability ($\delta W \geq 0$) are that

$$J \frac{dB_{x0}}{dz} \geq 0 \qquad (11.48)$$

and

$$B_{x0} \frac{dJ}{dz} \leq 0. \qquad (11.49)$$

Anzer establishes that these conditions are also necessary.

Observations appear to confirm Equation (11.48), but they are not yet sophisticated enough to judge whether or not Equation (11.49) holds. However, several deductions can be made. For instance, if the prominence mass is supported by the Lorentz force alone, one has

$$\rho \, \mathrm{d}g = B_{x0} J, \qquad (11.50)$$

and the fact that the left-hand side is positive implies $B_{x0} J > 0$. This may then be combined with Equation (11.49) to give $\mathrm{d}J^2/\mathrm{d}z \leq 0$, so that the magnitude of the current must decrease with height for a stable prominence.

The derivative of Equation (11.50) with respect to z is

$$\frac{\mathrm{d}\rho}{\mathrm{d}z} \mathrm{d}g = J \frac{\mathrm{d}B_{x0}}{\mathrm{d}z} + B_{x0} \frac{\mathrm{d}J}{\mathrm{d}z},$$

where Equations (11.48) and (11.49) imply that the two terms on the right have opposite signs. Thus, the mass density in the prominence may either decrease or increase with height, depending which of the two dominates.

Now, suppose, without loss of generality, that B_{x0} and J are both positive, and denote the value of B_x at the edge ($x = \tfrac{1}{2}d$) of the prominence by B_{xd}. Then, if the prominence currents are so small that $B_{xd} \simeq B_{x0}$, Equation (11.48) implies $\partial B_{xd}/\partial z > 0$. If, furthermore, the field outside the prominence is current-free ($j = 0$),

$$\frac{\partial B_{zd}}{\partial x} - \frac{\partial B_{xd}}{\partial z} = 0,$$

and so $\partial B_{zd}/\partial x > 0$. B_z must therefore increase with x for a stable configuration, which means that the field lines must have a curvature that is concave upward.

In the future, it is important to attempt a full stability analysis, including the internal structure of the prominence and the external field, line-tied in the photosphere. A start has been made on the problem by Brown (1958), who considered purely horizontal or purely vertical displacements. Also, Nakagawa (1970) has tested the stability of a horizontal, sheared field that is supporting prominence material.

11.2.5. Helical Structure

Anzer and Tandberg-Hanssen (1970) gave a simple model for the helical structure which may exist in some prominences and which becomes visible when they erupt (Figure 1.33). They assume the prominence is circular in cross-section and its magnetic configuration consists of the superposition of uniform fields (B_{x0} and B_{y0}) in the (horizontal) x- and y-directions, together with a purely azimuthal pinch field (3.19) having circular field lines in the (vertical) x–z plane. Thus

$$\mathbf{B} = B_{x0}\hat{\mathbf{x}} + B_{y0}\hat{\mathbf{y}} + B_\phi \hat{\boldsymbol{\phi}},$$

where

$$B_\phi = \begin{cases} \dfrac{\mu I R}{2\pi a^2}, & R < a, \\ \dfrac{\mu I}{2\pi R}, & R > a, \end{cases}$$

with I and a constant and $R = (x^2 + z^2)^{1/2}$. The resulting field lines depend on the value of the parameter $C = \mu I/(2\pi a B_{x0})$. For $C < 1$ the field lines exhibit a dip, as in the Kippenhahn–Schlüter model, whereas for $C > 1$ there are some closed field lines in the x–z plane.

11.3. Support in Configurations with Helical Fields

11.3.1. Support in a Current Sheet

Kuperus and Raadu (1974) consider how a prominence that is formed in a current sheet (Section 11.1.3) may be supported. The initial stages of the tearing mode lead to the formation of current filaments, which may coalesce during the nonlinear phase to form a single filament and so give the configuration shown in Figure 11.13(a). Support is provided by the Lorentz force, which acts upwards because the field lines fan out at the photosphere where they are tied. The configuration may be regarded as the sum of a vertical (current sheet) field together with a current filament field which has a series of closed field lines, as shown in Figure 11.13(b). The field of the

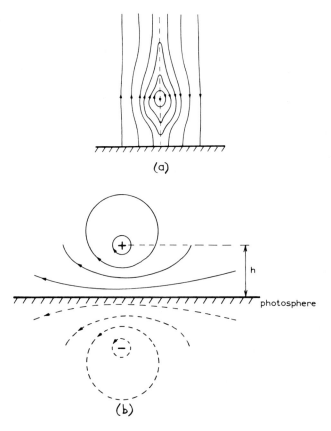

Fig. 11.13. The support of a prominence formed in a current sheet. (a) The magnetic field configuration due to a current filament in a neutral sheet. (b) The field (solid) due to a filament located a distance h above the photosphere (from Kuperus and Raadu, 1974).

current filament may in turn be regarded as coming from two line currents, namely I at a height h above the photosphere and I at a distance h below.

The supporting force is simply the force of repulsion between the two line currents, namely $\mu I^2/(4\pi h)$, where $I = 2B_\phi R\pi/\mu$, in terms of the field (B_ϕ) at a distance R from the current centre. The force can support a prominence mass of $m = \pi R^2 \rho$ per unit length if

$$\frac{\mu I^2}{4\pi h} = mg, \tag{11.51}$$

or, after substituting for I and m,

$$\frac{B_\phi^2}{\mu h} = \rho g. \tag{11.52}$$

With a density (ρ) of 10^{-10} kg m^{-3} and a height (h) of 10 000 km, this equation yields the reasonable field strength of $B_\phi \simeq 6$ G.

In order to demonstrate the feasibility of the above model convincingly, it would be necessary to follow the condensation process through to its final steady state. A potential difficulty is that, after a plasma compression by a factor of a hundred, the field lines near the photosphere are likely to slope *towards* the neutral sheet rather than away from it. The field would therefore provide no support at first and the plasma would fall down until it is supported by closed field lines near the base of the sheet, as in the Kippenhahn–Schlüter model. Also, the current in the configuration of Figure 11.13(a) appears to exceed the initial current in the sheet before condensation.

Kuperus and Raadu also consider small vertical perturbations (z) of the current filament from its equilibrium position. The equation of motion is simply $\rho \ddot{z} = -(B_\phi^2/(\mu h^2))z$, or, after using Equation (11.52), $\ddot{z} = -(g/h)z$, which shows that the filament is in stable equilibrium. It can oscillate with a period of $2\pi(h/g)^{1/2}$, which is about 20 min when $h = 10\,000$ km.

Van Tend and Kuperus (1978) and Kuperus and van Tend (1981) *Solar Phys.* **71**, 125, have extended this qualitative model to show that *non-equilibrium* occurs if the filament current (or its height) exceeds a critical value. The Equation (11.51) for vertical prominence equilibrium is generalised to

$$\frac{\mu I^2}{4\pi h} = IB + mg, \tag{11.53}$$

so that it includes the Lorentz force (IB) due to the current I in a background horizontal field ($B(h)$). This is solved for I and sketched as a function of h after writing the variation of B with h for a model active region in the form

$$B(h) = \begin{cases} B_0, & h < h_1, \\ B_0 \dfrac{h_1}{h}, & h_1 < h < h_2, \\ B_0 \dfrac{h_1 h_2^2}{h^3}, & h_2 < h. \end{cases}$$

The result is that I has a maximum value at h_2, which is taken as typically a few $\times\, 10^4$ km. The suggestion is therefore that, if I exceeds this maximum, the fila-

11.3.2. SUPPORT IN A HORIZONTAL FIELD

Lerche and Low (1980) have generalised the model of Kuperus and Raadu (Section 11.3.1) so as to analyse the magnetostatics of a cylindrical prominence supported by a horizontal field and bounded below by the photosphere. The adopted magnetic field has the form

$$\mathbf{B} = \left(\frac{\partial A}{\partial z}, B_y, -\frac{\partial A}{\partial x}\right), \tag{11.54}$$

with the component (B_y) along the prominence axis assumed uniform. The prominence has a radius R_0, and its axis is located at $x = 0$, $z = h$, while the photosphere is at $z = 0$.

Outside the prominence, the magnetic field is potential with $\nabla^2 A = 0$, subject to the following boundary conditions at infinity, at the photosphere and on the prominence interface:

$$B_x \to B_0, \; B_z \to 0, \quad \text{as } z \to +\infty, \tag{11.55}$$

$$B_z = 0, \quad \text{on } z = 0, \tag{11.56}$$

$$B_R, B_\phi \text{ continuous at } (y, z) = (R_0 \sin \phi, h + R_0 \cos \phi). \tag{11.57}$$

The boundary conditions (11.55) and (11.56) are satisfied automatically by taking A as the potential due to the horizontal field $(B_0 \hat{\mathbf{x}})$, together with line currents at $z = \pm a$ on the z-axis: $A = F_0 + B_0 z$, where

$$F_0 = F_* \log_e \frac{(z-a)^2 + x^2}{(z+a)^2 + x^2}.$$

The constant a is chosen to be $a = (h^2 - R_0^2)^{1/2}$, so that the line current at $(x = 0, z = a)$ lies inside the prominence cylinder, and F_0 is constant on its surface. The topology of the field lines depends on the value of $B_0 R_0 / F_*$, as indicated in Figure 11.14.

Inside the prominence, the force balance (11.27) may be written

$$\nabla^2 A = \mu J(R, \phi), \tag{11.58}$$

where (R, ϕ) are cylindrical coordinates centred on the prominence axis and

$$J = -\frac{\partial p}{\partial A}, \tag{11.59}$$

as shown in Section 3.6 (Equation (3.65)). In particular, Lerche and Low (1980) express J as

$$J = \sum_{n=0}^{\infty} j_n R^n \cos n\phi. \tag{11.60}$$

They then solve Equation (11.58) and use the boundary conditions (11.57) to deduce

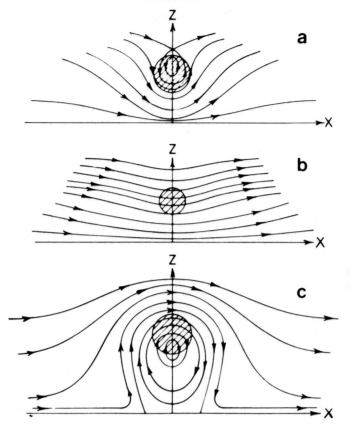

Fig. 11.14. Models for a cylindrical prominence with (a) a neutral point above the prominence ($B_0 R/F_* < -1$), (b) no neutral point above ($-1 < B_0 R_0/F_* < 0$), and (c) a neutral point below ($0 < B_0 R/F_* < 1$) or no neutral point below ($1 < B_0 R_0/F_*$ (from Lerche and Low, 1980).

the constants j_n, with the result that

$$J = \frac{4F_*(1 - 2\psi_0^2 R^2)}{R_0^2(1 + \psi_0^2 R^2 + 2\psi_0 R \cos\phi)^2},$$

where

$$\psi_0 = \frac{h - (h^2 - R_0^2)^{1/2}}{R_0^2}.$$

The pressure follows by integrating Equation (11.59), and it is necessary to take $h^2 \geq 9R_0^2/8$ so as to ensure that J does not vanish; otherwise there would be locations where $\partial p/\partial A$ is zero and the plasma cannot be supported.

However, the current is not uniquely determined by the boundary conditions, since it may be written more generally than Equation (11.60) as

$$J = \sum_{m,n=0}^{\infty} j_{mn} R^m \cos n\phi,$$

with no extra conditions to determine the constants (j_{mn}). In other words, there is more than one internal current distribution corresponding to an external field.

Low (1981) has cleverly extended the analysis to model support in a magnetic arcade. The magnetic field is imposed as the sum of an ambient arcade field, a diffuse line current (to represent the filament) and an image line current (to preserve line-tying). The pressure is deduced from magnetohydrostatic equilibrium normal to **B**, while the density and temperature follow from the perfect gas law and hydrostatic equilibrium along **B**. A field component along the filament is necessary to produce prominence-like temperatures.

11.4. Coronal Transients

Coronal transients accompany filaments that erupt, both in the quiet Sun and (along with two-ribbon flares) in active regions. About a third of them are in the form of loops, whose properties have been described in Section 1.4.3F, although several features are not yet well-established. For instance, their orientations relative to the filaments are not well-known: low down they may be in a plane perpendicular to the filament, but high up they appear to be inclined within 20° of it. The magnetic field strength is uncertain, but it is probably in the range 1 to 10 G. Also, the three-dimensional geometry is not known with certainty, although it is generally believed to represent a loop rather than an arcade or a shell.

In this section we shall summarise the theories for *loop transients* by extending somewhat a previous review (Priest, 1982b). The main properties in need of explanation are as follows:

(1) the details of the driving force, which is probably magnetic since the plasma beta is 0.1 or less;
(2) the uniform speed, which is observed for the transient summit between $3R_\odot$ and $5 R_\odot$ (typically 300 to 400 km s^{-1} for those associated with quiescent filaments and 700 to 800 km s^{-1} for those accompanying two-ribbon flares near plage filaments);
(3) the fixed angle which the transient legs subtend at the solar centre;
(4) the increase in loop width (h) with distance (R) like $R^{0.8}$;
(5) the increase in the radius of curvature (R_c) of the loop summit like $R^{1.6}$;
(6) the density enhancement at the loop summit by a factor of 5 to 10 and the density deficit beneath the transient by a factor of 2 compared with ambient coronal values.

11.4.1. Twisted Loop Models

An order-of-magnitude model for a loop transient as a twisted flux tube was presented by Mouschovias and Poland (1978). The plasma pressure is neglected, and the summit of the transient is regarded as a flux loop with a longitudinal field (B_l), surrounded by an azimuthal field B_{az} (Figure 11.15(a)). Since the speed of the transient is roughly constant, the balance between magnetic and gravitational forces may be written

$$\frac{B_{az}^2}{\mu R_c} - \frac{B_l^2}{\mu R_c} = \frac{nnGM_\odot}{R^2}, \qquad (11.61)$$

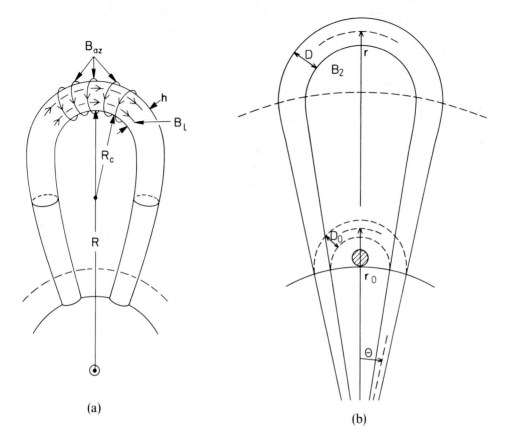

Fig. 11.15. Transient geometry for the models of (a) Mouschovias and Poland (1978) and (b) Pneuman (1980).

where the first term on the left represents the outwards magnetic pressure gradient associated with B_{az} and the second term gives the restoring tension force associated with B_l. Conservation of longitudinal flux, azimuthal flux, and transient mass implies

$$B_l h^2 = \text{const.}, \tag{11.62}$$

$$B_{az} h R = \text{const.}, \tag{11.63}$$

and

$$n h^2 R = \text{const.}, \tag{11.64}$$

since the loop length is roughly proportional to R.

In order to complete the above set of relations for B_{az}, B_l, n, R_c, h, Mouschovias and Poland assume that B_{az}/B_l is constant ($= 1.2$), and they are then able to deduce that

$$h \sim R, \tag{11.65}$$

$$R_c \sim R, \tag{11.66}$$

$$B_l \sim R^{-2}. \tag{11.67}$$

The relation (11.65) for the summit width agrees quite well with observations, but the theoretical radius of curvature (11.66) is too small, possibly due to the neglect of the background field which would exert a drag and so flatten the top of the loop. Finally, Equation (11.67) gives a field strength of 1.0 G at $R = 2R_\odot$ (where $n = 3.9 \times 10^{13}$ m^{-3} and $R_c = 0.4 R_\odot$) and 0.3 G at $R = 5 R_\odot$ (where $n = 4.1 \times 10^{12}$ m^{-3} and $R_c = 1.8 R_\odot$). The corresponding Alfvén speeds are 360 km s^{-1} and 330 km s^{-1}, respectively. The result that B_l declines as R^{-2} means that it will dominate more and more the background field, which falls off roughly like R^{-3}.

A difficulty with this simple model is that, although *filaments* often appear twisted, there is no indication of twist in loop transients. Also, Equation (11.61) implies that B_{az} needs to exceed B_l, which means that the twist is so great that the flux tube may well be subject to the helical kink instability as it rises (Sections 7.4.1., 7.5.3, and 10.2.3).

Anzer (1978) has modelled a transient by a circular *ring current* and has studied the initial acceleration as well as the later phase of constant velocity. The ring current forms an arc of a circular torus with major radius r (initially r_0) and minor radius d, such that the feet of the arc are at a constant distance $2r_0$ apart. Initially, $r = r_0$ so that the loop consists of a semicircle. The equation of motion of the loop summit in terms of its distance (R) from the solar centre is

$$m\frac{d^2R}{dt^2} = F_r - \frac{mM_\odot G}{R^2}, \qquad (11.68)$$

where m is the mass per unit length, and F_r is the magnetic force in the radial direction. If the total loop mass remains constant, m is given by

$$m = m_0 \frac{r_0 f_1(r_0)}{r f_1(r)},$$

where r is a given function of R and $f_1(r)$ increases from $\frac{1}{2}$ at $r = r_0$ to 1 as $r \to \infty$. The Lorentz force can be written as

$$F_r = \frac{\varphi^2}{\mu r} \frac{\partial}{\partial r}\left(\frac{1}{L}\right)$$

in terms of the flux (φ) through a whole circle and the self-inductance

$$L = 4\pi r [\log_e 8r/d - \tfrac{7}{4}].$$

Assuming that the flux φ_1 through the part of the circle modelling the transient is conserved, φ may be written $\varphi = \varphi_1/f(r)$, where $f(r)$ increases from $\frac{1}{2}$ at $r = r_0$ to 1 as $r \to \infty$.

The solutions of Equation (11.68) depend on the parameters φ_1^2/m_0 and r_0. If φ_1^2/m_0 is small so that the magnetic driving force is small, the loop reaches a maximum speed and then decelerates. At intermediate values of φ_1^2/m_0, the speeds reach roughly constant values between $2R_\odot$ and $6R_\odot$, in agreement with observations. At larger values the loop continues to accelerate. For the intermediate values, Anzer deduces a field strength of 0.5 G when $r_0 = \frac{1}{2}R_\odot$, $d/r = 0.2$ and $n_0 = 10^{14}$ m^{-3}. He also stresses that loop transients need not be current loops, since any driving force which falls off approximately like R^{-2} can produce similar velocity profiles.

11.4.2. UNTWISTED LOOP MODELS

Pneuman (1980, 1981) has modelled a loop transient as a simple loop lying above a filament, whose axis is orientated perpendicular to the overlying field. He supposes that the transient is driven outwards by the lifting of the filament, which increases the magnetic pressure under the helmet. The assumed geometry is shown in Figure 11.15(b), where the dashed curves denote the circular boundary of the flux tube in its equilibrium position when its width is D_0 and its top is at a distance r_0 from the solar centre. The solid curves show the tube at some later time with width D and displacement r. B_2 is the driving field behind the transient, B is the field in centre of the flux tube at its summit, and θ is the constant half-angle between the legs. The initial location of the filament, seen edgewise, is shown as the shaded region.

Neglecting any plasma pressure gradients, the equation of motion of the field line (1) which forms the upper boundary of the transient loop may be written in order of magnitude as

$$\rho \frac{d^2 r_1}{dt^2} = \frac{B^2}{\mu D} - \frac{B^2}{\mu R_c} - \frac{GM_\odot}{r^2}\rho, \qquad (11.69)$$

where

$$R_c = \frac{r \tan \theta}{1 + \tan \theta}$$

is the radius of curvature, and the first and second terms on the right-hand side of Equation (11.69) represent the magnetic pressure and tension, respectively, assuming a negligible field strength ahead of the transient. Similarly, the equation of motion for the field line (2), which forms the lower boundary of the loop, is

$$\rho \frac{d^2 r_2}{dt^2} = -\frac{B^2 - B_2^2}{\mu D} - \frac{B^2}{\mu R_c} - \frac{GM_\odot}{r^2}\rho, \qquad (11.70)$$

where $r = \frac{1}{2}(r_1 + r_2)$ and $D = r_1 - r_2$, in terms of the positions (r_1 and r_2) of the upper and lower field lines, respectively.

Pneuman assumes conservation of magnetic flux and mass for the loop, so that the magnetic field and density are given in terms of the initial equilibrium values (B_0 and ρ_0) by

$$B = B_0 \frac{D_0^2}{D^2}, \qquad \rho = \rho_0 \frac{D_0^2 r_0}{D^2 r}, \qquad (11.71)$$

Furthermore, the field below the transient is assumed to conserve its flux as it expands, so that

$$B_2 = B_{20} \frac{r_0^2}{r^2}, \qquad (11.72)$$

in terms of the initial field strength (B_{20}). The equilibrium field strength (B_{20}) and loop width (D_0) are given by setting the left-hand sides of Equations (11.69) and (11.70) to zero as $B_{20} = \sqrt{2} B_0$ and

$$D_0 = r_0 \bigg/ \left(\frac{1 + \tan \theta}{\tan \theta} + \frac{\mu \rho_0 GM_\odot}{r_0 B_0^2} \right).$$

Pneuman adopts as typical equilibrium values $B_0 = 5$ G, $\rho_0 = 1.7 \times 10^{-12}$ kg m^{-3}, $\theta = 20°$, $r_0 = 1.2\, R_\odot$, and he deduces that $B_{20} = 7$ G and $D_0 = 0.24\, R_\odot$. The equilibrium is neutrally stable, and so, in order to simulate the increased field produced by the lifting prominence, he raises the value of B_{20} to 8 G. Then he solves Equations (11.69) and (11.70) for r_1 and r_2 (or, equivalently, r and D), subject to the initial conditions $r = r_0$, $D = D_0$, $dr/dt = dD/dt = 0$. The results show that the transient is accelerated rapidly for about $2\, R_\odot$, and then it approaches a uniform speed of about 750 km s^{-1}. Asymptotically, $v \sim (1 - K/r)^{1/2}$, and the transient width increases linearly ($D \sim r$), in reasonable agreement with observations.

The magnetic forces fall off as r^{-2}, just like the gravitational force. Thus, once the lifting prominence has initiated the outward expansion, the driving force always slightly exceeds the restoring forces and the transient coasts out to infinity.

Pneuman has also modelled a transient as an arcade, so that the conservation laws (11.71) are replaced by

$$B = B_0 \frac{D_0 r_0}{Dr}, \qquad \rho = \rho_0 \frac{D_0 r_0^2}{Dr^2}.$$

The solutions are similar to those for a loop, except that the terminal velocity is smaller (about 400 km s^{-1}). Another point is that he suggests that the top of a transient should flatten because the component of gravity normal to the loop is largest at the top. The above solutions result when a relatively small driving force is applied over a large distance, but Pneuman shows that a similar behaviour (with the transient approaching a constant velocity at infinity) can also be produced if the prominence and its driving field do not follow the transient to infinity but instead provide a larger force over a shorter distance.

The model has been made more sophisticated by Anzer and Pneuman (1982), who suggest that the driving force is an increase in magnetic pressure underneath the rising filament due to reconnection. Equations similar in form to (11.69) and (11.70) are set up for the displacements (r_3 and r_4) of the top and bottom of the erupting prominence as well as the displacements (r_1 and r_2) of the leading and trailing edges of the transient loop. The rate of rise of the cusp-type neutral point P (Figure 10.20) is imposed from observations of rising 'post'-flare loops during the 29 July 1973 flare (Figure 1.38). The resulting velocity profiles and transient widths agree well with observations of transients from white-light coronagraphs, provided two parameters (representing the properties of the reconnected flux) are suitably chosen. The speeds are naturally higher when mass is allowed to fall out of the prominence than when it is conserved. Also, as expected, the speeds may be increased by adding either a larger amount of reconnected flux below the transient or adding flux at a faster rate. The one problem with the model is that there does not yet seem to be firm evidence of closed field lines below the transient.

11.4.3. NUMERICAL MODELS

The Huntsville group have performed a series of two-dimensional, time-dependent, numerical experiments, in an attempt to model the coronal transient phenomenon. The driving force is thermodynamic rather than magnetic, and the experiments follow the way the magnetic field inhibits and channels the expansion produced by a flare-associated temperature enhancement at the base of the corona. Nakagawa et al. (1978) consider a temperature pulse of magnitude 1.5×10^7 K for 5 s in an initially isothermal atmosphere at 1.5×10^6 K. They find that the propagation of finite-amplitude slow and fast waves produces a bubble-like density enhancement in an open potential magnetic field. By contrast, in a closed dipole field the density enhancement takes the form of a pair of horns near the base. The analysis was repeated for an equatorial plane by Wu et al. (1978) with a full energy equation and a plasma beta of 0.7 at the base. More details of the shock development and decay were given, as well as a comparison with the non-magnetic case. The transient itself was identified as a density enhancement between an MHD shock and a contact (piston) surface.

Steinolfson et al. (1978) consider motion in a meridional plane for an initially polytropic atmosphere and for beta values at the base of 0.1 and 1.0. They follow the response to a 5-min temperature enhancement (by a factor 4.17) and a density enhancement (by a factor 1.2) at the equator. In each case a disturbance propagates upward with its leading edge in the shape of an expanding loop. When the initial potential magnetic field is closed, the outward motion is retarded by a build-up of Lorentz forces. None of the ejected material reaches the outer corona, since it simply rises up and then falls back down. When the initial magnetic configuration is open, the ejected material is allowed to flow out along field lines, and so it does reach the outer corona with little lateral motion. The disturbance is preceded by a fast wave when $\beta = 0.1$ and a fast shock when $\beta = 1.0$, both of which weaken rapidly with latitudinal distance from the equator. A more realistic model has been developed by Steinolfson (1982) with a streamer configuration as the initial state; he finds that the type of pulse makes little difference to the subsequent evolution.

11.4.4. Conclusion

The analytical and numerical models to some extent complement one another. The analytical models of Sections 11.4.1 and 11.4.2 suppose the transient is driven by a magnetic force, and they all give acceleration up to a roughly constant velocity (except for the Mouschovias and Poland model, which analyses only the constant-velocity phase). However, they are at a fairly rudimentary stage of development and describe the motion of just the loop summit. There is a need to describe the whole loop, to include pressure gradients, and to compare the predicted plasma densities with observation.

The numerical models of Section 11.4.3 consider a thermodynamic driving force, and do not take account of Alfvén waves. Also, the resulting wave front would tend to form a shell in three dimensions rather than a loop. Nevertheless, the agreement with observations in the more recent models is impressive. In future, it is important to try and include the magnetic effects in a more active way, by incorporating magnetic reconnection and a magnetic driver.

Much remains to be done in this new subject. For example, it is not clear whether the filament eruption causes the transient or whether the transient event stimulates the filament to erupt. The effect of the ambient magnetic field has not been treated in detail. It may well minimise the drag by making the transient rotate to become aligned with the background field as it rises. Furthermore, the early stages of transient formation need to be understood, including the mass flow within the structure. For example, Fisher *et al.* (1981) have described how the transient loop forms by filling up the legs from below and how a dark loop develops below the transient and expands to become a cavity. The dark loop is first observed moving up at 80 km s^{-1} and later accelerates to 200 km s^{-1}. It is first seen when the prominence starts to rise very slowly before also accelerating up to 200 km s^{-1}. Transients associated with flares develop faster, become brighter and move much faster than those associated with quiescent prominence eruption. Finally, it should be remembered that the loop event is not the only type of transient, since one also finds types described as 'filled bottles', 'streamer separations', 'clouds', 'injections into streamers', and 'unclassified'!

CHAPTER 12

THE SOLAR WIND

12.1 Introduction

Variations in the Earth's magnetic field were first observed in the nineteenth century as a sudden increase (by about 10^{-3} G) followed by a slow decrease. These *geomagnetic storms* were sometimes found to occur one or two days after large solar flares, and further evidence for a link between the two phenomena came with the discovery of an 11-yr periodicity in both flare and geomagnetic activity. However, since then we have learnt that the interaction is an exceedingly complex one.

In the early 1900s Birkeland performed experiments to try and model the production of aurora by electrons ejected from the Sun. Later, Chapman (1929) suggested that a geomagnetic storm is caused by a stream of plasma ejected from a solar flare and travelling at 1000 km s^{-1} through the vacuum that was presumed to exist between the Sun and the Earth. More recently, Biermann (1951) proposed that the Sun is emitting 'solar corpuscles' continuously (rather than sporadically) and so is making comet tails point away from the Sun.

The first *theory* of the extended solar corona was by Chapman (1957), who considered a static atmosphere, with energy transfer by conduction alone, and who deduced that the corona extends to the Earth and beyond. For a steady, spherically symmetric state, the heat flux across a sphere of radius r is constant, so that

$$4\pi r^2 \kappa \frac{dT}{dr} = \text{const.}, \tag{12.1}$$

where $4\pi r^2$ is the surface area, $\kappa dT/dr$ is the heat flux density, and the coefficient of thermal conduction is $\kappa = \kappa_0 T^{5/2}$. Thus Equation (12.1) may be integrated to give

$$T = T_0 \left(\frac{R_\odot}{r}\right)^{2/7}, \tag{12.2}$$

after imposing the boundary conditions that $T = T_0$ at $r = R_\odot$ and that T vanishes at infinity. Taking $T_0 = 10^6$ K at the Sun, Equation (12.2) gives a value for the temperature at 1 AU as about 10^5 K, which suggests that the Earth is enveloped by an extremely hot plasma. Furthermore, from the equations of hydrostatic equilibrium

$$\frac{dp}{dr} = -\frac{GM_\odot mn}{r^2}$$

and state $p = nk_B T$ with $n = n_0$, $p = p_0$ at $r = R_\odot$, Chapman was able to deduce the pressure as

$$p = p_0 \exp\left[\frac{7GM_\odot mn_0}{5p_0 R_\odot}\left[\left(\frac{R_\odot}{r}\right)^{5/7} - 1\right]\right]. \tag{12.3}$$

It can be seen that, as r approaches infinity, so p tends to a constant value, but one difficulty with the model is that this constant value far exceeds any reasonable interstellar pressure. Another problem is that the density becomes indefinitely large at great distances.

Parker (1958) resolved the inconsistency in a truly classic paper by suggesting that the corona cannot be in static equilibrium and instead must be continuously expanding outwards. In other words, in the absence of a 'lid' (a strong pressure at infinity) to hold in the corona, it must stream outwards as the 'solar wind'. In his customary skillful way, Parker modelled this steady expansion and found solutions for which the pressure becomes vanishingly small at large distances. However, his ideas were not widely accepted at the time, and they led to a great controversy, which was only resolved in 1959 when the solar wind was directly observed by the statellites Lunik III and Venus I. Its properties began to be studied in detail by the American Venus probe, Mariner II, in 1962, and they have been investigated with increasingly greater precision ever since.

The main features of the solar wind have been outlined in Section 1.3.4C, and many more details can be found in the books by Parker (1963b), Brandt (1970), Hundhausen (1972), and Zirker (1977). Here the aim is simply to describe the classical theory of a steady, spherically symmetric expansion (Sections 12.2 and 12.3) and then to summarise some more recent work on coronal holes (Section 12.4).

12.2. Parker's Solution

The equations of mass continuity and motion for the steady, spherically symmetric flow of an isothermal plasma (having total number density n and speed v) are

$$4\pi r^2 nv = \text{constant},$$

$$mnv \frac{dv}{dr} = -\frac{dp}{dr} - \frac{GM_\odot mn}{r^2},$$

where

$$p = nk_B T. \qquad (12.6)$$

Thus, the outflowing plasma is assumed to be acted on by a pressure gradient (pushing it away from the Sun) and the force of gravity.

Since T is assumed uniform, n may be eliminated between Eqs. (12.4) and (12.5) to give

$$\left(v - \frac{v_c^2}{v}\right)\frac{dv}{dr} = \frac{2v_c^2}{r} - \frac{GM_\odot}{r^2}, \qquad (12.7)$$

where $v_c = (k_B T/m)^{1/2}$ is the *isothermal sound speed*, which is somewhat less than the (adiabatic) sound speed. Equation (12.7) possesses a *critical point* (A), where dv/dr is undefined. It occurs when $v = v_c$, $r = r_c \equiv GM_\odot/(2v_c^2)$ so that both the coefficient of dv/dr and the right-hand side of the equation vanish.

Equation (12.7) is so simple that it may be integrated analytically to give $v(r)$

implicitly from

$$\left(\frac{v}{v_c}\right)^2 - \log_e\left(\frac{v}{v_c}\right)^2 = 4\log_e\frac{r}{r_c} + \frac{2GM}{rv_c^2} + C, \tag{12.8}$$

where C is part of the constant of integration and the other part has been incorporated in the log terms for convenience. The solutions have the form shown schematically in Figure 12.1. The critical point (A) is a saddle point, and dv/dr becomes infinite at $v = v_c (r \neq r_c)$, while dv/dr vanishes at $r = r_c (v \neq v_c)$.

Several types of solution are present in Figure 12.1, depending on the value of C. Types I and II are unacceptable because they are double-valued and do not connect the solar surface ($r \ll r_c$) continuously with locations far from the Sun ($r \gg r_c$). Type III possesses supersonic speeds at the Sun which are not observed, and so the only allowable solutions are types IV and V.

The 'solar-wind' solution (type IV) passes through the critical point (A) and corresponds to the value $C = -3$, obtained by putting $v = v_c$ and $r = r_c$ in Equation (12.8). At large distances where $v \gg v_c$ the velocity behaves like $v \sim (\log_e r)^{1/2}$, for which the first and third terms are dominant in Equation (12.8), and the density falls off like $n \sim r^{-2}(\log_e r)^{-1/2}$, so that the pressure vanishes at infinity, as required. For a temperature of 10^6 K the predicted flow speed at 1 AU is about 100 km s^{-1}.

Solutions of type V are everywhere subsonic and are often referred to as *solar (or stellar) breezes*. (The term can also be used to denote wind solutions with zero total energy.) At large distances where $v \ll v_c$, the second and third terms balance in (12.8), and so the flow speed falls off to zero like $v \sim r^{-2}$. At the same time, the density and pressure approach constant values, which may seem to exclude these solutions, except for the fact that, when a sufficiently rapid fall-off in temperature is incorporated,

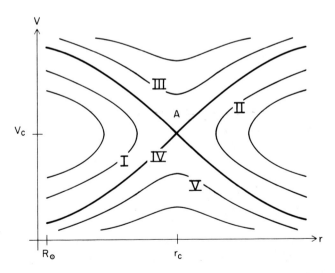

Fig. 12.1. A sketch of Parker's isothermal solutions, showing the different classes I, II, III, IV, V. Type IV (solar wind) passes through the critical point (A), where $v = v_c$ and $r = r_c$. Type V represents the subsonic, solar breeze solutions.

the pressure does go to zero. The predicted speed at the Earth's orbit is typically only 10 km s^{-1}.

When these solutions were proposed, it was unclear which was the most relevant to the Sun, but the question was resolved when Mariner II found speeds of several hundred kilometres per second near the Earth, and so it confirmed the solar wind solution as the appropriate one. The effect on the model of increasing the temperature is to enhance the isothermal sound speed (v_c) and to lower the critical-point radius (r_c), until, when the temperature exceeds 4×10^6 K, r_c falls below the solar surface. The main difficulties with the solution are that the outflowing plasma is expected to depart significantly from isothermality and the predicted density at the Earth (namely, 5×10^8 m^{-3} (500 cm^{-3})) is too high by a factor of a hundred, so that the mass flux is too high. Since the solution must pass through the critical point, the flow speed (v_0) at Sun is not arbitrary but is determined by the solution as typically 10 km s^{-1}. More realistic, spherically symmetric models (Section 12.3.1) with a lower mass flux give values for v_0 of 1 to 2 km s^{-1}, whereas coronal hole models (Section 12.4.2) whose base areas cover about $\frac{1}{4}$ of the solar surface possess a base speed of 10 to 20 km s^{-1}.

12.3. Models for a Spherical Expansion

Parker's elegant analytical solution was important in delineating the qualitative features of solar wind flow. It provides a classic example of how a physical idea may be modelled simply by incorporating only the dominant effects. Subsequently, as the reader may well imagine, the basic model was developed in numerous ways by incorporating many extra effects in an attempt to make the model more realistic. One of the technical difficulties in these often-numerical treatments is to find the solution which passes through the (sometimes-unknown) critical point (or points). We shall here outline only some of the more important extensions that apply to a steady, spherically expanding wind, and then will discuss in Section 12.4 the departures from spherical symmetry.

12.3.1. Energy equation

Spatial variations in temperature may be included by supplementing Equations (12.4) to (12.6) by an integrated energy equation, such as

$$nmvA\left(\tfrac{1}{2}v^2 + \frac{5p}{2nm} - \frac{GM_\odot}{r}\right) = A\kappa \frac{dT}{dr} + E_\infty, \qquad (12.9)$$

where $\kappa = \kappa_0 T^{5/2}$, $A = r^2$ and E_∞ is the energy at infinity (see Equation (2.40b)). The three terms on the left-hand side represent the flux of kinetic energy, enthalpy and gravitational energy, while the first term on the right is the conductive flux (assuming classical electron conduction). Its differential form (after substituting for $GM_\odot/r$ from Equation (12.5)) is

$$nv\left(\tfrac{3}{2}k_B \frac{dT}{dr} - \frac{k_B T}{n}\frac{dn}{dr}\right) = \frac{1}{A}\frac{d}{dr}\left(A\kappa\frac{dT}{dr}\right), \qquad (12.10)$$

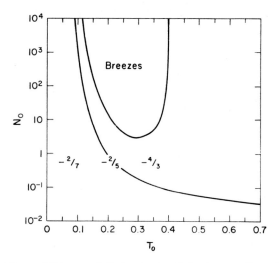

Fig. 12.2. The dependence of the asymptotic temperature behaviour on the surface density (n_0) and temperature (T_0) in units of $\kappa_0 (GM_\odot R_0 T_0^5)^{1/2} k_B^{-1}$ and $Gm_p M_\odot/(k_B R_\odot)$, respectively. Wind solutions with $T \sim r^{-2/7}$ are separated from those with $T \sim r^{-4/3}$ by a curve on which $T \sim r^{-2/5}$. The breeze solutions have zero total energy (after Roberts and Soward, 1972).

which follows from Equation (2.28d) when $p = nk_B T$ and the only contribution to L in Equation (2.32) is the conduction term. The topology of the solution curves is similar to Figure 12.1.

Several solutions have been discovered which traverse the critical point and have a vanishing temperature at infinity. They possess the following asymptotic behaviour: $T \sim r^{-2/7}$ (Parker, 1964), $T \sim r^{-2/5}$ (Whang and Chang, 1965), $T \sim r^{-4/3}$ (Durney, 1971). The first solution arises when the conductive flux is constant and dominates Equation (12.9) at large distances (as in Chapman's analysis of Section 12.1). It corresponds closely to the Parker isothermal solution, whereas the second solution occurs when the kinetic energy flux is the largest item. For the third type of solution the conductive flux is negligible, and so the flow is essentially adiabatic. In other words, $p/n^{5/3}$ is constant while $n \sim r^{-2}$, and so the temperature behaves like $p/n \sim n^{2/3} \sim r^{-4/3}$.

In a paper that is especially useful for *stellar* wind theory, Roberts and Soward (1972) showed how the type of solution depends on the surface values of the density (n_0) and temperature (T_0). Figure 12.2 shows that the Whang–Chang solution is a rather singular one that is unlikely to occur in practice. It separates conductively dominated solutions ($T \sim r^{-2/7}$) in cool, low-density winds from adiabatic solutions ($T \sim r^{-4/3}$) in hot, high-density winds. The values of n_0 and T_0 for the Sun probably lie just above the Whang–Chang solution.

12.3.2. Two-fluid model

In practice, the electrons and protons do not exchange enough energy to make their temperatures (T_p and T_e) the same. Hartle and Sturrock (1968) therefore replaced the single-fluid model by a two-fluid model, in which the electron fluid and proton

422   CHAPTER 12

fluid possess both the same density (by charge neutrality) and also the same velocity (since the Sun is electrically neutral).

The equations of mass continuity and momentum are

$$4\pi r^2 nv = \text{const.}, \tag{12.11}$$

$$nmv\frac{dv}{dr} = -\frac{d}{dr}(nk_B(T_e + T_p)) - \frac{GM_\odot mn}{r^2}, \tag{12.12}$$

where v and n are the common flow speed and number density. The separate energy equations for electrons and protons are

$$nv\left(\frac{3}{2}k_B\frac{dT_e}{dr} - \frac{k_B T_e}{n}\frac{dn}{dr}\right) = \frac{1}{r^2}\frac{d}{dr}\left(r^2\kappa_e\frac{dT_e}{dr}\right) - \frac{3}{2}vnk_B(T_e - T_p), \tag{12.13}$$

$$nv\left(\frac{3}{2}k_B\frac{dT_p}{dr} - \frac{k_B T_p}{n}\frac{dn}{dr}\right) = \frac{1}{r^2}\frac{d}{dr}\left(r^2\kappa_p\frac{dT_p}{dr}\right) + \frac{3}{2}vnk_B(T_e - T_p), \tag{12.14}$$

where v is a constant and $\kappa_e \approx (m_i/m_e)^{1/2}\kappa_p \approx 43\kappa_p$. Each of these two equations has the same form as Equation (12.10), except for the addition of the last term, which is proportional to the temperature difference between electrons and protons; it simulates the energy exchange by, for instance, Coulomb collisions between the two fluids. This coupling is important only near the Sun, where the density is high.

A numerical integration of the equations with T_e and T_p both equal to 2×10^6 K and n equal to 3×10^{13} m^{-3} at the base of the corona gives the temperature profiles shown in Figure 12.3. A single-fluid model (the limit as $v \to \infty$) gives a temperature of

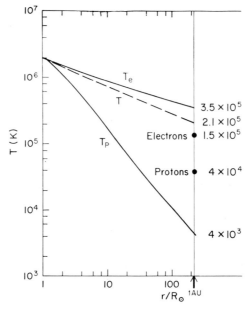

Fig. 12.3. The temperatures of electrons (T_e) and protons (T_p) as functions of distance for the two-fluid solar wind model of Hartle and Sturrock. The single-fluid profile is shown dashed and the observed values of low-speed streams are indicated by dots (after Hundhausen, 1972).

2.1×10^5 K at 1 AU, whereas the two-fluid model gives a somewhat higher electron temperature (3.5×10^5 K) and a much lower proton temperature (0.04×10^5 K). Because the electron conduction coefficient (κ_e) is much larger than the proton coefficient (κ_p), the electron temperature decreases with distance less rapidly than the proton temperature. The protons cool almost adiabatically, while the electrons remain hot because they are no longer forced to share their energy with the protons.

The Hartle–Sturrock model was very useful as a first attempt to allow for different electron and proton temperatures, but it does not agree quantitatively with observations of high-speed streams, which may be regarded as 'normal' solar wind. The main difficulty is that both the conductive flux and electron temperature are too high because the classical thermal conductivity is too high (Section 12.5). One way of lowering this value is to model the presence of a spiral magnetic field by allowing conduction only parallel to the field, but the most effective way is to introduce a collisionless conductivity. Another difficulty with the model is that it makes the flow speed (250 km s^{-1}) and proton temperature at 1 AU too low, especially for high-speed streams, which have $T_p \approx 2 \times 10^5$ K and $T_e \approx 10^5$ K. An extra energy source for the protons (such as collisionless damping of magnetohydrodynamic waves) can enhance the flow speed and proton temperature significantly if it is present above the critical point. Also, the difference between T_p and T_e can be reduced by enhancing the energy-exchange rate (v) above the Coulomb value.

12.3.3. Magnetic field

The magnetic field has several important effects on the properties of the solar wind. It causes differences in some parameters parallel and perpendicular to the field, such as the temperature. It also supports waves and produces micro-instabilities, but the only effect we shall consider here is the way it causes plasma close to the Sun to corotate with the Sun, almost like a solid body.

Parker (1958) considered qualitatively how a frozen-in magnetic field is dragged out by the solar wind, assuming a small magnetic energy compared with kinetic energy, so that the magnetic field just acts as a tracer and does not affect the wind speed. Also, Dessler (1967) has given a good discussion of the physics of the process in both fixed and rotating frames of reference. For a radial flow, the rotation of the Sun makes the magnetic field twist up into a spiral (Figure 12.4), just like a rotating gramophone record with its grooves corresponding to magnetic field lines and the motion of its needle corresponding to the solar wind outflow (provided the record is rotating backwards!).

Suppose the magnetic field is inclined at an angle ψ to the (radial) plasma velocity. Then the component of **v** perpendicular to **B** (namely $v \sin \psi$) must equal the speed of the (frozen-in) field line in that direction. But the field line rotates with the Sun, and so its speed relative to the solar surface is $\Omega(r - R_\odot)$ normal to the radius vector. Thus

$$v \sin \psi = \Omega(r - R_\odot) \cos \psi, \tag{12.15}$$

or

$$\tan \psi = \frac{\Omega(r - R_\odot)}{v}, \tag{12.16}$$

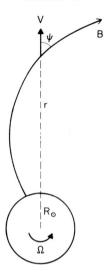

Fig. 12.4. A spiral magnetic field line attached to the Sun, which is rotating with angular speed Ω and is viewed from above the north pole. The solar wind is here assumed to dominate the magnetic field energetically, and it moves radially with speed v inclined at ψ to the magnetic field.

which gives $\psi \approx \tfrac{1}{4}\pi$ at the Earth. We note from Equation (12.16) that ψ vanishes at $r = R_\odot$, so that the magnetic field has been implicitly assumed normal to the solar surface there.

If the kinetic energy is so small compared with the magnetic energy that the flow speed at Earth is less than about 100 km s^{-1}, the magnetic field winds up into a very tight spiral and prevents the flow altogether at low latitudes. In practice, the plasma does have sufficient energy to overcome the field, but it does not completely dominate it either, so that both the radial and azimuthal flow components are affected by the field. An important parameter is the *Alfvén radius* ($r = r_A$), at which the flow speed becomes Alfvénic. Within that radius the magnetic field is so strong that it tends to keep the wind rotating with the Sun, and so increases the plasma's angular momentum as it moves out. Beyond the Alfvén radius the magnetic field has little effect on the solar wind, which therefore conserves its angular momentum.

Weber and Davis (1967) modelled this interaction between the solar wind and a magnetic field that is both radial and uniform at the solar surface. In spherical polars the magnetic field and fluid velocity in the equatorial plane are assumed to have components $(B_r, 0, B_\varphi)$ and $(v_r, 0, v_\varphi)$ that depend on r alone. The equation div $\mathbf{B} = 0$ then implies that

$$B_r = \frac{B_0 R_\odot^2}{r^2}, \tag{12.17}$$

where B_0 is the field strength at the solar surface. Furthermore, the equations of motion, continuity, energy, state and induction for steady, polytropic flow may be written

$$nm(\mathbf{v}\cdot\nabla)\mathbf{v} = -\nabla\left(p + \frac{B^2}{2\mu}\right) + (\mathbf{B}\cdot\nabla)\frac{\mathbf{B}}{\mu} - \frac{GM_\odot nm}{r^2}\hat{\mathbf{r}}, \tag{12.18}$$

THE SOLAR WIND

$$nv_r r^2 = \text{const.}, \tag{12.19}$$

$$\frac{p}{n^\alpha} = \text{const.}, \tag{12.20}$$

$$p = nk_B T, \tag{12.21}$$

$$0 = \nabla \times (\mathbf{v} \times \mathbf{B}). \tag{12.22}$$

The polytropic law (12.20) is adopted as an energy equation on grounds of simplicity, but the new feature is the presence of the Lorentz force in the momentum Equation (12.18).

The induction Equation (12.22) may be integrated to yield

$$v_r B_\varphi - v_\varphi B_r = \frac{C}{r}, \tag{12.23}$$

where C is constant and the term on the right is simply the θ-component of the electric field. Assuming $B_\varphi = 0$ and $v_\varphi = \Omega R_\odot$ at $r = R_\odot$, where Ω is the solar angular speed, Equation (12.23) gives $C = -\Omega R_\odot^2 B_0$, or, using Equation (12.17), $C = -\Omega r^2 B_r$. Thus Equation (12.23) determines the azimuthal field as

$$B_\varphi = \frac{v_\varphi - r\Omega}{v_r} B_r. \tag{12.24}$$

Next, the φ-component of Equation (12.18) becomes

$$nmv_r \frac{d}{dr}(rv_\varphi) = B_r \frac{d}{dr}(rB_\varphi),$$

where nv_r/B_r is constant by Equations (12.17) and (12.19). Thus, after integrating, we find

$$rv_\varphi - \frac{B_r}{nmv_r} rB_\varphi = L, \tag{12.25}$$

where the constant L is the *total angular momentum* per unit mass carried in both the plasma motion and the magnetic stresses. As the wind carries angular momentum away from the Sun, the Lorentz force transmits a torque to the solar surface, and it has probably slowed the Sun down significantly during its lifetime. Equation (12.24) may be used to eliminate B_φ from Equation (12.25) and so determine the azimuthal flow speed as

$$v_\varphi = \Omega r \frac{M_A^2 L/(r^2 \Omega) - 1}{M_A^2 - 1} \tag{12.26}$$

in terms of the radial *Alfvén Mach number*

$$M_A = \frac{v_r (\mu nm)^{1/2}}{B_r}.$$

The *Alfvén critical point* ($r = r_A$), where $M_A = 1$, was believed to be located at typically $20\,R_\odot$, but measurements from the Helios spacecraft suggest a mean value of $12\,R_\odot$. In order that v_φ remain finite there, it can be seen from Equation (12.26) that

$$L = \Omega r_A^2. \tag{12.27}$$

In other words, the total angular momentum density (L) is just what you would obtain from a solid-body rotation out to r_A.

Finally, the radial component of the equation of motion (12.18) is

$$nmv_r \frac{dv_r}{dr} - \frac{nmv_\varphi^2}{r} = -\frac{dp}{dr} - \frac{B_\varphi}{r}\frac{d}{dr}(rB_\varphi) - \frac{GM_\odot nm}{r^2},$$

which Weber and Davis solve, with v_φ, B_φ, n and p given by Equation (12.26), (12.24), (12.19) and (12.20), respectively. The $v_r - r$ phase-plane (Figure 12.5) is complicated by the presence of three critical points, where the flow speed attains successively the characteristic values for slow, Alfvén and fast waves. Since the plasma pressure is much smaller than the magnetic pressure far from the Sun, the Alfvén and fast magneto-acoustic speeds are very close together, while the slow speed equals the sound speed approximately. If the rotation is set equal to zero the Alfvén and fast-mode critical points coincide exactly.

Only the two solutions I and II which pass through all three critical points have a vanishing pressure at infinity. Solution II must, however, be ruled out, because it predicts much too small a flow speed at the Earth (9 km s^{-1}). Type I gives a radial speed there of 425 km s^{-1} and an azimuthal speed (v_φ) of 1 km s^{-1}. Although the predicted temperature and density do not agree well with observations, the analysis is important in delineating the topology of the solutions. Brandt et al. (1969) repeated the analysis with a full energy equation. They integrated away from the critical points and found $n = 6 \times 10^6$ m^{-3}, $v_r = 315$ km s^{-1}, $T = 3.2 \times 10^5$ K, $v_\varphi = 2.5$ km s^{-1}, and $\psi = 55°$ at 1 AU. This agrees reasonably with observations, except that the measurement of v_φ is difficult and somewhat uncertain at the present time. It used to be thought that $v_\varphi \approx 10$ km s^{-1}, but recent measurements from Helios seem to give a possibly much smaller value (0 to 10 km s^{-1}).

Sakurai (1982) has generalised the Weber and Davis solutions by allowing different values for the surface gravity (g_0) and surface temperature (T_0). When T_0 is high

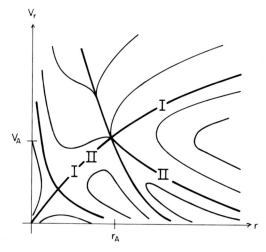

Fig. 12.5. A sketch of the solution topology for a spherical expansion from a rotating magnetised Sun (after Weber and Davis, 1967).

enough, the resulting stellar winds are thermally driven, whereas, when g_0 is small enough, they are centrifugally driven. On the other hand, when T_0 is small and g_0 is large, there are no wind solutions and the model atmosphere becomes hydrostatic (neglecting other effects such as radiation pressure).

12.4. Streamers and Coronal Holes

One assumption in the models of the previous sections is that the solar wind expansion is spherically symmetric, but this is a poor approximation close to the Sun where most of the acceleration takes place. On both eclipse and soft X-ray photographs (e.g., Figures 1.3(d), 1.4, 1.15 and 12.6(a)) one sees evidence of some regions with a predominantly open magnetic field and others with closed fields. The former are known as *coronal holes* (Section 1.3.4C), and the latter consist of *coronal loops* (Sections 1.3.4B and 6.5). Most of the solar wind is probably escaping from the solar surface as high-speed streams along the open fields of coronal holes, but it is possible that a significant mass flux dribbles across closed fields with the aid of small-scale interchange instabilities (Section 7.5.1) and produces low-speed streams.

The large-scale closed structures surmounted by open fields and known as *coronal streamers* (Section 1.3.4A) have been modelled by Pneuman and Kopp (1971), as described in (Section 12.4.1). They contain a high-pressure, high-density plasma because the closed magnetic field is strong enough to hold the plasma down. By contrast, the density of an open region falls off rapidly with height because the plasma can easily escape and because conductive losses lower the temperature and scale-height there. An additional factor is that the heating is likely to be more efficient in the closed regions and so make the temperature higher than in coronal holes. Since the total pressure (plasma plus magnetic) must remain continuous across the interface between closed and open regions, the fall-off of pressure with height in the open region means that there is a jump in magnetic field strength at the interface. The interface is therefore a curved current sheet, across which the tangential flow speed is discontinuous. The detailed structure of such a viscous-resistive boundary layer has not yet been studied. Pneuman and Kopp solve the isothermal magnetohydrodynamic equations self-consistently, and they deduce the radial variation of the coronal hole area. A complementary approach is to modify the standard solar wind models of Sections 12.2 and 12.3 by introducing a prescribed area variation; this sometimes gives rise to extra critical points (Section 12.4.2), as first pointed out by Parker (1963b, Appendix A).

12.4.1. Pneuman–Kopp Model

12.4.1A. Basic Model

Pneuman and Kopp (1971) have solved numerically the magnetohydrodynamic equations for steady, axisymmetric, coronal expansion from a sphere whose surface magnetic field is dipolar. The equations of motion, induction and continuity are

$$mn(\mathbf{v} \cdot \nabla)\mathbf{v} = -\nabla p - \frac{GM_\odot mn}{r^2}\hat{\mathbf{r}} + \mathbf{j} \times \mathbf{B}, \tag{12.28}$$

$$0 = \nabla \times (\mathbf{v} \times \mathbf{B}), \tag{12.29}$$

$$\nabla \cdot (n\mathbf{v}) = 0, \tag{12.30}$$

where m is the mean particle mass, n is the total particle density, T is uniform ($= 1.56 \times 10^6$ K), and

$$p = nk_B T, \tag{12.31}$$

$$\mathbf{j} = \operatorname{curl} \mathbf{B}/\mu, \tag{12.32}$$

$$\nabla \cdot \mathbf{B} = 0. \tag{12.33}$$

All variables are functions of r and θ alone (in spherical polars), and the boundary conditions are that n and B_r be prescribed at the solar surface as $n = n_0$, $B_r = B_{0p} \cos \theta$, where B_{0p} is the field strength at the poles (taken to be 1 G) and n_0 is a constant such that $n_0 k_B T = B_{0p}^2/(2\mu)$.

The equations are solved iteratively by starting with a current distribution and calculating $\mathbf{B}$ from Equation (12.32). The area $A(s)$ of each flux tube follows from Equation (12.33) as

$$A(s) = \frac{A_0 B_0}{B(s)}$$

in terms of the base values A_0 and B_0 and the distance (s) along each flux tube. Next, Equation (12.30) and the component of Equation (12.28) parallel to $\mathbf{B}$ give $n(s)v(s)A(s) = n_0 v_0 A_0$ and

$$\tfrac{1}{2}(v^2 - v_0^2) + \frac{k_B T}{m} \log_e \frac{p}{p_0} + \frac{GM_\odot}{R_\odot} \left[1 - \frac{R_\odot}{r} \right] = 0,$$

Fig. 12.6(a). A total eclipse (30 June, 1973), showing the outer corona in white light (courtesy High Altitude Observatory, NCAR, Boulder) and the inner corona in soft X-rays (courtesy American Science and Engineering, Cambridge, Mass.).

which are solved for $n(s)$ and $v(s)$, with v_0 determined to give the critical solar wind solution on each flux tube. Finally, a new current is determined from the component of Equation (12.28) normal to **B**, and the procedure is repeated until the solutions converge.

The resulting magnetic structure is shown in Figure 12.6b, with an appearance similar to that of a large helmet streamer. The field lines are dragged out by the solar wind to form a current sheet at the equator with a closed region below it extending out to $2.5\,R_\odot$. At large distances the field lines become radial. On polar field lines the flow becomes sonic at about $4\,R_\odot$ and then Alfvénic at $5\,R_\odot$, whereas near the equator where the magnetic field is weaker the flow first becomes Alfvénic (near the neutral point at $2.5\,R_\odot$) and then sonic at about $5\,R_\odot$. The polar field lines below the neutral point expand less rapidly than those near the closed region, and so, although the polar flow is slower than the low-latitude flow below $2\,R_\odot$, it is faster beyond $2\,R_\odot$. Above the cusp the current density increases from zero at the poles to a maximum at the equator.

The above values of 2.5 to $5\,R_\odot$ for the location of the Alfvén radius (r_A) are much smaller than the mean value ($12\,R_\odot$) deduced from Helios measurements. One reason for the discrepancy could be that the value of less than 1 G adopted for the mean base field strength is too small, and that, say, 8 G is more reasonable. This is the value obtained by assuming a field of 4×10^{-5} G at 1 AU and extrapolating back to coronal holes occupying 20% of the Sun's surface area.

In the future, it will be interesting to see how the Pneuman–Kopp model is modified by replacing the isothermal approximation with a full energy equation. Also, it is

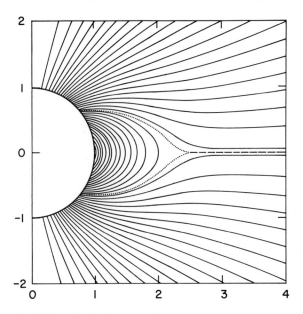

Fig. 12.6(b). Magnetic field lines for the Pneuman–Kopp model of isothermal, axisymmetric expansion from a dipolar surface field. The dotted curve marks the vortex-current sheet separating open and closed regions, while the dashed line indicates the current sheet extending away from the cusp-type neutral point (after Pneuman and Kopp, 1971).

important to decide what is the most relevant latitudinal variation of the base pressure to impose as a boundary condition. Instead of the iterative technique of Pneuman and Kopp, it would be preferable to solve the time-dependent MHD equations along the lines of Endler (unpublished thesis). Observationally, there is some evidence that the centre of a coronal hole produces a smooth, hot, rare, fast flow at large distances, whereas near the edges of the hole the flow is irregular, cool, and dense, and it possesses a lower mass flux density. Another point is that, although the Pneuman–Kopp solution is constrained to pass through the sonic critical point (Section 12.3.3), it is not forced to traverse the Alfvén critical point, and so it may have been inaccurate at large distances.

The axisymmetric Pneuman–Kopp model has a disc-like current sheet extending out from $2.5\,R_\odot$, and it is likely to represent qualitatively the structure of the inner solar wind for three-quarters of the solar cycle, especially near sunspot minimum, when the polar coronal holes dominate. Since the sheet is in practice often inclined to the plane of the Earth's orbit, the Earth consecutively samples plasma from the north polar hole and then from the south polar hole. Also, in reality the current sheet is usually warped like the rim of a sombrero due to departures from axisymmetry in the surface magnetic field (Figure 12.7). This means that the solar wind flow past the Earth is probably much more varied than it is at higher latitudes. Near solar maximum the polar holes no longer dominate and indeed may be absent, but coronal holes are present at mid-latitudes so that the global magnetic field is highly complex, with a current sheet that may meander far from the equator and may consist of several topologically distinct sheets (Hundhausen et al., 1981).

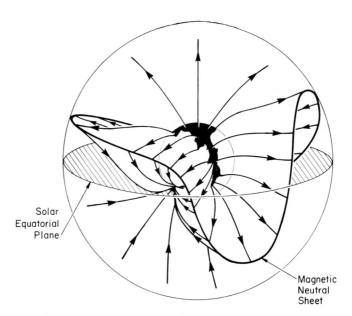

Fig. 12.7(a). A schematic drawing of the warped neutral current sheet that was present during the summer of 1973, when Skylab observed a large coronal hole stretching across the equator (Section 1.3.4C). The neutral sheet probably crosses the solar equatorial plane four times (after Hundhausen, 1981).

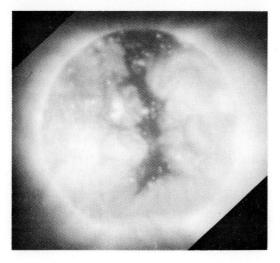

Fig. 12.7(b). A soft X-ray picture taken on June 1, 1973, of the coronal hole.

12.4.1B. Angular Momentum Loss

The effect of solar rotation on a spherically symmetric wind has been described in Section 12.3.3. For the axisymmetric Pneuman–Kopp model it has also been evaluated by Priest and Pneuman (1974). Departures from spherical symmetry reduce the angular momentum loss-rate by two effects, namely the presence of closed magnetic field regions with no loss and also the latitude variation in the location (r_A) of the Alfvénic point: since the Alfvén radius is closer to the Sun for field lines near the equator, the angular-momentum density L (see Equation (12.27)) is smaller there. The net result of both effects is that the angular momentum loss-rate is only 15% of that for a monopole field.

12.4.1C. Current Sheet

A model for steady flow in the current sheet which extends from the top of the helmet streamer has been considered by Pneuman (1972a) and Priest and Smith (1972). They suppose that the current sheet interior possesses a transverse magnetic field component, so that the field lines form stationary loops connected to the solar surface. Inside the sheet the density is enhanced and the flow is slower than in the ambient solar wind because of the additional Lorentz force. The sheet is thin enough that the plasma is able to slip outwards across the magnetic loops by magnetic diffusion. Solutions of the resistive magnetohydrodynamic equations are sought by separating the variables r and θ. It is found that the flow speed in the current sheet is less than half the ambient solar wind speed at the cusp and falls off rapidly to 10^{-8} G at $10\,R_\odot$. Incorporating a non-classical diffusion would modify the detail of the results considerably.

12.4.2. CORONAL HOLE MODELS

A full treatment of the magnetohydrodynamic interaction has so far been possible only for a simple geometry (Section 12.4.1), but a simple coronal hole model may be established by solving the solar wind equation in a prescribed magnetic geometry and neglecting the reaction of the plasma on the magnetic field. Observationally, one finds that the coronal hole density is lower than that of the average Sun by at least a factor of 3, and coronal holes have now been identified as the source of high-speed streams (with a typical particle flux of 3×10^{12} m^{-2} s^{-1}). Even for a spherically symmetric outflow this implies enormous speeds close to the Sun: e.g., a density of, say, 3×10^{11} m^{-3} at $2 R_\odot$ gives a speed of 120 km s^{-1}.

Kopp and Holzer (1976) considered polytropic flow in an open field of area $A(r)$ by solving the equations of continuity, momentum and energy in the form

$$nvA = \text{const.}, \tag{12.34}$$

$$mnv\frac{dv}{dr} = -\frac{dp}{dr} - \frac{GM_\odot mn}{r^2}, \qquad \frac{p}{n^\alpha} = \text{const.}, \tag{12.35}$$

where $p = nk_B T$ and α is a constant (taken to be 1.1.). These may be combined to give an equation for the Mach number

$$M = \frac{v}{c_s}, \tag{12.36}$$

where $c_s^2 = \alpha p/(nm)$. It takes the form

$$\frac{M^2 - 1}{2M^2}\frac{dM^2}{dr} = \frac{1}{2}\left[\frac{\alpha+1}{\alpha-1}\right]\left[1 + \left[\frac{\alpha-1}{2}\right]M^2\right]\frac{1}{g}\frac{dg}{dr}, \tag{12.37}$$

where $g(r) = A(r)^{2(\alpha-1)/(\alpha+1)}(E + GM_\odot/r)$.
Also,

$$E = \tfrac{1}{2}v^2 + \frac{\alpha}{\alpha-1}\frac{p}{nm} - \frac{GM_\odot}{r}$$

is the total energy, which is taken to be 1.8×10^{11} J kg^{-1} so as to produce the flow speed of 600 km s^{-1} at infinity that is characteristic of high-speed streams. The corresponding base temperature is 2×10^6 K.

Just as for Parker's solar wind solutions, Equation (12.37) may be integrated analytically to give $v(r)$ implicitly, and it possesses a critical point where $M^2 = 1$ and $dg/dr = 0$. However, unlike the case of spherical expansion ($A(r) \sim r^2$), there may be more than one critical point, and they may be either X-type or O-type in nature, depending on whether $g(r)$ possesses a minimum or a maximum, respectively.

The area function is written as $A(r) = A_\odot(r/R_\odot)^2 f(r)$, where $f(r)$ is a function that increases monotonically from 1 at $r = R_\odot$ to $f_{\max}$ (> 1) at large distances, most of the increase occurring between $1.4 R_\odot$ and $1.6 R_\odot$. Thus $A(r)$ increases like r^2 at low heights, much faster than r^2 near $1.5 R_\odot$, and then like r^2 again in the outer corona. By contrast, the area derived by Munro and Jackson (Figure 1.17(a)) increases 7 times faster than radially from the surface to $3 R_\odot$, and thereafter it is very nearly radial.

The flow topology depends on $f_{\max}$, as indicated in Figure 12.8. A Parker-type

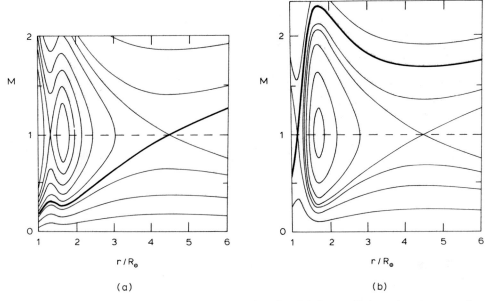

Fig. 12.8. The Mach number ($M = v/c_s$) as a function of radial distance (r) from the solar centre for polytropic flow in a coronal hole whose net non-radial divergence (f_{max}) is (a) 3 and (b) 12 (Kopp and Holzer, 1976).

critical point is always present at 4.5 $R_\odot$, but when f_{max} exceeds 2 an extra pair of critical points appears between $1R_\odot$ and $2R_\odot$, the inner one being X-type and the outer one O-type. When the coronal hole does not diverge too much ($f_{max} < 7.5$), the flow is close to spherical and the only solar wind flow (with a continuous connection of subsonic values at the Sun to supersonic values at large distances) is similar to Parker's solution: it is accelerated to a local maximum at the radius where the inner critical point is located, and then it remains subsonic until it crosses the outer critical point at 4.5 $R_\odot$, as exemplified in Figure 12.8(a). When the coronal hole expands rapidly ($f_{max} > 7.5$), the solar wind flow is rapidly accelerated to supersonic speeds at the inner critical point low down in the corona (1.15 $R_\odot$ for $f_{max} = 12$ in Figure 12.8(b)). The flow speed then attains a local maximum at the middle critical point, and thereafter it remains supersonic. As f_{max} is increased above 7.5, so the base flow becomes faster and the density falls everywhere, especially in the low corona where the rapid acceleration is occurring.

The main point of the Kopp–Holzer paper is that it explains qualitatively the low densities that are observed near the base of coronal holes. However, the rapid field divergence does not account for the high speeds that are observed at Earth, since all of their solar wind solutions, regardless of the value of f_{max}, possess the same asymptotic flow speed at large distances. Any polytropic model, even with a spherical expansion, can provide a high enough speed at 1 AU if the base temperature is taken high enough (2×10^6 K for the Kopp–Holzer model), but when a full energy equation is employed such high speeds are no longer produced.

The other contribution of Kopp and Holzer was to elucidate the topology of the solutions, with the two basic types shown in Figure 12.8. The analysis of Munro and Jackson (Section 1.3.4C, Figure 1.17(b)) gave high speeds low down in a coronal hole (200 km s^{-1} at 2.5 $R_\odot$), which suggests that a flow similar to Figure 12.8(b) (with rapid acceleration through the lower critical point) is the relevant type.

The above analysis was later extended by several authors. For instance, Steinolfson and Tandberg-Hanssen (1977) showed that similar features are present in a conductive model. Also, Leer and Holzer (1979) and Holzer and Leer (1980) have used the mass flux density at 1 AU and the coronal base pressure to deduce that the temperature maximum in the subsonic part of coronal holes is probably less than 2.5 × 10^6 K. For a conductive model with an energy equation of the form (12.10), they find that the solar wind speed at 1 AU (v_e) cannot be increased to high-speed values by either a rapidly diverging geometry or collisionless inhibition of heat conduction. The main effects of a rapid divergence are to lower the mass flux density at 1 AU and to increase it at the base (for fixed values of n_0 and T_0). Holzer (1977) considers a full energy equation including the addition of heat and momentum by, for instance, waves. The momentum addition can produce additional critical points that may be nodes or foci instead of saddle-points and O-type points. Furthermore, Leer and Holzer (1980) show that the addition of energy in the region of subsonic flow increases the solar wind flux but cannot significantly increase v_e. By contrast, energy addition in the supersonic flow region does not affect the mass flux but it significantly increases v_e. High-speed solar wind streams may be produced in three possible ways:

(i) subsonic momentum loss by, for instance, the magnetic field;
(ii) the transfer of energy by waves from the subsonic region to the supersonic region;
(iii) the propagation of energy (as Alfvén waves or fast waves) from the coronal base and its deposition in the supersonic region. (For a 10 G base field a velocity amplitude of 20 km s^{-1} would be needed to make $v_e = 700$ to 800 km s^{-1}.)

Ultimately, one would like to see a self-consistent solution of the plasma and magnetic field equations based on observed boundary conditions at the Sun and reproducing the properties at the Earth. As a start, Durney and Pneuman (1975) have taken a step towards a more realistic modelling of a solar wind by solving the flow equations along a potential field based on the observed photospheric field. Also, Seuss *et al.* (1977) have modelled three-dimensional, polytropic, magnetohydrodynamic flow for a coronal hole by assuming a small departure from spherical symmetry and expanding about a spherically symmetric state; the resulting hole boundary compares very favourably with the observed boundary of Munro and Jackson (Figure 1.17(a)).

12.5. Extra Effects

Solar wind theory is an immense subject, and so there are many effects which we have not covered. One possible source of the heating and momentum that is necessary to drive high-speed streams is outward propagating Alfvén waves, probably generated at the Sun. These have been studied thoroughly by Hollweg (1974, 1978 and 1981c). In particular, he accounts for the observation that the proton temperature and velocity at 1 AU tend to vary together in a systematic way, with faster flows having hotter protons. Adding more Alfvén wave energy can explain this correlation since it makes the flow both faster and hotter. Indeed, at 1 AU the dominant component of fluctuations is observed to be in the form of non-compressive Alfvén waves, with no changes in density or magnetic field magnitude. (The fast and slow compressional waves generated at the Sun are presumably either damped before they reach the Earth or are reflected at the transition region.) Hollweg shows that Alfvén wave pressure can accelerate the flow mainly in the supersonic region, and that it decreases like $\rho_0^{1/2}$ inside the Alfvén radius and like $\rho_0^{3/2}$ beyond it. (A more detailed analysis of wave-momentum transfer has been presented by Jacques (1977).) Hollweg also sets up a two-fluid solar wind model with Alfvén wave heating of the protons mainly in the supersonic region, and he obtains good agreement with observations at 1 AU. For instance, the temperature–velocity correlation is reproduced quite well by varying only the wave energy flux at the base. Future improvements would be to explain the nonlinear mechanism for damping the waves and to raise the density near the base by incorporating a non-radial geometry. Another possible source of heating and acceleration is the presence of fast magnetoacoustic waves (Barnes, 1979), which dissipate more easily than Alfvén waves and can propagate across the magnetic field.

The models that have been described here are fluid ones, but beyond about 0.1 AU, the quasi-fluid behaviour and transport coefficients should be derived from the microphysics of small-scale plasma instabilities rather than from Coulomb collisions (e.g., Schwarz, 1980, 1981). The steep temperature gradients that are predicted in fluid models invalidate the classical Spitzer conductivity law beyond 2 $R_\odot$, and they produce several plasma instabilities which reduce the conduction coefficient and transfer energy from electrons to protons. Also, the proton temperature is not isotropic, with the parallel temperature being held to only about twice the perpendicular temperature by, for example, the fire-hose instability or the mirror instability.

Other processes that need to be included in steady-state models are the effect of viscosity, the inclusion of He^{++} in the energetics, and the breakdown in ionisation equilibrium that first arises low down in coronal holes. Features that we have not touched in this chapter are: transverse spatial variations in the rotating solar wind structure, such as the interactions between slow and fast streams; time variations associated with the ejecta from flares and erupting prominences; interactions of the solar wind with the various planets and its termination at perhaps 100 AU.

Although solar wind models are reasonable to within a factor of two, more precise values for the flow parameters are not yet well-determined everywhere. In particular, there is a need for an accurate measurement both of the boundary conditions (on temperature, flow speed and wave flux) near the Sun and of the properties out of the ecliptic plane. The *Solar Polar Mission* and a *Lyman-α coronagraph* should help greatly, and, judging by past experiences, they are likely to produce some new surprises. It is important to investigate the contribution of plasma that is escaping from closed-field regions and from small-scale open-field regions rather than coronal holes. Also, there is a need to understand the recurrence of preferred longitudes and to model the solar-cycle variation of coronal holes and solar wind properties. Finally, it must be borne in mind that the plasma may not be escaping uniformally from the surface of a coronal hole, but instead it may possibly be ejected from many small-scale features such as ephemeral active regions.

APPENDIX I

UNITS

The question of what units to adopt is one that has traditionally complicated the subject of electromagnetism. It would clearly be advantageous if a single system could be accepted by all, even though the transition to standard units may be a painful one. I myself was reared on the gaussian system, and some years ago I suffered the trauma of changing to rationalised mks, which is part of the now internationally adopted SI system.

In the present book we have employed mks units, and have commonly quoted magnetic fields (**B**) in gauss (G), which is acceptable in SI. In formulae, however, the tesla (T) has been used*, such that

$$1\,T = 10^4\,G.$$

Also, in the text lengths have often been quoted in megametres (Mm), such that

$$1\,Mm = 10^3\,km = 10^6\,m,$$

although lengths in formulae are measured in m*. One of the most helpful discussions of units can be found in Boyd and Sanderson (1969), which has been used as a basis for this section.

At one time, it appeared that each book on electromagnetism used a different set of units. The reasons for confusion are as follows.

(i) A set of electromagnetic quantities may be measured in either electrostatic units (esu) or electromagnetic units (emu) or a combination of the two.

(ii) The magnetic permeability (μ) and dielectric constant (ε) may be either dimensionless or dimensional. In the former case, mass, length and time are fundamental, whereas in the latter case some extra fundamental unit is introduced.

(iii) Mass and length may be measured in either centimetres and grams (cgs) or metres and kilograms (mks).

(iv) A system may be 'rationalised' to remove the factor 4π from the equations.

The two main sets of units are *Gaussian cgs* (for which the defining equations are simpler, so it is a 'natural' system) and *rationalised mks* (for which Maxwell's equations are simpler, so it is easier to use).

Gaussian cgs

e, **E** and **D** are measured in esu, while **j**, **B** and **H** are in emu. Also, ε and μ are dimensionless. Length, mass and time are fundamental, and then other units are defined in terms of them by the appropriate equations. For instance: Newton's second law gives force; the force of repulsion between charges e_1 and e_2 separated by a distance r

* unless otherwise stated

gives charge; the force per unit length between currents I_1 and I_2 separated by r gives current; the magnetic induction at a distance r from a straight wire carrying a current I gives magnetic induction.

Quantity	Defining equation	Unit
length	–	1 cm
mass	–	1 g
time	–	1 s
force	$F = m\ddot{x}$	1 dyne = 1 g cm s^{-2}
charge	$F = \dfrac{e_1 e_2}{r^2}$	1 statcoulomb = 1 cm (dyne)$^{1/2}$
electric field	$E = \dfrac{e}{r^2}$	1 statvolt cm^{-1} = 1 (dyne)$^{1/2}$ cm^{-1}
electric displacement	$\mathbf{D} = \varepsilon \mathbf{E}$	1 statvolt cm^{-1}
current	$\dfrac{F}{l} = \dfrac{2 I_1 I_2}{r}$	1 abamp = 1 (dyne)$^{1/2}$
magnetic induction	$B = \dfrac{2I}{r}$	1 gauss = 1 abamp cm^{-1}
magnetic field	$\mathbf{H} = \mathbf{B}/\mu$	1 oersted = 1 gauss

If the current is measured in esu rather than emu, it is defined as the time rate of change of charge, and so has units of 1 statamp = 1 statcoulomb s^{-1}. The ratio of the values of I measured in the two units is,

$$\frac{I \text{ in esu}}{I \text{ in emu}} = c, \quad \text{say.}$$

The constant of proportionality c has units of cm s^{-1} and the approximate value 3×10^{10} cm s^{-1}.

Maxwell's equations in this system are

$$\text{curl } \mathbf{E} = -\frac{1}{c}\frac{\partial \mathbf{B}}{\partial t}, \quad \text{div } \mathbf{D} = 4\pi\rho,$$

$$\text{curl } \mathbf{H} = \frac{1}{c}\frac{\partial \mathbf{D}}{\partial t} + 4\pi \mathbf{j}, \quad \text{div } \mathbf{B} = 0.$$

If $\mathbf{j}$ is measured in esu rather than emu, $4\pi\mathbf{j}$ is replaced by $4\pi\mathbf{j}/c$.

Rationalised mks

Length, mass, time and current (amp) are fundamental, and other units are defined in terms of them by the appropriate equations. μ_0 is defined to be $4\pi \times 10^{-7}$ N amp^{-2} and is the vacuum value of μ. The unit of current is defined such that two wires carrying currents of 1 amp and separated by 1 m feel a force per unit length of 2×10^{-7} N m^{-1}. ε_0 is the vacuum value of ε and has units of A^2 s^2 N^{-1} m^{-2}. Its value is determined (as approximately 8.854×10^{-12}) from Coulomb's law ($F = e_1 e_2/(4\pi\varepsilon_0 r^2)$) such that two charges of 1 C separated by 1 m feel a repulsive force of $(4\pi\varepsilon_0)^{-1}$ N.

Quantity	Defining equations	Unit
length	–	1 m
mass	–	1 kg
time	–	1 s
force	$F = m\ddot{x}$	1 newton (N) = 1 kg m s^{-2}
current	$\dfrac{F}{l} = \dfrac{\mu_0}{4\pi} \cdot \dfrac{2I_1 I_2}{r}$	1 amp (A)
magnetic induction	$B = \dfrac{\mu_0}{4\pi} \cdot \dfrac{2I}{r}$	1 tesla = 1 N A^{-1} m^{-1}
magnetic field	$H = B/\mu$	1 A m^{-1}
charge	$I = \dfrac{dq}{dt}$	1 coulomb (C) = 1 A s
electric field	$E = \dfrac{1}{4\pi\varepsilon_0} \dfrac{e}{r^2}$	1 V m^{-1} = 1 N C^{-1}
electric displacement	$\mathbf{D} = \varepsilon \mathbf{E}$	1 C m^{-2}

Maxwell's equations are

$$\text{curl } \mathbf{E} = -\frac{\partial \mathbf{B}}{\partial t}, \qquad \text{div } \mathbf{D} = \rho,$$

$$\text{curl } \mathbf{H} = \frac{\partial \mathbf{D}}{\partial t} + \mathbf{j}, \qquad \text{div } \mathbf{B} = 0.$$

The wave equation in a vacuum becomes

$$\left(\nabla^2 - \varepsilon_0 \mu_0 \frac{\partial^2}{\partial t^2} \right) \mathbf{E} = 0,$$

and so $c^2 = (\varepsilon_0 \mu_0)^{-1}$.

Transformation of Symbols in Equations

A comparison of, for instance, Coulomb's law in the two sets of units shows that, in going from cgs to mks, e becomes $e(4\pi\varepsilon_0)^{-1/2}$. The transfer for other symbols is as follows.

Quantity	Gaussian	Rationalised mks
speed of light	c	$(\varepsilon_0 \mu_0)^{-1/2}$
charge	e	$e(4\pi\varepsilon_0)^{-1/2}$
current	j	$j(4\pi\varepsilon_0)^{-1/2}$
electric field	E	$E(4\pi\varepsilon_0)^{1/2}$
magnetic induction	B	$B(4\pi/\mu_0)^{1/2}$
electric displacement	D	$D(4\pi/\varepsilon_0)^{1/2}$
magnetic field	H	$H(4\pi\mu_0)^{1/2}$
dielectric constant	ε	$\varepsilon/\varepsilon_0$
magnetic permeability	μ	μ/μ_0
electrical conductivity	σ	$\sigma(4\pi\varepsilon_0)^{-1}$

Relation Between Units

Quantity	Gaussian	Rationalised mks
length	1 cm =	10^{-2} m
mass	1 g =	10^{-3} kg
time	1 s =	1 s
force	1 dyne =	10^{-5} N
energy	1 erg =	10^{-7} joule (J)
	(1 calorie =	4.185 J)
power	1 erg s^{-1} =	10^{-7} watt (W)
charge	1 statcoul =	$\frac{1}{3} \cdot 10^{-9}$ C
electric field	1 statvolt cm^{-1} =	3×10^4 V m^{-1}
displacement	1 statvolt cm^{-1} =	$\frac{1}{12\pi} \times 10^{-5}$ C m^{-2}
current	1 statamp =	$\frac{1}{3} \times 10^{-9}$ A
current density	1 statamp cm^{-2} =	$\frac{1}{3} \times 10^{-5}$ A m^{-2}
electrical conductivity	1 s^{-1} =	$\frac{1}{9} \times 10^{-9}$ mho m^{-1}
magnetic induction	1 G =	10^{-4} tesla (T)
magnetic field	1 oersted =	$\frac{1}{4\pi} \times 10^3$ A m^{-1}

APPENDIX II

USEFUL VALUES AND EXPRESSIONS

Physical Constants

Speed of light	$c = 2.998 \times 10^8$ m s^{-1}
Electron charge	$e = 1.602 \times 10^{-19}$ C
Electron mass	$m_e = 9.109 \times 10^{-31}$ kg
Proton mass	$m_p = 1.673 \times 10^{-27}$ kg
Mass ratio	$m_p/m_e = 1837$
Electron volt	$1\text{ eV} = 1.602 \times 10^{-19}$ J $= 11\,605$ K
Boltzmann constant	$k_B = 1.381 \times 10^{-23}$ J deg^{-1}
Gravitational constant	$G = 6.672 \times 10^{-11}$ N m^2 kg^{-2}
Gas constant	$\tilde{R} = 8.3 \times 10^3$ m^2 s^{-2} deg^{-1}
Permeability of free space	$\mu_0 = 4\pi \times 10^{-7} = 1.257 \times 10^{-6}$ henry m^{-1}
Permittivity of free space	$\varepsilon_0 = 8.854 \times 10^{-12}$ farad m^{-1}

Plasma Properties (n_e in m^{-3}, B in Gauss, T in °K, v in m s^{-1}, l in m)

Sound speed	$c_s = 152\, T^{1/2}$ m s^{-1}
Alfvén speed	$v_A = 2.8 \times 10^{12}\, B\, n^{-1/2}$ m s^{-1}
Plasma beta	$\beta = 3.5 \times 10^{-21}\, n\, T\, B^{-2}$
Magnetic Reynolds number	$R_m = 2 \times 10^{-9}\, l\, v\, T^{3/2}$
Scale-height	$\Lambda = 50\, T$ m for $\tilde{\mu} = 0.6$
Electrical conductivity	$\sigma = 10^{-3}\, T^{3/2}$ mho m^{-1}
Magnetic diffusivity	$\eta = 10^9\, T^{-3/2}$ m^2 s^{-1}
Thermal conductivity	$\kappa_\parallel = 10^{-11}\, T^{5/2}$ W m^{-1} deg^{-1}
Electron plasma frequency	$\omega_{pe} = 5.64\, n_e^{1/2} \times 10^7$ rad s^{-1}
Electron gyro-frequency	$\Omega = 1.76\, B \times 10^7$ rad s^{-1}
Proton gyro-frequency	$\Omega_p = 9.58\, B \times 10^3$ rad s^{-1}
Debye length	$\lambda_D = 6.9\, (T_e/n_e)^{1/2} \times 10$ m
Electron gyro-radius	$r_e = 2.38 \times 10^{-2}\, (T_e/11\,605)^{1/2}\, B^{-1}$ m
Proton gyro-radius	$r_p = 1.02(T_p/11\,605)^{1/2}\, B^{-1}$ m
Electron thermal speed	$v_{Te} = (k_B T_e/m_e)^{1/2} = \Omega r_e = 4.19 \times 10^5 (T_e/11\,605)^{1/2}$ m s^{-1}
Proton thermal speed	$v_{Ti} = (k_B T_p/m_p)^{1/2} = \Omega_p r_p = 9.79 \times 10^3 (T_p/11\,605)^{1/2}$ m s^{-1}

Solar Properties

Age	$= 4.5 \times 10^9$ yr
Mass	$M_\odot = 1.99 \times 10^{30}$ kg
Radius	$R_\odot = 696$ Mm $= 6.96 \times 10^8$ m $= 109$ earth radii
Surface gravity	$g_\odot = 274$ m s^{-2}
Escape velocity	$= 618$ km s^{-1}
Luminosity	$L_\odot = 3.86 \times 10^{26}$ W
Equatorial rotation period	$= 26$ days
Angular momentum	$= 1.7 \times 10^{41}$ kg m^2 s^{-1}
Mass loss rate	$= 10^9$ kg s^{-1}
Surface distances	1 arcsec $\equiv 726$ km
Distance from earth	1 AU $= 1.50 \times 10^{11}$ m $= 215$ $R_\odot$

Typical Sizes (1 Mm = 1000 km = 10^6 m)

depth of convection zone	200 Mm
depth of photosphere	500 km
height of chromosphere-corona transition	2–10 Mm
width of granule	1 Mm
width of supergranule	30 Mm
width of giant cell	300 Mm
width of sunspot	30 Mm

Temperature

Effective	5785 K
Solar centre	1.5×10^7 K
Base of photosphere	6000 K
Temperature minimum	4300 K
Chromosphere	10^4 K
Transition region	10^5 K
Corona	2×10^6 K
At 1 AU	10^5 K

Density

Solar centre	10^{32} m^{-3}
Photosphere	10^{23} m^{-3}
Transition region	10^{15} m^{-3}
Corona	5×10^{14} m^{-3}
At 1 AU	10^7 m^{-3}

Cylindrical Polar Coordinates (R, ϕ, z)

$$\nabla A = \frac{\partial A}{\partial R} \hat{\mathbf{R}} + \frac{1}{R} \frac{\partial A}{\partial \phi} \hat{\boldsymbol{\phi}} + \frac{\partial A}{\partial z} \hat{\mathbf{z}}.$$

$$\nabla \cdot \mathbf{B} = \frac{1}{R} \frac{\partial}{\partial R} (R B_R) + \frac{1}{R} \frac{\partial B_\phi}{\partial \phi} + \frac{\partial B_z}{\partial z}.$$

$$\nabla \times \mathbf{B} = \left(\frac{1}{R} \frac{\partial B_z}{\partial \phi} - \frac{\partial B_\phi}{\partial z} \right) \hat{\mathbf{R}} + \left(\frac{\partial B_R}{\partial z} - \frac{\partial B_z}{\partial R} \right) \hat{\boldsymbol{\phi}} +$$

$$+ \left(\frac{1}{R} \frac{\partial}{\partial R} (RB_\phi) - \frac{1}{R} \frac{\partial B_R}{\partial \phi} \right) \hat{\mathbf{z}}.$$

$$\nabla^2 A = \frac{1}{R} \frac{\partial}{\partial R} \left(R \frac{\partial A}{\partial R} \right) + \frac{1}{R^2} \frac{\partial^2 A}{\partial \phi^2} + \frac{\partial^2 A}{\partial z^2}.$$

$$(\mathbf{B} \cdot \nabla)\mathbf{B} = \left(B_R \frac{\partial B_R}{\partial R} + \frac{B_\phi}{R} \frac{\partial B_R}{\partial \phi} - \frac{B_\phi^2}{R} + B_z \frac{\partial B_R}{\partial z} \right) \hat{\mathbf{R}} +$$

$$+ \left(\frac{B_R}{R} \frac{\partial}{\partial R} (RB_\phi) + \frac{B_\phi}{R} \frac{\partial B_\phi}{\partial \phi} + B_z \frac{\partial B_\phi}{\partial z} \right) \hat{\boldsymbol{\phi}} +$$

$$+ \left(B_z \frac{\partial B_z}{\partial z} + B_R \frac{\partial B_z}{\partial R} + \frac{B_\phi}{R} \frac{\partial B_z}{\partial \phi} \right) \hat{\mathbf{z}}.$$

Spherical Polar Coordinates (r, θ, φ)

$$\nabla A = \frac{\partial A}{\partial r} \hat{\mathbf{r}} + \frac{1}{r} \frac{\partial A}{\partial \theta} \hat{\boldsymbol{\theta}} + \frac{1}{r \sin \theta} \frac{\partial A}{\partial \phi} \hat{\boldsymbol{\phi}}.$$

$$\nabla \cdot \mathbf{B} = \frac{1}{r^2} \frac{\partial}{\partial r} (r^2 B_r) + \frac{1}{r \sin \theta} \frac{\partial}{\partial \theta} (\sin \theta \, B_\theta) + \frac{1}{r \sin \theta} \frac{\partial B_\phi}{\partial \phi}.$$

$$\nabla \times \mathbf{B} = \frac{1}{r \sin \theta} \left(\frac{\partial}{\partial \theta} (\sin \theta \, B_\phi) - \frac{\partial B_\theta}{\partial \phi} \right) \hat{\mathbf{r}} + \left(\frac{1}{r \sin \theta} \frac{\partial B_r}{\partial \phi} - \frac{1}{r} \frac{\partial}{\partial r} (rB_\phi) \right) \hat{\boldsymbol{\theta}} +$$

$$+ \left(\frac{1}{r} \frac{\partial}{\partial r} (rB_\theta) - \frac{1}{r} \frac{\partial B_r}{\partial \theta} \right) \hat{\boldsymbol{\phi}}.$$

$$\nabla^2 A = \frac{1}{r^2} \frac{\partial}{\partial r} \left(r^2 \frac{\partial A}{\partial r} \right) + \frac{1}{r^2 \sin \theta} \frac{\partial}{\partial \theta} \left(\sin \theta \frac{\partial A}{\partial \theta} \right) + \frac{1}{r^2 \sin^2 \theta} \frac{\partial^2 A}{\partial \phi^2}.$$

$$(\mathbf{B} \cdot \nabla)\mathbf{B} = \left(B_r \frac{\partial B_r}{\partial r} + \frac{B_\theta}{r} \frac{\partial B_r}{\partial \theta} - \frac{B_\theta^2 + B_\phi^2}{r} + \frac{B_\phi}{r \sin \theta} \frac{\partial B_r}{\partial \phi} \right) \hat{\mathbf{r}} +$$

$$+ \left(B_r \frac{\partial B_\theta}{\partial r} + \frac{B_\theta}{r} \frac{\partial B_\theta}{\partial \theta} + \frac{B_r B_\theta}{r} + \frac{B_\phi}{r \sin \theta} \left(\frac{\partial B_\theta}{\partial \phi} - \cos \theta \, B_\phi \right) \right) \hat{\boldsymbol{\theta}} +$$

$$+ \left(B_r \frac{\partial B_\phi}{\partial r} + \frac{B_r B_\phi}{r} + \frac{B_\phi}{r \sin \theta} \frac{\partial B_\phi}{\partial \phi} + \frac{B_\theta}{r \sin \theta} \frac{\partial}{\partial \theta} (B_\phi \sin \theta) \right) \hat{\boldsymbol{\phi}}.$$

Orthogonal Curvilinear coordinates (u_1, u_2, u_3)

$$\mathbf{B} = B_1 \mathbf{e}_1 + B_2 \mathbf{e}_2 + B_3 \mathbf{e}_3.$$

$$\nabla A = \frac{1}{h_1} \frac{\partial A}{\partial u_1} \mathbf{e}_1 + \frac{1}{h_2} \frac{\partial A}{\partial u_2} \mathbf{e}_2 + \frac{1}{h_3} \frac{\partial A}{\partial u_3} \mathbf{e}_3.$$

$$\nabla \cdot \mathbf{B} = \frac{1}{h_1 h_2 h_3} \left[\frac{\partial}{\partial u_1} (h_2 h_3 B_1) + \frac{\partial}{\partial u_2} (h_3 h_1 B_2) + \frac{\partial}{\partial u_3} (h_1 h_2 B_3) \right].$$

$$\nabla \times \mathbf{B} = \frac{1}{h_2 h_3} \left[\frac{\partial}{\partial u_2} (h_3 B_3) - \frac{\partial}{\partial u_3} (h_2 B_2) \right] \mathbf{e}_1 +$$

$$+ \frac{1}{h_1 h_3} \left[\frac{\partial}{\partial u_3}(h_1 B_1) - \frac{\partial}{\partial u_1}(h_3 B_3) \right] \mathbf{e}_2 +$$

$$+ \frac{1}{h_1 h_2} \left[\frac{\partial}{\partial u_1}(h_2 B_2) - \frac{\partial}{\partial u_2}(h_1 B_1) \right] \mathbf{e}_3.$$

$$\nabla^2 A = \frac{1}{h_1 h_2 h_3} \left[\frac{\partial}{\partial u_1}\left(\frac{h_2 h_3}{h_1} \frac{\partial A}{\partial u_1}\right) + \frac{\partial}{\partial u_2}\left(\frac{h_3 h_1}{h_2} \frac{\partial A}{\partial u_2}\right) + \frac{\partial}{\partial u_3}\left(\frac{h_1 h_2}{h_3} \frac{\partial A}{\partial u_3}\right) \right].$$

e.g. Toroidal coordinates (u, v, ϕ)

$$x = \frac{a \sinh v \cos \phi}{\cosh v - \cos u}, \quad y = \frac{a \sinh v \sin \phi}{\cosh v - \cos u}, \quad z = \frac{a \sin u}{\cosh v - \cos u}.$$

$$h_1^2 = h_2^2 = \frac{a^2}{(\cosh v - \cos u)^2}, \quad h_3^2 = \frac{a^2 \sinh^2 v}{(\cosh v - \cos u)^2}.$$

Special Functions

Gamma function

$$\Gamma(x) = \int_0^\infty e^{-u} u^{x-1} \, du.$$

Beta function

$$B(x_1, x_2) = \int_0^1 u^{x_1 - 1}(1 - u)^{x_2 - 1} \, du.$$

Bessel functions $J_v(x)$ and $J_{-v}(x)$ satisfy

$$x^2 y'' + x y' + (x^2 - v^2) y = 0.$$

Hypergeometric function $F(a, b, c, x)$ satisfies

$$(x^2 - x)y'' + ((1 + a + b)x - c)y' + ab y = 0.$$

Error function

$$\operatorname{erf}(x) = 2\pi^{-1/2} \int_0^x e^{-u^2} \, du.$$

Legendre functions $P_v(x)$ and $Q_v(x)$ satisfy

$$(1 - x^2) y'' - 2x y' + v(v + 1) y = 0.$$

APPENDIX III

NOTATION

Latin Alphabet

a	flux tube radius
$A(s)$	cross-sectional area
$A(x, z)$	flux function
B_e	equipartition field
B	magnetic induction (or, loosely, magnetic field)
c	speed of light
c_v	specific heat at constant volume
c_p	specific heat at constant pressure
c_s	sound speed
c_T	tube speed or cusp speed
d	shock damping-length
$d(x)$	footpoint displacement
D	electric displacement
e	internal energy per unit mass; charge of an electron.
e_{ij}	rate of strain tensor
E	Ekman number
$E(k)$	energy spectrum function
E_H	heating
E	electric field
$\mathbf{E}_0$	total electric field
f	number of sunspots; fraction of ions ionised
F	wave flux; magnetic flux; strength of flux tube
$F(k)$	helicity spectrum function
$\mathbf{F}_c$	conductive flux
$\mathbf{F}_d$	driving force
$\mathbf{F}_g$	gravitational force
$\mathbf{F}_v$	viscous force
g	gravitational acceleration; number of sunspot groups
$g_\odot$	gravity at solar surface
G	gravitational constant
G	Gauss
H	heating
Ha	Hartmann number
H_v	viscous dissipation rate
H_w	wave heating
H	magnetic field

NOTATION

I	current
J	current
J_n	Bessel function
$\mathbf{j}$	current density
k_B	Boltzmann constant
$\mathbf{k}$	wavenumber vector
l	mixing length
L	total angular momentum density; luminosity
$L_\odot$	total solar luminosity
L_r	net radiation
$\mathscr{L}$	energy loss function
m	mean particle mass
M	Mach number
$M(r)$	mass inside sphere
M_A	Alfvén Mach number
$M_\odot$	mass of Sun
n	number density
n_+	number density of positive ions
n_-	number density of negative ions
n_a	number density of neutral atoms
n_e	electron number density
n_H	number density of hydrogen atoms
n_p	proton number density
$\hat{\mathbf{n}}$	principal normal
N	Brunt frequency
$\tilde{N}$	number of degrees of freedom
N_D	dynamo number
Nu	Nusselt number
p	plasma pressure
Pr	Prandtl number
P_m	magnetic Prandtl number
p_{ij}	stress tensor
P_l	Legendre polynomial
P_l^m	associated Legendre polynomial
$\mathbf{q}$	heat flux
$\mathbf{q}_r$	radiative flux
$Q(T)$	temperature dependent part of radiative loss function (Section 2.3.3)
Q	Chandrasekhar number
r	radial distance
r_A	Alfvén radius
r_c	critical-point radius
$\mathbf{r}$	Eulerian position vector
$\mathbf{r}_0$	Lagrangian position vector
R	radial distance from an axis
R	Wolf number
$\tilde{R}$	gas constant

Ra	Rayleigh number
R_c	radius of curvature
Re	Reynolds number
Ri	Richardson number
R_m	magnetic Reynolds number
Ro	Rossby number
$R_\odot$	radius of Sun
s	entropy per unit mass
T	temperature
T	Tesla
$\mathcal{T}$	Taylor number
T_c	central temperature
T_e	electron temperature
T_0	base temperature
T_1	summit temperature
u	electron conduction speed
U	shock speed
v_A	Alfvén speed
v_c	isothermal sound speed
v_i	inflow speed to current sheet
$\mathbf{v}$	plasma velocity
$\mathbf{v}_g$	group velocity
$\mathbf{v}_p$	phase velocity
W	magnetic energy
W	turbulent energy density
X	mass fraction; compression ratio; dynamo number

Greek Alphabet

α	coefficient of volume expansion; constant in radiative loss function; polytropic constant; constant for linear force-free field; kinematic dynamo coefficient
β	ratio of plasma to magnetic pressure
γ	ratio of specific heats
ε	nuclear energy generation rate
ε_0	permittivity of free space
η	magnetic diffusivity
$\tilde{\eta}$	eddy magnetic diffusivity
$\bar{\eta}$	fractional shock compression
θ	pinch ratio; temperature difference
θ_B	inclination of $\mathbf{k}$ and $\mathbf{B}_0$
θ_g	inclination of $\mathbf{k}$ and $\mathbf{g}$
θ_Ω	inclination of $\mathbf{k}$ and $\mathbf{\Omega}$
κ	thermal diffusivity
$\tilde{\kappa}$	opacity
κ_r	radiative conductivity

$\boldsymbol{\kappa}$	thermal conduction tensor
$\kappa_\parallel$	thermal conduction along $\mathbf{B}$
$\kappa_\perp$	thermal conduction normal to $\mathbf{B}$
κ_e	electron conduction coefficient
κ_p	proton conduction coefficient
λ	wavelength; damping length
λ_D	Debye length
Λ	scale-height
Λ_B	modified scale-height
$\tilde{\mu}$	mean atomic weight
μ_0	magnetic permeability of free space
ν	coefficient of kinematic viscosity
ξ	displacement from equilibrium
ρ	density
ρ_c	central density
ρ^*	charge density
σ	electrical conductivity
σ^*	anomalous conductivity
σ_1	direct conductivity
σ_2	Hall conductivity
σ_3	Cowling conductivity
σ_s	Stefan–Boltzmann constant
τ	optical depth; wave period; correlation time
τ^*	anomalous collision time
τ_A	Alfvén travel time
τ_c	conduction time
τ_d	diffusion time
τ_{ei}	electron-ion collision interval
τ_{en}	electron-neutral collision interval
τ_{ii}	ion-ion collision interval
τ_{in}	ion-neutral collision interval
τ_G	gravitational time scale
τ_{rad}	radiative time scale
Φ	twist; gravitational potential
Φ_{ij}	spectrum tensor
χ	constant in radiative loss function; inclination of flux tube boundary to vertical
Ψ	stream function; scalar magnetic potential
ω	wave frequency
$\boldsymbol{\omega}$	relative vorticity
ω_A	Alfvén frequency
ω_{pe}	electron plasma frequency
Ω	electron gyro-frequency
Ω_i	ion gyro-frequency

REFERENCES

Acheson, D. J.: 1978, *Phil. Trans. Roy. Soc.* **A289**, 459.
Acheson, D. J.: 1979a, *Solar Phys.* **62**, 23.
Acheson, D. J.: 1979b, *Nature* **277**, 41.
Acheson, D. J. and Gibbons, M. P.: 1978, *J. Fluid Mech.* **85**, 743.
Acheson, D. J. and Hide, R.: 1973, *Rep. Prog. Phys.* **36**, 159.
Adams, J. and Pneuman, G. W.: 1976, *Solar Phys.* **46**, 185.
Altschuler, M. D. and Newkirk, G.: 1969, *Solar Phys.* **9**, 131.
Ando, H. and Osaki, Y.: 1975, *Publ. Astron. Soc. Japan* **27**, 581.
Antia, H. M. and Chitre, S. M.: 1979, *Solar Phys.* **63**, 67.
Antiochos, S.: 1979, *Astrophys. J.* **232**, L125.
Anzer, U.: 1968, *Solar Phys.* **3**, 298.
Anzer, U.: 1969, *Solar Phys.* **8**, 37.
Anzer, U.: 1972, *Solar Phys.* **24**, 324.
Anzer, U.: 1978, *Solar Phys.* **57**, 111.
Anzer, U. and Pneuman, G. W.: 1982, *Solar Phys.* **79**, 129.
Anzer, U. and Tandberg–Hanssen, E.: 1970, *Solar Phys.* **11**, 61.
Athay, R. G.: 1976, *The Solar Chromosphere and Corona: Quiet Sun*, D. Reidel, Dordrecht, Holland.
Athay, R. G.: 1981a, in F. Orrall (ed.), *Skylab Active Region Workshop Proc.*, Colo. Ass. Univ. Press, Ch. 4.
Athay, R. G.: 1981b, Preprint.
Athay, R. G. and White, O. R.: 1977, *Proc. OSO8 Workshop, Nov. 1977*, University of Colorado.
Athay, R. G., White, O. R., Lites, B. W. and Bruner, E. C.: 1980, *Solar Phys.* **66**, 357.
Babcock, H. W.: 1961, *Astrophys. J.* **133**, 572.
Backus, G. E.: 1958, *Ann. Phys.* **4**, 372.
Bahng, J. and Schwarzschild, M.: 1963, *Astrophys. J.* **137**, 901.
Barbosa, D.: 1978, *Solar Phys.* **56**, 55.
Barnes, A.: 1979, in C. F. Kennel, L. J. Lanzerotti, and E. N. Parker (eds.), *Solar System Plasma Physics*, Vol. I, North Holland & Amsterdam, p. 249.
Barnes, C. W. and Sturrock, P. A.: 1972, *Astrophys. J.* **174**, 659.
Basri, G. S., Linsky, J. L., Bartoe, J. D. F., Brueckner, G. E., and Van Hoosier, M. E.: 1979, *Astrophys. J.* **230**, 924.
Batchelor, G. K.: 1967, *An Introduction to Fluid Dynamics*, Cambridge University Press, England, p. 555.
Baum, P. J. and Bratenahl, A.: 1977, *J. Plasma Phys.* **18**, 257.
Bazer, J. and Ericson, W. B.: 1959, *Astrophys. J.* **129**, 758.
Beckers, J. M.: 1972, *Ann. Rev. Astron. Astrophys.* **10**, 73.
Beckers, J. M.: 1976a, *Astrophys. J.* **203**, 739.
Beckers, J. M.: 1976b, in D. J. Williams (ed.), *Physics of Solar Planetary Environments*, American Geophysical Union, p. 89.
Beckers, J. M. and Schultz, R. B.: 1972, *Solar Phys.* **27**, 61.
Beckers, J. M. and Tallant, P.: 1969, *Solar Phys.* **7**, 351.
Bénard, H.: 1900, *Revue Générale des Sciences Pures et Appliquées.* **11**, 1261.
Bernstein, I. B., Frieman, E. A., Kruskal, M. D. and Kulsrud, R. M.: 1958, *Proc. Roy. Soc.* **A244**, 17.
Biermann, L.: 1941, *V. Astr. Ges.* **76**, 194.
Biermann, L.: 1946, *Naturwissenschaften* **33**, 118.
Biermann, L.: 1951, *Z. Astrophys.* **29**, 274.
Birn, J., Goldstein, H. and Schindler, K.: 1978, *Solar Phys.* **57**, 81.

REFERENCES

Birn, J. and Schindler, K.: 1981, in E. R. Priest (ed.), *Solar Flare Magnetohydrodynamics*, Gordon and Breach, London, Ch. 6.
Bohlin, J. D.: 1976, in D. J. Williams (ed.), *Physics of Solar Planetary Environments*, American Geophysical Union, p. 114.
Bonnet, R. M. and Dupree, A. K.: 1981, *Solar Phenomena in Stars and Stellar Systems*, D. Reidel, Dordrecht, Holland.
Boström, R.: 1973, *Astrophys. Space Sci.* **22**, 353.
Boyd, T. J. M. and Sanderson, J. J.: 1969, *Plasma Dynamics*, Nelson, London.
Braginsky, S. I.: 1965, *Rev. Plasma Phys.* **1**, 205.
Braginsky, S. I.: 1965, *Sov. Phys. JETP* **20**, 726.
Brandt, J. C.: 1970, *An Introduction to the Solar Wind*, W. H. Freeman, San Francisco.
Brandt, J. C., Wolf, C. and Cassinelli, J. P.: 1969, *Astrophys. J.* **156**, 1117.
Bray, R. J. and Loughhead, R. E.: 1964, *Sunspots*, Chapman and Hall, London.
Bray, R. J. and Loughhead, R. E.: 1974, *The Solar Chromosphere*, Chapman and Hall, London.
Brown, A.: 1958, *Astrophys. J.* **128**, 646.
Brown, J. C. and Smith, D. F.: 1980, *Rep. Prog. Phys.* **43**, 125.
Brueckner, G. E.: 1980, *Proc. XVII General Assembly of the IAU*.
Bruzek, A. and Durrant, C. J.: 1977, *Illustrated Glossary for Solar and Solar-terrestrial Physics*, D. Reidel, Dordrecht, Holland.
Burgers, J. M.: 1969, *Flow Equations for Composite Gases*, Academic Press, New York.
Busse, F. M.: 1973, *J. Fluid Mech.* **57**, 529.
Busse, F. M.: 1975a, *Geophys. J. Roy. Astron. Soc.* **42**, 437.
Busse, F. M.: 1975b, *J. Fluid Mech.* **71**, 193.
Callebaut, D. K. and Voslamber, D.: 1962, *Phys. Rev.* **127**, 1857.
Canfield, R. C., Priest, E. R. and Rust, D. M.: 1974, in Y. Nakagawa and D. M. Rust (eds.), *Flare-related Magnetic Field Dynamics*, NCAR, Boulder, U.S.A.
Cargill, P. J. and Priest, E. R.: 1980, *Solar Phys.* **65**, 251.
Cargill, P. J. and Priest, E. R.: 1982a, *Solar Phys.* **76**, 357.
Cargill, P. J. and Priest, E. R.: 1982b, *Geophys. Astrophys. Fluid Dynamics* **20**, 227.
Chandrasekhar, S.: 1961, *Hydrodynamic and Hydromagnetic Stability*, Cambridge University Press, England.
Chandrasekhar, S. and Kendall, P. C.: 1957. *Astrophys. J.* **126**, 457.
Chapman, S.: 1929, *Proc. Roy. Soc.* **A95**, 61.
Chapman, S.: 1957, *Smithsonian Contrib. Astrophys.* **2**, 1.
Chen, C. J. and Lykoudis, P. S.: 1972, *Solar Phys.* **25**, 380.
Cheng, A. F.: 1979, *J. Geophys. Res.* **84**, 2129.
Chew, G. F., Goldberger, M. L. and Low, F. F.: 1956, *Proc. Roy. Soc.* **A236**, 112.
Childress, S.: 1969, in S. Runcorn (ed.), *The Application of Modern Physics to the Earth and Planetary Interiors*, Wiley, London, p. 629.
Childress, S.: 1979, *Phys. Earth Planet. Inter.* **20**, 172.
Chiu, Y. C. and Wentzel, D. G.: 1972, *Astrophys. Space Sci.* **16**, 465.
Chiu, Y. T. and Hilton, H. H.: 1977, *Astrophys. J.* **212**, 873.
Chiuderi, C.: 1981, in R. M. Bonnet and A. K. Dupree (eds.), *Solar Phenomena in Stars and Stellar Systems*, D. Reidel, Dordrecht, Holland p. 269.
Chiuderi, C., Einaudi, G. and Torricelli–Ciamponi, G.: 1981, *Astron. Astrophys.* **97**, 27.
Chiuderi, C. and Giovanardi, C.: 1979, *Solar Phys.* **64**, 27.
Chiuderi, C. and Van Hoven, G.: 1979, *Astrophys. J.* **232**, L69.
Choe, J., Tataronis, J. A., and Grossmann, W.: 1977, *Plasma Physics* **19**, 117.
Christensen–Dalsgaard, J.: 1980, *Monthly Notices Roy. Astron. Soc.* **190**, 765.
Claverie, A., Isaak, G. R., McLeod, C. P., van der Raay, H. B., and Roca Cortes, T.: 1979, *Nature* **282**, 591.
Comfort, R. H., Tandberg-Hanssen, E. and Wu, S. T.: 1979, *Astrophys. J.* **231**, 927.
Cook, A. E. and Roberts, P. H.: 1970, *Proc. Camb. Phil. Soc.* **68**, 547.
Coppi, B., Galvao, R. Pellat, R., Rosenbluth, M. N., and Rutherford, P. H.: 1976, *Sov. J. Plasma Phys.* **2**, 533.
Courant, R and Hilbert, D.: 1963, *Methods of Mathematical Physics* Vol. 2, New York, Interscience, p. 367.

Cowley, S. W. H.: 1974a, *J. Plasma Phys.* **12**, 319.
Cowley, S. W. H.: 1974b, *J. Plasma Phys.* **12**, 341.
Cowling, T. G.: 1934, *Monthly Notices Roy. Astron. Soc.* **94**, 39.
Cowling, T. G.: 1946, *Monthly Notices Roy. Astron. Soc.* **106**, 446.
Cowling, T. G.: 1953, in G. P. Kuiper (ed.), *The Sun*, University of Chicago Press, p. 532.
Cowling, T. G.: 1976, *Magnetohydrodynamics*, 2nd edn., Adam Hilger, Bristol, England.
Cowling, T. G.: 1981, *Ann. Rev. Astron. Astrophys.* **19**, 115.
Cox, D. P. and Tucker, W. H.: 1969, *Astrophys. J.* **157**, 1157.
Craig, I. J. D.: 1981, in E. R. Priest (ed.), *Solar Flare Magnetohydrodynamics*, Gordon and Breach, London, Ch. 5.
Craig, I. J. D., Robb, T. D. and Rollo, M. D.: 1982, *Solar Phys.* **76**, 331.
Danielson, R. E.: 1961, *Astrophys. J.* **134**, 289.
Davis, R.: 1972, *Bull. Am. Phys. Soc.* **17**, 527.
Defouw, R. J.: 1976, *Astrophys. J.* **209**, 266.
Deinzer, W.: 1965, *Astrophys. J.* **141**, 548.
Deinzer, W. and Stix, M.: 1971, *Astron. Astrophys.* **12**, 111.
Dessler, A. J.: 1967, *Rev. Geophys.* **5**, 1.
Deubner, F. L.: 1973, in R. G. Athay (ed.), 'Chromospheric Fine Structure', *IAU Symp.* **56**, 263.
Deubner, F. L.: 1975, *Astron. Astrophys.* **44**, 371.
Deubner, F. L.: 1976, *Astron. Astrophys.* **51**, 189.
Deubner, F. L., Ulrich, R. K., and Rhodes, E. J.: 1979, *Astron. Astrophys.* **72**, 177.
Dicke, R. H.: 1979, *New Scientist* **83**, No. 1162, p. 12.
Dungey, J. W.: 1953, *Phil. Mag.* **44**, 725.
Durney, B. R.: 1971, *Astrophys. J.* **166**, 669.
Durney, B. R. and Pneuman, G. W.: 1975, *Solar Phys.* **40**, 461.
Durrant, C. J. and Roxburgh, I. W.: 1977, in A. Bruzek and C. J. Durrant (eds.), *Illustrated Glossary for Solar and Solar-Terrestrial Physics*, D. Reidel, Dordrecht, Holland, p. 1.
Duvall, T. L.: 1979, *Solar Phys.* **63**, 3.
Eddy, J. A.: 1976, *Science* **192**, 1189.
Evershed, J.: 1909, *Monthly Notices Roy. Astron. Soc.* **69**, 454.
Ferraro, V. C. A. and Plumpton, C.: 1961, *An Introduction to Magnetofluid Mechanics*, Oxford University Press, Oxford.
Field, G. B.: 1965, *Astrophys. J.* **142**, 531.
Finn, J. M.: 1975, *Nucl. Fusion* **15**, 845.
Fisher, R., Garcia, C. J., and Seagraves, P.: 1981, *Astrophys. J.* **246**, L161.
Forbes, T. G., Priest, E. R., and Hood, A.W.: 1982. *J. Plasma Phys.* **27**, 157.
Foukal, P. V.: 1975, *Solar Phys.* **43**, 327.
Foukal, P. V.: 1976, *Astrophys. J.* **210**, 575.
Frieman, F. and Rotenberg, M.: 1960, *Rev. Mod. Phys.* **32**, 898.
Furth, H. P., Killeen, J., and Rosenbluth, M. N.: 1963, *Phys. Fluids* **6**, 459.
Furth, H. P., Rutherford, P. H., and Selberg, H.: 1973, *Phys. Fluids* **16**, 1054.
Gabriel, A. H.: 1976, *Phil. Trans. Roy. Soc. Lond.* **A. 281**, 339.
Galeev, A. A., Rosner, R., Serio, S. and Vaiana, G. S.: 1981, *Astrophys. J.* **234**, 301.
Galloway, D. J. and Moore, D. R.: 1979, *Geophys. Astrophys. Fluid Dyn.* **12**, 73.
Galloway, D. J., Proctor, M. R. E., and Weiss, N. O.: 1977, *Nature* **266**, 686.
Galloway, D. J., Proctor, M. R. E., and Weiss, N. O.: 1978, *J. Fluid Mech.* **87**, 243.
Galloway, D. J. and Weiss, N. O.: 1981, *Astrophys. J.* **243**, 945.
Giachetti, R., Van Hoven, G., and Chiuderi, C.: 1977, *Solar Phys.* **55**, 371.
Gibson, E. G.: 1973, *The Quiet Sun*, NASA, Washington.
Gilman, P. A.: 1969, *Solar Phys.* **8**, 316: **9**, 3.
Gilman, P. A.: 1970, *Astrophys. J.* **162**, 1019.
Gilman, P. A.: 1976, in V. Bumba and J. Kleczek (eds.), 'Basic Mechanisms of Solar Activity' *IAU Symp.* **71**, 207.
Gilman, P. A.: 1977, *Geophys. Astrophys. Fluid Dynamics* **8**, 93.
Gilman, P. A.: 1978, *Geophys. Astrophys. Fluid Dynamics* **11**, 157.

Gilman, P. A. and Miller, J.: 1981, *Astrophys. J. Suppl.* **246**, 555.
Gingerich, O., Noyes, R. W., Kalkofen, W., and Cuny, Y.: 1971, *Solar Phys.* **18**, 347.
Giovanelli, R. G.: 1972, *Solar Phys.* **27**, 71.
Glatzmaier, G.: 1980, PhD thesis, Colorado University.
Goldreich, P. and Keeley, D. A.: 1977, *Astrophys. J.* **212**, 243.
Golub, L., Davis, J. M. and Krieger, A. S.: 1979, *Astrophys. J.* **229**, L145.
Golub, L., Krieger, A. S., Silk, J. K., Timothy, A. F., and Vaiana, G. S. 1974: *Astrophys. J.* **189**, L93.
Golub, L., Maxson, C., Rosner, R., Serio, S., and Vaiana, G. S.: 1980, *Astrophys. J.* **238**, 343.
Graff, P.: 1976, *Astron. Astrophys.* **49**, 299.
Graham, E.: 1975, *J. Fluid Mech.* **70**, 689.
Grec, C., Fossat, É., and Pomerantz, M.: 1980, *Nature* **288**, 541.
Green, R. M.: 1965, *IAU Symp.* **22**, 398.
Green, R. M. and Sweet, P. A.: 1966, *Astrophys. J.* **147**, 1153.
Greenspan, H. P.: 1968, *The Theory of Rotating Fluids*, Cambridge University Press, England.
Grossman, W. and Tataronis, J.: 1973, *Z. Phys.* **261**, 217.
Habbal, S. R. and Rosner, R.: 1979, *Astrophys. J.* **234**, 1113.
Habbal, S. R., Leer, E., and Holzer, T. E.: 1979, *Solar Phys.* **64**, 287.
Hartle, R. E. and Sturrock, P. A.: 1968, *Astrophys. J.* **151**, 1155.
Harvey, J. W.: 1969, PhD thesis, University of Colorado.
Harvey, J. W.: 1974, in Y. Nakagawa and D. M. Rust (eds.), *Flare-Related Magnetic Field Dynamics* NCAR- Boulder, U.S.A.
Hasegawa, A. and Chen, L.: 1974, *Phys. Rev. Lett.* **32**, 454.
Hathaway, D.: 1980, PhD thesis, Colorado University.
Heasley, J. N. and Mihalas, D.: 1976, *Astrophys. J.* **205**, 273.
Herzenberg, A.: 1958, *Phil. Trans. Roy. Soc.* **A250**, 543.
Heyvaerts, J.: 1974, *Astron. Astrophys.* **37**, 65.
Heyvaerts, J. and Priest, E. R.: 1976, *Solar Phys.* **47**, 223.
Heyvaerts, J., Priest, E. R. and Rust, D. M.: 1977, *Astrophys. J.* **216**, 123.
Heyvaerts, J., Lasry, J. M., Schatzman, M., and Witomsky, G.: 1980, *Lecture Notes Math.* **782**, 160.
Heyvaerts, J. and Schatzman, E.: 1980, *Proc. Japan-France Seminar on Solar Phys.* (ed. F. Moriyama & J. C. Henoux), p. 77.
Hildner, E.: 1974, *Solar Phys.* **35**, 123.
Hill, H. A. and Stebbins, R. T.: 1975, *Astrophys. J.* **200**, 471.
Hollweg, J. V.: 1974, *J. Geophys. Res.* **79**, 1539.
Hollweg, J. V.: 1978, *Rev. Geophys. Space Phys.* **16**, 689.
Hollweg, J. V.: 1979, *Solar Phys.* **62**, 227.
Hollweg, J. V.: 1981a, in F. Q. Orrall (ed.), *Solar Active Regions*, Ch. 10, Colo. Ass. Univ. Press.
Hollweg, J. V.: 1981b, *Solar Phys.* **70**, 25.
Hollweg, J. V.: 1981c, in S. Jordan (ed.), *The Sun as a Star*, NASA/CNRS, Ch. 15.
Hollweg, J. V.: 1982, *Astrophys. J.* **254**, 806.
Hollweg, J. V., Jackson, S., and Galloway, D.: 1982, *Solar Phys.* **75**, 35.
Hollweg, J. V. and Roberts, B.: 1981, *Astrophys. J.* **250**, 398.
Holzer, T. E.: 1977, *J. Geophys. Res.* **82**, 23.
Holzer, T. E. and Leer, E.: 1980, *J. Geophys. Res.* **85**, 4665.
Hood, A. W. and Priest, E. R.: 1979a, *Astron. Astrophys.* **77**, 233.
Hood, A. W. and Priest, E. R.: 1979b, *Solar Phys.* **64**, 303.
Hood, A. W. and Priest, E. R.: 1980a, *Astron. Astrophys.* **87**, 126.
Hood, A. W. and Priest, E. R.: 1980b, *Solar Phys.* **66**, 113.
Hood, A. W. and Priest, E. R.: 1981a, *Solar Phys.* **73**, 289.
Hood, A. W. and Priest, E. R.: 1981b, *Geophys. Astrophys. Fluid Dynamics* **17**, 297.
Hood, A. W.: 1983, *Solar Phys.* **87**, 279; **89**, 235.
Howard, R.: 1976, *Astrophys. J.* **210**, L159.
Howard, R.: 1977, *Ann. Rev. Astron. Astrophys.* **15**, 153.
Howard, R. and Labonte, B. J.: 1980, *Astrophys. J.* **239**, L33.
Howard, R. and Svestka, Z.: 1977, *Solar Phys.* **54**, 65.

Hundhausen, A. J.: 1972, *Coronal Expansion and Solar Wind*, Springer-Verlag, New York.
Hundhausen, A. J.: 1981, in *McGraw-Hill Encyclopedia of Science and Technology*, 5th Edn.
Hundhausen, A. J., Hansen, R. T., and Hansen, S. F.: 1981, *J. Geophys. Res.* **86**, 2079.
Hundhausen, J. R., Hundhausen, A. J., and Zweibel, E. G.: 1981, *J. Geophys. Res.* **86**, 11117.
Hunter, J. H.: 1970, *Astrophys. J.* **161**, 451.
Imshennik, V. S. and Syrovatsky, S. I.: 1967, *Sov. Phys. JETP* **25**, 656.
Ionson, J. A.: 1978, *Astrophys. J.* **226**, 650.
Ito, K.: 1980, *Earth Planet. Sci. Lett.* **51**, 451.
Ivanov, L. N. and Platov, Y. V.: 1977, *Solar Phys.* **54**, 35.
Jacques, S. V.: 1977, PhD thesis NCAR, Boulder, Colo.
Jakimiec, J.: 1965, *Astron. Astrophys.* **15**, 145.
Jeffrey, A.: 1966, *Magnetohydrodynamics*, Oliver and Boyd, Edinburgh.
Jeffrey, A. and Taniuti, T.: 1964, *Nonlinear Wave Propagation*, Academic Press, New York.
Jeffrey, A. and Taniuti, T.: 1966, *Magnetohydrodynamic Stability and Thermonuclear Containment*, Academic Press, New York.
Jensen, E., Maltby, P., and Orrall, F. Q. (eds.): 1979, 'Physics of Solar Prominences', *IAU Colloquium* **44**.
Jepps, S. A.: 1975, *J. Fluid Mech.* **67**, 625.
Jette, A. D.: 1970, *J. Maths. Anal. Appl.* **29**, 109.
Jockers, K.: 1978, *Solar Phys.* **56**, 37.
Jones, C. A.: 1982, in A. M. Soward (ed.), *Stellar and Planetary Magnetism*, Gordon and Breach.
Jordan, C.: 1975, in S. R. Kane (ed.), Solar Gamma, X-, and EUV Radiation', *IAU Symp.* **68**, 109.
Jordan, C.: 1977, in A. Bruzek and C. J. Durrant (eds.), *Illustrated Glossary for Solar and Solar-Terrestrial Physics*, D. Reidel, Dordrecht. Holland, p. 35.
Jordan, C.: 1981, ed. of *Proc. 3rd European Solar Meeting*, Oxford.
Kaburaki, O. and Uchida, Y.: 1971, *Pub. Astron. Soc. Japan* **23**, 405.
Kahler, S. W. and Kreplin, R. W.: 1970, *Solar Phys.* **14**, 372.
Kane, S. R.: 1974, *Space Sci. Lab. Series* **14**, Issue 76.
Kantrowitz, A. and Petschek, H. E.: 1966, in W. B. Kunkel (ed.) *Plasma Physics in Theory and Application*, McGraw-Hill, p. 147.
Kiepenheuer, K.: 1957, *The Sun*, University of Michigan Press.
Kippenhahn, R. and Schlüter, A.: 1957, *Zs. Ap.* **43**, 36.
Kirkland, K. and Sonnerup. B. U. Ö.: 1979, *J. Plasma Phys.* **22**, 289.
Knobloch, E.: 1977, *J. Fluid Mech.* **83**, 129.
Knobloch, E., Weiss, N. O., and Da Costa, L. N.: 1981, *J. Fluid Mech.* **113**, 153.
Kopp, R. A. and Kuperus, M.: 1968, *Solar Phys.* **4**, 212.
Kopp, R. A. and Holzer, T. E.: 1976, *Solar Phys.* **49**, 43.
Kopp, R. A. and Pneuman, G. W.: 1976, *Solar Phys.* **50**, 85.
Kraichnan, R. H.: 1973, *J. Fluid Mech.* **59**, 745.
Kraichnan, R. H.: 1976, *J. Fluid. Mech.* **77**, 753.
Krause, F. and Rädler, K. H.: 1980, *Mean-Field Magnetohydrodynamics and Dynamo Theory*, Pergamon, Oxford.
Kruskal, M. D.: 1954, U.S. Atomic Energy Commission Report No. NYO-6015.
Kruskal, M. D. and Oberman, C. R.: 1958, *Phys. Fluids* **1**, 275.
Kruskal, M. D. and Schwarzschild, M.: 1954, *Proc. Roy. Soc.* **223A**, 348.
Kuperus, M.: 1965, *Rech. Astron. Obsevr. Utrecht* **17**, 1.
Kuperus, M.: 1969, *Space Sci. Rev.* **9**, 713.
Kuperus, M. and Chiuderi, C.: 1976, *IAU Colloq.* **36**, 223.
Kuperus, M., Ionson, J. A. and Spicer, D. S.: 1981, *Ann. Rev. Astron. Astrophys.* **19**, 7.
Kuperus, M. and Raadu, M. A.: 1974, *Astron. Astrophys.* **31**, 189.
Kuperus, M. and Tandberg-Hanssen, E.: 1967, *Solar Phys.* **2**, 39.
Landman, D. A. and Finn, G. D.: 1979, *Solar Phys.* **63**, 221.
Latour, J., Spiegel, E. A., Toomre, J., and Zahn, J. P.: 1976, *Astrophys. J.* **207**, 233.
Laval, G., Mercier, C., and Pellat, R.: 1965, *Nuclear Fusion* **5**, 156.
Layzer, D., Rosner, R., and Doyle, H. T.: 1979, *Astrophys. J.* **229**, 1126.
Lee, M. A., Roberts, B., and Rae, I. C.: 1984, *Astrophys. J.*, submitted.

Leer, E. and Holzer, T. E.: 1979, *Solar Phys.* **63**, 143.
Leer, E. and Holzer, T. E.: 1980, *J. Geophys. Res.* **85**, 4681.
Leibacher, J. and Stein, R. F.: 1971, *Astrophys. Lett.* **7**, 191.
Leighton, R. B.: 1960, *IAU Symp.* **12**, 321.
Leighton, R. B.: 1963, *Ann. Rev. Astron. Astrophys.* **1**, 19.
Leighton, R. B.: 1964, *Astrophys. J.* **140**, 1559.
Leighton, R. B.: 1969, *Astrophys. J.* **156**, 1.
Leighton, R. B., Noyes, R. W., and Simon, G. W.: 1962, *Astrophys. J.* **135**, 474.
Lerche, I. and Low, B. C.: 1977, *Solar Phys.* **53**, 385.
Lerche, I. and Low, B. C.: 1980, *Solar Phys.* **66**, 285.
Leroy, B: 1980, *Astron. Astrophys.* **91**, 136.
Leroy, J. L.: 1979, in E. Jensen, P. Maltby and F. Q. Orrall (eds.), 'Physics of Solar Prominences', *IAU Colloq.* **44**, 56.
Levine, R. H.: 1974, *Astrophys. J.* **190**, 457.
Levine, R. H. and Nakagawa, Y.: 1974, *Astrophys. J.* **190**, 703.
Levine, R. H., Schulz, M., and Frazier, E. N.: 1982, *Solar Phys.* **77**, 363.
Levine, R. H. and Withbroe, G. L.: 1977, *Solar Phys.* **51**, 83.
Lighthill, M. J.: 1953, *Proc. Roy. Soc.* **A211**, 564.
Lighthill, M. J.: 1967, *IAU Symp.* **28**, 440.
Lin, R. P.: 1974, *Space Sci. Rev.* **16**, 189.
Lites, B. W., Bruner, E. C., Chipman, E. G., Shine, R. A., Rothman, G.J., White, O. R., and Athay, R. G.: 1976, *Astrophys. J.* **210**, L111.
Lock, R. C.: 1955, *Proc. Roy. Soc. A.* **233**, 105.
Low, B. C.: 1973, *Astrophys. J.* **181**, 209.
Low, B. C.: 1974, *Astrophys. J.* **193**, 243.
Low, B. C.: 1975, *Astrophys. J.* **197**, 251.
Low, B. C.: 1977, *Astrophys. J.* **212**, 234.
Low, B. C.: 1981, *Astrophys. J.* **246**, 538.
Low, B. C. and Nakagawa, Y.: 1975, *Astrophys. J.* **199**, 237.
Lorentz, E. N.: 1963, *J. Atmos. Soc.* **20**, 130.
Lundquist, S.: 1951, *Phys. Rev.* **83**, 307.
Lüst, R. and Schlüter, A.: 1954, *Z. Astrophys.* **34**, 263.
Lynn, Y. M.: 1966, *Phys. Fl.* **9**, 314.
McLellan, A. and Winterberg, F.: 1968, *Solar Phys.* **4**, 401.
MacQueen, R. M.: 1980, *Phil. Trans. Roy. Soc.* **297**, 605.
McWhirter, R. W. P., Thoneman, P. C., and Wilson, R.: 1975, *Astron. Astrophys.* **40**, 63.
Malkus, W. V. R. and Proctor, M. R. E.: 1975, *J. Fluid Mech.* **67**, 417.
Martin, S.: 1973, *Solar Phys.* **31**, 3.
Martin, S. F. and Harvey, K. L.: 1979, *Solar Phys.* **64**, 93.
Martres, M. J., Michard, R., Soru-Iscovici, I., and Tsap, T.: 1968, in K. O. Kiepenheuer (ed.), 'Structure and Development of Solar Active Regions', *IAU Symp.* **35**, 318.
Martres, M. J., Soru-Escaut, I., and Rayrole, J.: 1971, R. Howard (ed.), 'Solar Magnetic Fields', *IAU Symp.* **43**, 435.
Mattig, W.: 1958, *Z. Astrophys.*, **44**, 280.
Mein, N. and Schmieder, B.: 1981, *Astron. Astrophys.* **97**, 310
Mein, P.; Mein, N. and Schmieder, B.: 1980, *Proc. Japan-France Seminar on Solar Phys.* (ed. F. Moriyama and J. C. Henoux) p. 70.
Mercier, C. and Heyvaerts, J.: 1980, *Solar Phys.* **68**, 151.
Meyer, F. and Schmidt, H.: 1967, *Z. Astrophys.* **65**, 274.
Meyer, F. and Schmidt, H. U.: 1968, *Z. Angew. Math. Mech.* **48**, 218.
Meyer, F., Schmidt, H. U., Simon, G. W., and Weiss, N. O.: 1979, *Astron. Astrophys.* **76**, 35.
Meyer, F., Schmidt, H. U., and Weiss, N. O.: 1977, *Monthly Notices Roy. Astron. Soc.* **179**, 741.
Meyer, F., Schmidt, H. U., Weiss, N. O., and Wilson, P. R.: 1974, *Monthly Notices Roy. Astron. Soc.* **169**, 35.
Mikhailovsky, A. B.: 1974, *Theory of Plasma Instabilities*, Vols. 1 and 2, Consultants Bureau, New York.

Miller, G. and Turner L.: 1981, *Phys. Fluids* **24**, 363.
Milne, A. M. and Priest, E. R.: 1981, *Solar Phys.* **73**, 157.
Milne, A. M., Priest, E. R., and Roberts, B.: 1979, *Astrophys. J.* **232**, 304.
Mitchell, H. G. and Kan, J. R.: 1978, *J. Plasma Phys.* **20**, 31.
Moffatt, H. K.: 1978, *Magnetic Field Generation in Electrically Conducting Fluids*, Cambridge University Press, England.
Molodensky, M. M.: 1974, *Solar Phys.* **39**, 393.
Molodensky, M. M.: 1975, *Solar Phys.* **43**, 311.
Molodensky, M. M.: 1976, *Solar Phys.* **49**, 279.
Moore, D. W. and Spiegel, E. A.: 1964, *Astrophys. J.* **139**, 48.
Moore, R. L.: 1967, *IAU Symp.* **28**, 405.
Moore, R. L. and Tang F.: 1975, *Solar Phys.* **41**, 81.
Mouschovias, T. C. and Poland, A. I.: 1978, *Astrophys. J.* **220**, 675.
Muller, R.: 1973, *Solar Phys.* **32**, 409.
Munro, R. H. and Jackson, B. V.: 1977, *Astrophys. J.* **213**, 874.
Musman, S. and Rust, D. M.: 1970, *Solar Phys.* **13**, 261.
Nakagawa, Y.: 1970, *Solar Phys.* **12**, 419.
Nakagawa, Y. and Malville, J. M.: 1969, *Solar Phys.* **9**, 102.
Nakagawa, Y. and Raadu, M. A.: 1972, *Solar Phys.* **25**, 127.
Nakagawa, Y., Raadu, M. A., Billings, D. E., and McNamara, D.: 1971, *Solar Phys.* **19**, 72.
Nakagawa, Y., Raadu, M. A., and Harvey, J. W.: 1973, *Solar Phys.* **30**, 421.
Nakagawa, Y., Wu, S. T., and Han, S. M.: 1978, *Astrophys. J.* **219**, 314.
Neidig, D. F.: 1978, AFGC report TR-78-0194.
Neidig, D. F.: 1979, *Solar Phys.* **61**, 121.
Newcomb, W. A.: 1960, *Ann. Phys.* **10**, 232.
Noci, G.: 1981, *Solar Phys.* **69**, 63.
November, L. J., Toomre, J., Gebbi, K. B., and Simon, G. W.: 1981, *Astrophys. J.* **245**, L123.
Nye, A. H. and Thomas, J. H.: 1974, *Solar Phys.* **38**, 399.
Nye, A. H. and Thomas, J. H.: 1976a, *Astrophys. J.* **204**, 573.
Nye, A. H. and Thomas, J. H.: 1976b, *Astrophys. J.* **204**, 582.
Orrall, F. Q.: 1981, *Proc. Skylab Active Region Workshop*, Colo. Ass. Univ. Press.
Orrall, F. Q. and Zirker, J. B.: 1961, *Astrophys. J.* **134**, 72.
Osherovich, V.: 1979, *Solar Phys.* **64**, 261.
Osherovich, V.: 1982, *Solar Phys.* **77**, 63.
Osterbrock, D. E.: 1961, *Astrophys. J.* **134**, 347.
Parker, E. N.: 1953, *Astrophys. J.* **117**, 431.
Parker, E. N.: 1955a, *Astrophys. J.* **121**, 491.
Parker, E. N.: 1955b, *Astrophys. J.* **122**, 293.
Parker, E. N.: 1958, *Astrophys. J.* **128**, 664.
Parker, E. N.: 1963a, *Astrophys. J.* **138**, 552.
Parker, E. N.: 1963b, *Interplanetary dynamical processes*, Interscience, New York.
Parker, E. N.: 1964, *Astrophys. J.* **139**, 93.
Parker, E. N.: 1966, *Astrophys. J.* **145**, 811.
Parker, E. N.: 1972, *Astrophys. J.* **174**, 499.
Parker, E. N.: 1973, *J. Plasma Phys.* **9**, 49.
Parker, E. N.: 1974a, *Solar Phys.* **36**, 249.
Parker, E. N.: 1974b, *Astrophys. J.* **189**, 563; **190**, 429.
Parkatt, E. N.: 1974c, *Astrophys. J.* **191**, 245.
Parker, E. N.: 1975, *Astrophys. J.* **198**, 205.
Parker, E. N.: 1977, *Ann. Rev. Astron. Astrophys.* **15**, 45.
Parker, E. N.: 1978, *Astrophys. J.* **221**, 368.
Parker, E. N.: 1979a, *Cosmical Magnetic Fields*, Oxford University Press, England.
Parker, E. N.: 1979b, *Astrophys. J.* **230**, 905.
Parker, E. N.: 1979c, *Astrophys. J.* **230**, 914.
Parker, E. N.: 1979d, *Astrophys. J.* **231**, 250.

Parker, E. N.: 1979e, *Astrophys. J.* **231**, 270.
Parker, E. N.: 1979f, *Astrophys. J.* **232**, 282.
Parker, E. N.: 1979g, *Astrophys. J.* **233**, 1005.
Parker, E. N.: 1979h, *Astrophys. J.* **232**, 291.
Parker, E. N.: 1979i, *Astrophys. J.* **234**, 333.
Parker, E. N.: 1981, *Astrophys. J.* **244**, 631.
Peres, G., Rosner, R., Serio, S., and Vaiana, G. S.: 1982, *Astrophys. J.* **252**, 791.
Petschek, H. E.: 1964, *AAS–NASA Symp. on Solar Flares*, NASA SP-50 p. 425.
Petschek, H. E. and Thorne, R. M.: 1967, *Astrophys. J.* **147**, 1157.
Piddington, J. H.: 1973, *Astrophys. Space Sci.* **24**, 259.
Piddington, J. H.: 1977, *Astrophys. Space Sci.* **47**, 319.
Piddington, J. H.: 1978, *Astrophys. Space Sci.* **55**, 401.
Pikel'ner, S. B.: 1971, *Solar Phys.* **17**, 44.
Pneuman, G. W.: 1972a, *Solar Phys.* **23**, 223.
Pneuman, G. W.: 1972b, *Astrophys. J.* **177**, 793.
Pneuman, G. W.: 1980, *Solar Phys.* **65**, 369.
Pneuman, G. W.: 1981, in E. R. Priest (ed.), *Solar Flare Magnetohydrodynamics*, Gordon and Breach, London, Ch. 7.
Pneuman, G. W. and Kopp, R. A.: 1971, *Solar Phys.* **18**, 258.
Pneuman, G. W. and Kopp, R. A.: 1977, *Astron. Astrophys.* **55**, 305.
Pneuman, G. W. and Kopp, R. A.: 1978, *Solar Phys.* **57**, 49.
Pouquet, A. Frisch, U., and Leorat, J.: 1976, *J. Fluid Mech.* **77**, 321.
Pouquet, A. and Patterson, G. S.: 1978, *J. Fluid Mech.* **85**, 305.
Priest, E. R.: 1976, *Solar Phys.* **47**, 41.
Priest, E. R.: 1978, *Solar Phys.* **58**, 57.
Priest, E. R.: 1981a, *Solar Flare Magnetohydrodynamics*, Gordon and Breach, London.
Priest, E. R.: 1981b, in E. R. Priest (ed.), *Solar Flare Magnetohydrodynamics*, Gordon and Breach, London, Ch. 3.
Priest, E. R.: 1981c, in F. Orrall (ed.), *Solar Active Regions*, Ch. 9, Colo. Ass. Univ. Press.
Priest, E. R.: 1982a, in C. Jordan (ed.), Proc. 3rd European Solar Meeting, Oxford, 1981.
Priest, E. R.: 1982b, *Fundamentals of Cosmic Physics.* **7**, 363.
Priest, E. R. and Milne, A. M.: 1980, *Solar Phys.* **65**, 315.
Priest, E. R. and Pneuman, G. W.: 1974, *Solar Phys.* **34**, 231.
Priest, E. R. and Raadu, M.: 1975, *Solar Phys.* **43**, 177.
Priest, E. R. and Smith, D. F.: 1972, *Astrophys. Lett.* **12**, 25.
Priest, E. R. and Smith, E. A.: 1979, *Solar Phys.* **64**, 267.
Priest, E. R. and Sonnerup, B. U. Ö.: 1975, *Geophys. J. Roy. Astron. Soc.* **41**, 405.
Priest, E. R. and Soward, A. M.: 1976, V. Bumba and J. Kleczek (eds.), 'Basic Mechanisms of Solar Activity' *IAU Symp.* **71**, 353.
Proctor, M. R. E.: 1977, *J. Fluid Mech.* **80**, 769.
Proctor, M. R. E. and Galloway, D. J.: 1979, *J. Fluid Mech.* **90**, 273.
Raadu, M. A.: 1972, *Solar Phys.* **22**, 425.
Raadu, M. A. and Kuperus, M.: 1973, *Solar Phys.* **28**, 77.
Rae, I. C. and Roberts, B.: 1981, *Geophys. Astrophys. Fluid Dynamics*, **18**, 197.
Rayleigh, Lord: 1916, *Phil. Mag.* **32**, 529.
Raymond, J. C. and Smith, B. W.: 1977, *Astrophys. J. Suppl.* **35**, 419.
Reid, J. and Laing, E. W.: 1979, *J. Plasma Phys.* **21**, 501.
Ribes, E. and Unno, W.: 1980, *Astron. Astrophys.* **91**, 129.
Riesebieter, W. and Neubauer, F. M.: 1979, *Solar Phys.* **63**, 127.
Robbins, K. A.: 1976, *Proc. Nat. Acad. Sci. U.S.A.* **73**, 4297.
Roberts, B.: 1976, *Astrophys. J.* **204**, 268.
Roberts, B.: 1979, *Solar Phys.* **61**, 23.
Roberts, B.: 1981a, *Solar Phys.* **69**, 27.
Roberts, B.: 1981b, *Solar Phys.* **69**, 39.
Roberts, B.: 1981c, in L. E. Cram and J. H. Thomas (eds.), *The Physics of Sunspots*, p. 369.

Roberts, B. and Frankenthal, S.: 1980, *Solar Phys.* **68**, 103.
Roberts, B. and Priest, E. R.: 1975, *J. Plasma Phys.* **14**, 417.
Roberts, B. and Webb, A. R.: 1978, *Solar Phys.* **56**, 5.
Roberts, B. and Webb, A. R.: 1979, *Solar Phys.* **64**, 77.
Roberts, G. O.: 1970, *Phil. Trans. Roy. Soc. A.* **266**, 535.
Roberts, G. O.: 1972, *Phil. Trans. Roy. Soc.* **A271**, 411.
Roberts, P. H.: 1967, *An Introduction to Magnetohydrodynamics*, Longmans, London.
Roberts, P. H.: 1972, *Phil. Trans. Roy. Soc. A.* **272**, 663.
Roberts, P. H. and Soward, A. M.: 1972, *Proc. Roy. Soc.* **A328**, 185.
Roberts, P. H. and Soward, A. M.: 1978, *Rotating Fluids in Geophysics*, Academic Press, London.
Roberts, P. H. and Stewartson, K.: 1977, *Astron. Nachr.* **298**, 311.
Roberts, P. H. and Stix, M.: 1972, *Astron. Astrophys.* **18**, 453.
Rosenbluth, M. N., Krall, N. A. and Rostoker, N.: 1962, *Nuc. Fusion Suppl.* **1**, 143.
Rosner, R., Tucker, W. H. and Vaiana, G. S.: 1978, *Astrophys. J.* **220**, 643.
Roxburgh I. W.: 1976, in V. Bumba and J. Kleczek (eds.), 'Basic Mechanisms of Solar Activity', *IAU Symp.* **71**, 453.
Roxburgh, I. W.: 1981, in R. M. Bonnet and A. K. Dupree (ed.), *Solar Phenomena in Stars and Stellar Systems*, D. Reidel, Dordrecht, Holland, p. 399.
Rust, D. M.: 1967, *Astrophys. J.* **150**, 313.
Rust, D. M.: 1968, in K. O. Kiepenheuer (ed.), 'Structure and Development of Solar Active Regions', *IAU Symp.* **35**, 77.
Rust, D. M.: 1972, *Solar Phys.* **25**, 141.
Rust, D. M.: 1973, *Solar Phys.* **33**, 205.
Rust, D. M.: 1976, *Phil. Trans. Roy. Soc. Lond.* **A281**, 427.
Rust, D. M. and Bar, V.: 1973, *Solar Phys.* **33**, 445.
Rust, D. M. and Bridges, C. A.: 1975, *Solar Phys.* **43**, 129.
Sakurai, T.: 1976, *Publ. Astron. Soc. Japan* **28**, 177.
Sakurai, T.: 1982a, *Solar Phys.* **76**, 301.
Sakurai, T.: 1982b, *Astrophys. J.*, submitted.
Sakurai, T. and Levine, R. H.: 1981, *Astrophys. J.* **248**. 817.
Sakurai, T. and Uchida, Y.: 1977, *Solar Phys.* 52, 397.
Schatzman, E.: 1949, *Ann. Astrophys.* **12**, 203.
Schatzman, E.: 1956, *Ann. Astrophys.* **19**, 45.
Schatzman, E.: 1963, *Ann. Astrophys.* **26**, 166.
Schatzman, E.: 1965, *IAU Symp.* **22**, 337.
Schatzman, E. and Souffrin, P.: 1967, *Ann. Rev. Astron. Astrophys.* **5**, 67.
Scheuer, M. A. and Thomas, J. H.: 1981, *Solar Phys.* **71**, 21.
Schmahl, E. J.: 1979, in E. Jensen, P. Maltby, and F. Q. Orrall (eds.), 'Physics of Solar Prominences', *IAU Colloq.* **44**, 102
Schmidt, H. U.: 1964, W. Hess (ed.), *NASA Symp. on Phys. of Solar Flares*, NASA SP-50, p. 107.
Schmidt, H. and Zirker, J.: 1963, *Astrophys. J.* **138**, 1310.
Schlüter, A. and Temesvary, S.: 1958, *IAU Symp.* **6**, 263.
Schnack, D. D. and Killeen, J.: 1978, in *Theoretical and Computational Plasma Physics*, Int. Atom. Energy Agency, Vienna, p. 337.
Schnack, D. D. and Killeen, J.: 1979, *J. Nucl. Fusion* **19**, 877.
Schüssler, M.: 1977, *Astron. Astrophys.* **56**, 439.
Schüssler, M.: 1980, Nature **288**, 150.
Schussler, M.: 1981, *Astron. Astrophys.* **94**, L17.
Schwarz, S. J.: 1980, *Rev. Geophys. Space Phys.* **18**, 313.
Schwarz, S. J.: 1981, in R. M. Bonnet and A. K. Dupree (eds.), *Solar Phenomena in Stars and Stellar Systems*, D. Reidel, Dordrecht, Holland, p. 319.
Schwarzschild, M.: 1948, *Astrophys. J.* **107**, 1.
Sedlacek, Z.: 1971, *J. Plasma Phys.* **5**, 239.
Semel, M.: 1967, *Ann. d'Astrophys.* **30**, 513.
Serio, S., Peres, G., Vaiana, G. S., Golub, L. and Rosner, R.: 1981, *Astrophys. J.* **243**, 288.
Seuss, S. T., Richter, A. K., Winge, C. R., and Nerney, S. F.: 1977, *Astrophys. J.* **217**, 296.

Shafranov, V. D.: 1957, *J. Nuc. Energy III* **5**, 86.
Sheeley, N. R.: 1980, *Solar Phys.* **65**, 229.
Sheeley, N. R. and Harvey, J. W.: 1975, *Solar Phys.* **45**, 275.
Simon, G. and Leighton, R.: 1964, *Astrophys. J.* **140**, 1120.
Smith, E. A. and Priest, E. R.: 1977, *Solar Phys.* **53**, 25.
Somov, B. V.: 1978, *Solar Phys.* **60**, 315.
Somov, B. V. and Syrovatsky, S. I.: 1972, *Sov. Phys. JETP* **34**, 332.
Somov, B. V. and Syrovatsky, S. I.: 1980, *Sov. Astron. Lett.* **6**, 311.
Somov, B. V. and Syrovatsky, S. I. 1982, *Solar Phys.* **75**, 237.
Sonnerup, B. U. Ö: 1970, *J. Plasma Phys.* **4**, 161.
Sonnerup, B. U. Ö. and Priest, E. R.: 1975, *J. Plasma Phys.* **14**, 283.
Souffrin, P.: 1967, *IAU Symp.* **28**, 459.
Soward, A. M.: 1972, *Phil. Trans. Roy. Soc.* **A272**, 431.
Soward, A. M.: 1974, *Phil. Trans. Roy. Soc.* **A275**, 611.
Soward, A. M. and Priest, E. R.: 1977, *Phil. Trans. Roy. Soc. Lond.* **A284**, 369.
Soward, A. M. and Priest, E. R.: 1982, *J. Plasma Phys.* **28**, 335.
Spicer, D. S.: 1976, PhD thesis, University of Maryland (NRL report 8036).
Spicer, D. S.: 1977, *Solar Phys.* **53**, 305.
Spicer, D. S.: 1980 in P. A. Sturrock (ed.), *Solar Flares*, Colo. Ass. Univ. Press, Ch. 3.
Spicer, D. S.: 1981, *Solar Phys.* **70**, 149.
Spicer, D. S. and Brown, J. C.: 1981, in S. Jordan (ed.), *The Sun as a Star*, NASA/CNRS, Ch. 18.
Spiegel, E. A.: 1957, *Astrophys. J.* **126**, 202.
Spies, G. O.: 1974, *Phys. Fluids* **17**, 1188.
Spitzer, L.: 1962, *Physics of Fully Ionized Gases*, Interscience, New York.
Spruit, H. C.: 1974, *Solar Phys.* **34**, 277.
Spruit, H. C.: 1977, *Solar Phys.* **55**, 3.
Spruit, H. C.: 1979, *Solar Phys.* **61**, 363.
Spruit, H. C.: 1981a, *Space Sci. Rev.* **28**, 422.
Spruit, H. C.: 1981b, *Astron. Astrophys.* **98**, 155; in R. M. Bonnet and A. K. Dupree (eds.), *Solar Phenomena in Stars and Stellar Systems*, D. Reidel, Dordrecht, p. 289.
Spruit, H. C.: 1981c, in S. D. Jordan (ed.), *The Sun as a Star*, NASA/CNRS, p. 385.
Spruit, H. C.: 1981d, *Phys. of Sunspots* (ed. L. E. Cram & J. H. Thomas), p. 98, p. 359.
Spruit, H. C. and van Ballegooijen, A. A.: 1982, *Astron. Astrophys.* **106**, 58.
Spruit, H. C. and Zweibel, E. G.: 1979, *Solar Phys.* **62**, 15.
Steenbeck, M., Krause, F., and Rädler, K. H.: 1966, *Z. Naturforsch.* **21a**, 369.
Steenbeck, M. and Krause, F.: 1969, *Astron. Nachr.* **291**, 49.
Stein, R. F. and Leibacher, J.: 1974, *Ann. Rev. Astron. Astrophys.* **12**, 407.
Steinolfson, R. S.: 1982, *Astrophys. J.* **255**, 730; *Astron. Astrophys.* **115**, 39.
Steinolfson, R. S. and Tandberg-Hanssen, E.: 1977, *Solar Phys.* **55**, 99.
Steinolfson, R. S., Wu, S. T., Dryer, M., and Tandberg-Hanssen, E.: 1978 *Astrophys. J.* **225**, 259.
Stix, M.: 1970, *Astron. Astrophys.* **4**, 189.
Stix, M.: 1972, *Astron. Astrophys.* **20**, 9.
Stix, M.: 1973, *Astron. Astrophys.* **24**, 275.
Stix, M.: 1976, *Astron. Astrophys.* **47**, 243; *IAU Symp.* **71**, 367.
Stix, M.: 1981, *Solar Phys.* **74**, 79
Stuart, J. T.: 1954, *Proc. Roy. Soc.* **A221**, 189.
Stuart, J. T.: 1963, in L. Rosenhead (ed.), *Laminar Boundary Layers*, Oxford, England, p. 492.
Sturrock, P. A.: 1966, *Nature* **211**, 695.
Sturrock, P. A.: 1980, *Solar Flares*, Colo. Ass. Univ. Press, Boulder, U.S.A.
Sturrock, P. A. and Woodbury, E. T.: 1967, *Plasma Astrophys.*, New York, Academic Press, p. 155.
Sturrock, P. and Uchida, Y.: 1981, *Astrophys. J.* **246**, 331.
Suydam, B. R.: 1959, *Proc. 2nd U.N. Intern. Conf. Peaceful Uses of Atomic Energy* **31**, 157.
Svestka, Z.: 1976a, in D. J. Williams (ed.) *Physics of Solar Planetary Environments*, American Geophys. Union, p. 129.
Svestka, Z.: 1976b, *Solar Flares*, D. Reidel, Dordrecht, Holland.

Svestka, Z.: 1977, *Solar Phys.* **52**, 69.
Svestka, Z.: 1981, in E. R. Priest (ed.), *Solar Flare Magnetohydrodynamics*, Gordan and Breach, London, Ch. 2.
Sweet, P. A.: 1958, *IAU Symp.* **6**, 123.
Sweet, P. A.: 1969, *Ann. Rev. Astron. Astrophys.* **7**, 149.
Syrovatsky, S. I.: 1966, *Sov. Astron.* **10**, 270.
Syrovatsky, S. I.: 1971, *Sov. Phys. JETP* **33**, 933.
Syrovatsky, S. I.: 1976, *Sov. Astron. Letts* **2**, 35.
Syrovatsky, S. I.: 1978, *Solar Phys.* **58**, 89.
Syrovatsky, S. I.: 1982, *Solar Phys.* **76**, 3.
Tanaka, K. and Nakagawa, Y.: 1973, *Solar Phys.* **33**, 187.
Tandberg-Hanssen, E.: 1974, *Solar Prominences*, D. Reidel, Dordrecht, Holland.
Tandberg-Hanssen, E. and Anzer, U.: 1970, *Solar Phys.* **15**, 158.
Tasso, H.: 1977, *Plasma Phys.* **19**, 177.
Tayler, R. J.: 1957, *Proc. Phys. Soc. (London)* **B70**, 1049.
Taylor, J. B.: 1974, *Phys. Rev. Letters* **33**, 1139.
Taylor, J. B.: 1976, in D. E. Evans (ed.), *Pulsed High Beta Plasmas*, Pergamon Press, Oxford, p. 59.
Thomas, C. L.: 1976, in M. B. Hooper (ed.), *Computational Methods in Classical and Quantum Physics*, Advance publications, p. 178.
Thomas, J. H.: 1978, *Astrophys. J.* **225**, 275.
Thomas, J. H., Clark, P., and Clark, A.: 1971, *Solar Phys.* **16**, 51.
Topka, K., Moore, R., La Bonte, B., and Howard, R.: 1982, *Solar Phys.* **79**, 231.
Tucker, W. H.: 1973, *Astrophys. J.* **186**, 285.
Tucker, W. H. and Koren, M.: 1971, *Astrophys. J.* **168**, 283.
Tur, T. J. and Priest, E. R.: 1976, *Solar Phys.* **48**, 89.
Tur, T. J. and Priest, E. R.: 1978, *Solar Phys.* **58**, 181.
Uchida, Y.: 1965, *Astrophys. J.* **142**, 335.
Uchida, Y. and Kaburaki, O.: 1974, *Solar Phys.* **35**, 451.
Uchida, Y. and Low, B. C.: 1981, *Ind. J. Astrophys. Astron.* **2**, 405.
Ugai, M. and Tsuda, T.: 1977, *J. Plasma Phys.* **17**, 337.
Ulmschneider, P.: 1971, *Astron. Astrophys.* **12**, 297; **14**, 275.
Ulmschneider, P.: 1974, *Solar Phys.* **39**, 327.
Ulmschneider, P.: 1976, *Solar Phys.* **49**, 249.
Ulmschneider, P.: 1979, *Space Sci. Rev.* **24**, 21.
Ulmschneider, P. and Kalkofen, W.: 1977, *Astron. Astrophys.* **57**, 199.
Ulrich, R.: 1970, *Astrophys. J.* **162**, 993.
Van Hoven, G.: 1976, *Solar Phys.* **45**, 95.
Van Hoven, G.: 1981, in E. R. Priest (ed.), *Solar Flare Magnetohydrodynamics*, Gordon and Breach, London, Ch. 4.
Van Hoven, G. and Cross, M. A.: 1973, *Phys. Rev.* **A7**, 1347.
Van Tend, W. and Kuperus, M.: 1978, *Solar Phys.* **59**, 115.
Vasyliunas, V. M.: 1975, *Rev. Geophys. Space Phys.* **13**, 303.
Vernazza, J. E., Avrett, E. H., and Loeser, R.: 1981, *Astrophys. J. Suppl.* **245**, 350.
Vesecky, J. F., Antiochos, S. K., and Underwood, J. H.: 1979, *Astrophys. J.* **233**, 987.
Waddell, B. V., Rosenbluth, M. N., Monticello, D. A., and White, R. B.: 1976, *Nucl. Fusion* **16**, 528.
Waddell, B. V., Carreras, B., Hicks, H. R., and Holmes, J. A.: 1979, *Phys. Fluids* **22**, 896.
Webb, A. W. and Roberts, B.: 1978, *Solar Phys.* **59**, 249.
Webb, A. W. and Roberts, B.: 1980, *Solar Phys.* **68**, 71 and **68**, 87.
Weber, E. J. and Davis, L.: 1967, *Astrophys. J.* **148**, 217.
Weber, W. J.: 1979, *Solar Phys.* **61**, 345.
Weiss, N. O.: 1966, *Proc. Roy. Soc.* **A293**, 310.
Weiss, N. O.: 1974, in L. Mestel and N. O. Weiss (eds.), *Magnetohydrodynamics*, Swiss Soc. Astron. Astrophys., Geneva, p. 185.
Weiss, N. O.: 1981, *J. Fluid Mech.* **108**, 247; **108**, 273.
Weiss, N. O.: 1982, in A. M. Soward (ed.), *Planetary and Stellar Magnetism*, Gordon and Breach.

Wentzel, D. G.: 1974, *Solar Phys.* **39**, 129.
Wentzel, D. G.: 1976, *Solar Phys.* **50**, 343.
Wentzel, D. G.: 1977, *Solar Phys.* **52**, 163.
Wentzel, D. G.: 1978, *Solar Phys.* **58**, 307.
Wentzel, D. G.: 1979, *Astron. Astrophys.* **76**, 20.
Whang, Y. C. and Chang, C. C.: 1965, *J. Geophys. Res.* **70**, 4175.
Whitham, G. B.: 1974, *Linear and Nonlinear Waves*, Wiley Interscience, New York.
Wilson, A.: 1774, *Phil. Trans. Roy. Soc.* **64**, 1.
Wilson, O. C., Vaughan, A. H., and Mihalas, D.: 1981, *Scientific American* **244**, No 2, 104.
Withbroe, G. L. and Noyes, R. W.: 1977, *Ann. Rev. Astron. Astrophys.* **15**, 363.
Woltjer, L.: 1958, *Proc. Nat. Acad. Sci.* **44**, 489.
Wragg, M. A. and Priest, E. R.: 1981a, *Solar Phys.* **70**, 293.
Wragg, M. A. and Priest, E. R.: 1981b, *Solar Phys.* **69**, 257.
Wragg, M. A. and Priest, E. R.: 1982, *Solar Phys.* **80**, 309.
Wu, S. T., Dryer, M., Nakagawa, Y., and Han, S. M.: 1978, *Astrophys. J.* **219**, 324.
Yang, C. K. and Sonnerup, B. U. Ö.: 1976, *Astrophys. J.* **206**, 570.
Yeh, T. and Axford, W. I.: 1970, *J. Plasma Phys.* **4**, 207.
Yeh, T. and Pneuman, G. W.: 1977, *Solar Phys.* **54**, 419.
Yoshimura, H.: 1975, *Astrophys. J. Suppl.* **29**, 467.
Yoshimura, H.: 1977, *Solar Phys.* **52**, 41.
Yoshimura, H.: 1978, *Astrophys. J.* **220**, 692; **226**, 706.
Yoshimura, H.: 1979, *Astrophys. J.* **227**, 1047.
Yun, H. S.: 1968, PhD thesis, Ind. Univ., U.S.A.
Zeldovich, Ya. B. and Ruzmaikin, A. A.: 1980, *Soviet Scientific Review*, New York.
Zirin, H. and Stein, A.: 1972, *Astrophys. J.* **178**, L85.
Zirin, H.: 1974, *Vistas. Astron.* **16**, 1.
Zirin, H. and Tanaka, K.: 1973, *Solar Phys.* **32**, 173.
Zirker, J. B.: 1977, *Coronal Holes and High Speed Wind Streams*, Colo Ass. Univ. Press, Boulder.
Zwaan, C.: 1978, *Solar Phys.* **60**, 213.
Zweibel, E.: 1980, *Solar Phys.* **66**, 305.
Zweibel, E. G. and Hundhausen, A. J.: 1982, *Solar Phys.* **76**, 261.

INDEX

Absorption, resonant 187, 223–225
Absorption line 2, 7
Acoustic-gravity wave 170–173, 178
Acoustic (or hydrodynamic) shock wave 89, 189–192, 195–197, 213–216, 381
Acoustic (or sound) wave 89, 157, 170, 178–180, 189, 213–216
Active prominence 56
Active region (or plage) 21–22, 30–31, 37–45, 209
  ephemeral 21, 23–24, 32, 343
  remnant 39, 63
Active-region (or plage) filament (or prominence) 38, 56, 58–61, 345, 375, 377, 382–385, 390
Active-region loop 42, 222, 234–235, 241
Active-region motions 44–45
Active-region streamer 42
Adiabatic behaviour 17, 85, 243–244, 421, 423
Adiabatic temperature gradient 17
Alfvén critical point (or radius) 35, 424–426, 429–431
Alfvén Mach number 94
Alfvén radius (or critical point) 35, 424–426, 429–431
Alfvén resonance 225
Alfvén speed 83, 92, 440
Alfvén surface wave 225, 322
Alfvén wave 22, 36, 89, 157, 163, 168, 217, 352, 434
  compressional 158, 162–163, 170
  kinetic 225
  nonlinear 220–223
  shear 159–162
  torsional 161
Alpha-(α-) effect 330
Alpha-omega (α–ω) dynamo 336–338
Ambipolar diffusion 77
Analogy, vorticity magnetic field 96
Angular momentum 4, 425, 431, 441
Annihilation, magnetic 97, 347
Anomalous conductivity 79
Antidiffusion mode 234
Arcsec 4, 441
Arcade, coronal 34, 39, 59, 138–139, 145–152, 367–368, 374, 386–390, 410, 414
Arch filament 38, 46, 68

Assumptions, MHD 92
Astronomical unit 441
Atmospheric model 207–213
Atomic weight 82–83
Attraction, hydrodynamic 311–313
Attractor, strange 339–341

Bennet's relation 124
Bessel function 443
Beta, plasma 24, 83, 94, 440
Beta limitation 351
Body wave 321
Boltzmann constant 440
Boltzmann equations 73
Boussinesq approximation 281, 283, 346
Breeze, solar 419, 421
Bright point, X-ray 24, 30, 32, 38, 343, 357
Brunt frequency 164
Buckling 270
Buffeting, granular 319
Buoyancy 16
  magnetic 291–299, 325, 330, 339–340
Butterfly diagram 54–55, 338

Calorie 439
Cavity
  coronal 59
  resonant 178–179
Centrifugal force 154
Chandrasekhar number 95
Chromosphere xv, 7, 9, 26–29, 207
Chromospheric model 26
C–N cycle 13
Collision time 76, 78
Collisionless shock wave 194
Compact (or simple-loop) flare 67, 235, 356–366
Complex of activity 39
Complex variable 231, 403
Compression (or density) ratio 201
Compression wave 190
Compressional Alfvén wave 158, 162–163, 170
Conduction, thermal 85, 207–208, 312, 417, 440
Conductivity
  anomalous 79
  Cowling 76

direct 77
electrical 78, 440
Hall 77
radiative 16
Conservation of flux 411
Conservation (or jump or Rankine–Hugoniot) relations 192, 195, 197, 199–202
Constant-α magnetic field 125, 133, 135, 137–140
Contact discontinuity 203
Continuity, mass 80–81
Continuous spectrum 183, 185, 187
Convection
  inhibition 298
  magneto- 280–290, 314
  overturning 282
  penetrative 181
Convection zone 6, 15–18, 441
Convective dynamo 341–342
Convective (or thermal) instability 165, 280–290
Convective overstability 181, 276–277, 282–284, 299
Convective roll 276, 308
Cool loop core 239, 384
Core, solar 5, 13
Coriolis force 154, 166, 297, 329–330
Corona 2, 11, 29–36, 417
Coronagraph 2, 29, 58
Coronal arcade 34, 39, 59, 138–139, 145–152, 367–368, 374, 386–390, 410, 414
Coronal condensation 42
Coronal evolution 145–149
Coronal hole 22, 32–34, 209, 343, 420, 431–434
Coronal loop 32, 42–44, 70–71, 222, 234–245, 362, 384, 413, 427
Coronal loop flow 240–245
Coronal rain 45–46
Coronal streamer 29, 427–431
Coronal transient 61–64, 66, 410–416
Correlation time 334
Couette flow 271
Coulomb logarithm 78–79
Cowling conductivity 76
Cowling's theorem 106, 327–328
Current, ring 412
Current density 78
Current (or magnetic) dissipation 225–234
Current filament 226, 233–234
Current-free (or potential) magnetic field 130–133, 231, 403, 408
Current sheath 226–227
Current sheet 97–98, 113–116, 226–233, 345–357, 390–395, 406–408, 429
  heliomagnetic 34, 427, 430–431
Cusp resonance 225
Cusp (or tube) speed 170, 184, 318

Cut-off 171
Cyclotron frequency 170
Cylindrical polar coordinates 442
Cylindrical tearing mode (or resistive kink) instability 363–366

Damping (or dissipation) length 89
Debye length 75, 440
Density 5–6, 15, 25, 29, 33, 82, 441
Diagnostic diagram 171–172, 178
Differential rotation 18, 328–329, 341, 343
Diffusion
  ambipolar 77
  resistive 144–145
  turbulent 331, 343
Diffusion region 348–352
Diffusion time 5, 96, 325
Diffusivity
  eddy 80, 312–313, 334–335, 338
  magnetic 77–78, 312, 440
  thermal 87
Dimensionless parameters 93–96
Discontinuity
  contact 203
  rotational 36, 205
  tangential 36, 114, 203
Dispersive wave 156
Dissipation
  magnetic (or current) 225–234
  ohmic 77, 90, 194, 219, 227, 312
  shock 214–216, 218–220
  topological 230
  viscous 89
Double tearing mode instability 365–366
Downflow (or downdraft) 21–22, 44, 213, 244, 316
Dynamo 325–343
  α–ω 336–338
  convective 341–342
  kinematic 326, 331–338
  MHD 338–342
  nearly symmetric 331–332
  turbulent 332–334
Dynamo number 335–336
Dynamo wave 335–336

Eclipse 1–2, 12, 29–30, 37, 427, 428
Eddy diffusivity 80, 312–313, 334–335, 338
Effective temperature 4, 48, 441
Ekman number 94
Electric field 75, 78
Electrical conductivity 78, 440
Electrical neutrality 75
Electrical resistivity 77
Electron charge 440

Electron mass 440
Electron volt 440
Ellerman bomb 44, 50
Emerging flux 24, 38, 67–69
Emerging flux model 228–230, 356–360
Energy
  electromagnetic 90
  flare 66
  internal 84
  potential 246–247, 257
Energy balance 14, 235–240, 358–359, 398
Energy equation 84, 351, 420–421, 434
Energy loss, atmospheric 209
Energy loss function 85
Energy (or variational) method 257–265, 306, 404
Energy spectrum function 333
Enthalpy 85
Entropy 85, 90, 192, 195–196
Ephemeral active region 21, 23–24, 32, 343
Equation
  energy 84, 351, 420–421, 434
  Euler–Lagrange 261–263, 362
  heat 84
  induction 77–80, 96–101
  Klein–Gordon 324
  mechanical energy 90
  motion 81–82
Equations
  Boltzmann 73
  conservation 192, 195, 197, 199–202
  Maxwell 73
  slender flux tube 316
  summary of 91–93
Error function 443
Erupting prominence (or filament) 61–64, 67, 367–368, 375–377, 402, 408
Eruptive instability 367–368, 375–377
Escape velocity 4, 441
Euler–Lagrange equation 261–263, 362
Evanescent wave 171, 177–178, 183
Evaporation 71, 244, 381
Evershed flow 2, 44, 50, 243, 309
Evolutionary condition 193
Exchange of stability 248
Expansion coefficient 281
Expansion wave 190
Exploding granule 19
Explosive instability 248, 360
Expulsion, magnetic flux 285–290

Faculae 21
Fast magnetoacoustic shock wave 36, 193, 202–203, 218, 434–435
Ferraro isorotation law 107
Fibril 27, 42, 50, 59, 68

Field line, magnetic 109
Filament (or prominence) 2–3, 37, 56–64, 377, 382–410
Filament
  active-region (or plage) 38, 56, 58–61, 345, 375, 377, 382–385, 390
  arch 38, 46, 68
  current 226, 233–234
  quiescent 39, 56–61, 63, 377, 382, 390
  winking 61
Filament channel 59
Filigree 20–21
First-order smoothing 333
Five-minute oscillation 3, 19, 175–182
Flare 2, 50, 64–72, 106, 344–381, 417
  homologous 66
  simple-loop (or compact) 67, 235, 356–366
  sympathetic 66
  thermal 66, 68
  two-ribbon 41, 61, 67, 69–71, 345, 366–381, 410
Flare energy 66
Flare-induced coronal wave 175
Flare kernel 70
Flare trigger 358, 360
Flash phase of flare 65, 67, 356
Flow
  active-region 44–45
  coronal loop 240–245
  Evershed 2, 44, 50, 243, 309
  meridional 18, 343
  siphon 242–243, 309, 386, 390
Flow instability 270–272
Flute instability 256, 265–266, 314
Flux
  heat 86
  missing 298
  Poynting 90, 197, 201
Flux tube 108–113, 121–130, 186–187, 223–224, 231, 259, 289, 291, 297, 303, 305, 342, 362, 375
  intense 314–324
Flux tube strength 109–110
Force-free equilibria, multiple 368–375
Force-free magnetic field 69, 111, 119, 125–129, 133–149, 230, 232, 259, 263, 365, 368–376, 386
Frequency
  Brunt 164
  cyclotron 170
Frozen flux theorem 99

Gamma function 443
Gas constant 440
Gas law 14
Gaussian cgs units 436–437
Geomagnetic storm 417

Geometrical acoustics 214
Giant cell 17, 22, 441
Global solar oscillations 15, 177
Grain 19
Granular buffeting 319
Granulation 2, 17, 19–20, 338, 441
    umbral 49, 181
Granule, exploding 19
Gravitational acceleration 14
Gravitational force 81
Gravitational mode instability 274
Gravitational potential 108
Gravity wave 163–166, 178
Green's function 132, 142, 186, 333
Group velocity 156, 166–170
Gyro-frequency 440
Gyro-radius 440

Hα 7, 37
Hα kernel 68–69
Hα knot 68–69, 71
Hale cycle 54
Hall conductivity 77
Hartmann number 95, 277
Harvard–Smithsonian reference atmosphere 24–25, 83, 96, 180
Heat equation 84
Heat flux 86
Heating 89
    adiabatic 17, 85, 243, 421, 423
    atmospheric 206–245
Helical structure (prominence) 405–410
Helicity 330, 334, 346
Heliomagnetic current sheet 34, 427, 430–431
Helmet streamer 30, 59
High-speed stream 36, 432, 434
Homologous flare 66
Hydrodynamic attraction 311–313
Hydrodynamic (or acoustic) shock wave 89, 189–192, 195–197, 213–216, 381
Hydromagnetic inertial wave 296, 330
Hydromagnetic Rayleigh–Taylor (or Kruskal–Schwarzschild) instability 256, 267, 294
Hypergeometric function 443

Impulsive phase of flare 65, 356
Incompressible 92, 251, 253, 264, 322
Induction equation 77–80, 96–101
Inertial wave 166–168
Initial-value problem 186, 248
Inner network magnetic field 23–24
Instability 246–279
    convective (or thermal) 165, 280–290
    double tearing mode 365–366
    eruptive 367–368, 375–377

    exchange (or interchange) 256, 265–267, 304
    explosive 248, 360
    flow 270–272
    flute 256, 265–266, 314
    gravitational mode 274
    hydromagnetic Rayleigh–Taylor (or Kruskal–Schwarzschild) 256, 267, 294
    intense magnetic field 316–319
    interchange 256, 265–267, 304
    Kelvin–Helmholtz 271–272
    kink 112, 259–263, 269, 362–365, 377
    leak 277, 282
    magnetic buoyancy 293–296
    radiative 276, 277–279, 383
    Rayleigh–Taylor 251–257, 267–268
    resistive 272–276, 325
    resistive kink (or cylindrical tearing mode) 363–366
    rippling mode 274
    sausage 256, 268–269
    tearing mode 275, 288, 345–346, 363–366, 406
    thermal (convective) 165, 280–290
    thermal (radiative) 276, 277–279, 383
Intense flux tube 22, 220, 314–324
Intense magnetic field instability 316–319
Interchange (or exchange) instability 256, 265–267, 304
Interconnecting loop 31–32, 235, 241
Interface 183
Interior, solar 4–6, 13–18
Intermediate wave 169, 193, 205
Internal energy 84
Ionisation 17, 25, 280
    partial 78
Ion-sound turbulence 79
Isotropic turbulence 333

Jet 29
Joule mode 233
Jump (or conservation) relations 192, 195, 197, 199–202

Kappa mechanism 181
Kelvin–Helmholtz instability 271–272
Kernel
    Hα 68–69
    flare 70
Kinematic approximation 93, 285, 290
Kinematic dynamo 326, 331–338
Kinetic Alfvén wave 225
Kink instability 112, 259–263, 269, 362–365, 377
Kippenhahn–Schlüter prominence model 395–402, 404
Klein–Gordon equation 324

Knot
  Hα 68–69, 71
  magnetic 22, 46, 50, 304
Kruskal–Schwarzschild (or hydromagnetic Rayleigh–Taylor) instability 256, 267, 294
Kruskal–Shafranov limit 264, 363

Lagrangian position 249
Law
  Ampère's 78
  Ferraro's isorotation 107
  Newton's cooling 87
  Ohm's 75–76
  perfect gas 14
  scaling (for loop) 228, 237
  Sporer's 54
  sunspot polarity 53
Leak instability 277, 282
Legendre function 443
Length
  damping 89
  Debye 75, 440
  mixing 14, 17, 343
Lighthill mechanism 181
Line
  absorption 2, 7
  polarity-inversion 58, 69
Line-tying 259, 266, 270, 362, 366, 376–377, 393–395
Linear pinch 124, 268–269
Longitudinal tube wave 323
Loop
  active-region 42, 222, 234, 235, 241
  coronal 32, 42–44, 70–71, 222, 234–245, 362, 384, 413, 427
  interconnecting 31–32, 235, 241
  'post'-flare 67, 70–71, 235, 345, 367, 377–381
  quiet-region 32, 235
  sunspot 43
  thermally isolated 236, 238, 241
  twisted 111, 122, 125–126, 264, 270, 362, 376–377, 384–385, 410–412
Loop prominence 56, 69–70
Loop transient 410
Lorentz force 101–106
Luminosity 4, 14–15, 441
Lundquist number 93

Mach number 94, 195
Macrospicule 29
Magnetic annihilation 97, 347
Magnetic buoyancy 291–299, 325, 330, 339–340
Magnetic buoyancy instability 293–296
Magnetic diffusivity 77–78, 312, 440
Magnetic (or current) dissipation 225–234
Magnetic elements 22

Magnetic field xv–xvi, 22–24, 48, 58, 69, 211–212, 436
  constant-α 125, 133, 135, 137–140
  force-free 69, 111, 119, 125–129, 133–149, 230, 232, 259, 263, 365, 368–376, 386
  inner network 23–24
  photospheric 22–24
  potential (or current-free) 130–133, 231, 403, 408
  uniform twist 125–126, 129, 262–263, 363, 384
  variable twist 363, 384
Magnetic field line 109
Magnetic flux 42
Magnetic flux concentration 285–290
Magnetic flux conservation 128–129
Magnetic flux expulsion 285–290
Magnetic flux tube 108–113, 121–130, 186–187, 223–224, 231, 259, 289, 291, 297, 303, 305, 342, 362, 375
  intense 22, 220, 314–324
  slender 296, 304, 308, 314–316
Magnetic heating 217–234
Magnetic island 148, 152
Magnetic knot 22, 46, 50, 304
Magnetic neutral point 104, 231–232, 328, 348, 379, 409, 414, 429
Magnetic nonequilibrium 230, 407
Magnetic permeability 73, 440
Magnetic Prandtl number 94, 277
Magnetic pressure 101–105, 111, 395
Magnetic reconnection 68, 114–116, 228, 345–357, 414
Magnetic reconnection rate 346, 353, 357
Magnetic Reynolds number 93, 96–101, 440
Magnetic solar wind model 423–426
Magnetic tension 101–105, 111, 161, 163, 395
Magnetic (or Alfvén) wave 22, 36, 89, 157–163, 168, 217, 352, 434
Magnetoacoustic-gravity wave 173–175
Magnetoacoustic surface wave 185
Magnetoacoustic wave 168–170, 217, 434
Magnetoconvection 280–290, 314
Magnetograph 3
Magneto-gravity overstability 181
Magneto-gravity wave 175
Magnetohydrostatics 117–152, 299–306, 395–410
Magnetostatic nonequilibrium 230, 399, 407
Main phase of flare 65, 356
Marginal stability 246, 248, 285
Mass, solar 3, 441
Mass continuity 80–81
Mass fraction 15
Mass loss 441
Maunder minimum 53, 339, 340

Maxwell's equations 73
Mean atomic weight 82
Mean-field electrodynamics 332–334
Mean particle mass 82
Mechanical energy equation 90
Mechanism
 kappa 181
 Lighthill 181
 Petschek 351–355
Megametre 74, 436
Meridional flow 18, 343
Mesogranulation 17, 19
Metastability 248, 359, 360
MHD assumptions 92
MHD dynamo 338–342
Micropore 22
Microsurge 46
Minimum energy theorem 130, 135, 137
Missing sunspot flux 298–299
Mixing length 14, 17, 343
Moat 47, 299, 310
Mode
 antidiffusion 234
 intermediate 169, 193, 205
 Joule 233
 kink 320–323
 normal 186, 246, 251–257
 sausage 320–321
Mode coupling 366
Model
 atmospheric 207–213
 chromospheric 26
 emerging flux flare 228–230, 356–360
 energy balance loop 235–240
 Pneumann–Kopp 427–431
 spaghetti 304
Modelling xvi
Momentum, angular 4, 425, 431, 441
Motion
 active region 44–45
 equation of 81–82
 sunspot 50–52
Mottle 27
Moving magnetic feature 46–47

Network 3, 21–23, 26, 28
Neutral point 104–106
 magnetic 104, 231–232, 328, 348, 379, 409, 414, 429
Neutral sheet 115, 391–392
Neutral (or marginal) stability 246, 248, 285
Neutrality, electrical 75
Neutrino 13, 15
Newton's Law of cooling 87
Nonequilibrium

 magnetic 230, 399, 407
 thermal 238–239, 360–361, 383–388, 391–393
Nonlinear Alfvén wave 220–223
Nonlinear tearing mode instability 366
Nonthermal velocity 206
Normal mode 186, 246, 251–257
Notation xvi, 444
Number
 Alfvén Mach 94
 Chandrasekhar 95
 dynamo 335–336
 Ekman 94
 Hartmann 95, 277
 Lundquist 93
 Mach 94, 195
 magnetic Prandtl 94, 277
 magnetic Reynolds 93, 96–101, 440
 Nusselt 95
 Prandtl 95, 276
 Rayleigh 95, 276, 281
 Reynolds 94, 271
 Richardson 272
 Rossby 94
 sunspot 52–53
 Taylor 95, 276
Numerical experiment 414–415
Nusselt number 95

Oblique shock wave 194, 199–205
Ohm's law 75–76
Ohmic dissipation (or diffusion or heating) 77, 90, 194, 219, 227, 312
Opacity 14–15
Oscillation
 five-minute 3, 19, 175–182
 global solar 15, 177
 three-minute 181
 umbral 49, 182, 321
Overstability 248
 convective 181, 276–277, 282–284, 299
 magneto-gravity 181
Overturning convection 282

Parameters, dimensionless 93–96
Parcel argument 16
Passive medium 144
Penetrative convection 181
Penumbra 49–50, 308–309
Penumbral wave 50, 175, 182
Perfect gas law 14, 82
Perfectly conducting limit 99–101
Permeability 73, 440
Permittivity 73, 440
Perpendicular shock wave 194, 197–199
Petschek mechanism 351–355

Phase of flare
    flash 65, 67, 356
    impulsive 65, 356
    main 65, 356
    preflare 65, 67, 356
Phase velocity 156, 169
Photosphere 7–8, 19–26, 441
Photospheric magnetic field 22–24
Photospheric ringing 177
Photospheric velocity 69
Pinch
    diffuse 270
    linear 124, 268–269
    reversed field 137
Pinch ratio 137
Pinched discharge 268
Pitch 122
Plage (or active region) 21–22, 30–31, 37–45, 209
Plage (or active-region) filament 38, 56, 58–61, 345, 375, 377, 382–385, 390
Plasma xvi
Plasma beta 24, 83, 94, 440
Plasma frequency 440
Pneuman–Kopp model 427–431
Polar coordinates
    cylindrical 442
    spherical 442
Polar coronal hole 33–34
Polar crown 59
Polar diagram 159, 169–170
Polar plume 30
Polarity-inversion line 58, 69
Polytropic approximation 93, 425, 432, 433
Pore 20, 305
'Post'-flare loops 67, 70–71, 235, 345, 367, 377–381
Potential
    gravitational 108
    scalar 130
    vector 136
Potential energy 246–247, 257
Potential (or current-free) magnetic field 130–133, 231, 403, 408
Poynting flux 90, 197, 201
P–P cycle 13
Prandtl number 95, 276
Preflare phase 65, 67, 356
Prominence (or filament) 2–3, 37, 56–64, 377, 382–410
    active 56
    active-region (or plage) 38, 56, 58–61, 345, 375, 377, 382–385, 390
    erupting 61–64, 67, 367–368, 375–377, 402, 408
    loop 56, 69–70
    quiescent 39, 56–61, 63, 377, 382, 390
Prominence feet 402
Prominence formation 240, 382–395
Prominence oscillation 407
Prominence properties 56, 58
Prominence stability 404–405
Prominence support 395–410
Prominence thread 59–61
Proton mass 430
Pseudo-isotropic turbulence 332–334

Quasi-linear approximation 333
Quiescent filament (or prominence) 39, 56–61, 63, 377, 382, 390
Quiet-region loop 32, 235
Quiet Sun 4, 13–36, 209

Radiative conductivity 16
Radiative cooling time 278
Radiative damping 87
Radiative instability 276, 277–279, 383
Radiative loss 87–89, 208
Radio burst
    type II 66
    type III 66
    type IV 66
Radius
    Alfvén 35, 424–426, 429–431
    solar 3, 441
Rain, coronal 45–46
Rankine–Hugoniot (or jump) relations 192, 195, 197, 199–202
Rationalised mks units 437–438
Ray path 219–220
Rayleigh criterion 271
Rayleigh number 95, 276, 281
Rayleigh–Taylor instability 251–257, 267–268
Reconnection, magnetic 68, 114–116, 228, 345–357, 414
Resistive diffusion 144–145
Resistive instability 272–276, 325
Resistive kink (or cylindrical tearing mode) instability 363–366
Resistivity, electrical 77
Resonance 180–181, 319
    Alfvén 225
    cusp 225
Resonance overlap 366
Resonant absorption 187, 223–225
Resonant cavity 178–179
Reversed-field pinch 137
Reynolds number 94
    critical 271
Richardson number 272
Ring current 412

Ringing, photospheric 177
Rippling mode instability 274
Roll, convective 276, 308
Rossby number 94
Rotation 296, 441
 differential 18, 328–329, 341, 343
Rotational discontinuity 36, 205

Sausage instability 256, 268–269
Scalar potential 130
Scale height 118–119, 121, 440
Scaling law 228, 237
Schwarzschild condition (or criterion) 17, 165
Sector boundary 35–36
Self-inductance 412
Self-similar (or similarity) solution 301, 313, 353
Shear 148, 387, 399, 402
Sheet
 current 97–98, 113–116, 226–233, 345–357, 390–395, 406–408, 429
 neutral 115, 391–392
Shock wave 189–205
 acoustic 89, 189–192, 195–197, 213–216, 381
 collisionless 194
 fast magnetoacoustic 36, 193, 202–203, 218, 434–435
 oblique 194, 199–205
 perpendicular 194, 197–199
 slow magnetoacoustic 193, 202–203, 351–352, 381
 strong 189
 switch-off 194, 203–205
 switch-on 194, 203–205
 weak 189
Shock wave damping-length 216, 219, 222
Shock wave dissipation 214–216, 218–220
Shock wave propagation 214–216, 218–220
Shock width 192
Similarity (or self-similar) solution 301, 313, 353
Simple-loop (or compact) flare 67, 235, 356–366
Siphon flow 242–243, 309, 386, 390
Skylab 3, 34, 60–61, 64, 70, 344, 430
Slender flux tube 296, 304, 308, 314–316
Slow magnetoacoustic shock wave 193, 202–203, 351–352, 381
Solar age 441
Solar breeze 419, 421
Solar cycle 2, 32, 52–56, 335–338
Solar flare 2, 50, 64–72, 106, 344–381, 417
 simple-loop (or compact) 67, 235, 356–366
 two-ribbon 41, 61, 67, 69–71, 345, 366–381, 410
Solar mass 3, 441
Solar Maximum Mission 71, 299, 381
Solar model 13–15

Solar oscillation, global 15, 177
Solar properties 35–36, 441
Solar radius 3, 441
Solar wind 3, 32–36, 417–435
Solar wind model
 isothermal 418
 magnetic 423–426
 two-fluid 421–423, 435
Sound speed 92, 440
 isothermal 418
Sound (or acoustic) wave 89, 157, 170, 178–180, 189, 213–216
Spaghetti model of sunspots 304–305
Specific heat 83–84
Spectrum tensor 333
Speed
 Alfvén 83, 92, 440
 cusp 170, 184, 318
 light 73, 440
 sound 92, 440
 tube 170, 184, 318
Spicule 2–3, 26–27, 45, 213, 240, 243, 318–320, 324
Sporer's law 54
Spray 56
Stability
 exchange of 248
 marginal (or neutral) 246, 248, 285
 meta- 248, 359, 360
 over- 181, 248, 276–277, 282–284, 299
 prominence 404–405
 sunspot 306–308
Standing wave 177
Stochastic motion 228
Strange attractor 339–341
Streamer
 active-region 42
 coronal 29, 427–431
 helmet 30, 59
Strong field approximation 93
Subflare 69
Sunspot 1–3, 22, 38, 42, 46–56, 280–324, 441
 satellite 50, 67, 69
Sunspot cooling 298–299
Sunspot cycle 2, 32
Sunspot decay 311–314
Sunspot equilibrium 299–308
Sunspot evolution 309–314
Sunspot formation 309–311
Sunspot loop 43
Sunspot maximum 35, 53–54, 430
Sunspot model 139–140, 146, 304
Sunspot motion 50–52
Sunspot number 52–53
Sunspot polarity law 53

Sunspot stability 306–308
Supergranulation 17, 19, 22, 211–212, 338, 441
Surface gravity 4, 441
Surface wave 183–186, 225, 321–323
　magnetoacoustic 185
Surge 44, 50, 56, 67, 243
Suydam criterion 270
Switch-off shock wave 194, 203–205
Switch-on shock wave 194, 203–205
Sympathetic flare 66

Tangential discontinuity 36, 114, 203
Taylor number 95, 276,
Taylor–Proudman theorem 106–107
Taylor's hypothesis 137
Tearing-mode instability 275, 288, 345–346, 363–366, 406
　double 365–366
　nonlinear 366
Temperature 4–7, 15, 25, 441
　effective 4, 48, 441
Temperature minimum 24, 79
Terminal velocity 414
Theorem
　Cowling's 106, 327–328
　frozen-flux 99
　minimum energy 130, 135, 137
　Taylor–Proudman 106–107
　virial 107–108, 134
　Woltjer's 135
Thermal conduction 85, 207–208, 312, 417, 440
Thermal diffusivity 87
Thermal equilibrium 207
Thermal flare 66, 68
Thermal instability
　convective 165, 280–290
　radiative 220, 239
Thermal nonequilibrium 238–239, 360–361, 383–388, 391–393
Thermal speed 80, 440
Thermally isolated loop 236, 238, 241
Three-minute oscillation 181
Time
　collison 76, 78
　correlation 334
　diffusion 5, 96, 325
　radiative cooling 278
Tokamak 137
Topological dissipation 230
Toroidal coordinates 443
Torsional Alfvén wave 161–162
Torus 270, 412
Transient
　coronal 61–64, 66, 410–416
　loop 410

Transition region (or zone) 26, 29, 34, 59, 207, 441
Trapping, wave 178
Trigger, flare 65, 67, 356
Tube (or cusp) speed 170, 184, 318
Tube wave 112, 320–324
　longitudinal 323
　transversal 323
Tunneling 179
Turbulence
　ion-sound 79
　isotropic 333
　pseudo-isotropic 332–334
Turbulent diffusion 331, 343
Turbulent dynamo 332–334
Twist, flux tube (or loop) 111, 122, 125–126, 264, 270, 362, 376–377, 384–385, 410–412
Two-fluid solar wind model 421–423, 435
Two-ribbon flare 41, 61, 67, 69–71, 345, 366–381, 410

Umbra, sunspot 48–49
Umbral dot 49, 305
Umbral flash 49, 182
Umbral granulation 49, 181
Umbral oscillation 49, 182, 321
Uniform-twist magnetic field 125–126, 129, 262–263, 363, 384
Unipolar area (or region) 22, 34–35
Unipolar magnetic field 21, 36
Units 436–439
　Gaussian cgs 436–437
　rationalised mks 437–438

Variable-twist magnetic field 363, 384
Variational (or energy) method 257–265, 306, 404
Vector potential 136
Velocity
　escape 4, 441
　group 156, 166–170
　nonthermal 206
　phase 156, 169
　photospheric 69
　terminal 414
Virial theorem 107–108, 134
Viscosity, coefficient of 81
Viscous dissipation 89
Viscous force 81
Vorticity magnetic field analogy 96

Wave
　acoustic (or sound) 89, 157, 170, 178–180, 189, 213–216
　acoustic-gravity 170–173, 178
　Alfvén (or magnetic) 22, 36, 89, 157–163, 168, 217, 352, 434

Alfvén surface 225, 322
body 321
compression 190
compressional Alfvén 158, 162–163, 170
dispersive 156
dynamo 335–336
evanescent 171, 177–178, 183
expansion 190
flare-induced coronal 175
gravity 163–166, 178
hydromagnetic inertial 296, 330
inertial 166–168
intermediate 169, 193, 205
magnetoacoustic 168–170, 217, 434
magnetoacoustic-gravity 173–175
magnetoacoustic surface 185
magneto-gravity 175
penumbral 50, 175, 182
shear Alfvén 159–162

shock (see shock wave)
sound (or acoustic) 89, 157, 170, 178–180, 189, 213–216
standing 177
surface 183–186, 225, 321–323
torsional Alfvén 161
tube 112, 320–324
Wave generation 180
Wave pressure 213
Wave steepening 213–214
Wave trapping 178
Wilson effect 51–52, 301
Winking filament 61
WKB approximation 156, 173
Woltjer's theorem 135

X-ray bright point 24, 30, 32, 38, 343, 357
X-rays 30–32, 70–71

GEOPHYSICS AND ASTROPHYSICS MONOGRAPHS

Editor:

BILLY M. McCORMAC (Lockheed Palo Alto Research Laboratory)

Editorial Board:

R. GRANT ATHAY (High Altitude Observatory, Boulder)
W. S. BROECKER (Lamont-Doherty Geological Observatory, New York)
P. J. COLEMAN, Jr. (University of California, Los Angeles)
G. T. CSANADY (Woods Hole Oceanographic Institution, Mass.)
D. M. HUNTEN (University of Arizona, Tucson)
C. DE JAGER (the Astronomical Institute at Utrecht, Utrecht)
J. KLECZEK (Czechoslovak Academy of Sciences, Ondřejov)
R. LÜST (President Max-Planck-Gesellschaft zur Förderung der Wissenschaften, München)
R. E. MUNN (University of Toronto, Toronto)
Z. ŠVESTKA (The Astronomical Institute at Utrecht, Utrecht)
G. WEILL (Service d'Aéronomie, Verrieres-le-Buisson)

1. R. Grant Athay, *Radiation Transport in Spectral Lines.*
2. J. Coulomb, *Sea Floor Spreading and Continental Drift.*
3. G. T. Csanady, *Turbulent Diffusion in the Environment.*
4. F. E. Roach and Janet L. Gordon, *The Light of the Night Sky.*
5. H. Alfvén and G. Arrhenius, *Structure and Evolutionary History of the Solar System.*
6. J. Iribarne and W. Godson, *Atmospheric Thermodynamics.*
7. Z. Kopal, *The Moon in the Post-Apollo Era.*
8. Z. Švestka, *Solar Flares.*
9. A. Vallance Jones, *Aurora.*
10. C.-J. Allègre and G. Michard, *Introduction to Geochemistry.*
11. J. Kleczek, *The Universe.*
12. E. Tandberg-Hanssen, *Solar Prominences.*
13. A. Giraud and M. Petit, *Ionospheric Techniques and Phenomena.*
14. E. L. Chupp, *Gamma Ray Astronomy.*
15. W. D. Heintz, *Double Stars.*
16. A. Krüger, *Introduction to Solar Radio Astronomy and Radio Physics.*
17. V. Kourganoff, *Introduction to Advanced Astrophysics.*
18. J. Audouze and S. Vauclaire, *An Introduction to Nuclear Astrophysics.*
19. C. de Jager, *The Brightest Stars.*
20. J. L. Sérsic, *Extragalactic Astronomy.*
21. E. R. Priest, *Solar Magnetohydrodynamics.*
— V. A. Bronshten, *Physics of Meteoric Phenomena.*

NOTES

NOTES

NOTES

NOTES